10th International Conference on Aerophysics and Physical Mechanics of Classical and Quantum Systems (APhM 2016)

Journal of Physics: Conference Series Volume 815

Moscow, Russia
5 – 9 December 2016

Editors:

Sergey T. Surzhikov
Daniil Andrienko
Yacine Babou
Gianpiero Colonna
Alexey S. Dikaljuk Michael K. Ermakov
Igor A. Kryukov
Eduard V. Teodorovich

ISBN: 978-1-5108-3900-7
ISSN: 1742-6588

Printed from e-media with permission by:

Curran Associates, Inc.
57 Morehouse Lane
Red Hook, NY 12571

Some format issues inherent in the e-media version may also appear in this print version.

Printed by Curran Associates, Inc. (2017)

For permission requests, please contact the Institute of Physics at the address below.

Institute of Physics
Dirac House, Temple Back
Bristol BS1 6BE UK

Phone: 44 1 17 929 7481
Fax: 44 1 17 920 0979

techtracking@iop.org

Additional copies of this publication are available from:

Curran Associates, Inc.
57 Morehouse Lane
Red Hook, NY 12571 USA
Phone: 845-758-0400
Fax: 845-758-2633
Email: curran@proceedings.com
Web: www.proceedings.com

Table of contents

Volume 815

10th International Conference on Aerophysics and Physical Mechanics of Classical and Quantum Systems

5–9 December 2016, Moscow, Russian Federation

Accepted papers received: 21 February 2017
Published online: 6 April 2017

Theoretical and Computational Fluid Dynamics

Nonequilibrium Processes in Gas Dynamics

Preface

10th International Conference "Aerophysics and Physical Mechanics of Classical and Quantum Systems" – APhM–2016 was held from 5 to 9 December 2016 at the Institute for Problems in Mechanics of the Russian Academy of Sciences (IPMech RAS), Moscow, Russia.

The aim of the Conference was to bring together leading experience of new research findings, databases, computer codes, information on new instruments and methods of physical experiments in the field of physical and chemical kinetics in gas dynamics, plasma physics in the processes of combustion and explosion.

The Conference was organized as follow. Every working day started with a plenary session where lectures concerning some actual problems of astrophysics and aeromechanics were presented.
The following lectures were delivered by well-known experts in various fields of related problems:

- Prof. S.A. Rashkovsky (IPMech RAS) "Quantum mechanics: a theory of particles or a classic field theory?"
- Prof. Yu.D. Chashechkin (IPMech RAS) "Dynamics and structures of flows in astrophysical, aerophysical and geophysical hydrodynamics. Observations, laboratory and mathematical modeling".
- Prof. S.A. Isaev (St.Petersburg State University of Civil Aviation) "Numerical modeling of the turbulent tear flows with the emphasis on aerodynamics of thick profiles with vortex cells and an intensification of the heat transfer by the oval-trench wells".
- Prof. I.A. Znamenskaya (Moscow Lomonosov State University, Physical department) "Panoramic methods of experimental studies in aerophysics and physical mechanics".
- Dr. Y. Babou (Bioengineering and Aerospace engineering department University of Carlos III, Madrid) "Plasma airflow jets diagnosis by means of time-resolved tomography".
- Prof. E.V. Teodorovich (IPMech RAS) "Diffusion of a passive impurity in a random velocity field".

After the lectures the work carried out in 5 sections:
- Experimental physical mechanics
- Nonequilibrium Processes in Gas Dynamics
- Physical–Chemical Gas Dynamics
- Physics of Gas Discharge
- Hydro aeromechanics

The last two days were devoted to special talks and discussions.

Totally 75 oral reports devoted to the topics of a section were heard and discussed.
37 papers were submitted for publication in the JPCS. The others were published in the electronic journal «Physical-Chemical Kinetics in Gas Dynamics» by Moscow Lomonosov State University.

Figure. From left to right: Prof. S.A. Isaev (St.Petersburg State University of Civil Aviation); Prof. S.T. Surzhikov (IPMech RAS); Prof. I.A. Znamenskaya (Lomonosov Moscow State University, Physical department); Dr. Y.Babou (Bioengineering and Aerospace engineering department University of Carlos III, Madrid, Spain)

Most of the papers presented at the conference are performed under financial support from the Programs of basic researches of the Russian academy of sciences, the corresponding projects of the Russian Foundation for Basic Research, and of the Russian Science Foundation.

160 delegates from 19 organizations took part in the Conference.

The next APhM Conference will be held in December, 2017 at the Institute for Problems in Mechanics of the Russian Academy of Sciences, Moscow, Russia.

IOP Conf. Series: Journal of Physics: Conf. Series **815** (2017) 011001 doi:10.1088/1742-6596/815/1/011001

Founders

The Federal Agency for Scientific Organizations (FASO Russia).

Organisation

Chairmen

Dmitry M. Klimov
Sergey T. Surzhikov

Secretary

Olga V. Shalashova

Organising Committee

Sergey T. Surzhikov, Alexey S. Dikalyuk, Michael K. Ermakov, Alexandra L. Zheleznyakova, Igor A. Kryukov, Viktor V. Kuzenov, Larisa B. Ruleva, Olga V. Shalashova, Elena Ja. Sysoeva, Michael A. Kotov, Roman K. Seleznev, Dmitry A. Storozhev, Sergey I. Solodovnikov.

Program Committee

- Daniil Andrienko (USA),
- Yacine Babou (Spain),
- Gianpiero Colonna (Italy),
- Michael K. Ermakov (Russia),
- Dmitry M. Klimov (chairman, Russia),
- Anatoly F. Kolesnikov (Russia),
- Igor A. Kryukov (Russia),
- Joseph Shang (USA),
- Sergey T. Surzhikov (Russia),
- Valery A. Tovstonog (Russia).

Editors

- Editor: Sergey T. Surzhikov (Russia), Dr. Sci.

Invited Editors:

- Daniil Andrienko (USA), PhD,
- Yacine Babou (Spain), PhD,
- Gianpiero Colonna (Italy), PhD,
- Alexey S. Dikaljuk (Russia), PhD,
- Michael K. Ermakov (Russia), PhD,
- Igor A. Kryukov (Russia), PhD,
- Eduard V. Teodorovich (Russia), Dr. Sci.

We thank all the conference participants. We appreciate the efforts of the members of Program and Organizing Committee during the preparation and providing the Conference.

Dmitry M. Klimov
Sergey T.Surzhikov
Eduard V. Teodorovich

IOP Conf. Series: Journal of Physics: Conf. Series **815** (2017) 011002 doi:10.1088/1742-6596/815/1/011002

Peer review statement

All papers published in this volume of *Journal of Physics: Conference Series* have been peer reviewed through processes administered by the proceedings Editors. Reviews were conducted by expert referees to the professional and scientific standards expected of a proceedings journal published by IOP Publishing.

IOP Conf. Series: Journal of Physics: Conf. Series **815** (2017) 012001 doi:10.1088/1742-6596/815/1/012001

Numerical Investigation of Penning Discharge Characteristics using 2D/3V Particle-In-Cell Method

A S Dikalyuk[1,2,3] **and S E Kuratov**[1,2]

[1]All-Russia Research Institute of Automatics, Moscow, 127055, Russia
[2]A. Ishlinsky Institute for Problems in Mechanics of the Russian Academy of Sciences, Vernadsky prospekt 101(1), Moscow, 119526, Russia
[3]Moscow Institute of Physics and Technology, Dolgoprudny, 141701, Russia

aleks.dikalyuk@gmail.com

Abstract. Numerical model for the simulation of characteristics of Penning discharge is presented in the paper. The model is based on the 2D/3V axisymmetric electrostatic particle-in-cell algorithm implemented on unstructured triangular meshes. In order to simulate kinetic processes in the gas discharge plasma Monte-Carlo collision (MCC) method was incorporated into the model. Considered processes for the numerical experiments are elastic scattering of electrons on the neutral particles and ionization of neutral particles by electrons. Using this model various characteristics of Penning discharge such as anode and cathode currents, distribution of charged components in the discharge chamber have been calculated at various applied voltages and magnetic flux densities. The results of numerical experiments have been analyzed and discussed.

1. Introduction

Behavior of the plasma created in rarified gas in the presence of external magnetic field is not well understood. A known physical system that can be used for studying such processes is Penning discharge. The schematic view of Penning discharge chamber is shown in figure 1. It consists of cylindrical anode and two round cathodes placed coaxially at both ends of anode. The characteristic size of the chamber is 1 cm. An external magnetic field is applied to the system parallel to the axis of symmetry. The typical parameters of the Penning discharge are as follows: gas pressure of the order of 1 mTorr, voltage drop between electrodes is 1–10 kV, strength of the magnetic field ~ 0.1 T. At mentioned parameters combined effect of quadrupole electric field and axial magnetic field enlarges lifetime of the electrons in the discharge chamber and therefore raises the number of ionization events produced by single electron.

Penning gas discharge has been studied both experimentally and numerically [1–8]. In [1, 2] experimental measurements of the Penning discharge characteristics (mainly currents) have been made at various parameters (magnetic field, geometry, etc.) of the system. Usage of annular anode allowed visualizing the structure of gas discharge plasma. In [3, 4] the design of the Penning discharge chamber suitable for usage as ion source was proposed and studied experimentally. The corresponding volt-ampere characteristics of the Penning discharge were obtained for helium and molecular hydrogen at various pressure. Also measurements of extracted current versus anode voltage at different pressures were performed. In [5] numerical simulation of dissociation and ionization kinetics in the Penning discharge in hydrogen has been carried out based on the local thermodynamic equilibrium approach. Temporal behavior of neutral and charged atomic and molecular gas components has been obtained as well as kinetic parameters of discharge plasma (mobility of charged components, diffusion coefficients). The study has allowed determining the set of elementary processes that should correctly describe the composition of gas discharge plasma. The series of studies [6–8] is dedicated to numerical simulation of spatial structure of Penning discharge plasma using different formulations of drift-diffusion model. Development of the models of continuum mechanics for the description of the Penning discharge is important since they are much less computationally demanding than kinetic models. However the most detailed description of the plasma ignited in the rarified gas can be obtained using kinetic models. An example of these models is particle-in-cell model combined with Monte-Carlo collision method (PIC-MCC).

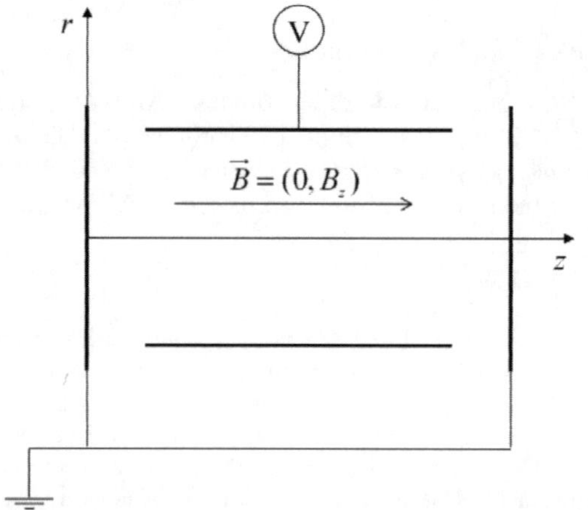

Figure 1. Schematic view of Penning discharge geometry.

In the study numerical simulation of two-dimensional spatial structure of Penning discharge plasma is performed using PIC-MCC approach. Dependence of the anode, cathode and extracted current on the applied magnetic field and voltage is calculated as well as energy distribution function of the charged particle in the gas discharge plasma. Results of numerical experiments are analyzed and discussed.

2. Description of the PIC-MCC method

Particle-in-cell method introduces a concept of macroparticle [9, 10]. Macroparticle is a computational particle that simulates behaviour of a large number of real plasma particles (electrons or ions). Macroparticles follow the trajectory of real particles in the plasma since the Lorentz force depends only on the charge-to-mass ratio [11]. The movement of macroparticles changes the distribution of electromagnetic field in the system. It means that the whole simulation process is self-consistent. It can be shown that actually PIC method is a mean for solution of Vlasov-Maxvell system of equations [12].

For the numerical solution of the Penning gas discharge problem 2D/3V axisymmetric electrostatic PIC–MCC model was used. In order to briefly characterize numerical methods and approaches that have been used in the computer implementation of the PIC-MCC method mentioned above let us consider a computational cycle shown in figure 2.

The first stage of a computational cycle is the solution of the Poisson equation. It is formulated for the axisymmetric geometry:

$$\Delta\varphi = \frac{1}{r}\left(\frac{\partial}{\partial r}\left(r\frac{\partial\varphi}{\partial r}\right) + \frac{\partial}{\partial z}\left(r\frac{\partial\varphi}{\partial z}\right)\right) = \frac{\rho(r,z)}{\varepsilon_0} \tag{1}$$

In equation (1) φ is the electrostatic potential, $\rho(r,z)$ is the charge density and ε_0 is the electric constant. For the boundary value problem the boundary conditions can be formulated in the following form:

$$\left.\left(\alpha_1\varphi+\alpha_2\frac{\partial\varphi}{\partial\vec{n}}\right)\right|_{\partial D}=g(r,z) \tag{2}$$

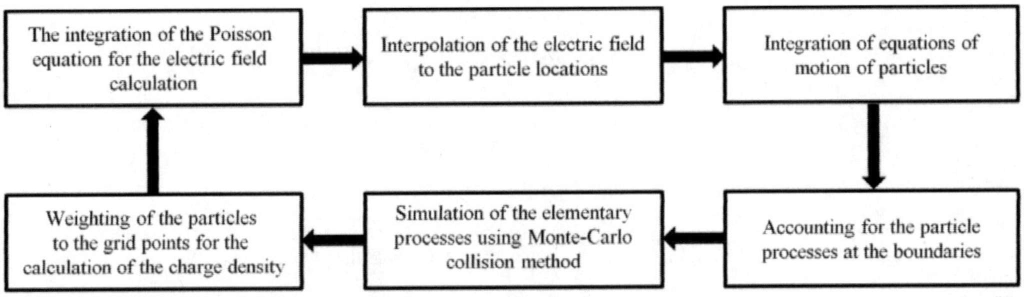

Figure 2. PIC–MCC computational cycle.

For the numerical solution of equation (1) the computational domain is subdivided into triangles. Usage of unstructured grids for the numerical simulation of gas discharges is necessary since discharge chamber may be of complex shape. Descritization of equation (1) is carried out using cell-centered finite volume method [13]. Geometrical multigrid method is used for the numerical solution of the system of linear algebraic equations resulted from the descritization process. Implementation of the multigrid procedure on the unstructured grid is performed using overset meshes approach [14].

Development of the particle-in-cell method on the unstructured grids requires special procedures of interpolation of the electric field to the particles position for the Lorentz force calculation and localization of macroparticles relative to the mesh elements (particle search algorithm) [12, 15]. These procedures constitute the second stage of the computational cycle. Both of these procedures can be easily implemented using local basis functions $N_n^k(z,r)$ constructed for the n-th node (local numbering of triangle nodes, varies from 1 to 3) of the k-th element (figure 3).

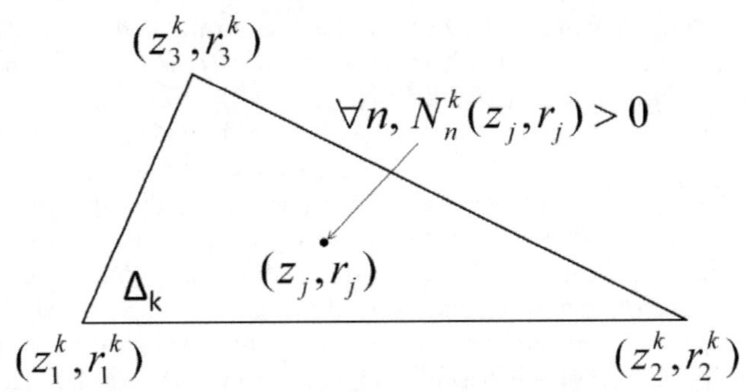

Figure 3. Local basis functions formalism used in the unstructured PIC method.

For the interpolation procedure the following property of the local basis function is important. If the values of interpolated function f are known in the k-th triangle nodes (z_m^k, r_m^k), then for every point with coordinates (z_j, r_j) that lies inside triangle:

$$f(z_j,r_j)=\sum_{n=1}^{3}N_n^k(z_j,r_j)f(z_n^k,r_n^k) \tag{3}$$

IOP Conf. Series: Journal of Physics: Conf. Series **815** (2017) 012001 doi:10.1088/1742-6596/815/1/012001

For the macroparticle search algorithm the following property is used [15]. If the macroparticle lies within the k -th triangle then the values of all local basis functions $N_n^k(z_p, r_p)$ are nonnegative, where z_p and r_p are the axial and radial coordinates of the considered macroparticle.

In the third stage of the computational cycle one has to solve equations of particles motion which for the nonrelativistic case in the cylindrical coordinate system can be written in the following form [16]:

$$\begin{aligned}
&\frac{dr}{dt} = v_r \\
&\frac{d\theta}{dt} = \frac{v_\theta}{r} \\
&\frac{dz}{dt} = v_z \\
&m\frac{dv_r}{dt} = q(E_r + v_\theta B_z - v_z B_\theta) + m\frac{v_\theta^2}{r} \\
&m\frac{dv_\theta}{dt} = q(E_\theta - v_r B_z + v_z B_r) - m\frac{v_r v_\theta}{r} \\
&m\frac{dv_z}{dt} = q(E_z + v_r B_\theta - v_\theta B_r)
\end{aligned} \tag{4}$$

In (4) r, θ, z are the radial, angular and axial particle coordinates, v_r, v_θ, v_z are the corresponding velocity components, E_r, E_θ, E_z are the projections of the vector of the electric field, B_r, B_θ, B_z are the projections of the vector of the magnetic field, m is the macroparticle mass, q is the macroparticle charge.

It is evident that system of equations (4) has singularity when particle with nonzero angular momentum approaches to the axis of symmetry [16]. If one is going to solve numerically this system it has to deal with the problem. Boris [11, 16] has proposed the numerical procedure which successfully overcomes mentioned above difficulty using Cartesian–cylindrical coordinate systems transformation.

Another problem is concerned with the Boris numerical method itself. Boris method is of the class of leap-frog methods, which means that velocities of the particles and their coordinates are not known in the same moments of time, however initial conditions are formulated in the same moment of time. In order to initialize calculation procedure correctly one has to obtain appropriate velocity of the particle. This problem was formulated and solved in [16].

On the next stage of computational cycle processes at the boundaries have to be considered. For example, plasma particles hitting the metal electrodes generate electric current. Dielectric boundaries are able to accumulate charge of plasma particles and impact on the distribution of the potential inside the discharge chamber [17]. Ions and electrons of plasma can hit surfaces of discharge chamber and generate fluxes of charge particles directed back to the gas discharge plasma due to the processes of secondary electron emission, ion-electron emission [18].

Accounting for the secondary emission processes at the boundaries can be implemented according to method described in [17]. Number of particles injected from the boundary into the computational area due to boundary-macroparticle collision is determined using the random number R uniformly distributed on the interval $[0,1]$. If $R < (\gamma - \mathrm{int}(\gamma))$ then number of particles to be injected into the computational area is $N_{inj} = \mathrm{int}(\gamma) + 1$ otherwise $N_{inj} = \mathrm{int}(\gamma)$. A model of the corresponding process has to be applied for the initialization of the new particle.

It is possible to account for elementary processes in the PIC algorithm using the Monte-Carlo collision approach based on the null collision method [19-21]. Monte-Carlo collision method consists of the following steps [21]. Calculation of the kinetic energy of the particle i of sort s :

$$\varepsilon_i = 0.5 m_s V_i^2 \tag{5}$$

Determining the magnitude of the total collision cross-section corresponding to this energy:

$$\sigma_{ttl}(\varepsilon_i) = \sigma_1(\varepsilon_i) + ... + \sigma_N(\varepsilon_i) \tag{6}$$

Here $\sigma_j(E_i)$ for $1 \le j \le N$ is the cross section of the j-th type of collision between the species s and the target species. The collision probability is calculated using the following formula:

$$P_i = 1 - \exp(-\Delta t V_i \sigma_{ttl}(\varepsilon_i) n_{trg}(\vec{x_i})) \tag{7}$$

In (7) V_i and $\vec{x_i}$ are the velocity and position of the i-th particle correspondingly, $n_{trg}(\vec{x_i})$ is the local density of the target species at the position of the i-th particle. A collision occur if a random number uniformly distributed on the interval [0,1] is less then P_i. If the collision occurs then one has to determine the collision type using another random number. After that new particles have to be initialized based on the model [19-22] of the process corresponding to the given collision type.

Disadvantage of the approach presented above is due to the necessity of calculation of P_i for each particle in the simulation. In order to overcome this difficulty the null collision method was proposed. Detailed description of the method is given in [21]. The idea of the method is to introduce the constant collision frequency ν in the following manner:

$$\nu = \max_{\vec{x},E}\left(n_{trg}\sigma_{ttl}V\right) = \max_{\vec{x}}\left(n_{trg}\right)\max_{E}\left(\sigma_{ttl}V\right) \tag{8}$$

Physically it means that additional collisional process is introduced with collision frequency of the special form. This collisional process is called null collision since no real interaction occurs. Then one can determine maximum fraction of the total number of particles of sort s which experience collision:

$$P_{null} = 1 - \exp(-\nu'\Delta t) \tag{9}$$

The colliding particles are chosen randomly among all the particles of the given sort s participating in the simulation (eliminating duplicates), and each particle is checked for the type of the collision.

The last stage of computational cycle is the calculation of the charge density of plasma particles. These fields are used for the construction of source term for the Poisson equation. In case of unstructured grid and cylindrical coordinate system one can utilize the formalism of local basis function [12, 15] in order to perform the calculations. The corresponding formula can be written as follows [12]:

$$q_n = \frac{1}{V_n}\sum_k\sum_j \frac{Q_j N_n^k(z_j, r_j)}{2\pi r_j} \tag{10}$$

In (10) summation is performed over all particles j that are in triangle k that contains the node n. z_j and r_j are the axial and radial coordinates of macroparticle, Q_j is the charge of macroparticle, in the two-dimensional case V_n is calculated according to the following formula:

$$V_n = \sum_k \frac{S_k}{3} \tag{11}$$

In (11) S_k is the area of the triangle k that contains node n.

3. Statement of the Penning gas discharge problem

In this section initial data for Penning gas discharge simulation are going to be presented. Geometry of the computational domain that represents the gas discharge chamber as well as unstructured mesh and boundary conditions for the Poisson equation are shown in figure 4. The dashed black line r=0 cm is the axis of symmetry with the corresponding boundary condition ($\partial\varphi/\partial r$=0). The thick black line represents the grounded cathode (φ=0 V). The part of the boundary at z=1.1 cm represents an orifice for the extraction of ions from the gas discharge plasma. The boundary condition for the orifice is assumed to be $\partial\varphi/\partial z$=0. The cylindrical anode (thin black line) is placed inside the gas discharge chamber with the applied voltage $\varphi = V_0$. Magnetic field is assumed to be uniform and parallel to the axis of symmetry of the system. The unstructured mesh shown in figure 4 consists of 2438 nodes and 4642 triangular elements.

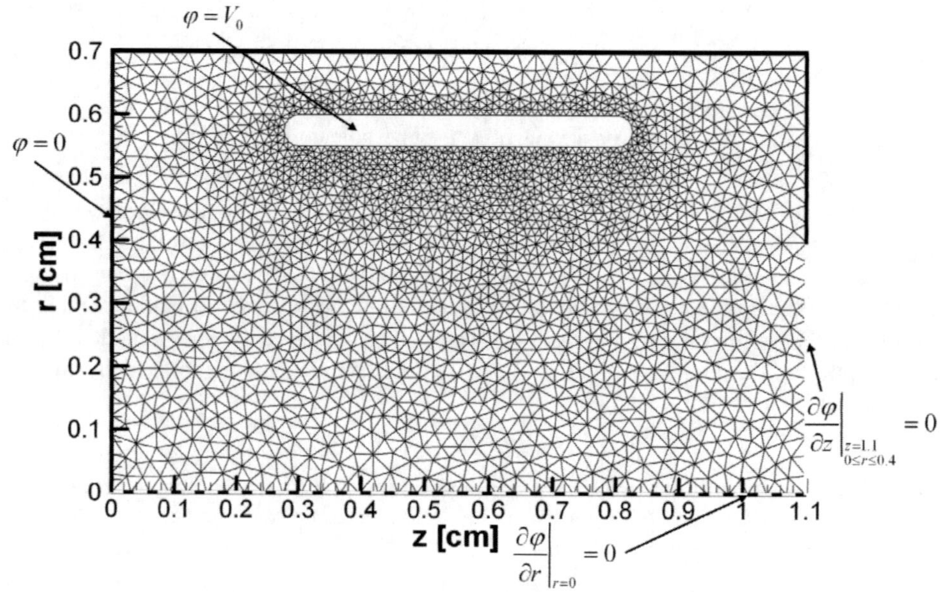

Figure 4. Geometry of the discharge chamber and computational mesh.

In the study Penning gas discharge in molecular hydrogen H_2 at pressure p=1.2 mTorr and temperature T=300 K is going to be considered. Elementary processes accounted for in the study are elastic electron scattering on H_2 molecules and ionization of H_2 molecules by electrons. The data on the cross-section for these processes were taken from [23]. Also ion-electron emission from the cathode has been taken into account. The data on the dependence of the electron yield per ion versus energy was taken from [24].

In the PIC–MCC model two species of macroparticles have been considered: electrons and H_2^+ ions. Primary electrons and ions were placed in the center of the computational domain. Initial density of the charged particles were $5\cdot10^8$ cm^{-3}. Initial velocities of the particles were sampled from Maxwell distribution with temperature 300 K.

In the following section results of numerical experiments carried out at different magnetic field strengths and anode voltages according to the PIC-MCC model described above are going to be presented and analyzed.

4. Numerical results of Penning gas discharge simulation

Structure of the Penning gas discharge has been investigated for the anode voltage V_0=2000 V at the following magnitudes of the axial projection of the magnetic field (others projections are assumed to be equal zero): B_z =0.05; 0.075; 0.1; 0.15 T. Numerical simulation has showed that gas discharge could not be ignited at B_z =0.05 T. However gas discharge is stable at slightly increased magnetic field B_z =0.075 T and above this value.

Distributions of the electron and ion number densities at B_z =0.075; 0.1; 0.15 T are shown in figures 5–10. It can be easily seen that the structure of the Penning gas discharge strongly depend on the magnitude of the magnetic field. Increasing magnetic field leads to the increase of the maximum of

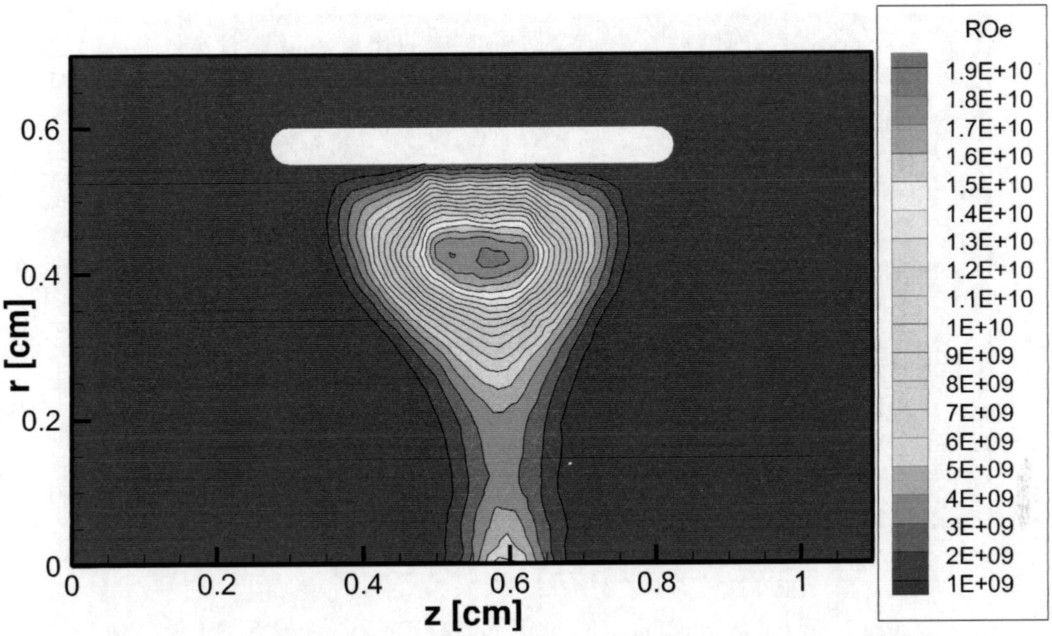

Figure 5. Spatial distribution of electron number density in cm^{-3} at $B_z = 0.075$ T.

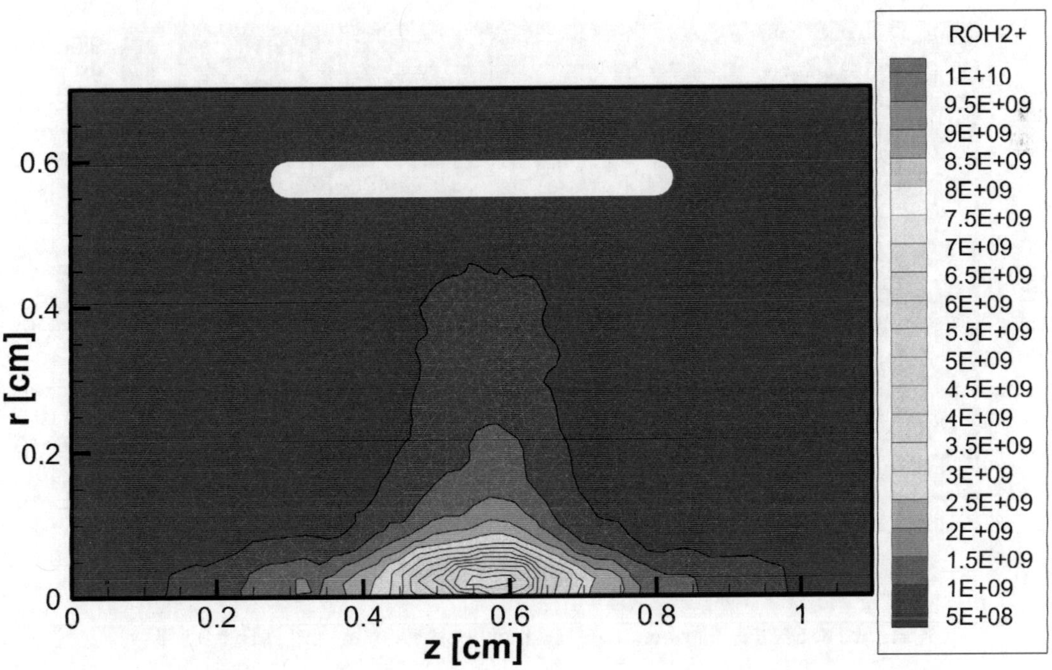

Figure 6. Spatial distribution of H_2^+ number density in cm^{-3} at $B_z = 0.075$ T.

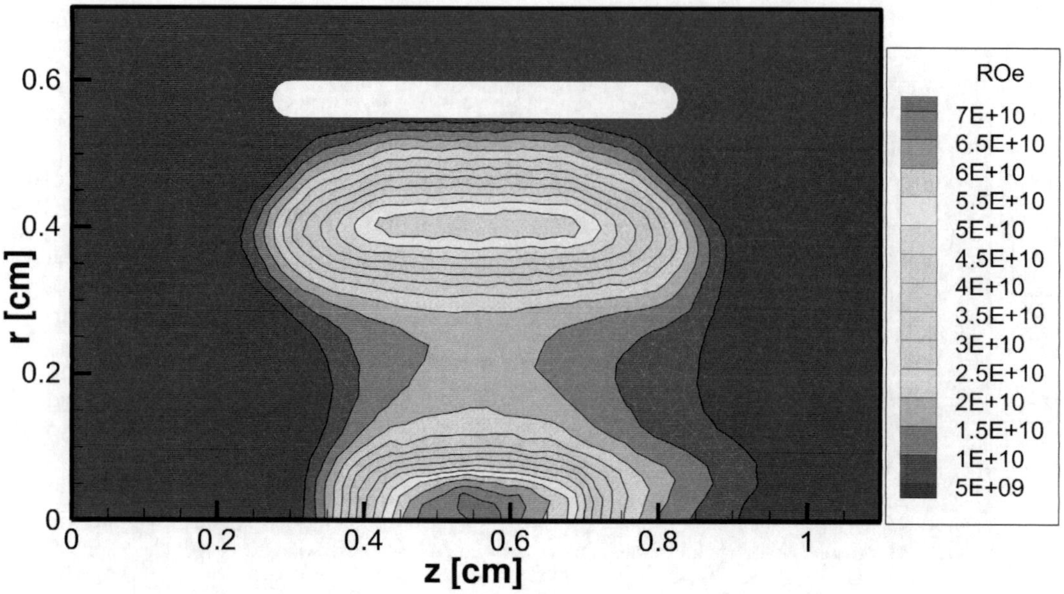

Figure 7. Spatial distribution of electron number density in cm^{-3} at $B_z = 0.1$ T.

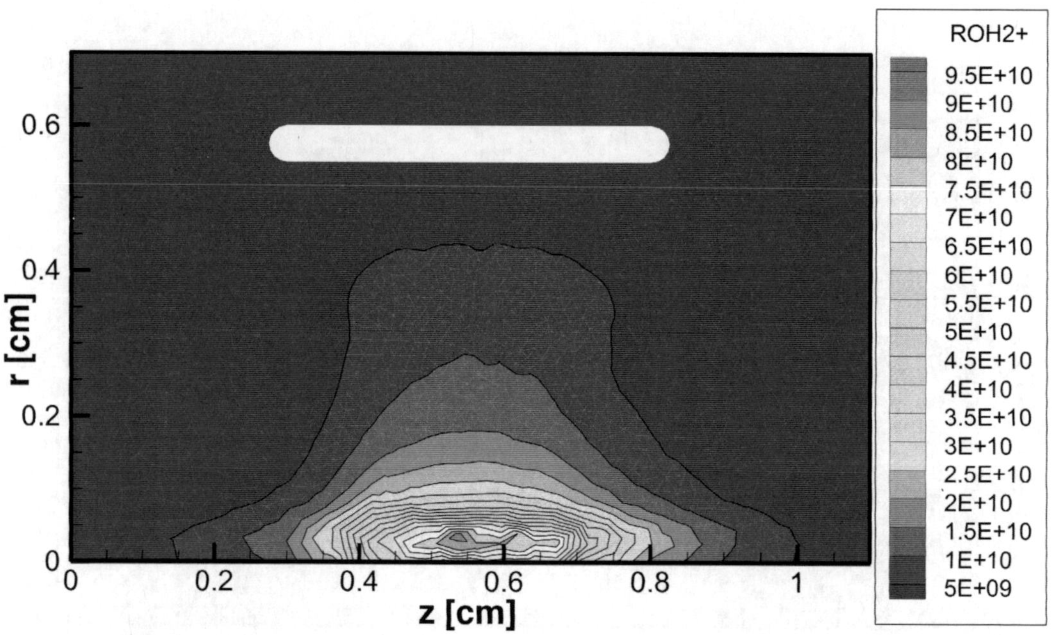

Figure 8. Spatial distribution of H_2^+ number density in cm^{-3} at $B_z = 0.1$ T.

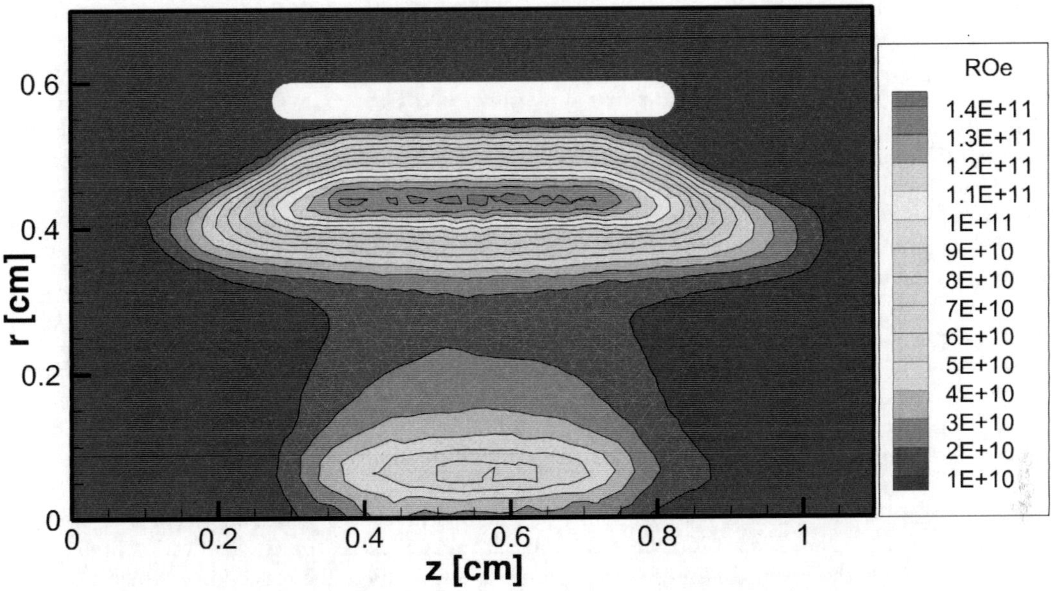

Figure 9. Spatial distribution of electron number density in cm^{-3} at $B_z = 0.15$ T.

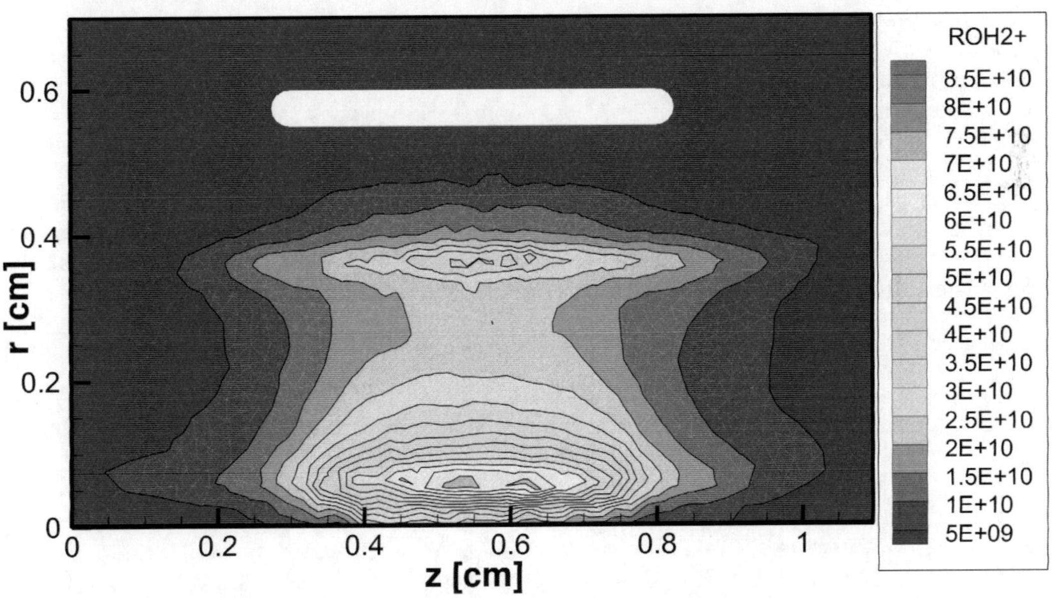

Figure 10. Spatial distribution of H_2^+ number density in cm^{-3} at $B_z = 0.15$ T.

electron number density in the gas discharge chamber. However maximum H_2^+ ion number density grows when magnetic field increases from 0.075 to 0.1 T and then becomes smaller when magnetic field further increases to 0.15 T. However total number of ions (as can be estimated using the number of macroparticles that represent H_2^+ ions) grows with increasing magnetic field. Figures 5–10 also show that the volume of the discharge chamber filled with gas discharge plasma increases with the increasing magnetic field.

In the figures 11, 13 and 15 the radial distributions of the number densities of electrons and ions in the cross section with axial coordinate z=0.55 cm are shown for B_z=0.075; 0.1; 0.15 T correspondingly. It can be noticed absence of the pronounced maximum in the electrons number density distribution in the vicinity of the axis of symmetry at B_z=0.075 T. In figures 13 and 15 one can observe that in the regions close to the axis of symmetry in the considered cross section number densities of the electrons and ions are very close to each other (plasma is closer to electroneutrality than in other regions of the discharge chamber). Also it can be noticed that the increasing magnetic field leads to shift of the maximum in the distribution of plasma particle number densities from the axis of symmetry to the anode as can be seen in figures 13 and 15. Figure 15 shows the presence of the second maximum in the distribution of ions number density.

In the figures 12, 14 and 16 the radial distributions of absolute value of axial projection of ion current density through the cathode, anticathode and orifice are shown for B_z=0.075; 0.1; 0.15 T correspondingly. The orifice is located in the range of radial coordinates $0 \le r \le 0.4$ cm. For all considered cases one can conclude that the radial distribution of ions leaving the Penning gas discharge chamber through orifice has the form of the relatively narrow beam. Also one can observe shift of the maximum of ion current density distribution from the axis of symmetry to the boundary of the orifice.

In the figures 17 and 18 energy distributions of macroparticles that represent electrons and ions at different values of B_z are shown. One can see that at greater values of B_z fraction of high energy electrons grows. On the other hand increasing magnetic field leads to the reduction of fraction of high energy ions. It is worth notecing that results presented in this paragraph should be considered preliminary and requiring conducting of additional numerical experiments and analysis.

In the figures 19 and 20 dependence of anode, cathode and extracted through orifice currents on the value of axial projection of magnetic field and on the anode voltage correspondingly are shown. It can be seen that variation of the magnetic field influences on the gas discharge currents stronger than the variation of anode voltage. Increase of magnetic field at 1.5 times (from 0.1 T to 0.15 T) leads to 2 time increase of anode current (from 2.7 mA to 5.45 mA). However increase of anode voltage at 1.7 times (from 1500 to 2500) leads to only 1.3 times increase of anode current (from 2.03 mA to 2.7 mA). Analysis of the gas discharge parameters such as electrons and ions number densities distributions, spatial distribution of the potential, axial and radial projections of electric field at different anode voltages (to be specific, at 1500, 2000, 2500 V) at fixed value of magnetic field B_z=0.1 T has shown that there is no qualitative influence of anode voltage on the structure of plasma of Penning gas discharge.

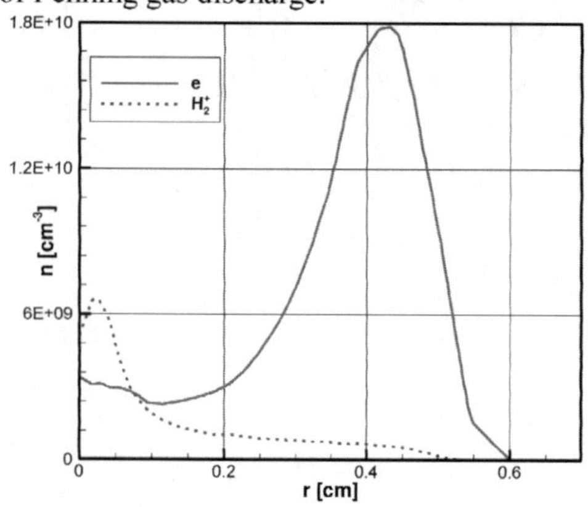

Figure 11. Radial distribution of electrons and ions number density in cross section with axial coordinate z=0.55 cm, B_z=0.075 T.

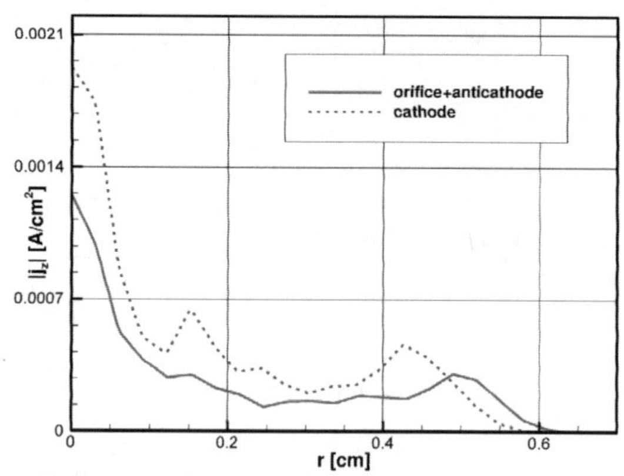

Figure 12. Radial distribution of absolute value of axial projection of ion current density through orifice, anticathode and cathode, B_z=0.075 T.

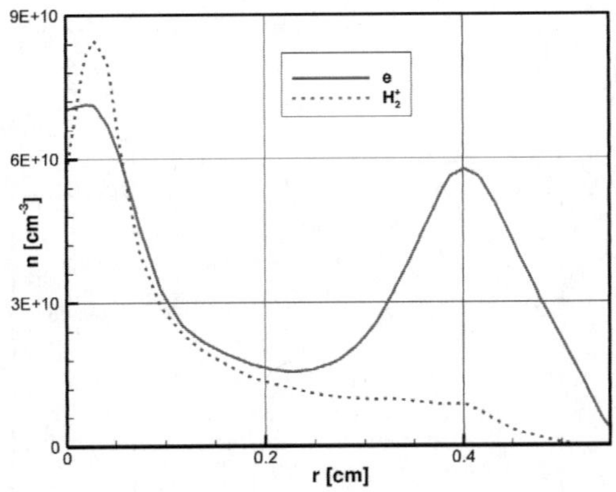

Figure 13. Radial distribution of electrons and ions number density in cross section with axial coordinate z=0.55 cm, B_z=0.1 T.

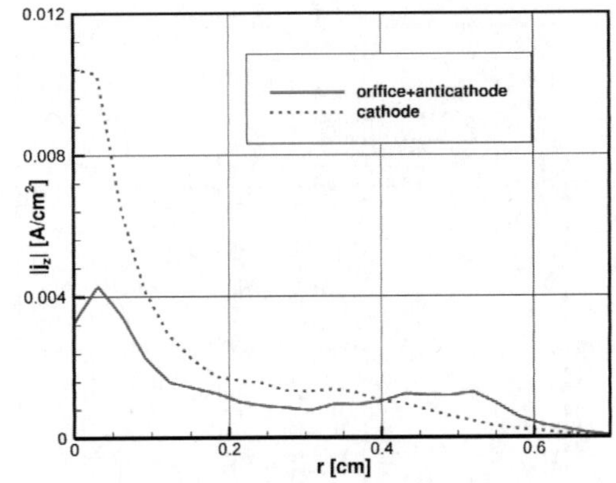

Figure 14. Radial distribution of absolute value of axial projection of ion current density through orifice, anticathode and cathode, B_z=0.1 T.

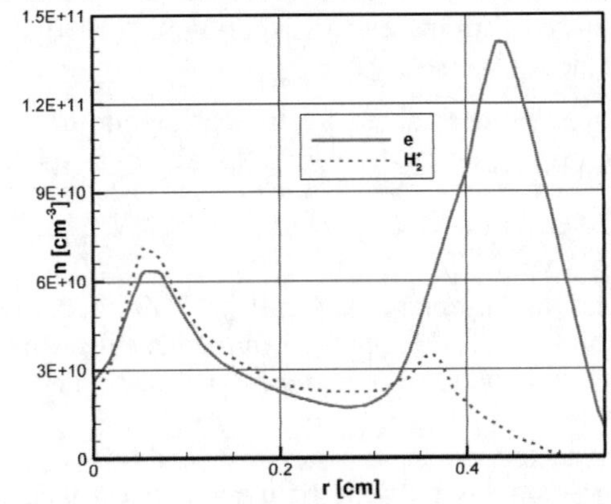

Figure 15. Radial distribution of electrons and ions number density in cross section with axial coordinate z=0.55 cm, B_z=0.15 T.

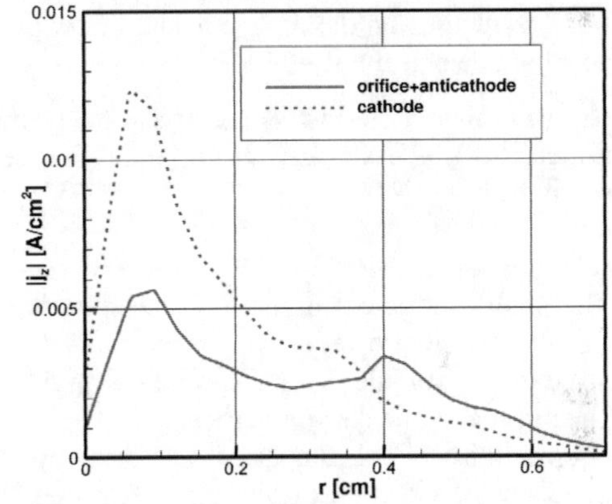

Figure 16. Radial distribution of absolute value of axial projection of ion current density through orifice, anticathode and cathode, B_z=0.15 T.

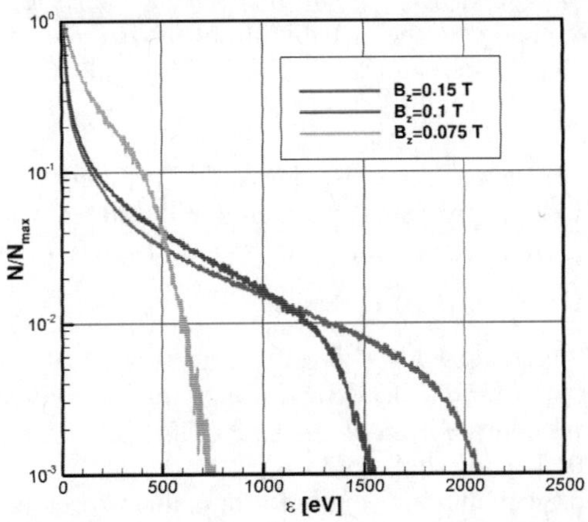

Figure 17. Energy distribution of macroparticles that represent electrons.

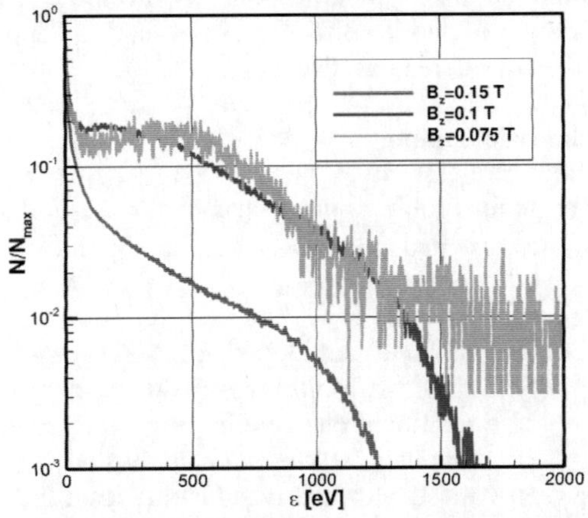

Figure 18. Energy distribution of macroparticles that represent ions H_2^+.

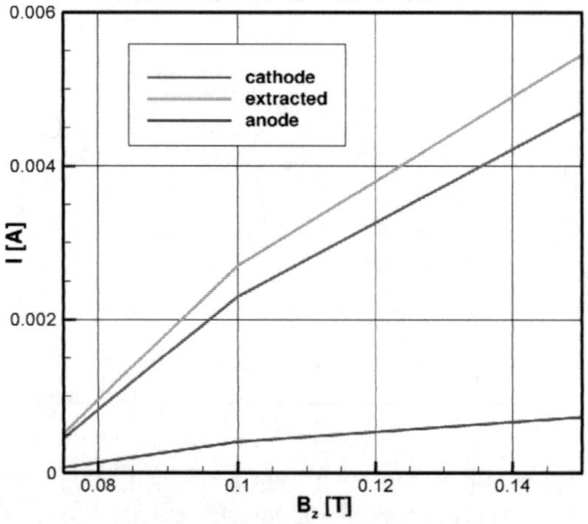

Figure 19. Dependence of anode, cathode and extracted currents on the magnetic field strength at fixed anode current V_0=2500 V.

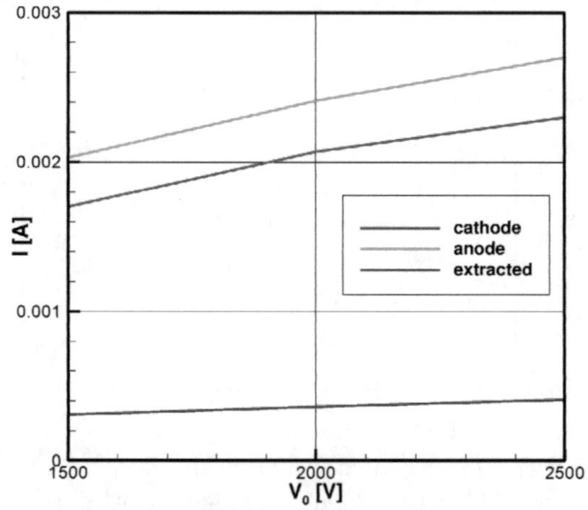

Figure 20. Dependence of anode, cathode and extracted currents on the anode voltage at fixed magnetic field strength B_z=0.1 T.

In general particle-in-cell method used in the study may be considered for the calculation of the parameters of the gas discharges relevant for the ignition of burning mixtures [25, 26] and high-speed flow control [27-29].

5. Conclusion

In the study numerical model for the simulation Penning gas discharge is described. The model is based on the electrostatic axisymmetric 2D/3V particle-in-cell (PIC) approach implemented on the unstructured triangular grid combined with Monte-Carlo collision (MCC) method for the accounting of the plasma elementary processes.

Using this model numerical experiments were conducted aimed at the investigation of the dependence of structure of plasma of Penning gas discharge on the magnetic field and anode voltage. Results of numerical experiments have shown that the variation of magnetic field has significant impact on the Penning gas discharge electrodynamic structure. It means that real properties of the magnetic system (permanent magnets, coils) responsible for creation of magnetic field in the system should be taken into account for accurate simulation of Penning gas discharge characteristics. Variation of anode voltage does not lead to the qualitative change in the distribution of the parameters of the considered gas discharge.

Acknowledgments

Authors are grateful to the prof. S.T. Surzhikov and staff of Laboratory of Radiative Gas Dynamics of Institute for Problems in Mechanics for useful discussions of the problems considered in the study.

The presented study was supported by the Russian Science Foundation project № 16–11–10275.

References

[1] Kreindel Yu E and Ionov A S 1964 *J. of Technical Physics* **34** 1199

[2] Schitov N, Kanshin I, Mamedov N 2015 Investigation of the dependency of penning ion source operational characteristics on its geometric parameters *Physical-Chemical Kinetics in Gas Dynamics* **16** 4 http://chemphys.edu.ru/issues/2015-16-4/articles/590/

[3] Rovey J L, Ruzic B P and Houlahan T J 2007 Simple penning ion source for laboratory research and development applications *Review of Scientific Instruments* **78** 106101

[4] Rovey J L 2008 Design parameter investigation of a cold-cathode Penning ion source for

general laboratory applications *Plasma Sources Sci. Technol.* **17** 035009
[5] Storozhev D 2014 Numerical simulation of the kinetcs of ionization and dissociation of hydrogen in Penning discharge plasma in the LTE approach *Physical-Chemical Kinetics in Gas Dynamics* **15** 3 http://chemphys.edu.ru/issues/2014-15-3/articles/229/
[6] Surzhikov S, Kuratov S 2014 Modified drift-diffusion model of the Penning discharge *Physical-Chemical Kinetics in Gas Dynamics* **15** 6 http://chemphys.edu.ru/issues/2014-15-6/articles/257/
[7] Surzhikov S T 2015 Application of the modified drift-diffusion theory to study of the two-dimensional structure of the penning discharge *Proc. 53rd AIAA Aerospace Sciences Meeting* AIAA 2015–1832
[8] Surhzikov S T 2016 Two-dimensional structure of Penning discharge in cylindrical chamber with axial magnetic field at pressure 1 Tor *Technical Physics Letter*
[9] Hockney R W and Eastwood J W 1988 *Computer simulation using particles* (New York: CRC Press)
[10] Birdsall C K and Langdon A B 1985 *Plasma Physics Via Computer Simulation* (New York: McGraw-Hill, Inc.)
[11] Berezin Yu A and Vshivkov V A 1980 *Particle Method in Dynamics of Rarified Plasma* (Novosibirsk: Nauka)
[12] Grigoriev Yu N, Vshivkov V A and Fedoruk M P 2004 *Numerical Simulation via Particle-In-Cell Methods* (Novosibirsk: Izdatelstvo SO RAN)
[13] Versteeg H K and Malalasekera W 2007 *An Introduction to Computational Fluid Dynamics: the Finite Volume Method* (Harlow: Pearson Education)
[14] Mavriplis D J 1995 *Multigrid Techniques for Unstructured Meshes* (Hampton: Institute for Computer Applications in Science and Engineering)
[15] Spirkin A M 2006 *A Three-Dimensional Particle-In-Cell Methodology on Unstructured Voronoi Grids with Applications to Plasma Microdevices* (Worcester: Worcester Polytechnic Institute)
[16] Delzanno G L and Camporeale E 2013 On particle movers in cylindrical geometry for particle-in-cell simulations *J. of Comp. Phys.* **253** pp 259–277
[17] Bromley B P 2001 *Computational Modeling of the Axial-Cylindrical Inertial Electrostatic Confinement Fusion Neutron Generator* (Urbana-Champaign: University of Illinois at Urbana-Champaign)
[18] Raizer Yu P 1991 *Gas Discharge Physics* (Berlin: Springer)
[19] Birdsall C K 1991 Particle-in-cell charged-particle simulations plus monte carlo collisions with neutral atoms PIC-MCC *IEEE Trans. on Plasma Sci.* **19** pp 65–85.
[20] Donko Z 2011 Particle simulation methods for studies of low-pressure plasma sources *Plasma Sources Sci. Technol.* **20** 024001
[21] Vahedi V and Surendra M 1995 A monte carlo collision model for the particle in cell method: applications to argon and oxygen discharges *Comp. Phys. Comm.* **87** pp 179–198.
[22] Nanby K 2000 Probability theory of electron-molecules, ion-molecule, molecule-molecule, and Coulomb collisions for particle modeling of materials processing plasmas and gases *IEEE Trans. on Plasma Sci.* **28** pp 971–990
[23] Tawara H, Itikawa Y, Nishimura H and Yoshino M 1990 Cross sections and related data for electron collisions with hydrogen molecules and ions *J. Phys. Chem. Ref. Data.* **19** pp 617–636
[24] Phelps A V and Petrovic Z Lj 1999 Cold-cathode discharges and breakdown in argon: surface and gas phase production of secondary electrons *Plasma Sources Sci. Technol.* **8** pp R21–R44
[25] Seleznev R K and Surzhikov S T 2015 Quasi-One-Dimensional and Two-Dimensional Numerical Simulation of Scramjet Combustors *Proc. 51th AIAA/ASME/SAE/ASEE Joint Propulsion Conference* AIAA 2015-4166

[26] Seleznev R K, Surzhikov S T and Shang J A 2016 Quasi-One-Dimensional Analysis of Hydrogen-Fueled Scramjet Combustors *Proc. 52nd AIAA/ASME/SAE/ASEE Joint Propulsion Conference* AIAA 2016-4569

[27] Surzhikov S T, Shang J S 2005 Viscous Interaction on a Flat Plate with a Surface Discharge in Magnetic Field *High Temperature* **43**(1) pp 019–030

[28] Surzhikov S T, Shang J S 2012 Coupled Radiation-Gasdynamic Model for Stardust Earth Entry Simulation *Journal of Spacecraft and Rockets* **49**(5) pp 875-888

[29] Surzhikov S T 2011 Radiative-Convective Heat Transfer of a Spherically Shaped Space Vehicle in Carbon Dioxide *High Temperature* **49**(1) pp 92–107

IOP Conf. Series: Journal of Physics: Conf. Series **815** (2017) 012002 doi:10.1088/1742-6596/815/1/012002

Numerical simulation of the kinetics of dissociation and ionization of molecular hydrogen in the penning discharge plasma with the use of the reduced kinetic model

D A Storozhev [1,2,3], **S E Kuratov** [1]

[1]N.L. Dukhov All-Russian Research Institute of Automatics (State Corporation «ROSATOM» company), Suschevskaya Street 22, Moscow 127055, Russia
[2]Moscow Institute of Physics and Technology (State University), 9 Institutskiy per., Dolgoprudny, Moscow Region, 141701, Russia
[3]A. Ishlinsky Institute for Problems in Mechanics of the Russian Academy of Sciences, Vernadsky Prospect 101(1), Moscow, 119526, Russia

dmitry.stor@gmail.com

Abstract. A simplified kinetic scheme for describing the kinetics of ionization and dissociation of hydrogen in the Penning discharge plasma is developed. The calculations of the component composition of the Penning discharge plasma are performed. Numerical simulation results are compared with the results provided by the detailed state-to-state kinetic scheme.

1. Introduction

In recent years the increasing interest in the study of the kinetics of molecular hydrogen and its isotopes in low-temperature plasma and Penning discharge has been primarily associated with the development of the electrical devices such as neutron generators and ion sources [1–3]. One of the most important characteristics that define the efficiency of these devices is a chemical composition of the Penning discharge plasma. Component composition of the discharge plasma depends on the characteristic energy of charged particles (electrons and ions) gained by them in the crossed electric and magnetic fields, and determines by the rate of energy exchange with the neutral gas molecules. The complexity of the determination of the component composition in the Penning discharge plasma is due to the need to take into account a large number of kinetic processes describing the collisions of an electron and molecules, molecules and atoms, molecules in various excited states, the complexity of solving the system of kinetic equations, as well as small amount of experimental data on the plasma chemical composition [4–6].

Available papers related to the numerical simulation of atomic-molecular composition of plasma in ion sources mainly concern the devices with high gas-discharge current used in fusion reactors, as well as sources of H^- ions [1,5]. At the same time many ion sources used in the gas-filled neutron sources operates at relatively low discharge current ($I = 0.1 \div 2$ mA) and the use of the detailed kinetic schemes in the compute-intensive 2D and 3D PIC models for calculation of the plasma chemical composition appears to be difficult.

In this paper we consider the Penning discharge at the pressure $p \approx 10^{-3} \div 10^{-2}$ Torr. The potential difference applied between the cathode and the anode is $\varphi = 1 \div 3$ kV. The current through the discharge is equal to $I = 0.1 \div 2\,\text{mA}$, and the ionization degree is $\alpha \approx 10^{-2}$. In the paper the component composition of the plasma in molecular hydrogen using the state-to-state and the simplified kinetic schemes is calculated. With the use of numerical simulation, the population of vibrational states of the hydrogen molecules, the discharge degree of ionization and dissociation, as well as the impact of various kinetic processes on the plasma component composition were defined

2. Numerical simulation results

A big number of plasma-chemical processes occur in the area of the Penning discharge [7–11]. First of all it is the processes of direct excitation of vibrationally excited states by electron impact, stepwise mechanism of excitation of vibrationally excited states occurred due to the excitation of the electronic states, the ionization of particles by electron impact, processes of dissociative ionization, dissociation of molecules flowing through the electronically excited states $B^1\Sigma_u^+$, $C^1\Pi_u$, $B'^1\Sigma_u^+$, $D^1\Pi_u$, $B''^1\Sigma_u^+$, $D'^1\Pi_u$, $b^3\Sigma_u^+$ and chemical reactions involving heavy molecules and atoms. In this paper, the kinetic scheme (see table 1) has been developed to describe the kinetics of ionization and dissociation of hydrogen in the Penning discharge plasma.

Calculation of the component composition of the Penning discharge plasma in the zero-dimensional approximation was carried out under the following assumptions:

1) The initial data set included the electron density N_e^0 and the concentration of neutral hydrogen molecules. In the calculations the electron density was taken as a constant.

2) The electron energy distribution function (EEDF) was considered either Maxwell distribution function or a combination of Maxwell EDF (simulating thermal electrons produced inside the main discharge volume) and the delta function (describes the fast electrons emitted from the cathode as the result of ion-electron emission). It was assumed that the share of fast electrons is 10%.

3) The energy of thermal electrons was varied within the range of 1÷50 eV, the energy of the fast electrons was considered to be equal to 1 keV, the temperature of neutral molecules and ions was taken to be equal to 300–500 K.

4) The physical time was considered to be equal to 20 microseconds. It corresponds to the pulse width in the ion source based on the Penning discharge.

A detailed analysis of different ionization and dissociation mechanisms in the Penning discharge plasma is given in figure 1. The figure shows that the main mechanisms of atomic ions production are the dissociative ionization processes №2–3. At the same time, with the increase of the electron temperature, the influence of the mechanism №2 on the concentration of atomic ions decreases, and the influence of the mechanism №3 increases. It should be also noted that the share of the process of ionization of neutral atoms by the electron impact in the formation of atomic ions is less than 3%. At low temperatures, the main mechanism of electron molecule dissociation is №10, flowing through the excited electronic state $b^3\Sigma_u^+$. With the increase of the electron temperature, hydrogen atoms predominantly produced by the processes of dissociative ionization of molecule №4 and №5, and the mechanism of dissociation №8.

A detailed analysis of the ionization and dissociation mechanisms in the case of non-Maxwell EDF used in the calculations is given in figures 2–3. On the bar charts the influence of various mechanisms of atomic ions production in the hydrogen plasma is shown. It should be noted that the inclusion of the monoenergetic electrons in the kinetic model leads to the increase of the influence of dissociative ionization mechanism №3 on the concentration of atomic ions. Numerical simulations show that in this case the fraction of atomic hydrogen ions produced by the mechanism № 3 in the kinetic scheme is equal to 75-85% (see figure 2). Figure 3 shows the influence of different mechanisms on the production of hydrogen atoms. In this case, as in calculations with Maxwell EEDF, it appears that the main mechanisms of dissociation are the mechanisms № 6 and № 10, as well as the processes of dissociative ionization № 2 – 3.

Table 1. Simplified kinetic scheme.

Molecule ionization		
1. $H_2\left(X^1\Sigma_g^+, v_i\right)+e \rightarrow H_2^+\left(X^2\Sigma_g^+\right)+e$	$v_i=0$	$0 \le \varepsilon \le 200$ eV
Dissociative ionization		
2. $H_2\left(X^1\Sigma_g^+, v_i\right)+e \rightarrow H_2^+\left(X^2\Sigma_g^+\right)+2e \rightarrow H(1s)+H^+ +2e$	$v_i=0$	$0 \le \varepsilon \le 200$ eV
3. $H_2\left(X^1\Sigma_g^+, v_i\right)+e \rightarrow H_2^+\left({}^2\Sigma_u^+\right)+2e \rightarrow H(1s)+H^+ +2e$	$v_i=0$	$0 \le \varepsilon \le 200$ eV
Dissociation		
4. $H_2\left(X^1\Sigma_g^+, v_i\right)+e \rightarrow H_2\left(B^1\Sigma_u^+\right)+e \rightarrow H(1s)+H(2l)+e$	$v_i=0$	$0 \le \varepsilon \le 200$ eV
5. $H_2\left(X^1\Sigma_g^+, v_i\right)+e \rightarrow H_2\left(C^1\Sigma_u^+\right)+e \rightarrow H(1s)+H(2l)+e$	$v_i=0$	$0 \le \varepsilon \le 200$ eV
6. $H_2\left(X^1\Sigma_g^+, v_i\right)+e \rightarrow H_2\left(B'^1\Sigma_u^+\right)+e \rightarrow 2H+e$	$v_i=0$	$0 \le \varepsilon \le 200$ eV
7. $H_2\left(X^1\Sigma_g^+, v_i\right)+e \rightarrow H_2\left(D^1\Pi_u\right)+e \rightarrow 2H+e$	$v_i=0$	$0 \le \varepsilon \le 200$ eV
8. $H_2\left(X^1\Sigma_g^+, v_i\right)+e \rightarrow H_2\left(B''^1\Sigma_u^+\right)+e \rightarrow 2H+e$	$v_i=0$	$0 \le \varepsilon \le 200$ eV
9. $H_2\left(X^1\Sigma_g^+, v_i\right)+e \rightarrow H_2\left(D'^1\Pi_u\right)+e \rightarrow 2H+e$	$v_i=0$	$0 \le \varepsilon \le 200$ eV
10. $H_2\left(X^1\Sigma_g^+, v_i\right)+e \rightarrow H_2\left(b^3\Sigma_u^+\right)+e \rightarrow H(1s)+H(1s)+e$	$v_i=0$	$0 \le \varepsilon \le 200$ eV
Atom ionization		
11. $H+e \rightarrow H^+ +e+e$	CS[12]	$0 \le \varepsilon \le 1000$ eV
Processes with participation of heavy particles		
12. $H_2^+ + H \rightarrow H_2 + H^+$	K= 9.D-16 cm^6s^{-1}[13]	
13. $H + H_2(v \ge 4) \rightarrow H + H_2^+$	K= 2.5D-15 cm^6s^{-1}[13]	
14. $H + H + H_2 \rightarrow H_2(X, v=14) + H_2$	K= $2.68*10^{-31}*T^{-0.6}$ cm^6s^{-1}[5]	
15. $H + H + H \rightarrow H_2(X, v=14) + H$	K= $8.04*10^{-31}*T^{-0.6}$ cm^6s^{-1}[5]	
16. $H_2^+ + H_2 \rightarrow H^+ + H + H_2$	CS[13]	
Recombination		
17. $e+e+H^+ \rightarrow e + H(n)$	$K = 6.34\times10^{-27} n^2 \times \dfrac{1}{T_e^3\left(E_n^{2.33}+4.38E_n^{1.72}+1.32E_n\right)}$ cm^6s^{-1} [12]	
18. $e+H^+ \rightarrow H + h\nu$	$K = 3.92\times10^{-14}\dfrac{1}{n}\times\dfrac{E_n^{1.5}}{E_n+0.35}$, cm^3s^{-1} [14]	
19. $e+e+H_2^+ \rightarrow e + 2H$	K= $8.75*10^{-27}T_e^{-4.5}$ cm^6s^{-1}[5]	
20. $e+H_2^+ \rightarrow H(1) + H(n)$	CS [12]	

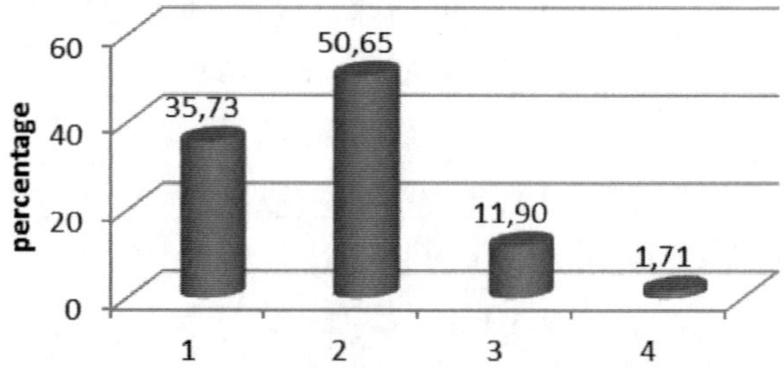

Figure 1. The fraction of atomic ions produced by the following reactions (EEDF is Maxwell, $T_e = 10\,\text{eV}$, $N_e^0 = 3*10^{17}\ \text{m}^{-3}$):

1. $H_2\left(X^1\Sigma_g^+, v_i\right) + e \rightarrow H_2^+\left(X^2\Sigma_g^+\right) + 2e \rightarrow H(1s) + H^+ + 2e$
2. $H_2\left(X^1\Sigma_g^+, v_i\right) + e \rightarrow H_2^+\left(^2\Sigma_u^+\right) + 2e \rightarrow H(1s) + H^+ + 2e$
3. $H + e \rightarrow H^+ + e + e$
4. $H_2^+ + H \rightarrow H_2 + H^+$

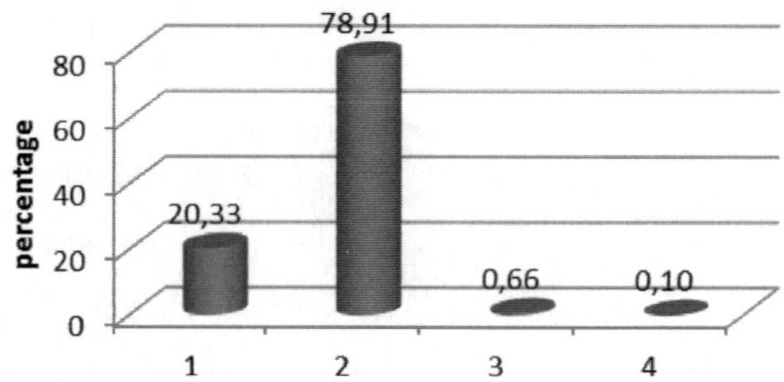

Figure2. The fraction of atomic ions produced by the following reactions (EEDF is Nonmaxwell, $T_e = 10\,\text{eV}$, $N_e^0 = 3*10^{16}\ \text{m}^{-3}$):

1. $H_2\left(X^1\Sigma_g^+, v_i\right) + e \rightarrow H_2^+\left(X^2\Sigma_g^+\right) + 2e \rightarrow H(1s) + H^+ + 2e$
2. $H_2\left(X^1\Sigma_g^+, v_i\right) + e \rightarrow H_2^+\left(^2\Sigma_u^+\right) + 2e \rightarrow H(1s) + H^+ + 2e$
3. $H + e \rightarrow H^+ + e + e$
4. $H_2^+ + H \rightarrow H_2 + H^+$

It is worth noting that the vibrational distribution function of hydrogen molecules, in contrast to the glow discharge, has no significant effect on the kinetics of plasma chemical processes in the Penning discharge. Most chemical processes (ionization, dissociation, excitation of the electronic states) take place with the participation of hydrogen molecules in the ground vibrational state. However, with the increase of the electron temperature and concentration, the number of vibrationally excited molecules increases and, consequently, its influence on the kinetics of plasma-chemical reactions also increases

Nevertheless, calculations carried out in this paper show that the relative share of neutral atoms and ions produced in reactions with participation of vibrationally excited molecules is less than 10%. It should be also noted that some dissociation mechanisms provided in the kinetic scheme, produce more than 40% of neutral atoms from vibrationally excited states, however, these reactions do not significantly affect the concentration of hydrogen atoms. Figure 4 gives the vibrational distribution function in the Penning discharge plasma.

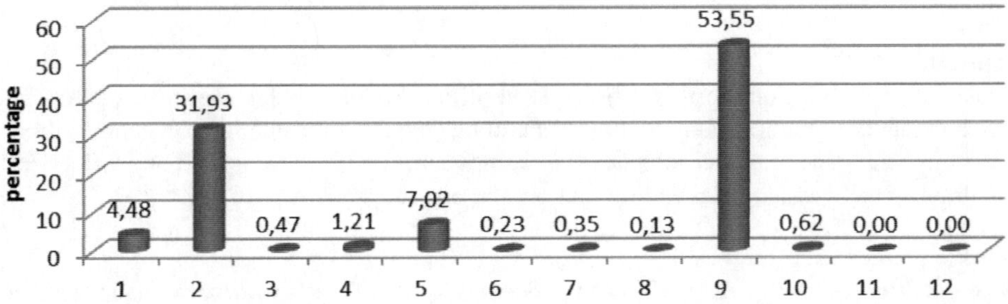

Figure **3.** The fraction of neutral atoms produced by the following reactions (EEDF is nonmaxwell, $T_e = 3$ ev, $N_e^0 = 3*10^{16}$ m^{-3})

1. $H_2(X^1\Sigma_g^+, v_i) + e \rightarrow H_2^+(X^2\Sigma_g^+) + 2e \rightarrow H(1s) + H^+ + 2e$
2. $H_2(X^1\Sigma_g^+, v_i) + e \rightarrow H_2^+({}^2\Sigma_u^+) + 2e \rightarrow H(1s) + H^+ + 2e$
3. $H_2(X^1\Sigma_g^+, v_i) + e \rightarrow H_2(B^1\Sigma_u^+) + e \rightarrow H(1s) + H(2l) + e$
4. $H_2(X^1\Sigma_g^+, v_i) + e \rightarrow H_2(C^1\Pi_u^+) + e \rightarrow H(1s) + H(2l) + e$
5. $H_2(X^1\Sigma_g^+, v_i) + e \rightarrow H_2(B'^1\Sigma_u^+) + e \rightarrow 2H + e$
6. $H_2(X^1\Sigma_g^+, v_i) + e \rightarrow H_2(D^1\Pi_u) + e \rightarrow 2H + e$
7. $H_2(X^1\Sigma_g^+, v_i) + e \rightarrow H_2(B''^1\Sigma_u^+) + e \rightarrow 2H + e$
8. $H_2(X^1\Sigma_g^+, v_i) + e \rightarrow H_2(D'^1\Pi_u) + e \rightarrow 2H + e$
9. $H_2(X^1\Sigma_g^+, v_i) + e \rightarrow H_2(b^3\Sigma_u^+) + e \rightarrow H(1s) + H(1s) + e$
10. $e + H_2^+ \rightarrow H(1) + H(n)$
11. $e + e + H^+ \rightarrow e + H(n)$
12. $e + H^+ \rightarrow H + hv$

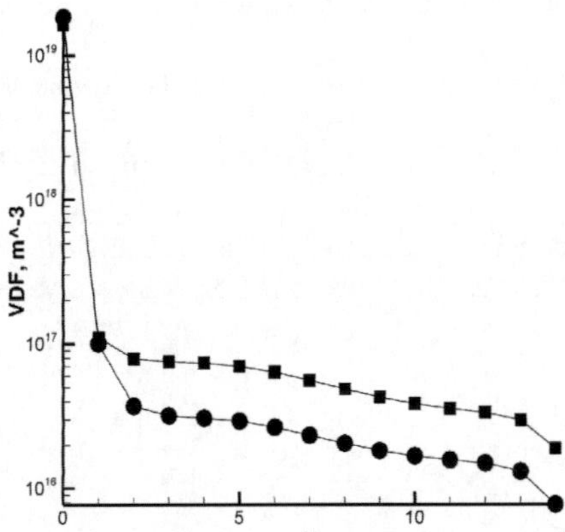

Figure 4. Vibrational distribution function (EEDF is Maxwell) for different electron temperature: $T_e = 50$ eV (squares), $T_e = 10$ eV (circles).

In the case the molecular ions of hydrogen can achieve sufficiently high energies (more than 100 eV) the process $H_2^+ + H_2 \rightarrow H^+ + H + H_2$ can have a significant impact on the plasma chemical composition.

3. Conclusion

On the basis of the work done in this article, a simplified kinetic model was developed. The model allows to calculate the chemical composition of Penning discharge plasma in molecular hydrogen. The developed simplified kinetic model includes five components and 20 reactions. The work is performed under the grant of the Russian Science Foundation, the project № 16–11–10275.

References

[1] Janev R 1995 *Atomic and molecular processes in fusion edge plasmas*(Springer Science & Business Media)

[2] Ikeda K, Agostinetti P, Brombin M, Fantz U, Geng S, Kaneko O and Sartori E 2015 Recent Studies of Hydrogen Negative Ion Source and Beam Production for NBI in Large Helical Device. *In 25th International Toki Conf.*

[3] Storozhev D and Surzhikov S 2015 Numerical simulation of two-dimensional structure of glow discharge in molecular hydrogen *45th AIAA Thermophysics Conf* 10.2514/6.2015-3108

[4] Laricchiuta A, Celiberto R, Esposito F and Capitelli M 2006 State-to-state cross sections for H2 and its isotopic variants *Plasma Sources Sci. Technol.* **15** pp S62–S66

[5] Matveyev A and Silakov V 1995 Kinetic processes in a highly-ionized non-equilibrium hydrogen plasma *Plasma Sources Sci. Technol* **4** pp 606–617

[6] Engelhardt A and Phelps A 1963 Elastic and inelastic collision cross sections in hydrogen and deuterium from transport coefficients *Phys. Rev.* **131** 5 2115

[7] Seleznev R, Surzhikov S and Shang J 2016 Quasi-one-dimensional analysis of hydrogen-fueled scramjet combustors *Proc. 52nd AIAA/ASME/SAE/ASEE Joint Propulsion Conference AIAA* 2016-4569

[8] Seleznev R and Surzhikov S 2015 Quasi-one-dimensional and two-dimensional numerical simulation of scramjet combustors *Proc.51th AIAA/ASME/SAE/ASEE Joint Propulsion Conference AIAA* 2015-4166

[9] Dikalyuk A and Surzhikov S 2010 The modeling of dust particles in a normal glow discharge: the comparison of two charged models *Proc. of 41st Plasmadynamics and Laser conference* AIAA 2010-4310

[10] Storozhev D and Surzhikov S 2013 Numerical simulation of glow discharge in a magnetic field through the solution of the boltzmann equation *J. of basic and applied phys.* **2** 3

[11] Storozhev D and Surzhikov S 2015 Numerical simulation of the two-dimensional structure of glow discharge in molecular nitrogen in light of vibrational kinetics *High Temp.* **53** 3 pp 307–318

[12] Janev R, Langer W, Evans K and Post D 1988 Elementary Processes in Hydrogen-Helium Plasmas *Nuclear Fusion* **28** 3 529

[13] Méndez I, Gordillo-Vázquez F, Herrero V and Tanarro I 2006 Atom and ion chemistry in low pressure hydrogen DC plasmas *J. Phys. Chem. A* **110** pp 6060–66

[14] Stubbers R, Kim H and Miley G 2002 Two-dimensional modeling of a radically-convergent cylindrical inertial electrostatic confinement (IEC) fusion device *5th US/Japan workshopIEC fusion Madison*

Quasi-stationary convection in a periodic-pulsed optical discharge in high pressure rare gas

V P Zimakov, V A Kuznetsov, N G Solovyov, A N Shemyakin, A O Shilov and M Yu Yakimov

A.Ishlinsky Institute for Problems in Mechanics RAS, 101-1 Vernadskogo pr., Moscow, Russia, 119526

yakimov@lantanlaser.ru

Abstract. Unusual convection flows were observed in stabilized pre-breakdown phase of the periodic-pulsed optical discharge (POD) called "quiet" POD. The discharge was a relatively weakly glowing plasma filament sustained by focused $\lambda = 1.064$ μm laser pulses with repetition rate of $f_r = 50 \div 100$ kHz at the intensity several times below than that required for the optical breakdown to occur. No strong shock waves or irregular turbulence around the discharge were observed, in contrast to breakdown types of POD. Significant laser beam refraction measured in the beam cross-section behind the discharge zone was explained by the gas heating in the discharge up to 10 kK, providing high gradients of gas density and refraction index. Intense convective flow was detected on the schlieren images as thermal traces of the laser-induced gas streams flowing from the discharge zone, directed mainly normally to the optical axis. Repeated relaxation of the gas expanding after being rapidly heated by the laser pulse is proposed to explain the effect. The periodic-pulsed discharge located in the elongated beam waist generates an anisotropic heated region with gas streams and vortices, which may form the observed regular convective flow at the late stages of expanding.

1. Introduction

Laser induced quasi-regular convection stream that could not be explained by thermo-gravity mechanism together with surprisingly strong laser beam refraction were two most interesting manifestations of the unusual optical discharge phenomenon of stable sustaining of pre-breakdown phase of the periodic-pulsed optical discharge (POD) observed by the authors of the paper in high pressure xenon in the vicinity of the focused beam waist.

The possibility of a non-breakdown plasma sustaining in a periodic-pulsed optical discharge (POD) in a focused periodic-pulsed Yb^{3+} laser beam ($\lambda = 1.07$ μm) with pulse frequency being above 2 kHz, radiation pulse length of 200 μs, and pulse power above 200 W in high-pressure ($p = 10 \div 20$ bar) xenon was found and investigated in [1]. The characteristics of this type of POD were found to be closer to those of continuous optical discharge (COD) than to those of an optical breakdown or laser spark [2, 3]. In the non-stationary phase POD plasma brightness and the intensity of its ion spectral lines considerably exceeded those of COD plasma under similar conditions [4] indicating higher temperature of POD plasma. After the end of each laser pulse POD plasma recombines in a characteristic time of 10 to 20 μs, with the gap between pulses up to 200 μs or even more, while plasma initiation in the next pulse is still possible. In 200 μs after the pulse end the concentration of free electrons or ions in the gas is not enough to re-initiate discharge. The initial laser radiation absorption in the next pulse occurs due to electronic transitions between energy levels of the excited xenon atoms that preserved the excited state to the front of the next pulse.

The current paper experiments deal with a different kind of POD, which may be characterized as pre-breakdown phase of the periodic-pulsed optical discharge. This kind of POD differs from those described in the above paragraph and [1] primarily by temporal characteristics of the lasers used [5, 6]. A Q-switched periodic-pulsed fiber laser of YLP series from IPG Photonics Corp. was used in present experiments [6]. The laser average power is $P = 50$ W, pulse repetition rate $f_r = 50 \div 100$ kHz, pulse energy up to $E = 1$ mJ, the pulse FWHM $\tau = 170$ ns. YLP series is characterized by high pulse-to-pulse stability at the level of a fraction of a percent. The laser radiation wavelength is $\lambda = 1.064$ μm.

Though the beam quality of the laser was high, the intensity of the focused beam in the range of $10^9 \div 10^{10}$ W/cm^2 was roughly an order of magnitude lower than that required for an optical breakdown at this wavelength in pure high-pressure noble gases [3]. Practically, when the laser beam is focused inside a quartz bulb under $p = 8 \div 16$ bar of xenon, the focal ratio being $F = f/d < 4$, where d is the beam diameter on the lens, f is the focal length (the numerical aperture $NA > 0.125$), with high pulse repetition rate and pulse energy close to the maximal $E = 1$ mJ, the laser breakdown will occur, but not in each pulse, since it needs initiation from the dust suspension particles produced by erosion of the auxiliary electrodes in the gas volume after electric spark or short-time arc discharge. If voltage of several hundred volts is applied to the electrodes, laser breakdowns will stop due to removing the dust particles by the ionic wind.

In experiments with the YLP laser, a new quasi-stationary POD mode, which can be characterized as a pre-breakdown POD mode, has been obtained by the authors. It could be possible mainly because of strict pulse-to-pulse repeatability of the laser beam characteristics. This kind of POD differs dramatically in its properties from the periodic-pulsed optical breakdown [7, 8] as well as from POD [1] and COD [2, 4], which means that one can speak about a self-contained phenomenon of pre-breakdown POD or “quiet” laser spark, or “quiet” POD.

The “quiet” spark does not occur in a focused beam by itself, initiation by casual ordinary optical breakdown flash or by electric spark or arc discharge flash is required for its occurrence. The “quiet” spark does not stand out with a bright light, a close look is required to notice it in a daylight, but it distorts the transmitted laser beam almost as COD does [4, 9]. It also creates laser-induced convective flows in the gas, clearly visible on schlieren images of the discharge region.

2. Experimental layout

Even though periodic-pulsed sustaining of quasi-stationary laser plasma is similar to POD with sub-millisecond long pulses in its nature, the reason in significant differences between POD and the “quiet” POD is applying high power short laser pulses in the latter case. In the experiments the laser pulse power was up to $P_L = 6.5$ kW, pulse width being $\tau_{FWHM} = 170$ ns. The laser pulse and discharge glow pulse waveforms are shown in figure 1.

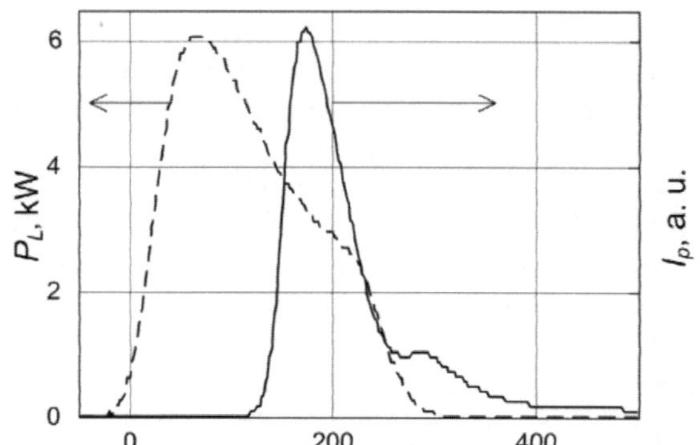

Figure 1. Waveforms of laser pulse power $P_L(t)$ (the dashed line) and integral discharge glow intensity $I_p(t)$ from the region of interaction between laser radiation and xenon, $p = 15 \pm 1$ bar, $F = 7$, $\tau_{FWHM} = 170$ ns, $f_r = 50$ kHz.

The pulse frequency is $f_r = 50$ kHz, the pulse energy is up to $E = 1$ mJ, the wavelength is $\lambda = 1.064$ μm. The beam shape is close to the ground transverse mode with propagation ratio $M^2 = 1.8$ and beam parameter product BPP = 0.5 mm·mrad, which gives estimation for beam diameter near the beam waist $d_0 \approx 20 \div 30$ μm, focal ratio being $F = f/d = 7$. The uncertainty of d_0 is particularly brought by optical distortion of the beam in the walls of the quartz bulb.

With $F = 7$ power density up to $I = 2\times10^9$ W/cm^2 could be achieved, which allowed the possibility of an optical breakdown (ordinary laser spark) only in presence of initiation centers (dust) at pulse repetition rate of $f_r = 50$ kHz. When using a shorter focus lens with $F = 4.5$ achievable power density raised up to $I = (5 \div 7)\times10^9$ W/cm^2, which provided ordinary laser spark flashes in a wider pulse energy range, but narrowed the "quiet" POD observation range at the same time.

Standard xenon short arc lamps of 50 W, 150 W and 1000 W rated input electric power with filling pressures of $p = 15$ bar, $p = 14$ bar and $p = 8$ bar respectively were used as xenon filled quartz bulbs. The pressure in the lamp was increased due to heating, the smaller the lamp was, the more it was increased. A laser beam of a diameter $d = 7$ mm at the output of the laser collimator was focused near the xenon bulb center with lenses of focal length $f = 30$ mm or $f = 50$ mm, which gave the focal ratios $F = f/d = 4.5$ and $F = 7$, respectively, or numerical apertures $NA = 0.11$ and $NA = 0.08$.

The lamp electrodes were used for initiating plasma with a spark or a short-term arc discharge; moreover, high continuous voltage up to $V = 1.5$ kV could be applied to the electrodes, which could be used to remove the dust and the charged particles from the lamp volume.

A scheme of the experimental set up, allowing us to sustain a quasi-stationary "quiet" POD with a periodic-pulsed laser, investigate the intensity distribution of the laser radiation passed through the plasma, and acquire schlieren images of the plasma region and convective flows generated by it, is shown in figure 2.

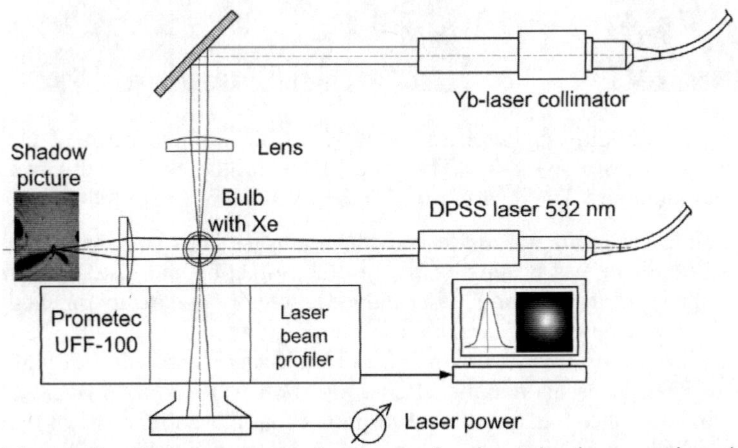

Figure 2. The experimental set up for investigating quasi-stationary plasma sustaining with nanosecond repeating pulses of pre-breakdown intensity in $p = 8 \div 16$ bar xenon. The focal ratio is $F = 4 \div 7$ ($\mathrm{NA} = 0.08 \div 0.12$), the laser pulse power is $P_L = 3 \div 6$ kW, the average power is $P = 25 \div 50$ W, the pulse length is $\tau_{FWHM} = 170$ ns, pulse repetition rate is $f_r = 50 \div 100$ kHz.

Prometec Laserscope UFF-100 laser beam profiler is used to measure the intensity distributions of the laser beam (measuring procedure described in [10]), and schlieren images are acquired in an expanded diode-pumped Nd^{3+} laser beam working at 532 nm wavelength. As the beam becomes astigmatic after passing the elliptical walls of the quartz bulb, a horizontal or vertical Foucault knife, placed in the corresponding beam waist behind the lens projecting schlieren images onto a watching screen, was used depending on the major orientation of the density gradients investigated in the gas.

3. Results of the experiments

3.1. Plasma appearance and observation conditions

While periodic-pulsed laser radiation with a wavelength of $\lambda = 1.064$ μm, a pulse repetition rate $f_r = 50$ kHz and pulse width $\tau_{FWHM} = 170$ ns being focused in xenon under the pressure $p = 8{\div}16$ bar to achieve an intensity in the focus $I = 1{\div}5$ GW/cm^2, a very regular glowing in the form of about 100 ns long flashes, repeating with the laser pulse repetition rate, could be observed in the vicinity of the beam focus after initiation procedure. The intensity threshold of the phenomenon was about $I_t = 1$ GW/cm^2 at xenon pressure $p = 14{\div}15$ bar and about $I_t = 2$ GW/cm^2 at xenon pressure $p = 8{\div}10$ bar.

The discharge observed, as opposed to the laser spark (figure 3), demonstrated relatively weak and quasi-stable glow near the beam waist without perturbations typical for a laser spark.

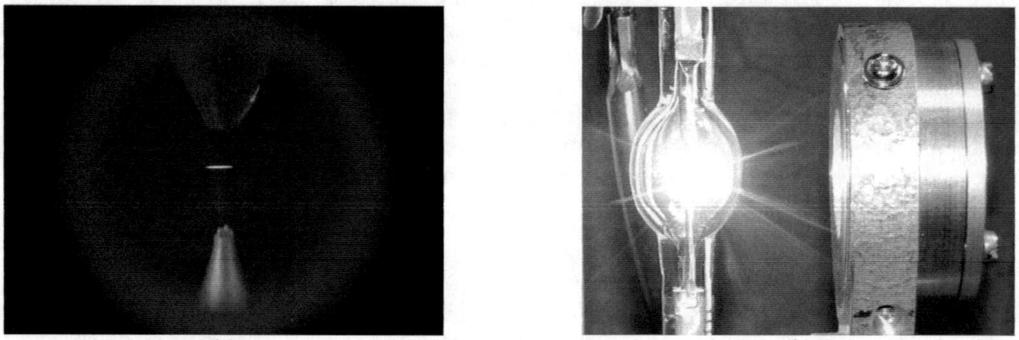

(a) (b)

Figure 3. Quasi-stationary periodic-pulsed glowing discharge zone observed near the waist of the focused laser beam (a) against common laser spark (b). The ratio of the brightness of these two discharges is about several orders of magnitude. The inter-electrode gap (a) is 2.5 mm.

The glow looked white; the visible spectrum was uniform with no strong lines, similar to the recombination spectrum of xenon. Significant gas heating, leading to the refraction and distortion of the transmitted laser beam, as well as formation of a laser-induced convective stream directed perpendicularly to the laser beam axis, were observed near the interaction region.

The time-average laser radiation absorption was less than 5% and so it was measured with high standard deviation more than 30%. The laser radiation and plasma radiation waveforms in figure 1 show that notable plasma glowing and thus the laser radiation absorption occur close to the end of the laser pulse. Low brightness of the plasma emission, low laser radiation absorption, and high pulse-to-pulse stability set the phenomenon observed apart from other periodic-pulsed optical discharge types. A stable distortion pattern of the laser beam passed through the interaction area, as well as quite stationary convective flows, allows us to talk about quasi-stationarity of the "quiet" POD itself.

An ordinary laser spark (figure 3, b) has also been observed in the same set up with the same laser at maximum laser radiation power. To obtain the laser spark, an accurate adjustment of the beam focusing was required, using a short-focus lens with $F = 4.5$ and passing the laser beam through the bulb wall area with minimal optical aberrations. Under these conditions relatively regular flashes of the laser spark could be observed, though the laser breakdown did not occur in each pulse. Since the intensity was set below the breakdown level, random factors contributed to the breakdown development, such as presence and migration of dust and active particles, ions and excited atoms inside the lamp bulb due to irregular flows created by shockwaves generated after laser spark flashes. High laser pulse repetition rate also contributed in the breakdown threshold reduction.

The occurrence of an ordinary laser spark was followed by the raise of laser radiation absorption, while the "quiet spark" pre-breakdown plasma could be observed, apart from a glow in the focus region, by a thermal lens, manifested itself in the distortion of the transmitted laser beams, both green laser beam used to obtain schlieren pictures and discharge-sustaining IR-beam.

Casual laser breakdown flashes did generally lead to initiation of the quasi-stationary glowing of the "quiet" POD, which was then stably sustained after decreasing the power density down to 1 GW/cm^2 under the pressure p = 14÷15 bar.

Figure 4, (a) shows the appearance of the glow observed in the focus area; the beam borders, the optical axis, and the focus position are also shown by lines. The glowing intensity distribution on the image in arbitrary units with real linear sizes is shown in figure 4, (b).

The laser pulse waveforms (figure 1) and the integral plasma glowing shown in figure 4 allow us to conclude that the discharge is an initial stage of an optical breakdown which does not develop due to the lack of the laser intensity and pulse width. The plasma size along the laser beam is ~1 mm (figure 4), the glowing channel diameter is of the order of magnitude of the laser beam waist (several tens microns), and the plasma recombination radiation pulse width is ~ 100 ns.

For a time of 20 μs between the pulses free electrons in the examined pressure range p = 8÷16 bar recombine and cannot be responsible for the laser radiation initial absorption. However, heated gas and some atoms in electron-excited states remain in the focus zone since the previous pulse. These atoms can provide the initial radiation absorption at 1.064 μm due to broadened 6s–6p (5d–6p, 6p–6d) spectral transitions. For example, the closest strong enough absorption line with a center at $\lambda_c = 1.053\,\mu\mathrm{m}$ corresponds to $5\mathrm{p}^5\left({}^2\mathrm{P}^0_{3/2}\right)6\mathrm{p}-5\mathrm{p}^5\left({}^2\mathrm{P}^0_{3/2}\right)6\mathrm{d}$ transition. The temperature estimation with an induced heat lens, the nature of the glowing, high stability and repeatability also tell us about a possibility of initial absorption with the 5p^{5}6s group meta-stable and resonance levels.

The authors have not succeeded in initiating a continuous optical discharge with the "quiet" spark, probably due to its bright phase being short-lived and, thus, average electron and electron-excited atom concentration being too low for COD initiation even when the provided laser radiation power is high enough.

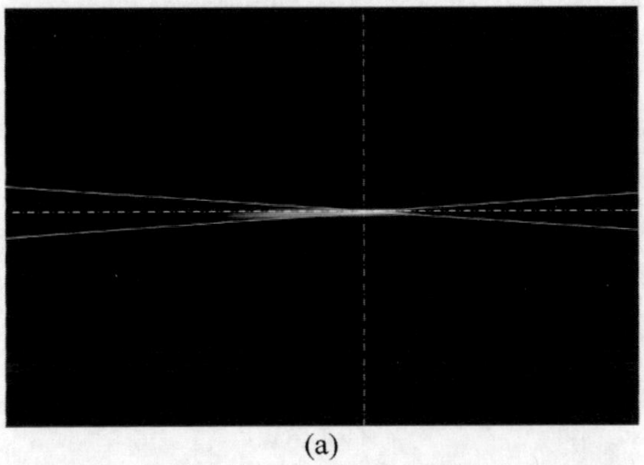

(a)

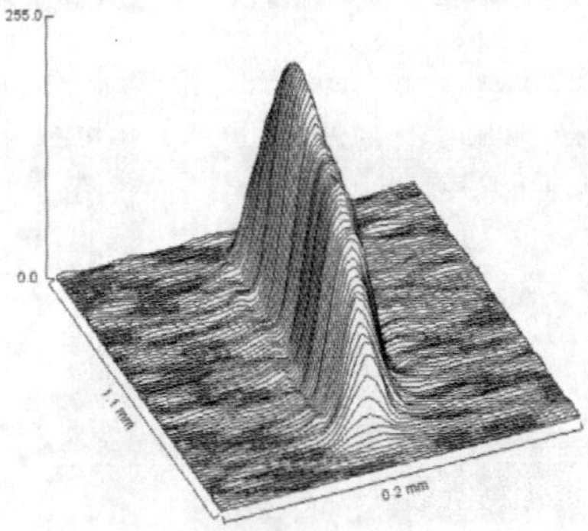

(b)

Figure 4. Spatial characteristics of the time-average discharge glow. (a) Pattern of a discharge glow. White lines show the optical axis, focus and laser beam borders (solid lines). (b) 2-D plot of the discharge glow. Plasma length is ~900 μm, its diameter is 30÷40 μm. Xenon, p = 15 bar, F = 7, P_L = 6 kW, P = 50 W, τ_{FWHM} = 170 ns, f_r = 50 kHz.

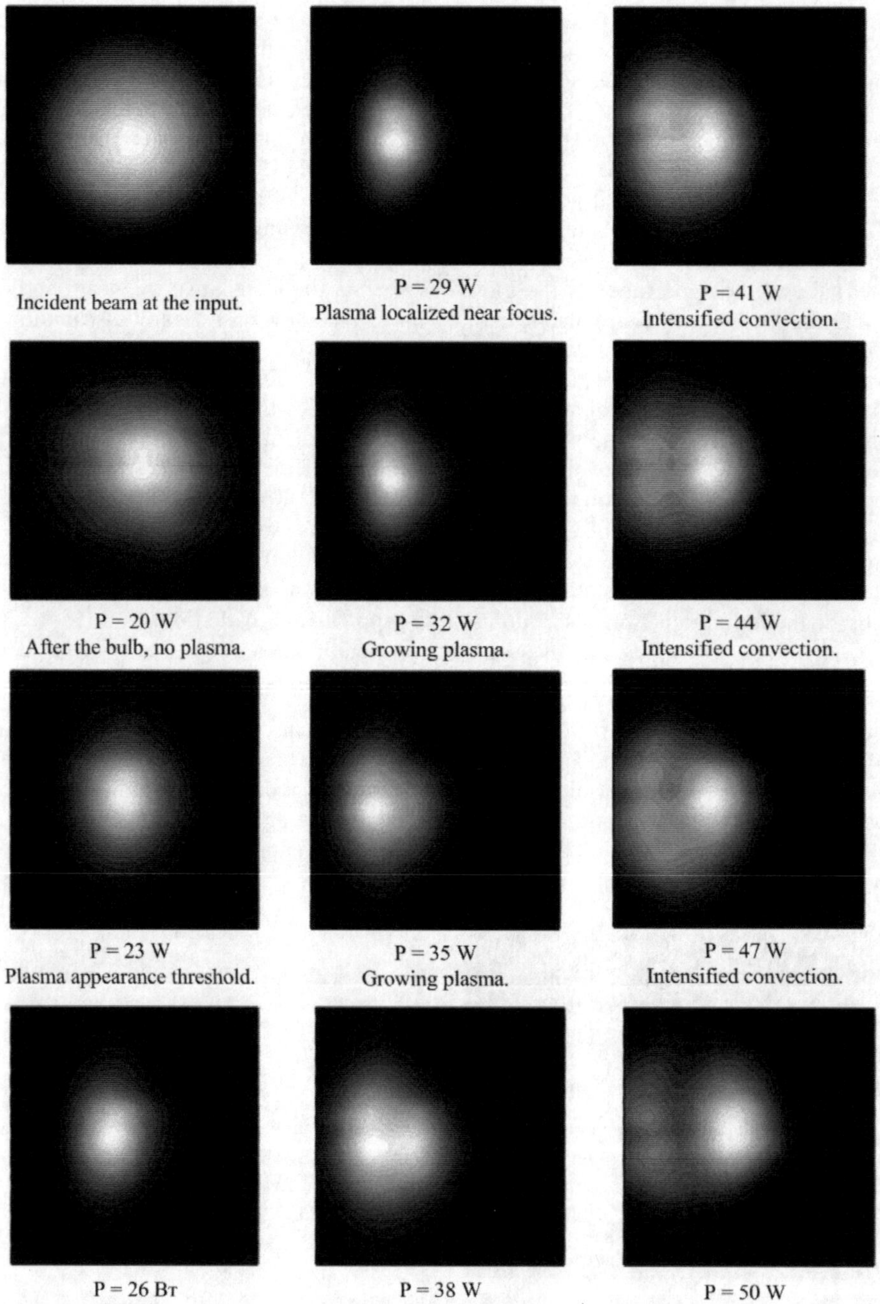

Figure 5. The distortion of the radiation intensity distributions of the YLP laser beam after passing through the interaction region. The average laser power P is varied. The beam was focused with $F = 7$ inside a bulb of a short-arc xenon lamp ($p \approx 16$ bar), $f_r = 50$ kHz. The measurements was made at a distance of 39 mm behind the focus point, the frame size is 6×6 mm^2. The symmetry axis of the pictures has no assigned direction. The real gravity vector is normal to the picture plane.

3.2 Laser beam distortion by an induced thermal lens

Figure 5 depicts the results of measurements of the laser beam average intensity distributions at a distance of 39 mm behind the lens focal plane, showing changes in the laser beam distortion by a "quiet" POD plasma, depending on the radiation average and pulse power.

With the occurrence and development of the glowing plasma, the laser beam diameter at 39 mm behind the focus is seen to become about 2 times smaller. Authors applied the method of a negative thermal lens analysis developed in [9], which presumes optical characteristics of the gas in the glowing region can be modeled as a cylindrical domain in which the refraction index is radially distributed according to parabolic law, with a minimum in the axis. Estimations followed by the model [9] show that the beam distortion observed in the cylindrical interaction region of 1 mm length and 0.1 mm diameter can be explained by assuming that the refraction index difference between the axis and the edge is 0.004, which corresponds to the temperature difference of 7 to 10 kK between the axis and the edge of the region examined, when filled with Xe, p = 15 bar. For this estimation, the temperature dependence of the refraction index at p = 15 bar pressure mentioned in [4] has been used.

The heated region diameter estimation of 0.1 mm comes from the 30 μm beam diameter, corresponding to the heating start region, according to further expanding of the gas heated zone for a 170 ns long pulse with the sound speed in xenon being 240 m/s. During the pulse, the heated gas region diameter expands up to about 100 μm diameter.

Estimation of the energy required for heating a cold gas, originally located in a waist volume of 30 μm diameter, 1 mm length, up to 10 kK temperature, gives the value of about 40 μJ, which is 4% of the laser pulse energy and corresponds to the observed laser radiation absorption in the "quiet spark" discharge. Time average energy balance of the "quite" POD was similar to that of COD [4] with thermal conductivity as dominating energy dissipation mechanism.

3.3 Laser–induced convection

The analysis of the obtained intensity distributions in the laser beam after passing the glowing zone (figure 5) and the plasma shadow images in the "quiet" laser spark mode (figures 6, 7) showed that changes are observed starting from approximately 0.5 to 0.6 mJ pulse energy, which can be interpreted as generation of quasi-stationary convective flows in a form of an expanding flat stream emitted from the centre of laser beam focus area mainly directed normally to the optical axis. This is not gravity convection, as the flow direction does not depend on the orientations of the laser beam and the bulb against the gravity force.

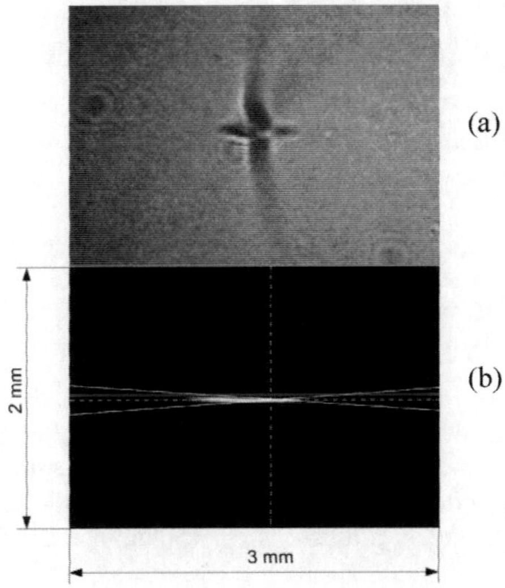

Figure 6. A schlieren image of the plasma with the convective flows (a) and the corresponding image of the plasma glowing (b) with beam border lines (solid) and symmetry lines (dash-dot) shown in the focus area (F = 7). The radiation is directed from the left to the right.

A typical schlieren image of the region around the "quiet" POD showing the convective flow observed is presented at figure 6, (a) together with a picture of the discharge glow (b). The flow is seen to come out of a small volume approximately corresponding to the glowing zone center in the form of the darker "wings" stretched normally to the optical axis on the schlieren image which indicates the heat build-up zone borders.

The convective flow image is not quite stationary and not quite symmetric. The flow evolutions are given in figure 7 in the form of a series of the discharge region schlieren images, made with 1 s intervals.

Streams of the heated gas, seen on schlieren images as dark stripes at one or both sides from the laser beam optical axis, do not flow from the discharge center symmetrically and almost normally to the optical axis, as it could be concluded from some photos. The flow periodically turns out to be seen as if it was flowing to one side, i.e. mainly down in figure 7.

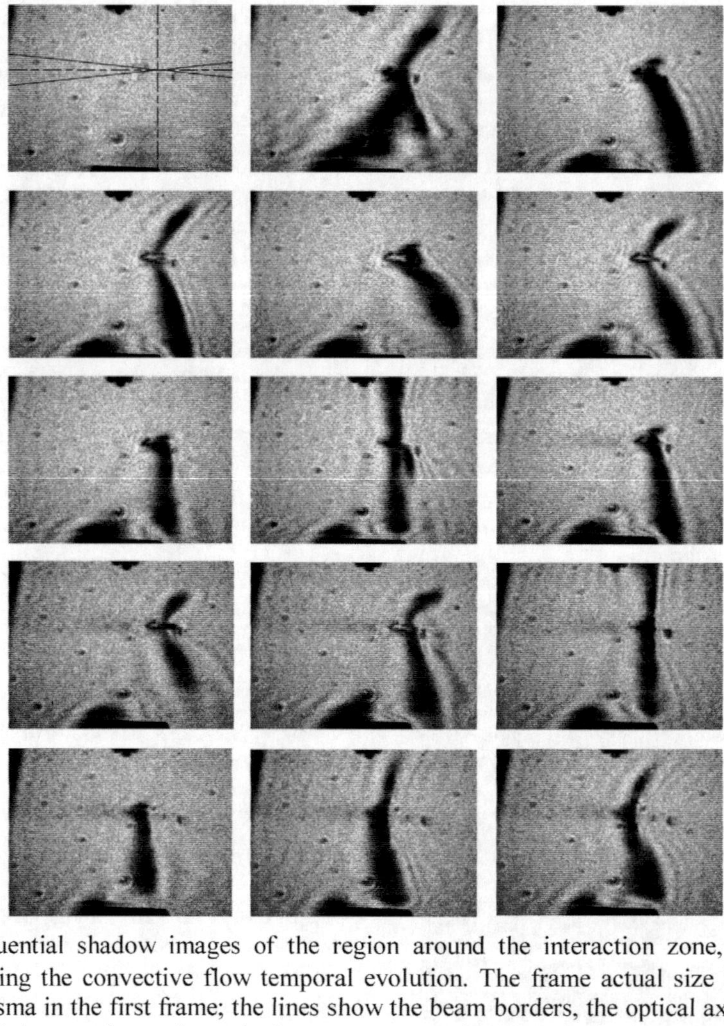

Figure 7. Sequential shadow images of the region around the interaction zone, taken with 1 s intervals, showing the convective flow temporal evolution. The frame actual size is 3.2×2.4 mm^2. There is no plasma in the first frame; the lines show the beam borders, the optical axis and the focus location. 150 W short arc lamp electrode tips are seen near the center on both upper and bottom sides of the frames, the inter electrode gap is 2.2 mm. The xenon pressure is $p \approx 15$ bar. The gravity force vector is directed upwards in this experiment.

This direction is not related to the gravity force direction, as the flow behavior is quite similar when the optical axis is oriented vertically.

Since the gravity force direction has practically no effect on the convective flow pattern, a conclusion can be made that the flow velocity is considerably higher than that of the gravity convection of a gas heated up to the same temperature.

In figure 5, changes of the passing beam refraction pattern are related to strong convective flows occurring when the average radiation power achieves $P = 35$ W (pulse energy is $E = 0.6$ mJ). The assigned axis of symmetry, alongside which the intensity distribution of transmitted laser beam expands and transforms further after the distortion in the interaction zone, appears simultaneously with the convective flow formation and generally coincide with the premature direction of the convection. The assigned direction of the convection flow appeared to occur due to random factors, such as mutual arrangement of the point of the discharge, electrodes and walls of the quartz bulb.

As it was found in the experiments, the convection flows formed in the "quiet" POD prevented stationary discharges (particularly, arc discharge and COD) from being sustained in combination with the "quiet" POD. When the authors tried to subject the stationary discharge plasma to the periodic-pulse laser radiation of the short high intensity pulses to obtain fast local heating of stationary plasma, the formation of "quiet" POD disturbed the stability of stationary discharges. All attempts to increase the laser pulse energy in the "quiet" POD caused growing plasma oscillations of stationary discharges until they went out.

4. Discussions of the results

Based on the obtained discharge heat trail schlieren images, an assumed convection flow pattern in the discharge zone is presented in figure 8 with arrows. The quasi-stationary gas flow comes into the discharge zone from opposite sides along the laser beam axis and spreads asides from the axis as a flat stream. This stream was not generally symmetric and had a dominant direction, apparently due to random factors, involving discharge volume geometry. The stream in the schlieren projection looks predominantly flat, just slightly widening, which can be explained by the interaction of the flow observed with toroidal or semi-toroidal vortical convection cells formed close to it and shown in figure 8 as couples of closed flow lines on both sides from the discharge heat trail located in the flow part coming out of the discharge zone.

Such a complicated gas flow pattern can appear as a result of periodic-pulsed heating and the following expanding and relaxation dynamics of the gas in POD. Just after the laser pulse when high-temperature small cylindrical gas volume stretched along the beam axis begins to expand, a cylindrical shock wave coming from the optical axis to the periphery is formed. The shock wave is followed by formation of a volume of moving heated gas core, which relaxation generates interacting gas flows of various directions. At long times of heated gas core relaxation, the flows can transform into directed jets and vortices.

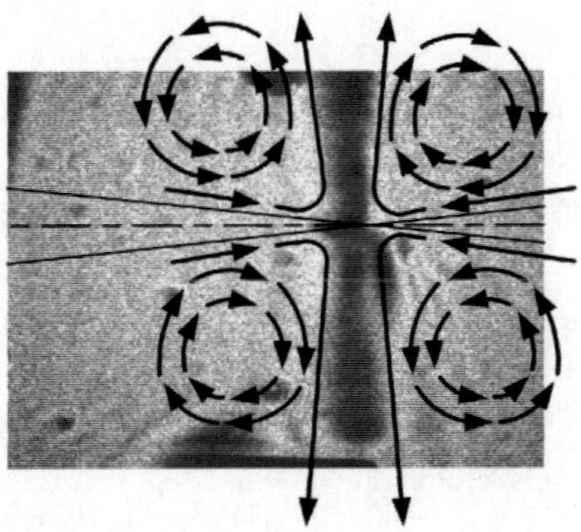

Figure 8. The assumed convection pattern in the discharge zone. The conditions and the schlieren pattern correspond to a frame from figure 7. The thinner solid lines show the laser beam conventional border, the dashed line shows the optical axis. Arrows show schematically the directions of the convective gas flows.

Calculations [11] show a flow pattern similar to that presented in figure 8, which can be formed within approximately 50 μs during relaxation of the plasma core produced in a single laser breakdown. In the authors' experiment, the pulse period was from 10 to 20 μs, and the flow picture observed was formed due to many successive pulses. The "quiet" POD mode allowed the flow pattern to be stationary and achieve the quasi-static convection mode, which in first approximation corresponds to the flow pattern at long relaxation times after a single laser breakdown.

It is stated in [11] that the flow pattern is difficult to observe at long relaxation times after a single laser breakdown due to gas cooling. Indeed, in many calculations and experiments, for example, in well-known papers [12-13], they generally just examine and register asymmetric axial flows and single vortices, developing within short time right after the laser breakdown, before the gas has cooled. The pattern of a flat expanding stream between a symmetric pair of vortices appearing at long expanding times has not been observed in experiments before present studies.

5. Summary and conclusion

The phenomenon of sustaining a stable glowing plasma filament in high pressure xenon under periodic-pulsed laser radiation with intensity lower than required for the optical breakdown of the gas has been observed in the experiments.

The phenomenon observed differs from well known optical breakdown or laser spark phenomena with relatively low plasma light emission, low laser radiation absorption, high stability and strict pulse-to-pulse repeatability. The laser intensity threshold of maintaining this new type of POD, which can be called "pre-breakdown" or "quiet" POD, is at least an order of magnitude lower than required for an optical breakdown to occur. It also should be emphasized that time average energy balance of pre-breakdown POD is similar to that of COD with predominant thermal conductivity losses.

The observed significant laser beam intensity distribution distortion after passing the discharge zone was used to estimate gas heating in the interaction area up to 10 kK, which could provide high gradients of gas density responsible for laser radiation refraction in the discharge zone.

The authors also have found that gas heating in "quiet" POD is followed by a laser-induced gas stream flowing out of the discharge zone, directed mainly normally to the optical axis, well seen on the schlieren images. The phenomenon cannot be explained through gravity convection mechanism, as the flow direction does not depend on the gravity force direction and the gas velocity is estimated to be significantly higher than that of the gravity convection. The pattern of convection flows generated by a "quiet" POD looks very similar to that formed at final relaxation stages after a single optical breakdown.

Actual realization mechanisms of the convection observed and ideas of possible applications require further studies.

Acknowledgements
The authors are grateful to S.E. Yakush for useful discussions.

References

[1] Rudoy I G, Solovyov N G, Soroka A M, Shilov A O and Yakimov M Yu 2015 Xenon Plasma Sustained by Pulse-Periodic Laser Radiation *Plasma Physics Reports* **41**(10) pp 858–861
[2] Raizer Yu P 1980 Optical discharges *Soviet Physics Uspekhi* **23** pp 789–806
[3] Raizer Yu P 1977 *Laser–induced discharge phenomena* (New York: Consultants Bureau)
[4] Zimakov V P, Kuznetsov V A, Solovyov N G, Shemyakin A N, Shilov A O and Yakimov M Yu 2016 Interaction of Near–IR Laser Radiation with Plasma of a Continuous Optical Discharge *Plasma Physics Reports* **42**(1) pp 68–73
[5] IPG Photonics Single-Mode Quasi-CW Ytterbium Fiber Lasers YLR-QCW: http://www.ipgphotonics.com/group/view/9/Lasers%2FQuasi_CW_Fiber_Lasers#[1_micron]
[6] IPG Photonics Ytterbium Nanosecond Pulsed Fiber Lasers YLPN: http://www.ipgphotonics.com/group/view/10/Lasers%2FNanosecond_Fiber_Lasers#[1.06_μm]

IOP Conf. Series: Journal of Physics: Conf. Series **815** (2017) 012003 doi:10.1088/1742-6596/815/1/012003

[7] Uhlenbusch J and Viol W 1990 H_β-line profile measurements in optical discharges *J. Quant. Spectroscopy and Radiation Transfer* **44** pp 47–56
[8] Tretyakov P K, Grachev G N, Ivanchenko A I, Krainev V L, Ponomarenko A G and Tishchenko V N 1994 Stabilization of an optical discharge in a supersonic argon flow *Physics-Doklady* **39**(6) pp 415–416
[9] Zimakov V P, Kedrov A Yu, Kuznetsov V A, Shemyakin A N, Solov'yov N G and Yakimov M Yu 2011 Laser sustained plasma ball lensing effect controlled by means of coaxial gas flow *Proc. SPIE 7913* p 791305
[10] Graham M P and Weckman D C 1995 A comparison of rotating-wire- and rotating-pinhole-type laser beam analysers when used to measure pulsed Nd:YAG laser beams *Meas. Sci. Technol.* **6**(10) pp 1492-1499
[11] Ghosh S and Mahesh K 2008 Numerical simulation of the fluid dynamic effects of laser energy deposition in air *J. Fluid Mech.* **605** pp 329–354
[12] Dors I G and Parigger C G 2003 Computational Fluid-Dynamic Model of Laser-Induced Breakdown in Air *Applied Optics* **42**(30) pp 5978–5985
[13] Glumac N G, Elliott G S and Boguszko M 2005 Temporal and Spatial Evolution of a Laser Spark in Air *AIAA Journal* **43**(9) pp 1984–1994

IOP Conf. Series: Journal of Physics: Conf. Series **815** (2017) 012004 doi:10.1088/1742-6596/815/1/012004

Numerical simulating the two-dimensional structure of the Penning discharge using the modified drift-diffusion model

S T Surzhikov

A. Ishlinsky Institute for Problems in Mechanics of the Russian academy of sciences, Vernadsky prospekt 101(1), Moscow, 119526, Russia
Dukhov All-Russian Scientific Research Institute of Automatics, Moscow, Russia

E-mail: surg@ipmnet.ru

Abstract. Spatial structure of the Penning discharge at pressures $p=0.1-0.01$ Torr is investigated within the framework of the modified drift-diffusion model. This modification includes three models of elementary physical processes, which take into account peculiarities of gas discharge processes at low pressures and large reduced electric fields. Presented results of numerical simulations are in reasonable agreement with available experimental data.

1. Introduction

Electric discharges, suggested by Penning, are investigated during more than 80 years [1–3]. The basic configuration of this discharge is defined as the discharge in cylindrical chamber which is equipped by two plane cathodes and hollow anode. An axial magnetic field is a significant peculiarity of the discharge. The typical voltage drop between electrodes in such a discharge is of $V\sim1000$ V and the magnetic field induction is of ~0.1 T.

In various applications the most popular are the Penning discharge chambers at the pressures of $p\sim10^{-2}-10^{-5}$ Torr filled with the gases H_2, D_2, Ar or Xe. At such a low pressures the magnetic field plays a significant role in maintaining a gas discharge.

In spite of wide field of application of these discharges a scientific literature contains very small amount of papers containing the description of spatial structure of such discharges. Bounded number of fragmentary experimental data does not allow perform a validation of computational codes that were developed. Note that modern methods of mathematical modeling allow predict a spatial structure of gas discharges of various kind, and, unfortunately, corresponding experimental data practically absent.

Among developments of numerical simulation methods which are intended for modeling spatial structure of gas discharges, we can stay on the methods of particles in cells (PIC). Application of the method for plasma in magnetic field, including the simulation of the Penning discharge, has been demonstrated in [4-6].

However, the well known advantages of the PIC methods for simulation of rarefied plasma become helpless at increasing the pressure in discharge chambers up to order of 1 Torr. In this case the models of the magneto hydrodynamics (MHD) appear to be efficient. In turn, the MHD models lost its advantage at low pressures. It is obvious that there is a need in developing the both classes of models.

The drift-diffusion models (DDM) are the variant of the MHD models. In previous papers by the author the DDM were used for investigating the normal direct current discharge in molecular hydrogen and nitrogen in transversal magnetic field at pressure $p\sim1$ Torr [7,8].

Two-dimensional model of the Penning discharge was investigated in [9-11]. Relatively high pressure of ~ 1 Torr was studied in [9,10]. This enables one to use the classical drift-diffusion model that is based on the Townsend formula. The first attempt to use the DDM for analysis of the Penning discharge at pressure p~10^{-3} Torr was made in [11], and this caused a necessity to modify the classical model. This modification concerns to two aspects.

Firstly, inside gas discharge chambers at low pressures (for example, at pressure p ~ 1 mTorr) the reduced field achieves high values of order $E/p=10^6$ V/cm×Torr. This means that, when calculating the mobilities it is necessary to take into account a change of physical mechanisms of interaction between charged and neutral particles under transition from low to high reduced fields [15,16].

Simple model for calculation of drift velocities of ions and electrons under transition from low to high reduced fields was proposed in [11]. The results of calculation of drift velocities using this model are shown in figure 1. It is seen that at $E/p \geq 100$ V/cm×Torr there is observable slowing down the drift velocity at increasing E/p. However, it is obvious that when $E/p > 10^6$ V/cm×Torr this model does not work.

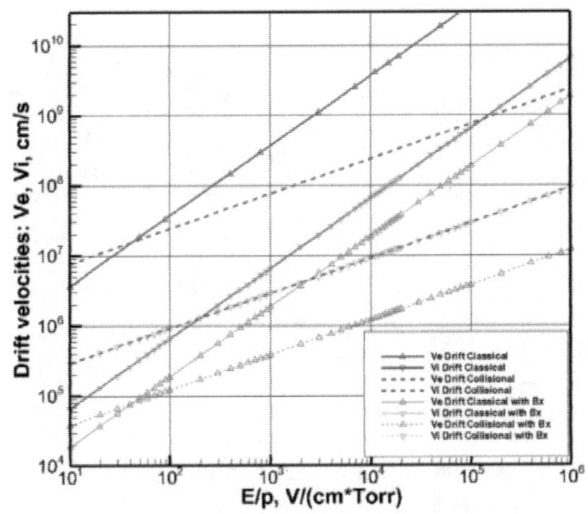

Figure 1. Drift velocities of electrons and ions versus reduce field.

The second modification relates to the method for calculation of ionizing processes at increasing E/p. It is well known that the Townsend formula gives acceptable results at $E/p < 1000$ V/cm×Torr (more precise, for molecular hydrogen for $150 < E/p < 600$ V/cm×Torr). The model of ionization coefficient proposed in [11], is illustrated in Fig.2. It is assumed that at energy of electrons less than potential ionization of hydrogen, the Townsend formula is correct. At higher energies the ionization coefficient α is calculated as follows:

$$\alpha = N\sigma_i \ ,$$

where σ_i is the cross-section of ionizing collisions predicted by the Tompson formula:

$$\sigma_i = 4\pi a_0^2 \left(\frac{I_{H_2}}{\varepsilon} \right)^2 \cdot \frac{\left(\varepsilon - I_{H_2} \right)}{I_{H_2}} \text{, cm}^2.$$

Here, I_{H_2} is the potential of ionization of H_2, $4\pi a_0^2 = 3.52 \times 10^{-16}$ cm^2; a_0^2 is the Bohr radius, and ε is the energy of electrons (as a rule, I_{H_2} and ε are measured in eV).

For cross-linking of numerical values of the first ionization coefficient final formula is wondered as:

$$\alpha = pA^{*}\exp\left[-B/\left(E/p\right)\right],$$

where A^{*}=2.75 1/cm×Torr; B = 130 V/cm×Torr.

According to figure 2 the ionization coefficients firstly increase and then decrease as E/p decreases.

So, in the Penning discharge used as the ions accelerator, there are two competitive mechanisms at applied voltage drop increasing, namely, a decreasing of ionization efficiency, and an increasing of charged particle energies.

In the present work the modified DDM is applied for analysis the Penning discharge at the pressure p = 0.01÷0.1 Torr. At these pressures the classical DDM cannot be applied for modeling the gas discharge chamber with size of ~ 1cm. Under these conditions typical reduced fields do not exceed the values of $E/p \sim 10^4$ V/cm×Torr.

Note, that reduced field in gas discharge chamber depends not only on applied voltage drop, but also on the configuration of volume charge.

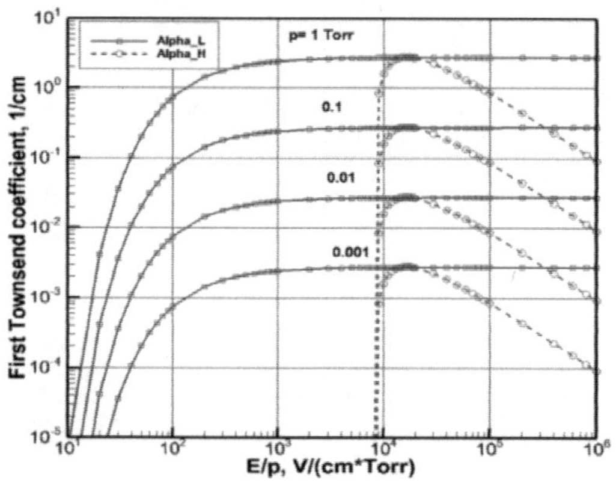

Figure 2. Ionization coefficient for modified DDM at pressure p=1.0÷0.001 Torr.

2. Governing equations

A cylindrical two-dimensional 2D glow discharge in molecular hydrogen between flat electrodes with a presence of axial magnetic field is considered (see figure.3).

A modified drift-diffusion model [10] (MDDM) is used for description of gas discharge processes. This theory is based on the continuity equations for electron and positive ion concentrations n_e and n_i together with the equations for the electro-static field $\mathbf{E} = -\mathrm{grad}\varphi$. The governing equations of the drift-diffusion theory with axial magnetic field have the following form:

$$\frac{\partial n_e}{\partial t} + \frac{\partial \Gamma_{e,x}}{\partial x} + \frac{1}{r}\frac{\partial r\Gamma_{e,y}}{\partial r} = \dot{\omega}_i = \alpha(E)\left|\mathbf{\Gamma}_e\right| - \beta n_i n_e \,, \tag{1}$$

$$\frac{\partial n_i}{\partial t} + \frac{\partial \Gamma_{i,x}}{\partial x} + \frac{1}{r}\frac{\partial r\Gamma_{i,y}}{\partial r} = \dot{\omega}_i = \alpha(E)\left|\mathbf{\Gamma}_e\right| - \beta n_i n_e \,, \tag{2}$$

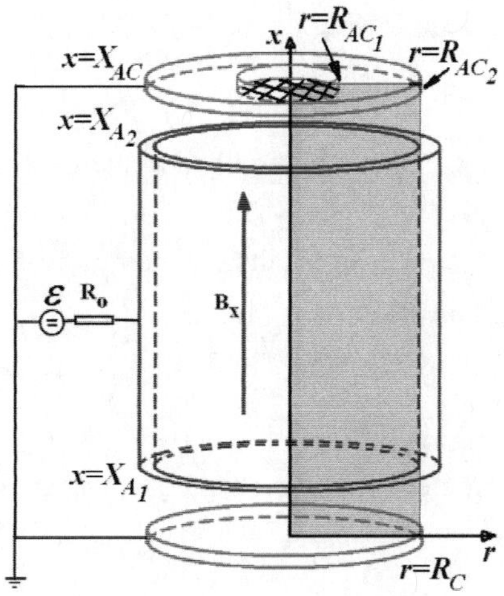

Figure 3. Schematic representation of a glow discharge in external magnetic field.

$$\frac{\partial^2\varphi}{\partial x^2}+\frac{1}{r}\frac{\partial}{\partial r}r\frac{\partial\varphi}{\partial r}=4\pi e(n_e-n_i)\ , \tag{3}$$

where

$$\mathbf{\Gamma}_e=n_e\mathbf{u}_e=-\hat{\mathbf{D}}_e\text{grad}n_e-n_e\hat{\boldsymbol{\mu}}_e\mathbf{E}\ ,\ \mathbf{\Gamma}_i=n_i\mathbf{u}_i=-\hat{\mathbf{D}}_i\text{grad}n_i+n_i\hat{\boldsymbol{\mu}}_i\mathbf{E}\ ,$$

$$\hat{\mathbf{D}}_e=\begin{pmatrix}D_{e,x} & 0\\ 0 & \dfrac{D_{e,r}}{1+b_e^2}\end{pmatrix},\ \hat{\mathbf{D}}_i=\begin{pmatrix}D_{i,x} & 0\\ 0 & \dfrac{D_{i,r}}{1+b_i^2}\end{pmatrix},\ \hat{\boldsymbol{\mu}}_e=\begin{pmatrix}\mu_{e,x} & 0\\ 0 & \dfrac{\mu_{e,r}}{1+b_e^2}\end{pmatrix},\ \hat{\boldsymbol{\mu}}_i=\begin{pmatrix}\mu_{i,x} & 0\\ 0 & \dfrac{\mu_{i,r}}{1+b_e^2}\end{pmatrix}, \tag{4}$$

$$\mathbf{j}=e(\mathbf{\Gamma}_i-\mathbf{\Gamma}_e),$$

$\alpha(E)$ and β are the ionization and recombination coefficients, Γ_e and Γ_i are the electron and ion flux densities, μ_e and μ_i are the electron and ion mobility, D_e and D_i are the electron and ion diffusion coefficients, φ is the potential of electric field, $\mathbf{u}_e$ and $\mathbf{u}_i$ are the averaged velocities of electrons and ions, $\mu_e=\dfrac{e}{m_e\nu_e}$ and $\mu_i=\dfrac{e}{m_i\nu_{in}}$ are the electrons and ions mobilities, $D_e=\dfrac{kT_e}{e}\mu_e$ and $D_i=\dfrac{kT_i}{e}\mu_i$ are the electrons and ions diffusion coefficients, ν_{en},ν_{ei} and ν_{in} are the frequencies of electron-neutral, electron-ion, and ion-neutral collisions, $\nu_e=\nu_{en}+\nu_{ei}$, T_e and T_i are the temperatures of electrons and ions.

Taking into account that the magnetic field has only *x*-component B_x (see figure 3), one can write:

$$\Gamma_{e,x}=n_eu_{e,x}=\left(-D_e\frac{\partial n_e}{\partial x}-\mu_en_eE_x\right),\quad \Gamma_{e,r}=n_eu_{e,r}=\frac{1}{1+b_e^2}\left(-D_e\frac{\partial n_e}{\partial r}-\mu_en_eE_r\right),$$

$$\Gamma_{e,\varphi}=n_eu_{e,\varphi}=-b_e\Gamma_{e,r},$$

$$\Gamma_{i,x}=n_iu_{i,x}=\left(-D_i\frac{\partial n_i}{\partial x}-\mu_in_iE_x\right),\quad \Gamma_{i,r}=n_iu_{i,r}=\frac{1}{1+b_i^2}\left(-D_i\frac{\partial n_i}{\partial r}+\mu_in_iE_r\right),$$

where $b_e=\dfrac{\mu_e}{c}B_x$ and $b_i=\dfrac{\mu_i}{c}B_x$ are the Hall parameters of electrons and ions.

The source term in the right hand side can be written as following

$$|\Gamma_e| = \sqrt{\Gamma_{e,x}^2 + \Gamma_{e,r}^2 + \Gamma_{e,\varphi}^2} \ . \tag{5}$$

Note that the azimuthal component of electronic flux $\Gamma_{e,\varphi}$ can be significant due to the condition $b_e >> 1$.

The introduced coefficients, which are taking into account a physically meaningful magnetic field, can be presented in the following form:

$$b_e = \frac{\mu_e B_x}{c} = \frac{\omega_e}{\nu_e}, \qquad b_i = \frac{\mu_i B_x}{c} = \frac{\omega_i}{\nu_{in}},$$

where

$\omega_e = \frac{eB_x}{m_e c}$ and $\omega_i = \frac{eB_x}{m_i c}$ are the Larmor frequencies of electrons and ions, m_e and m_i are the masses of electrons and ions.

The boundary conditions for charged particles and electrical potential are formulated as following:

$$x = X,\ r < R_{AC_1}: \quad \frac{\partial n_i}{\partial x} = 0, \Gamma_e = \gamma^* \Gamma_i,\ \varphi = 0,$$

$$x = 0; x = X,\ r > R_{AC_1}: \quad \frac{\partial n_i}{\partial x} = 0, \Gamma_e = \gamma \Gamma_i,\ \varphi = 0, \tag{6}$$

where γ is the coefficient of the secondly ion-electron emission, γ^* is the effective coefficient of the secondly ion-electronic emission, which is taken of the order $\gamma^* \sim (10^{-1} - 10^{-2})\gamma$.

For cylindrical surface of anode the following boundary conditions were used:

$$X_{A_1} < x < X_{A_2},\ r = R_C: \quad \frac{\partial n_e}{\partial r} = 0,\ n_i = 0,\ \varphi = V \ .$$

For cylindrical surface between anode and cathode:

$$X_{A_1} > x,\ r = R_C: \quad \frac{\partial n_e}{\partial r} = \frac{\partial n_i}{\partial r} = \frac{\partial \varphi}{\partial r} = 0 \ ,$$

$$X_{A_2} < x,\ r = R_C: \quad \frac{\partial n_e}{\partial r} = \frac{\partial n_i}{\partial r} = \frac{\partial \varphi}{\partial r} = 0 \ . \tag{7}$$

A quasi-neutral plasma cloud of a spherical shape in the center of the chamber is used as the initial condition.

The transport and thermo-physics properties do not depend on temperature (for H_2), therefore:

$$\mu_e(p) = \frac{3.7 \cdot 10^5}{p}, \frac{\text{cm}^2}{\text{V} \cdot \text{s}}, \quad \mu_i(p) = \frac{6.55 \cdot 10^3}{p}, \frac{\text{cm}^2}{\text{V} \cdot \text{s}},$$

$$D_e = \mu_e(p) T_e, \qquad D_i = \mu_i(p) T_i,$$

where p is the pressure in Torr.

The recombination coefficient β and electron temperature are taken as constants $\beta = 2 \times 10^{-7}$ cm^3/s and $T_e = 11610\text{K}$.

The ionization coefficient for H_2 is determined as follows (the 1st Townsend formula):

$$\alpha(E) = p^* A^* \exp\left[-\frac{B}{\left(|\mathbf{E}|/p^*\right)}\right], \text{cm}^{-1},$$

where $A^* = 2.75 \dfrac{1}{\text{cm} \cdot \text{Torr}}, \quad B = 130 \dfrac{\text{V}}{\text{cm} \cdot \text{Torr}}$.

Equations (1)–(3) are supplemented with the equation for the external electric circuit (see figure 1), which is written for a stationary current as

$$\varepsilon = V + IR_0 ,$$

where V is the voltage on the electrodes, I is the total discharge current, ε is the e.m.f. in power supply, and R_0 is the external resistance.

The modified drift-diffusion model includes two modifications of the classical DDM, namely, the modification of drift mobilities and the modification of velocity of ionization.

In Ref.[10] it was suggested to use the following relations for ion and electron mobilities

$$\mu_{i,eff} = \min\left\{\frac{\mu_i p}{p}, \frac{9.2 \cdot 10^4}{\sqrt{p_{Torr} E_V}}\right\}, \quad \mu_{e,eff} = \min\left\{\frac{\mu_e p}{p}, \frac{2.4 \cdot 10^6}{\sqrt{p_{Torr} E_V}}\right\}, \tag{8}$$

and for corresponding diffusion coefficients

$$D_{i,eff} = \mu_{i,eff} T_i, \; D_{e,eff} = \mu_{e,eff} T_e .$$

This gives a possibility to use classical definitions for drift velocities

$$\mathbf{V}_{dr,i} = \mu_{i,eff} \mathbf{E}, \; \mathbf{V}_{dr,e} = \mu_{e,eff} \mathbf{E}.$$

The ionization coefficient for H_2 is determined as follows:

$$\alpha_{low}(E) = pA^* \exp\left[-\frac{B}{\left(|\mathbf{E}|/p\right)}\right], \text{ cm}^{-1}, \text{ at low reduced fields} \tag{9}$$

and

$$\alpha_{high} = N_n \sigma_i \;, \text{cm}^{-1}, \text{ for high reduced fields,} \tag{10}$$

where the ionization cross section σ_i could be calculated using the Tomson formula

$$\sigma_i = 4\pi a_0^2 \left(\frac{I_{H_2}}{\varepsilon}\right)^2 \cdot \frac{\left(\varepsilon - I_{H_2}\right)}{I_{H_2}}, \text{ cm}^2.$$

Instead of the reduced fields we can use the dependence on the electron and ion energies, where kinetic energies are estimated as following:

$$\varepsilon_{[eV]} = 2.84 \cdot 10^{-14} V^2_{e[cm/s]}, \quad \varepsilon_{[eV]} = 1.05 \cdot 10^{-12} V^2_{i[cm/s]} .$$

Now in the MDD model we can formulate constitutive relations for electrons:

$$\Gamma_{e,x} = -D_{e,eff} \frac{\partial n_e}{\partial x} - \mu_{e,eff} n_e E_x ,$$

$$\Gamma_{e,r} = -\frac{D_{e,eff}}{1 + b^2_{e,eff}} \frac{\partial n_e}{\partial r} - \frac{\mu_{e,eff}}{1 + b^2_{e,eff}} n_e E_r ,$$

$$\Gamma_{e,\varphi} = -b_{e,eff}\Gamma_{e,r}, \quad b_{e,eff} = \frac{\mu_{e,eff} B_x}{c},$$

and for ions:

$$\Gamma_{i,x} = -D_{i,eff}\frac{\partial n_i}{\partial x} + \mu_{i,eff} n_i E_x ,$$

$$\Gamma_{i,r} = -\frac{D_{i,eff}}{1+b_{i,eff}^2}\frac{\partial n_i}{\partial r} + \frac{\mu_{e,eff}}{1+b_{i,eff}^2} n_i E_r ,$$

$$\Gamma_{i,\varphi} = -b_{i,eff}\Gamma_{i,r}, \quad b_{i,eff} = \frac{\mu_{i,eff} B_x}{c} .$$

3. Results of numerical simulation

The calculations were performed for the following initial data: the pressure in molecular hydrogen H_2 was of p=0.1 ÷ 0.01 Torr; e.m.f. of power supply E =200 V, R_0 = 3000 Ohm, $\gamma = 0.33$, $\gamma^* \sim 10^{-2}\gamma$, $B_x = 0.1$T, $R_C = R_{AC_2} = 0.55$ cm, $R_{AC_1} = 0.2$ cm, $X_{A_1} = 0.425$ cm, $X_{A_2} = 0.675$ cm, and $X_C = 1.1$ cm (this value was used as a spatial scale of the problem under consideration).

Figure 4 shows the electron and ion concentrations in discharge chamber at pressures p=0.1 and 0.01 Torr. The main feature of these distributions is high concentrations of charged particles in the central domain of the discharge chamber at high pressure and low concentrations when the pressure is low.

Distributions of electron and ion concentrations at pressures p=0.1 Torr (left picture) and p=0.01 Torr (right picture) are shown in figure 4. At low pressure the domain of volume charge is shifted to internal cylindrical surface of the anode.

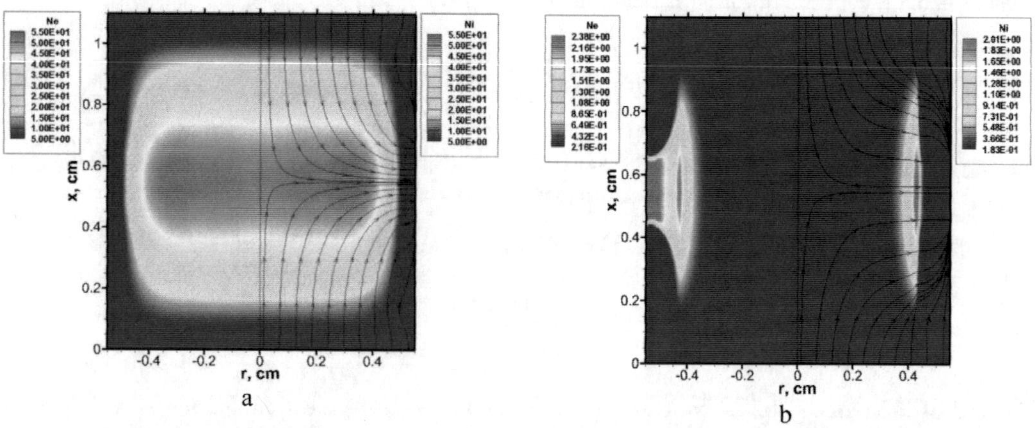

Figure 4. Concentration of (a) electrons Ne and (b) ions Ni (in 10^{10} cm^{-3}).

Decreasing of the pressure leads to significant change of the reduced field in the gas discharge chamber. From figure 5 one can see that near anode the reduced field E/p sharply increases, whereas in the volume the reduced field is distributed quite smoothly. The distribution of the ionization coefficient (see figure 6) corresponds to this distribution of the reduced field.

Differences between coefficients of ionization α_x and α_r, corresponding to the axial and radial fluxes of electrons points out to a "switching" from one dependence to another (see relations (9), (10)).

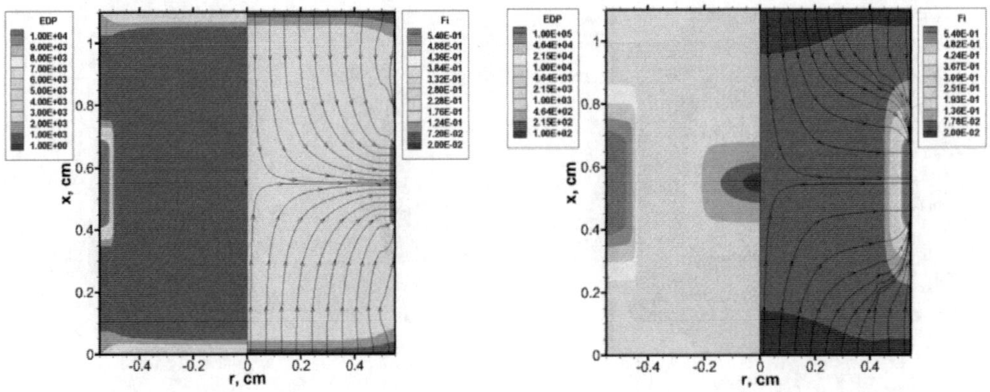

a b

Figure 5. (a) Reduced field (EDP=E/p in V/(cm*Torr)) and (b) electric potential (Fi = $\varphi/\mathcal{E}$) at p = 0.1 Torr and p = 0.01 Torr.

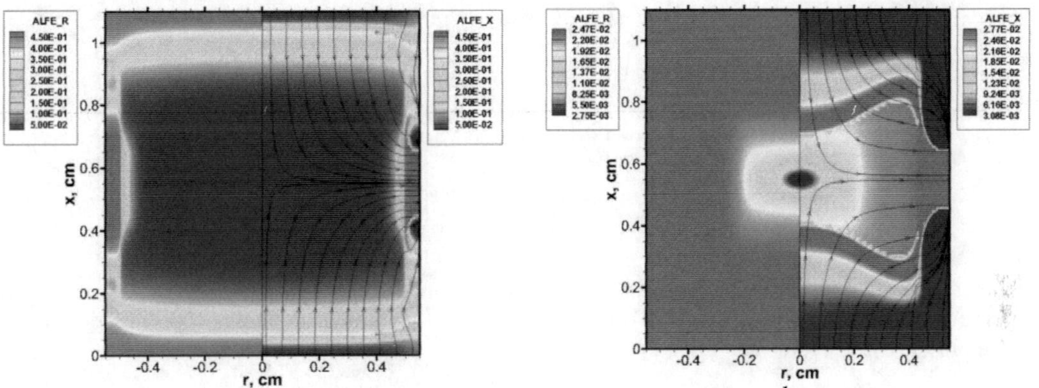

a b

Figure. 6. Coefficients of ionization (in 1/cm) at (a) p = 0.1 Torr and (b) at p = 0.01 Torr; ALFE_R=α_r is the coefficient of ionization calculated with E_r, ALFE_X=α_x is the coefficient of ionization calculated with E_x.

It is seen that at p = 0.1 Torr these differences are insignificant, whereas at p = 0.01 Torr the efficiency of ionization of neutral particles induced by electron collisions are different.
The source term $\dot{\omega}_i$ in Eqs.(1) and (2) is shown for two pressures in figure 7. This term is decreased by two order of the value when the pressure is decreased one order of the value. The volume of increased ionization is located near anode. So, such model of ionization in the Penning discharge gives low efficiency of ionization in the discharge volume. Nevertheless, presented results of numerical simulations are in reasonable qualitative agreement with available experimental data [1-3].

Axial distributions of ion and electron velocities in the Penning discharge at pressures p = 0.1 Torr and p = 0.01 Torr are shown in figure 8. The corresponding axial distributions of the reduced electric field and the energy of electrons, which were calculated using the axial velocity of electrons, are shown in figure 9.

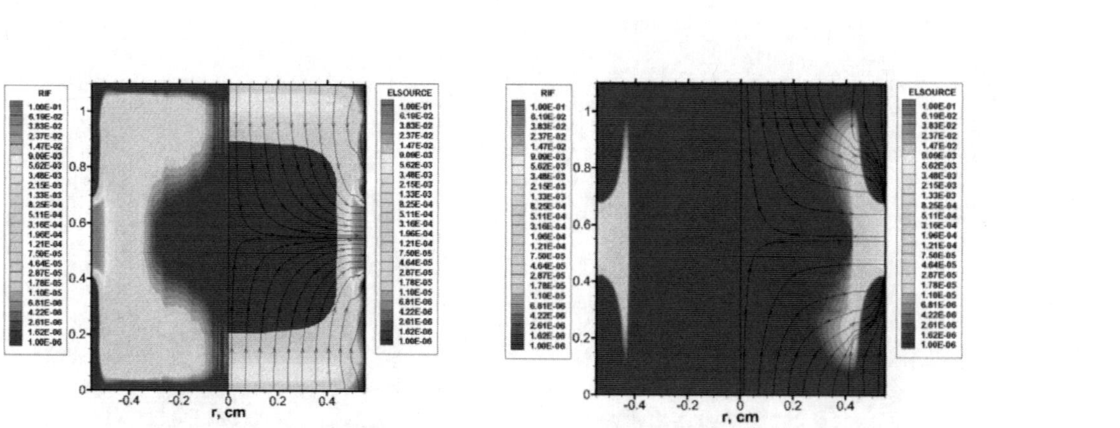

Figure. 7. Source of ionization ELSOURCE= $\dot{\omega}_i$ (in $10^{20}/(cm^3 * s)$) at (a) p=0.1 Torr and (b) p=0.01 Torr. Left parts of the figures show source of ionization due to azimuthal movement of electrons.

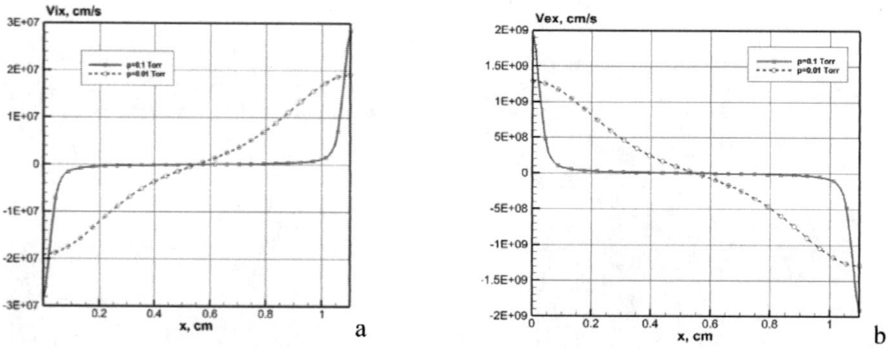

Figure. 8. Axial distributions of (a) ion and (b) electron average velocities in x-direction at the pressure p=0.1 Torr and 0.01 Torr.

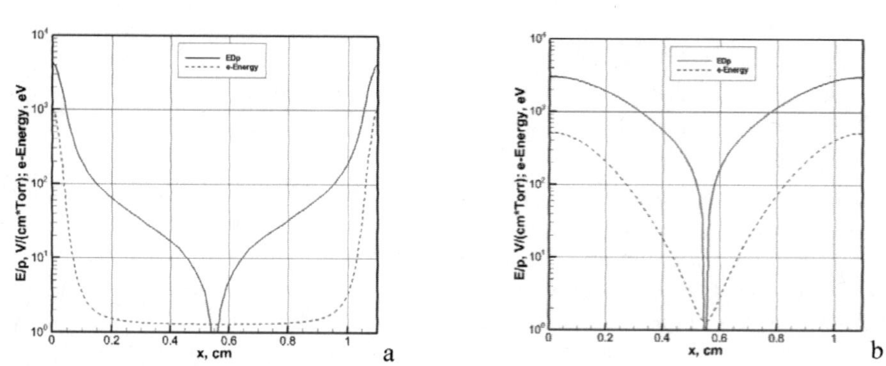

Figure 9. Axial distribution of reduced field (in V/(cm*Torr) and electron energy in eV at the pressure p=0.1 Torr and 0.01 Torr.

Conclusion

Spatial structure of the Penning discharge at pressures $p = 0.1 - 0.01$ Torr has been investigated within the framework of the modified drift-diffusion model. This modification includes three models of elementary physical processes, which take into account peculiarities of gas discharge processes at low pressures and large reduced electric fields. These are:

1) The non-linear model of behavior of ion and electron drifts depending on the reduced electric field in wide range of variation ;
2) The model of ionization of neutral particles in collisions with electrons in wide range of variation of the reduced electric field;
3) The anisotropic model of ionization by the axial, radial, and azimuthal fluxes of electrons.

The analysis of the presented predictions of spatial structure of the Penning discharge using the modified drift-diffusion model allows to suggest that the significant disadvantage of the model consists in an absent of the explicit source term connected with the low-energy electrons that were born in gas discharge volume after the first acts of ionizations by the high-energy electrons, being accelerated near cathode and anti-cathode. Presented MDD model should be improved in this aspect.

Acknowledgments
The work was supported by the Russian Science Foundation grant # 16-11-10275.

References
[1] Penning F M 1936 The Glow Discharge at Low Pressure Between Coaxial Cylinders in an Axial Magnetic Field *Fhysica III* (9) pp 873-894
[2] Hirsch E N 1964 On the Mechanizm of The Penning Discharge *Brit. J. Appl. Phys.* **15** pp 1535-1544
[3] Safronov B G , Konovalov V G 1974 Investigation of a Penning Discharge at Low Pressure. *NASA Technical Translation NASA TT F-15,523* p 14
[4] Birdsall C K , Langdon A B 1985 Plasma Physics, via Computer Simulation. *McGraw-Hill, New York* p 464
[5] Surzhikov S T 1999 Expansion of Multi-Charged Plasma Clouds Into Ionospheric Plasma With Magnetic Field. *Journal of Spacecraft and Rockets* **36**(6) pp 907-911
[6] Dikalyuk A S 2017 Development of Particle-in-Cell Solver for Numerical Simulation of Penning Discharge *AIAA 2017-0842* p 22
[7] Surzhikov S T, Shang J S 2014 Normal Glow Discharge in Axial Magnetic Field *Plasma Sources Sciences and Technology* **23**(054017) p 8
[8] Surzhikov S T 2016 Drift-Diffusion Model of Normal Glow Discharge in an Axial Magnetic Field *Doklady Physics* **61**(12) pp 596-600
[9] Surzhikov S T 2017 The Two-Dimensional Structure of the Penning Discharge in a Cylindrical Chamber with Axial Magnetic Field at Pressure of about 1 Torr *Technical Physics Letters* **43**(2) pp 169-172
[10] Surzhikov ST 2017 Numerical Study of Acceleration of the Hydrogen Ions in the Penning Discharge at Pressures Around 1 Torr *AIAA 2017-0155* p 13
[11] Surzhikov S T 2015 Application of the Modified Drift-Diffusion Theory to Study of the Two-Dimensional Structure of the Penning Discharge *AIAA 2015-1832* p 27
[12] Von Engel A, and Steenbeck M 1932 Elektrische Gasentladungen, Vol.II *Springer, Berlin* p 68
[13] Surzhikov S T 2013 Computational Physics of Electric Discharges in Gas Flows *Walter de Gruyter, Berlin/Boston* p 428
[14] Shang J S 2016 Computational Electromagnetic-Aerodynamics *Wiley, Hoboken, New Jersey* p 424
[15] Huxley L G H , Crompton R W 1974 The Diffusion and Drift of Electrons in Gases *John Wiley & Sons. New York* p 669
[16] McDaniel E W 1964 Collision Phenomena in Ionized Gases *John Wiley & Sons, Inc., New York* p 797 p
[17] McDaniel E W, Mason E A 1973 The Mobility and Diffusion of Ions in Gases *Willey and Sons, New York, London, Sydney, Toronto* p 381

IOP Conf. Series: Journal of Physics: Conf. Series **815** (2017) 012005 doi:10.1088/1742-6596/815/1/012005

Surface electromagnetic actuator in rarefied hypersonic flow

S T Surzhikov

A. Ishlinsky Institute for Problems in Mechanics of the Russian academy of sciences, Vernadsky prospekt 101(1), Moscow, 119526, Russia

Dukhov All-Russian Scientific Research Institute of Automatics, Moscow, Russia

E-mail: surg@ipmnet.ru

Abstract. Hypersonic flow past the surface with sharp edge is investigated. This surface forms an obtuse angle therefore the shock wave generated by the leading edge interacts with the surface. The effect of influence of the surface direct current discharge and a transverse magnetic field on the gas dynamic characteristics is investigated. To solve this problem a numerical simulation is used. The calculation model includes the set of the Navier – Stokes and energy conservation equations, as well as equations of electrodynamics in the ambipolar approximation, and the Poisson equation.
Results of numerical modelling the gas dynamics and electrodynamics of gas discharge with magnetic field show the change in the structure of the shock-wave interaction with surface far from location of the gas discharge. It is shown that the low-current glow discharge can be used as an electromagnetic actuator in hypersonic flows.

1. Introduction

In the late 90th of the last century, a series of experimental studies to provide data on the mass and heat fluxes for aerodynamic configurations typical for hypersonic vehicles [1] were carried out. The results of these experiments were thoroughly documented that allowed different groups to perform a validation of computational studies for author's computer codes developed by them. It is important that the results of the validation of these calculations were analyzed in detail by the group of highly qualified specialists in the field of computational aerodynamics (CFD), and the results of this analysis were published in the proceedings of a number of AIAA conferences. A review of these papers is presented in [1]. Among the studied aerodynamic configurations, special attention was focused on

- Blunt and sharp cones at Mach number $M = 8$,
- Blunt cone with flare at $M = 8 \div 10$,
- Hollow sharp cylinder with flare at $M = 9 \div 11$,
- Interaction of shock waves at Mach~10 (the fourth type of interaction in the Edney classification),
- Reflection of stationary shock waves with forming the regular configuration or the Mach configuration, as well as the hysteresis phenomena.

A series of experimental and theoretical works was devoted to the interaction of shock waves with the turbulent boundary layers in various configurations of the flow. Each of these configurations is of great interest for the study of hypersonic flows. In accordance of the purpose of this paper we draw attention to one of these tasks - hypersonic flow formed near surface which forms an obtuse angle with a sharp front edge. The detailed analysis of this problem was done in the papers [1–6], where the hollow-cylinder-flare flows were studied. Baseline data for one of the selected experiments (CUBRC Run # 14) are shown in table 1 (data are taken from table 1 [3]).

Among other above mentioned works we highlight the work [3], where not only the results of calculations using two computer codes (WIND [7] and DPLR [8]), but also accomplished study of the properties obtained from numerical solutions on different meshes are presented. It was found that the quality of computational meshes and the quality of the calculation method applied is a problem of particular importance. The good agreement between calculated and experimental data is obtained only on very detailed grids.

Table 1.

Freestream Conditions	Extended Hollow Cylinder Flare
CUBRC Run	14
Test gas	N_2
Mach Number	10.3
V_∞, **cm/s**	2.304×10^5
p_∞, **erg/cm³**	315.2
T_∞, **K**	120.4
ρ_∞, **g/cm³**	8.810×10^7
T_w, **K**	295.2

In this paper by the example of the problem of streamline of the sharp edge and surface, forming an obtuse angle, the impact of the glow discharge placed on the surface on the structure of the flow is examined. Schematic picture of the problem is shown in figure 1.

General formulation of the problem of creating plasma actuators was discussed in [9–12]. Earlier, the glow discharges localized on surface in the hypersonic flow was numerically studied in [13–15]. A detailed presentation of the numerical methods used to solve the problem of plasma actuators on the basis of the surface glow discharge of low power is given in [16, 17].

In the present paper an electromagnetic actuator formed by two electrodes made in the form of strips arranged transverse to the flow on the surface, as shown in figure 1. The subject of a numerical study is to identify perturbations in the pressure and heat flux distributions along the surface of the obtuse angle. As an additional control parameter of the plasma actuator a transverse magnetic field, which the polarization is shown in figure 1, is considered.

2. Gas dynamic model

We consider streamline of a sharp leading edge by a viscous, heat-conducting and partially ionized gas. The surface forms an obtuse angle before which the glow discharge is lighted, and magnetic field is applied in the cross-flow direction. The glow discharge is organized between two rectangular sections of the electrode at the surface, as shown in figure 1. An external electric circuit is taken into account. It consists of a power source and an ohmic resistance. In [9–12] it has been shown that the order in arrangement of electrodes in the flow is important. In present paper, it is assumed that the cathode is located upstream (ground electrode in figure 1).

The set of governing equations of the dynamics of a viscous, heat-conducting, and partially ionized gas takes into account the current flowing through that as well as the azimuthal magnetic field. It looks as follows:

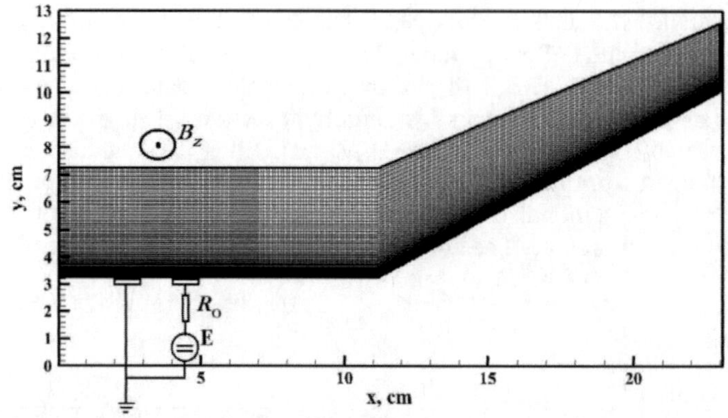

Figure 1.

$$\frac{\partial \rho}{\partial t}+\mathrm{div}(\rho\mathbf{V})=0\ , \quad (1)$$

$$\frac{\partial \rho u}{\partial t}+\mathrm{div}(\rho u\mathbf{V})=-\frac{\partial p}{\partial x}-\frac{2}{3}\frac{\partial}{\partial x}(\mu\,\mathrm{div}\mathbf{V})+\frac{\partial}{\partial y}\left[\mu\left(\frac{\partial u}{\partial y}+\frac{\partial v}{\partial x}\right)\right]+2\frac{\partial}{\partial x}\left(\mu\frac{\partial u}{\partial x}\right)+f_{M,x}, \quad (2)$$

$$\frac{\partial \rho v}{\partial t}+\mathrm{div}(\rho v\mathbf{V})=-\frac{\partial p}{\partial y}-\frac{2}{3}\frac{\partial}{\partial y}(\mu\,\mathrm{div}\mathbf{V})+\frac{\partial}{\partial x}\left[\mu\left(\frac{\partial u}{\partial y}+\frac{\partial v}{\partial x}\right)\right]+2\frac{\partial}{\partial y}\left(\mu\frac{\partial v}{\partial y}\right)+f_{M,y}, \quad (3)$$

$$\rho c_p\frac{\partial T}{\partial t}+\rho c_p\mathbf{V}\mathrm{grad}T=\frac{\partial p}{\partial t}+\mathrm{div}(\lambda\,\mathrm{grad}T)+\mathbf{V}\,\mathrm{grad}(p)+Q_\mu+Q_J, \quad (4)$$

$$\frac{\partial n}{\partial t}+\mathrm{div}(\mathbf{V}n)=\mathrm{div}\left(\mu_e^* D_a\mathrm{grad}n\right)+\dot{\omega}_e, \quad (5)$$

$$\mathrm{div}\left[(\mu_i+\tilde{\mu}_e)n\mathbf{E}+\left(\tilde{D}_e-D_i\right)\mathrm{grad}n\right]=0, \quad (6)$$

where x and y are the Cartesian coordinates connected with unit vectors $\boldsymbol{i}$ and $\boldsymbol{j}$; $\mathbf{V}=(u,v)$ is the velocity vector and its projections on axis x and y; ρ and p are the density and pressure; μ is the dynamic viscosity coefficient; c_p is the heat capacity at constant pressure; T is the temperature of the gas; λ is the heat conductivity coefficient; $f_{M,x}$, $f_{M,y}$ are the components of the ponderomotive force; Q_J is the heat release power due to the electric current; Q_μ is the dissipative function:

$$Q_\mu=\mu\left[2\left(\frac{\partial u}{\partial x}\right)^2+2\left(\frac{\partial v}{\partial y}\right)^2+\left(\frac{\partial v}{\partial x}+\frac{\partial u}{\partial y}\right)^2+\frac{2}{3}\left(\frac{\partial u}{\partial x}+\frac{\partial v}{\partial y}\right)^2\right], \quad (7)$$

$\dot{\omega}_e=\alpha(E)|\Gamma_e|-\beta_e n^2$ is the source term related to the ionization (the first summand) and to the recombination (the second summand); $\alpha(E)$ is the coefficient of ionization (the 1st Townsend coefficient); β_e is the recombination coefficient; $|\Gamma_e|$ is the module of the vector of electron flux; E is the module of the electric field strength; D_a is the ambipolar diffusivity; $\mathbf{j}=e(\mathbf{\Gamma}_i-\mathbf{\Gamma}_e)$ is the vector of current density; $\mathbf{\Gamma}_i$ and $\mathbf{\Gamma}_e$ are the vectors of the ion and electron flux densities.

An approximation of the quasineutral plasma is used, therefore $n=n_i=n_e$, where n_i, n_e are the concentrations of ions and electrons;

$$\tilde{\mu}_e = \frac{\mu_e}{1+b_e^2}, \quad \mu_e^* = \frac{\mu_e + \mu_i}{\left(1+b_e^2\right)\mu_i + \mu_e} \cong \frac{\mu_e}{\left(1+b_e^2\right)\mu_i + \mu_e}, \quad \tilde{D}_e = \frac{D_e}{1+b_e^2},$$

$b_e = \dfrac{\mu_e B_z}{c}$ is the Hall parameter, B_z is the magnetic field induction, c is the speed of light.

When one takes into account the magnetic field the term in the right hand side in equation (5) should be modified as follows:

$$\dot{\omega}_e = \alpha(E)\tilde{\mu}_e E - \beta n^2 .$$

Equation (6) is the Maxwell equation for conservation of electric current under steady-state conditions. Equation (5) expresses the continuity of the charged components of the partially ionized gas [16,17]. To obtain this equation, the following assumptions were used:

- velocities of the charged and neutral particles in partially ionized gas are equal, $\mathbf{V} = \mathbf{V}_e = \mathbf{V}_i$ ($\mathbf{V}$ – is the mean velocity of neutral particles, $\mathbf{V}_e$, $\mathbf{V}_i$ are the mean velocities of electrons and ions),
- the concentrations of ions and electrons are equal ($n = n_e = n_i$),
- charged particles diffuse with a factor of ambipolar diffusivity D_a.

Taking into account that the electric field strength is connected with the electric potential φ $\mathbf{E} = -\mathrm{grad}\varphi$, instead of equation (6) it is more preferable to use the following equation for potential:

$$\frac{\partial \varphi}{\partial x}\left(\frac{\partial b_e \tilde{\mu}_e n}{\partial x}\right) - \frac{\partial \varphi}{\partial y}\left(\frac{\partial b_e \tilde{\mu}_e n}{\partial y}\right) + \frac{\partial}{\partial x}\left[(\mu_i + \tilde{\mu}_e) n \frac{\partial \varphi}{\partial x}\right] + \frac{\partial}{\partial y}\left[(\mu_i + \tilde{\mu}_e) n \frac{\partial \varphi}{\partial y}\right] -$$

$$-\frac{\partial}{\partial x}\left[\left(\tilde{D}_e - D_i\right)\frac{\partial n}{\partial x}\right] - \frac{\partial}{\partial y}\left[\left(\tilde{D}_e - D_i\right)\frac{\partial n}{\partial y}\right] + \frac{\partial}{\partial x}\left[b_e \tilde{D}_e \frac{\partial n}{\partial y}\right] - \frac{\partial}{\partial y}\left[b_e \tilde{D}_e \frac{\partial n}{\partial x}\right] = 0 . \qquad (8)$$

Equations (1) – (6) are recorded in the Cartesian coordinate system. To calculate the structure of gas flow the new variables in the curved geometry are introduced $\xi = \xi(x, y)$, $\eta = \eta(x, y)$. These variables are connected with Cartesian coordinates by the one-to-one transformations. Now, the coordinate ξ is directed along surface, and the η -coordinate is directed normally to surface.

The boundary conditions for system (1) – (6) are presented in the following form (see figure 1):

$$\xi = 0: \quad u = V_\infty, \quad \upsilon = 0, \quad T = T_\infty, \quad p = p_\infty, \quad \rho = \rho_\infty, \quad n = n_\infty, \quad \frac{\partial \varphi}{\partial \xi} = 0;$$

$$\xi = L: \quad \frac{\partial u}{\partial \xi} = \frac{\partial v}{\partial \xi} = \frac{\partial T}{\partial \xi} = \frac{\partial n}{\partial \xi} = \frac{\partial \rho}{\partial \xi} = \frac{\partial \varphi}{\partial \xi} = 0 ;$$

$$\eta = 0, \quad \xi > \xi_{Edge}, \qquad u = \upsilon = 0, \quad \frac{\partial p}{\partial \eta} = 0 ;$$

$\eta = 0$, $\xi_{C_1} \le \xi \le \xi_{C_2}$ (the cathode section):

$$\frac{\partial n}{\partial \eta} = 0, \quad \varphi = 0;$$

$\eta = 0$, $\xi_{A_1} \le \xi \le \xi_{A_2}$ (the anode section): $\dfrac{\partial n}{\partial \eta} = 0, \quad \varphi = V_c$;

$\eta = 0$; $\xi_{Edge} < \xi < \xi_{C_1}$, $\xi_{C_2} < \xi < \xi_{A_1}$, $\xi > \xi_{A_2}$ (the dielectric surface):

$$n = 10^{-5} n_0, \quad \frac{\partial \varphi}{\partial \eta} = 0 ;$$

$$y = H, \quad x > 0: \quad u = V_\infty, \quad v = 0, \quad T = T_\infty, \; p = p_\infty, \quad \rho = \rho_\infty, \quad n = n_\infty, \quad \frac{\partial \varphi}{\partial \eta} = 0 .$$

Here: ξ_{Edge} is the longitudinal coordinate of the leading sharp edge; $\xi_{C_1}, \xi_{C_2}, \xi_{A_1}, \xi_{A_2}$ are the longitudinal coordinates of the cathode and anode boundaries; n_0 is the typical concentration of charged particles in glow discharge ($n_0 \sim 10^{10}$ см$^{-3}$); V_c – is the voltage drop between cathode and anode. Surface temperature T_w =300 K is assumed as constant for all calculations.

The electro-technical equation for external electric circuit is used in the following form:

$$IR_0 + V_c = \mathrm{E} , \tag{9}$$

from which the voltage drop V_c can be found, R_0 is the ohmic resistance, I is the total current in electric circuit, which is determined by integral over the surfaces of electrodes:

$$I = \int_0^L (\mathbf{jn})_c \mathrm{d}x = \int_0^L (\mathbf{jn})_a \mathrm{d}x ,$$

where $\mathbf{n}$ is the unit vector normal to cathode (c) and anode (a) surfaces; $\mathbf{j}$ is the vector of current density. All dimensions in the task are related to the lengths of the electrodes in z–direction.

Gas dynamic profiles without electric discharge, as well as a plasma cloud with charged particles concentration of above cathode sections were used as initial conditions for the calculations. The time-asymptotic method was used for integration of governing equations therefore the precise type of initial conditions was not very significant. Nevertheless, due to nonlinear character of the problem under consideration the initial conditions should be quite accurate.

In accordance with [16] the hypersonic flow of molecular nitrogen is considered. The constitutive relationships for N_2 are based on the molecular-kinetic theory [18]:

$$\mu_m = 2.67 \cdot 10^{-5} \sqrt{M_A T} \frac{1}{\sigma^2 \Omega^{(2,2)*}} , \; \text{g/(cm·s)},$$

is the dynamic viscosity,

$$\lambda = 8.334 \times 10^{-4} \sqrt{\frac{T}{M_A}} \frac{\left(0.115 + 0.354 \frac{c_p}{R_0} M_A\right)}{\sigma^2 \Omega^{(2,2)*}} , \; \text{W/(cm·K)},$$

is the heat conductivity,

$$c_p = 8.317 \times 10^7 \frac{7}{2} \frac{1}{M_A}, \quad \rho = \frac{p}{T} \frac{M_A}{R_0} , \; M_A = 28 \text{ g/mole},$$

$$R_0 = 8.314 \times 10^7 \text{ erg/(mole·K)}, \; \sigma = 3.68 \text{ Å}, \; \Omega^{(2,2)*} = 1.157 \left(T^*\right)^{-0.1472} ,$$

$$T^* = \frac{T}{(\varepsilon/k)}, \quad (\varepsilon/k) = 71.4 .$$

Coefficient of ionization was calculated as follows [16]:

$$\nu_i = \left(\frac{\alpha}{p^*}\right) p^* E \mu_e \left(p^*\right), \quad \left(\frac{\alpha}{p^*}\right) = A \exp\left[-\frac{B}{\left(E/p^*\right)}\right], \text{ (cm·Topp)}^{-1},$$

where: $p^* = p\frac{293}{T}$, $D_e = \mu_e\left(p^*\right)T_e$, $D_+ = \mu_+\left(p^*\right)T_+$, cm^2/s,

$\mu_e\left(p^*\right) = 4.2\times10^5\frac{1}{p^*}$, $\mu_+\left(p^*\right) = 1450\frac{1}{p^*}$, cm^2/(s·V),

$\beta = 2\times10^{-7}$ см^3/с, $T_e = 11610$ К, $A = 12$ (cm·Torr)$^{-1}$, $B = 342$ V/(cm·Torr),

μ_+ is the drift of ions.

Heat release due to the electric current (the Joule heat) is calculated by using the calculated current density and electric field strength:

$$Q_J = \eta(\mathbf{j}\cdot\mathbf{E}) = 1.6\times10^{-19}\eta\left[nE^2\left(\mu_+ + \tilde{\mu}_e\right) + \left(\tilde{D}_e - D_+\right)\mathbf{E}\,\mathrm{grad}n\right], \tag{10}$$

where η is the coefficient of efficiency of gas heating. Setting this ratio in the range $\eta \sim 0.1 \div 0.3$ allows partially take into account the shortcomings of the model used in the description of the kinetics of physical and chemical processes, as well as the processes of vibrational relaxation.

An additional force acted on gas due to the presence of magnetic field has two components:

$$\mathbf{F}_M = f_{M,x}\mathbf{i} + f_{M,y}\mathbf{j} = c^{-1}[\mathbf{jB}]. \tag{11}$$

These two projections of the magnetic field are taken into account in equations (2) and (3).The set of equations (1) – (6) describing a motion of a viscous, heat-conducting, and partially ionized gas along the surface with glow discharge at the presence of external magnetic field was solved by the type-asymptotic method up to achieving a numerical convergence with a relative error of 10^{-5}, which is calculated over the entire field of flow for functions u, v, p, T, n, φ. At each step the following equations were integrated numerically:

- the continuity equation and the Navier-Stokes equations,
- the energy conservation equation,
- the set of electrodynamic equations.

The AUSM method was used for integrating the gas dynamic equations. The energy conservation equation was integrated with the use of the implicit finite-difference method of the second order accuracy in time and in space. The same numerical approach was used for integration of the electrodynamic equations. This part of the numerical simulation procedure was the most time-consuming due to exponential character of the source of ionization.

Second internal iterative process (at each time step) consisted in achieving the mutual convergence of functions obtained at solving the Fourier-Kirchhoff equations (the energy conservation equation) and the set of electro-dynamic equations. The need in this iterative process was caused by the fact that the gas temperature is very strongly influenced by its ionization rate, which, in turn, produces strong perturbations in the right-hand side of equation (5).

3. Results of numerical simulation

Two configurations of mutual location of electrodes were investigated. Coordinates of electrodes for these configurations are shown in table 2. The choice of these two configurations follows from the obvious fact that creation of an artificial disturbance to the leading edge leads to significant consequences for the entire flowfield.

IOP Conf. Series: Journal of Physics: Conf. Series **815** (2017) 012005 doi:10.1088/1742-6596/815/1/012005

Figure 2 shows the distribution of the gas-dynamic parameters of the streamlined surface without an electrical discharge. It is assumed that the edge of the plate is located at a distance of 1 cm from the computational domain boundaries.

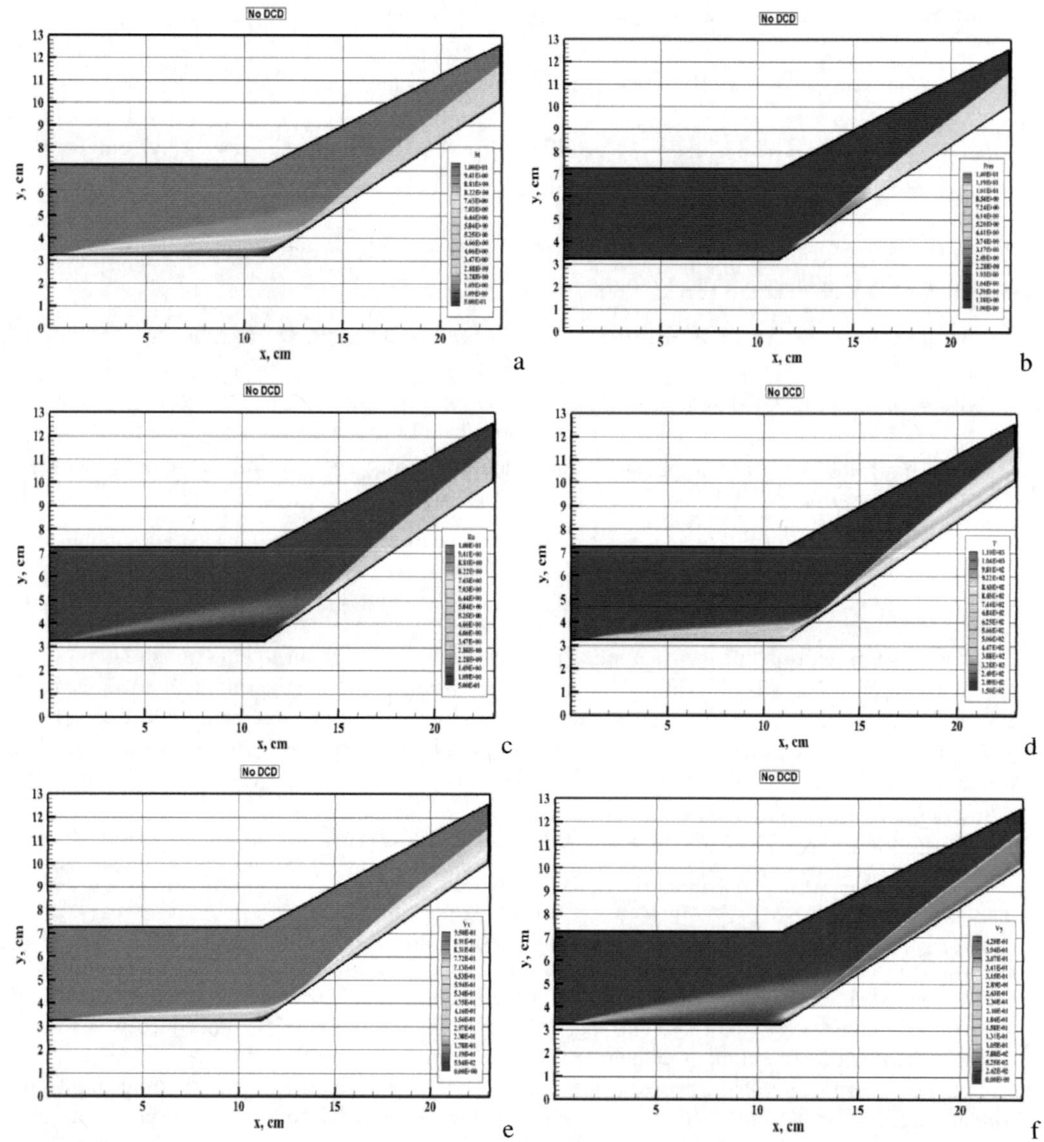

Figure 2. Gas dynamic parameters without gas discharge: *a* – Mach numbers; *b* – pressure (p/p_∞); *c* – density (ρ/ρ_∞); *d* – temperature (in K); *e* – longitudinal velocity (u/V_∞), *f* – transversal velocity (υ/V_∞).

Distributions of Mach number (figure 2a) and the longitudinal velocity (figure 2f) shows that near the sharp bend of the streamlined surface (cm) low speeds are observed. Calculations on detailed computational grids with high accuracy methods show that there is a fairly extensive area of the vortex motion of the gas. In present case, the length of this zone does not exceed of about 1 cm.

From figure 2*f* one can see that two areas of the flow spread are in the positive direction. The first area is from the leading edge, and the second area is from the sharp turn of the surface. The most high pressure is observed on the surface in place of the fall of the bow shock. In the vicinity of this place there is a noticeable increase in temperature (figure 2*d*) and, as a consequence, the increasing of the heat flux to the surface. Here there is also a natural increase in density (figure 2*c*).

The formulation of present task includes an examination of the possibility to modify the gas dynamic functions by the use of the surface glow discharge of low power. Figure 3 shows the distribution of the gas-dynamic parameters when the glow discharge is used for the first electrode configuration

Table 2.

	Configuration # 1	**Configuration #2**
x_{C_1}, **cm**	2	3
x_{C_2}, **cm**	3	4
x_{A_1}, **cm**	4	5
x_{A_2}, **cm**	5	6

The first configuration of the electrodes is considered with e.m.f. $E = 200\,\text{V}$ and the ohmic resistance of $R_0 = 300\,\text{k}\Omega$. In the place of the gas discharge plasma location one can observe heating of the gas up to $T \sim 1100\,\text{K}$. This temperature is commensurable with the temperature near place of the shock wave interaction with surface.

Perturbations in pressure and density distributions are insignificant, whereas changes in the longitudinal and transversal velocity are visible.

In the case of the second configuration of electrodes with e.m.f. E = 200 V and $R_0 = 300\,\text{k}\Omega$ the similar perturbation in distributions of gas dynamic functions close to the new arrangement of the electrodes are observed. In this case the results of numerical simulation are shown in figure 4.

Significantly more noticeable perturbations of gas-dynamic parameters are observed in the magnetic field. Corresponding numerical results are shown in figures 5 and 6.

A polarization of the magnetic field produces a force component $f_{M,y}$ towards the surface, i.e. pushing the gas to the streamlined surface.

Figures 5 and 6 show the distribution of the gas-dynamic parameters for the 1st and 2nd electrode configurations respectively.

Before analyzing the characteristics of gas-dynamic perturbations associated with the interaction between the surface glow discharge and external magnetic field, we consider the appropriate distribution of the electro-dynamic parameters shown in figures 7 and 8. Figures 7a and 8a show the distribution of electric potential, which clearly shows the location of the anode and cathode (the largest potential).

Figures 7b and 8b show axial component of the ponderomotive force. Near the cathode (or rather, near its right border) it is positive, and at the left border of the anode it is negative, which leads to a marked braking of gas flow. The normal component of the ponderomotive force is directed to the surface. The highest value of the heat release power occurs to be between the cathode and the anode (figures 7d and 8d). In the same area the greatest concentration of charged particles (figures 7e and 8e) is observed.

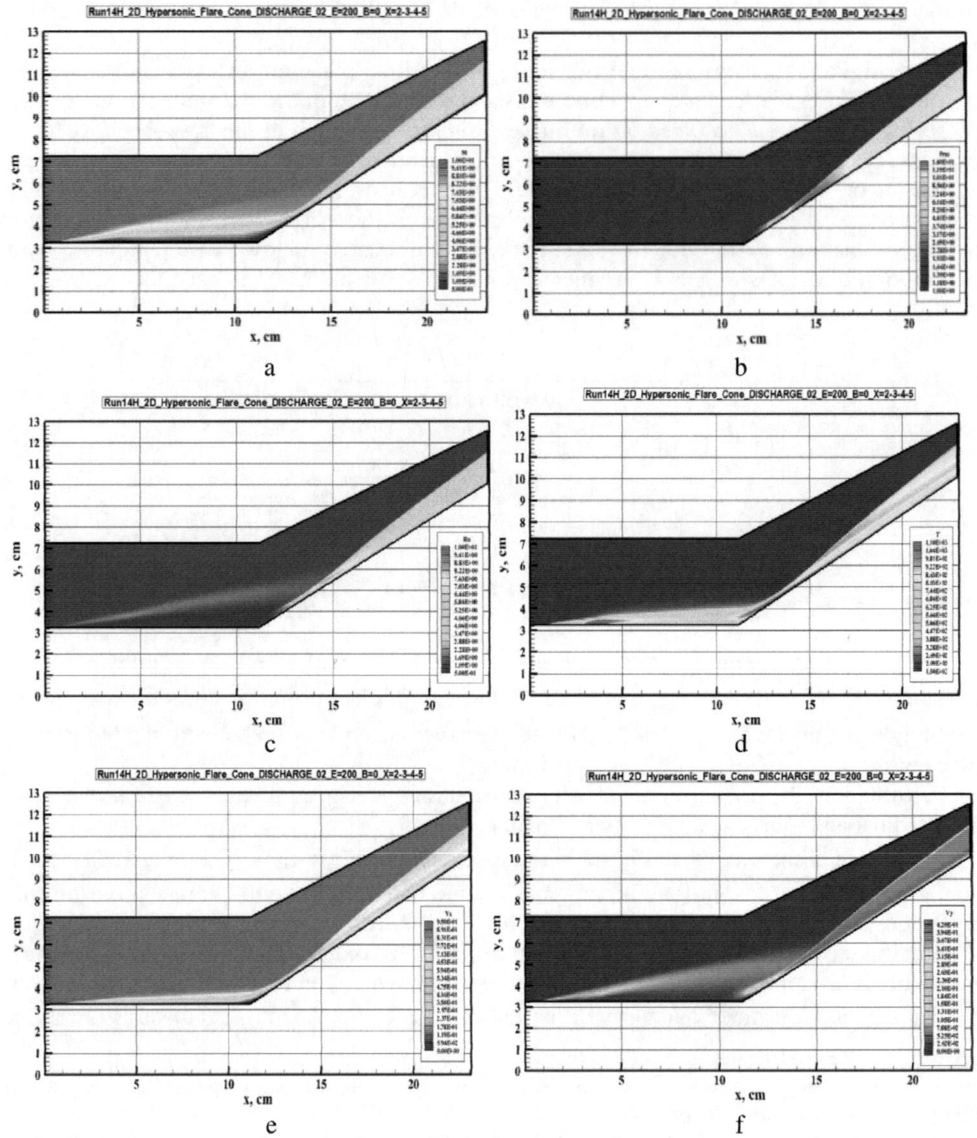

Figure 3. No magnetic field. Gas dynamic parameters with gas discharge of the first configuration: a – Mach numbers; b – pressure (p/p_∞); c – density (ρ/ρ_∞); d – temperature (in K); e – longitudinal velocity (u/V_∞), f – transversal velocity (v/V_∞).

Perturbations of the gas-dynamic parameters shown in figures 5 and 6 are conditioned by the presents of glow discharge with magnetic field which general characteristics are shown in figures 7 and 8.

The localized domain of deceleration of the gas flow between the cathode and the anode is visible in figures 5a,e and 6 a,e . In this domain the normal velocity component dramatically increases from the surface (figures 5f and 6f), as well as the local domain of high pressure and density are seen (figures 5 b and 6b, figures 5c and 6c). It is noteworthy that in this area the gas is heated to a temperature of ~2300 K, which is higher than that in the discharge without magnetic field.

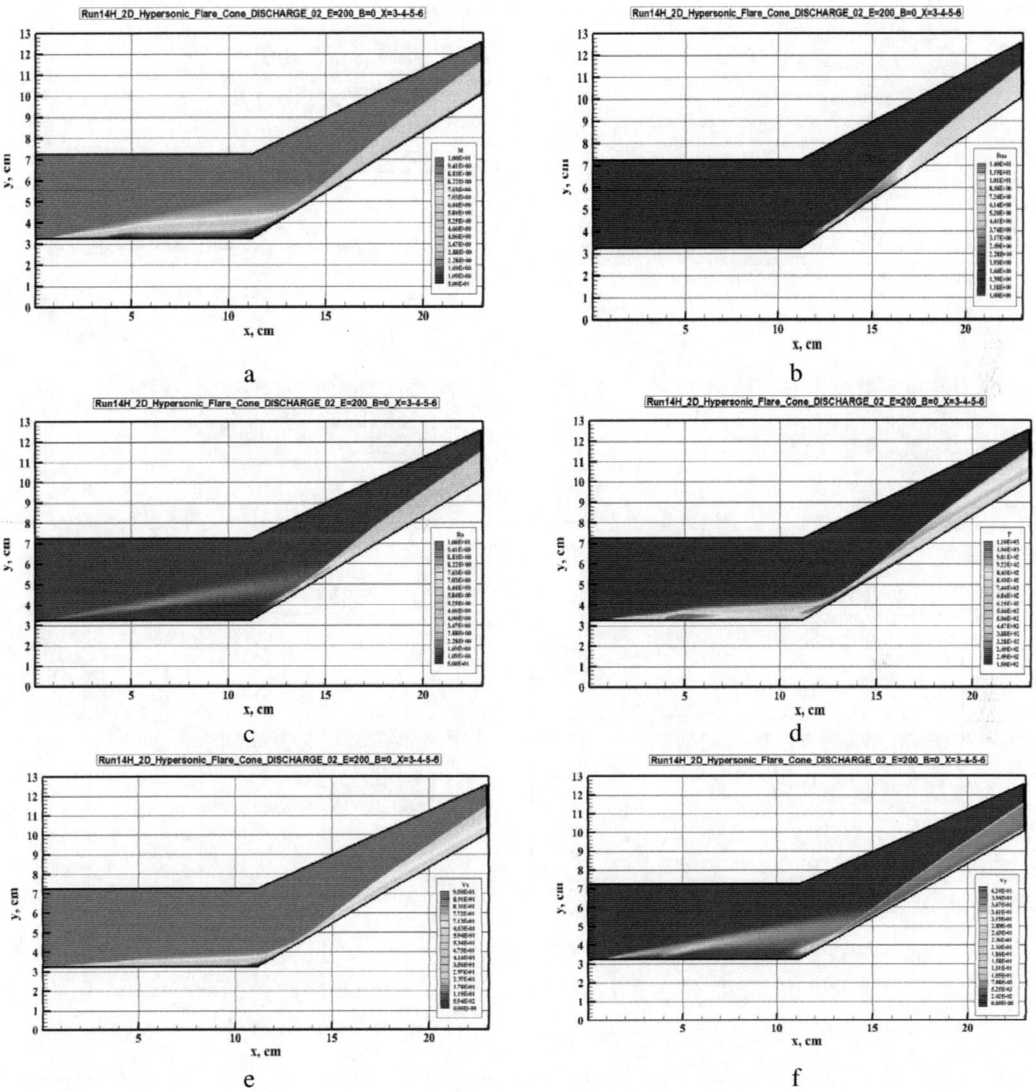

Figure 4. No magnetic field. Gas dynamic parameters with gas discharge of the second configuration: a – Mach numbers; b – pressure (p/p_∞); c – density (ρ/ρ_∞); d – temperature (in K); e – longitudinal velocity (u/V_∞), f – transversal velocity (υ/V_∞).

IOP Conf. Series: Journal of Physics: Conf. Series **815** (2017) 012005 doi:10.1088/1742-6596/815/1/012005

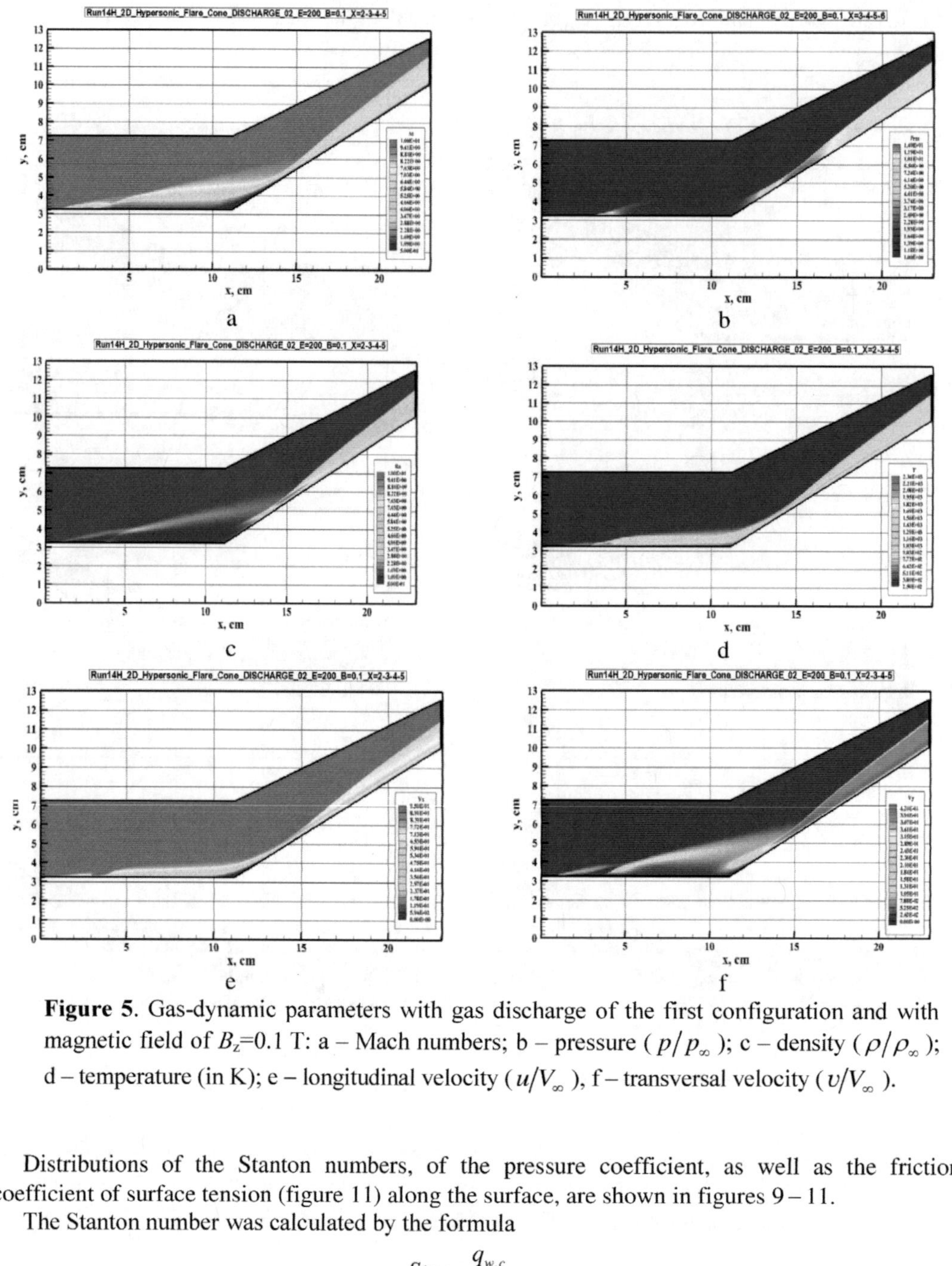

Figure 5. Gas-dynamic parameters with gas discharge of the first configuration and with magnetic field of B_z=0.1 T: a – Mach numbers; b – pressure (p/p_∞); c – density (ρ/ρ_∞); d – temperature (in K); e – longitudinal velocity (u/V_∞), f – transversal velocity (υ/V_∞).

Distributions of the Stanton numbers, of the pressure coefficient, as well as the friction coefficient of surface tension (figure 11) along the surface, are shown in figures 9 – 11.

The Stanton number was calculated by the formula

$$St = \frac{q_{w,c}}{\frac{1}{2}\rho_\infty V_\infty^3},$$

where $q_{w,c}$ is the convective heat flux.

The coefficients of the pressure and friction are calculated as following

$$c_p = \frac{p - p_\infty}{\frac{1}{2}\rho_\infty V_\infty^2}, \quad c_f = \frac{\tau_w}{\frac{1}{2}\rho_\infty V_\infty^2} = \frac{\left(\mu \dfrac{\partial u}{\partial y}\right)_w}{\frac{1}{2}\rho_\infty V_\infty^2} .$$

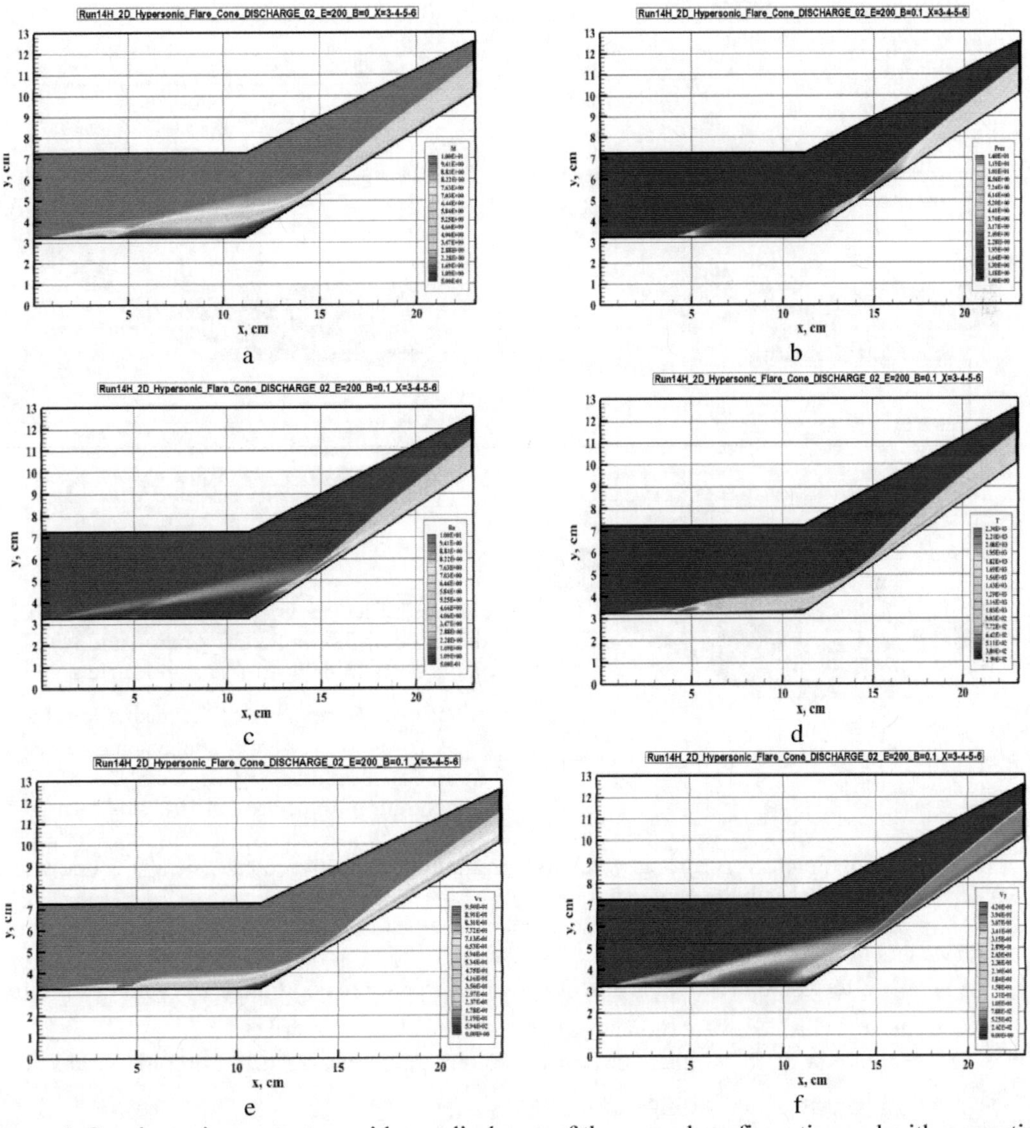

Figure 6. Gas-dynamic parameters with gas discharge of the second configuration and with magnetic field of B_z=0.1 T: a – Mach numbers; b – pressure (p/p_∞); c – density (ρ/ρ_∞); d – temperature (in K); e – longitudinal velocity (u/V_∞), f – transversal velocity (v/V_∞).

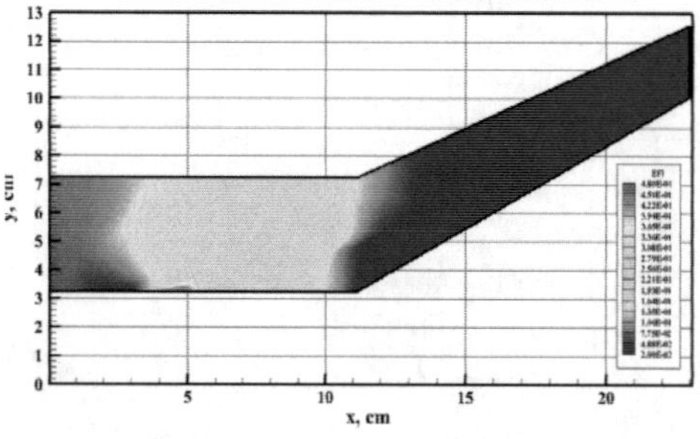

a

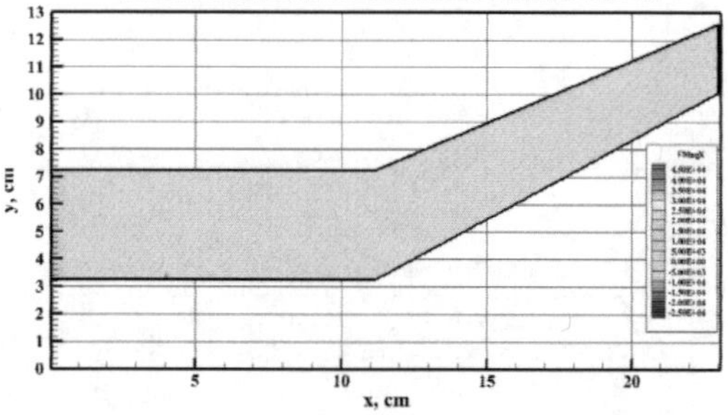

b

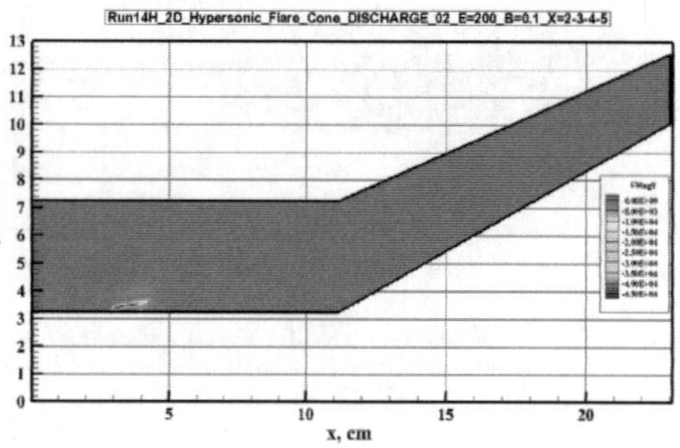

c

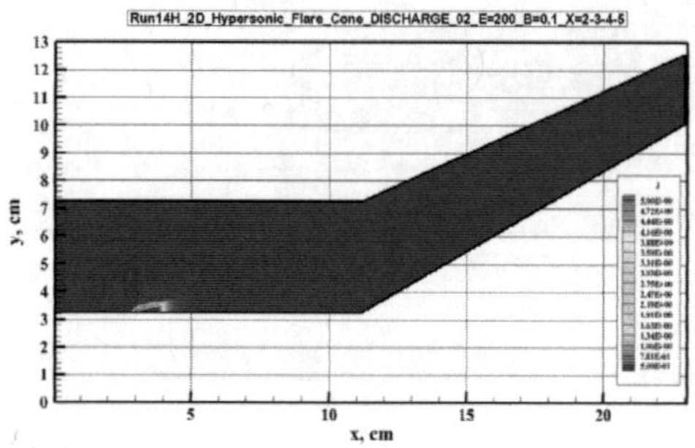

d

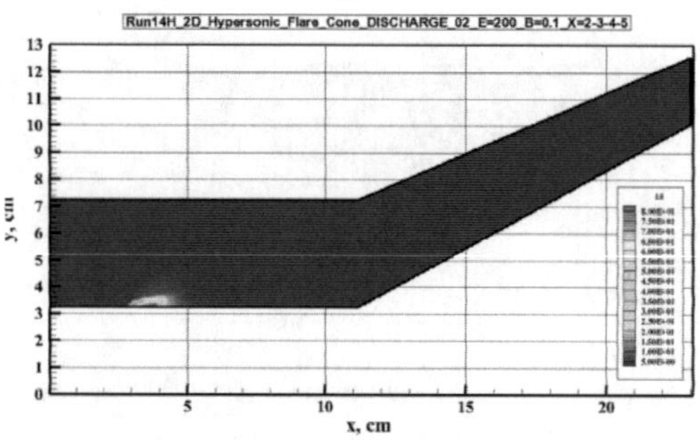

e

Figure 7. Electrodynamic parameters with gas discharge of the first configuration and with magnetic field of B_z=0.1 T: a – electric potential (φ/Ɛ), b – *x*-component of the magnetic force (dyn), c – *y*-component of the magnetic force (dyn), d – current density (mA/cm^2), e – concentration of electron in 10^{10} cm^{-3}.

The Stanton coefficient very strongly increases in the domain between cathode and anode without discharge, especially with the magnetic field

An important effect of the electric discharge and magnetic field is a shift of the heating domain by the amount of ~3 cm along the inclined surface.

A similar effect is observed in the distribution of the pressure coefficient c_p with the difference that it increases at displacement of the maximal pressure along the surface.

The distribution of the surface friction coefficient c_f points to increasing the vortex in gas discharge with magnetic field (figures 11 a, b).

IOP Conf. Series: Journal of Physics: Conf. Series **815** (2017) 012005 doi:10.1088/1742-6596/815/1/012005

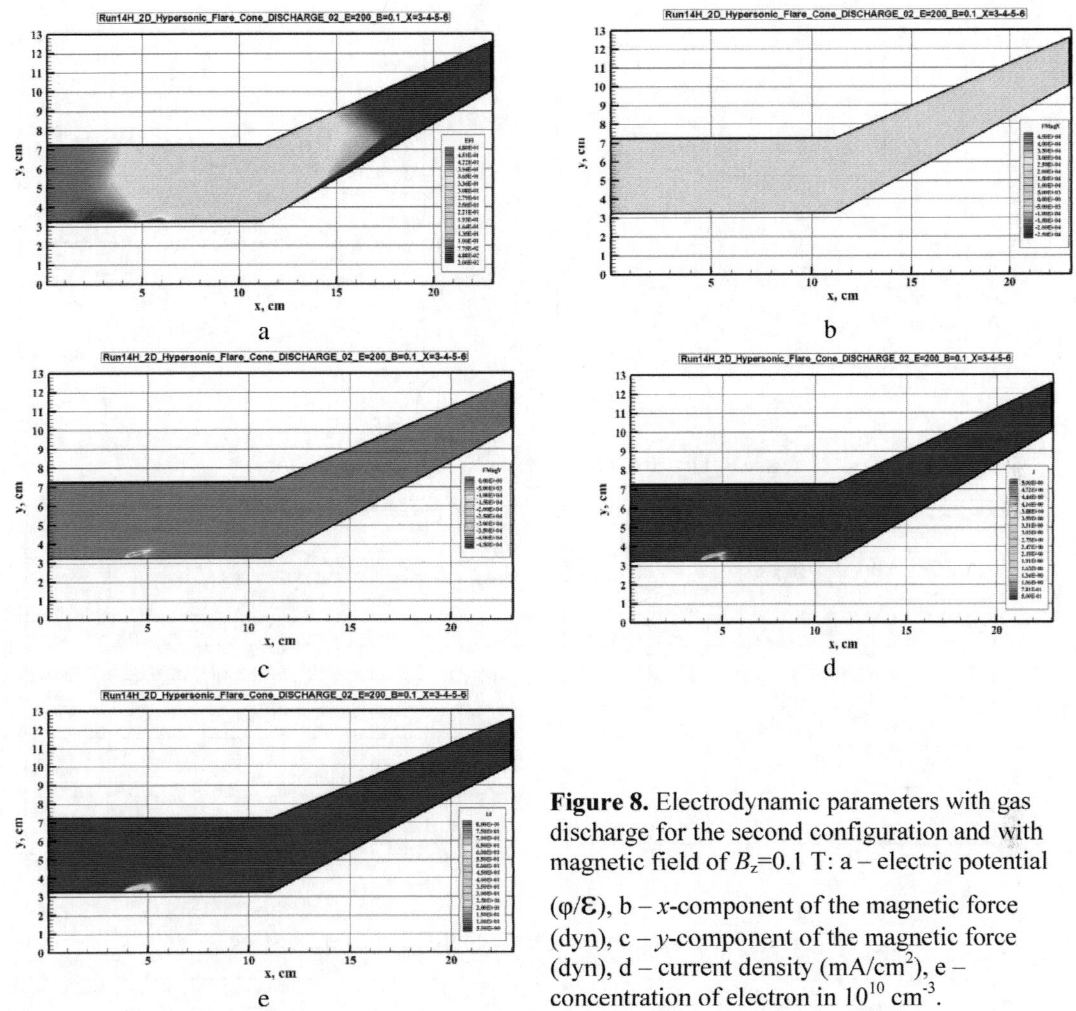

Figure 8. Electrodynamic parameters with gas discharge for the second configuration and with magnetic field of B_z=0.1 T: a – electric potential (φ/ℇ), b – *x*-component of the magnetic force (dyn), c – *y*-component of the magnetic force (dyn), d – current density (mA/cm^2), e – concentration of electron in 10^{10} cm^{-3}.

Conclusion

The problem of hypersonic flow past the surface with sharp edge has been considered. This surface forms an obtuse angle therefore the shock wave generated by the leading edge interacts with the surface.

The effect of influence of the surface direct current discharge and a transverse magnetic field on the gas dynamic characteristics is investigated. To solve this problem a numerical simulation model has been used. This model is based on set of the Navier – Stokes and energy conservation equations, as well as on equations of electrodynamics in the ambipolar approximation, and the Poisson equation.

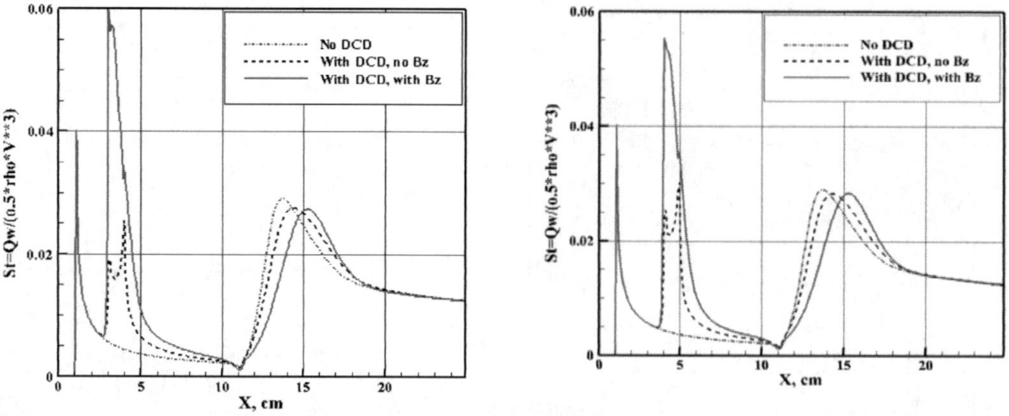

a b

Figure 9. Distribution of the St-number along surface.

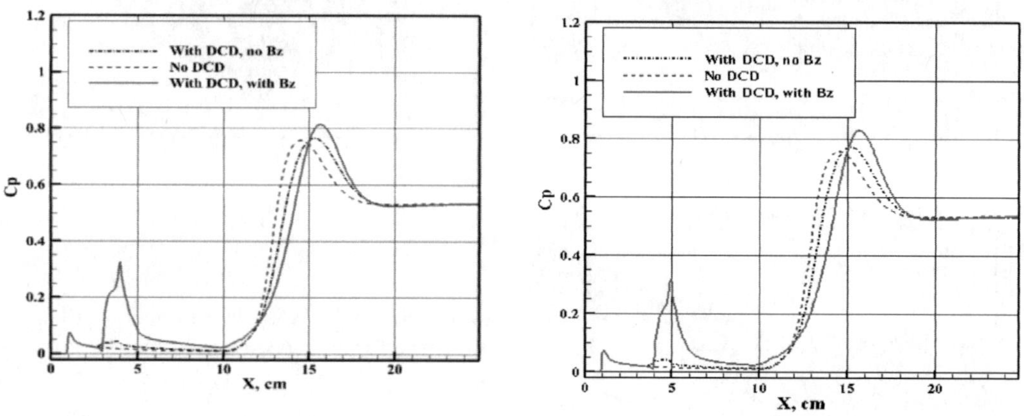

a b

Figure 10. Distribution of the C_p-coefficient along surface.

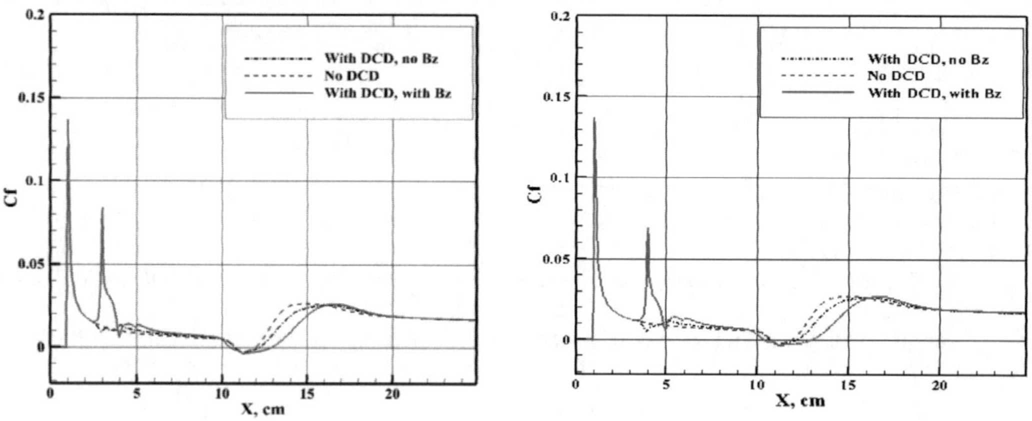

a b

Figure 11. Distribution of Cf-coefficient along surface.

The results of numerical modeling the gas dynamics and electrodynamics of gas discharge with magnetic field point out to a change in the structure of the shock-wave interaction with surface far from location of the gas discharge.

So, by the calculations it is shown that the low-current glow discharge can be used as an electromagnetic actuator in hypersonic flows.

Acknowledgements
This work was performed under support of the Russian Science Foundation project # 16–11–10275.

References

[1] Knight D 2006 Test Cases for CFD Validation of Hypersonic Flight *RTO-TR-AVT-007-V3*
[2] MacLean M and Holden M 2004 Validation and Comparison of WIND and DPLR Results for Hypersonic Laminar Problems *AIAA 2004-0529* p 17
[3] Holden M S, Wadhams T P, Harvey J K and Candler G V 2006 Comparison between Measurements in Regions of Laminar Shock Wave Boundary Layer Interaction in Hypersonic Flows with Navier-Stockes and DSMC Simulations *RTO-TR-AVT-007-V3* pp 4.1–4.56
[4] Gnoffo P A 2001 CFD Validation Studies for Hypersonic Flow Prediction *AIAA 2001-1025* p 13
[5] Wang W–L and Boyd I D 2002 Continuum Breakdown in Hypersonic Viscous Flows *AIAA 2002–0651* p 11
[6] Kato H and Tannehill J C 2001 Computation of Hypersonic Laminar Separated Flows using an Iterated PNS Algoritm *AIAA 2001–1028* p 16
[7] *NPARC WIND CFD Code, Version 5* http://www.grc.nasa.gov/www/winddocs
[8] Wright M J, Candler G V and Bose D 1998 Data Parallel Line Relaxation Method for the Navier-Stockes Equations *AIAA Journal* **36** (9) pp 1603–1609
[9] Surzhikov S T 2009 Rarefied Hypersonic Flow in a Plane Channel with a Surface Glow Discharge *Fluid Dynamics* **44** (2) pp 302–313
[10] Surzhikov S T 2009 Glow Discharge in External Magnetic Field in Hypersonic Flow of Rarefied Gas *High Temperature* **47** (4) pp 459-471
[11] Surzgikov S T 2009 Interaction of a Glow Discharge with a Rarefied Hypersonic Flow in a Curvilinear Channel *Russian Journal of Physical Chemistry* B **3**(3) pp 383–390.
[12] Shang J S and Surzhikov S T 2005 Magnetoaerodynamic Actuator for Hypersonic Flow Control *AIAA Journal* **43** (8) pp 1633-1652
[13] Shang J S, Surzhikov S T, Kimmel R, Gaitonde D, Menart J and Hayes J 2005 Mechanisms of plasma actuators for hypersonic flow control *Progress in Aerospace Sciences* **41** pp 642–668
[14] Shang S, Chang C and Surzhikov S 2007 Simulating Hypersonic Magneto-Fluid Dynamic Compression in Rectangular Inlet *AIAA Journal* **45** (11) pp 2710 –2720
[15] Shang J, Kimmel R L, Menart J and Surzhikov S T 2008 Hypersonic Flow Control Using Surface Plasma Actuator *Journal of Propulsion and Power* **24** (5) pp 923–934
[16] Surzhikov S T 2013 *Computational Physics of Electric Discharges in Gas Flows* (Berlin/Boston: Walter de Gruyter GmbH) p 428
[17] Shang J S 2016 Computational Electromagnetic Aerodynamics *IEEE Series on RF and Microwave Technology* (New Jersey:Wiley & Sons, Inc., Hoboken,) p 424
[18] Hirschfelder J O, Curtis C F and Bird R B 1964 *Molecular Theory of Gases and Liquids* (New York:Wiley) p 1219

The spectral characteristic investigations of normal glow discharge

M A Kotov[1], P V Kozlov[1], L B Ruleva[1], S I Solodovnikov[1], S T. Surzhikov[1] and V A Tovstonog [2]

[1] A. Ishlinsky Institute for Problems in Mechanics of the Russian academy of sciences, Vernadsky prospekt 101(1), Moscow, 119526, Russia
[2] Bauman Moscow State Technical University, Baumanskaya 2-ya, 5, Moscow, 105005, Russia, Moscow, Russia

E-mail: ma_kotov@mail.ru, ruleva@ipmnet.ru

Abstract. Description of experimental installation created for investigation of the normal glow discharges in different gases at pressures of ~ 1 Torr is presented. It is assumed that systematic research of the current-voltage characteristics, spatial configuration of the normal glow discharges, as well as their spectral emissions will be investigated with the use of the installation.
Preliminary experimental investigations of the spectral emission in helium, nitrogen, and argon of the normal glow discharge have been conducted.

1. Introduction

The phenomenon of the normal glow discharge (NGD) has been open about 100 years and until the present is the significant object of the basic physics. Schematic of the normal glow discharge is shown in figure 1. Here ε and R0 are e.m.f. of power supply and ohmic resistance of the external electric circuit.

Normal glow discharge exists between the two "infinite" flat electrodes. The term "infinite" electrodes used in order to emphasize that the boundary effects on the electrodes do not affect the structure of the NGD. As a rule, such discharges permanently exist with parameters close to the minimum of the Paschen breakdown curve. As an example, for molecular nitrogen N_2, voltage drop between electrodes reaches only a few hundred volts, and the total current through the discharge gap is measured by several milliamps.

Law "normal current density" states: with the increase of the total current through a glow discharge the current density in the center of the current spot remains practically unchanged. Simple, but very effective, the one-dimensional method for calculating of the parameters of the normal glow discharge was created by Engel and Steenbeck [1]. The two-dimensional hybrid fluid-dynamic model [2] has demonstrated the real physics of the phenomenon.

The paper presents preliminary studies of the spectral characteristics of normal glow discharge which were performed on the new experimental installation. The installation is shown in figure 2. It includes the cylindrical vacuum chamber with internal diameter of 230 mm and with height of 195 mm. This chamber is mounted on the laboratory table together with Baratron capacitance gauge and Vacom pressure sensor to control rarefaction level. Spectrograph is installed in front of acrylic window to register discharge emission spectra. Vacuum pump and the gas supply pipes are connected to the chamber via the valve system. High Voltage Power Supply is mounted under the laboratory table.

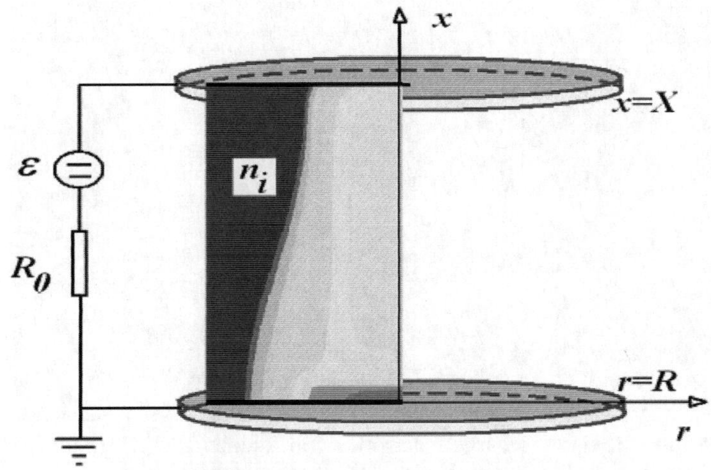

Figure 1. Schematic of the normal glow discharge.

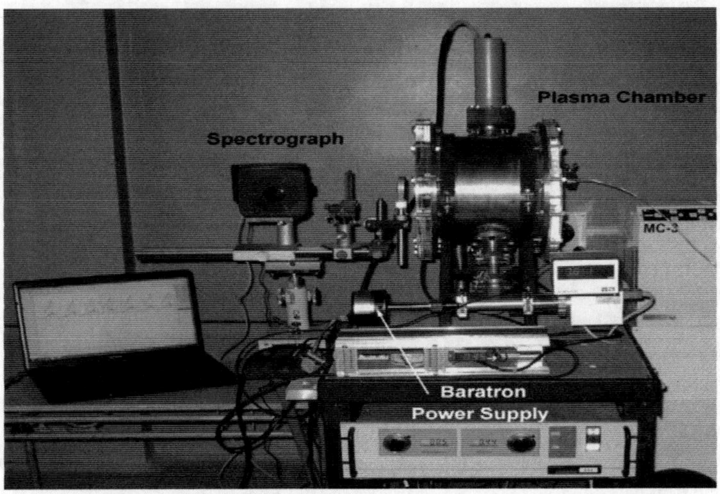

Figure 2. The normal glow discharge (NGD) experimental facility.

As it was mentioned above the general goal of the experiments is the investigation of the normal glow discharges. But it is well known that channel of current (or several channels) of the normal glow discharge between flat electrodes is unstable. As a rule it can stochastically change their location. To provide a stabilization of location of investigated channel of current the slightly convex electrodes are used (see figure 3).

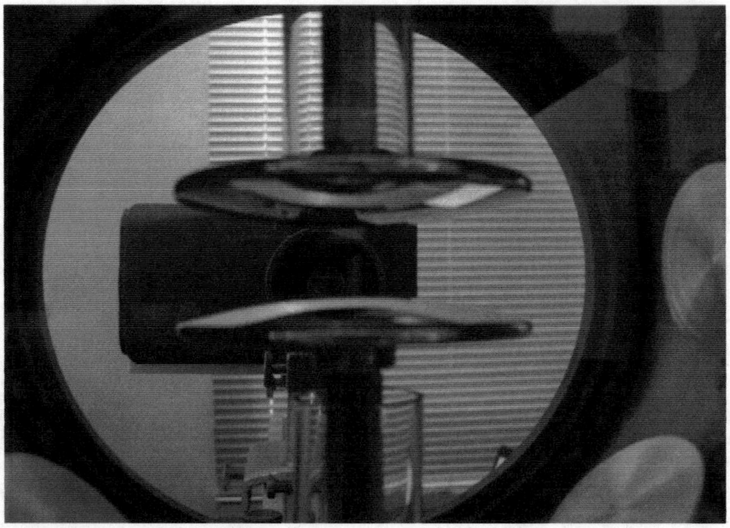

Figure 3. Gas discharge electrodes with diameter of 100 mm for creation of the normal glow discharge. Observation window diameter is of 320 mm.

2. Research methodology at the NGD facility

Figure 4 shows schematic of the experimental installation.

Laser device and spectrograph with collimating lens can be moved in parallel relative to the discharge gap. To register discharge emission spectra the Horiba spectrograph CP140-1824 with the gated CCD line Hamamatsu S11156-2048 of spectral resolution 1.25 nm is used. The spectra registration was carried out in the range of 374-1000 nm, the short-range boundary was determined by emission passing through the acrylic window, which allows alignment laser to monitor the spatial position of the area between the electrodes and take the discharge pictures (figure 3).

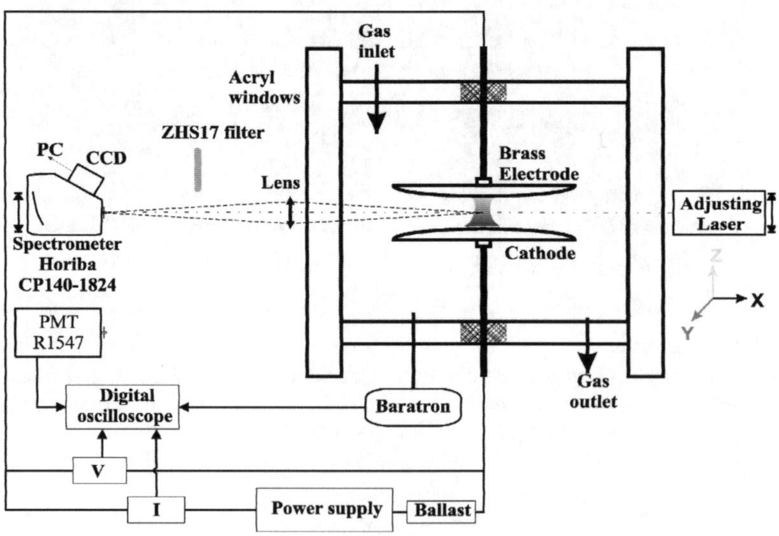

Figure 4. NGD facility schematic.

As a source of high-voltage discharge power the Spellman SA10PN4 is used (with constant and regulated voltage and current: 0,4A 10kV). The discharge voltage information was shot through the divider directly from the electrodes, and the current information from the reference point of power supply. The ballast resistance (1-100 kOm) is connected in the discharge power supply circuit.

The experimental procedure includes the initial evacuation of the discharge chamber oil-free fasting to 10^{-5} Torr pressure control wide-range vacuum gauge type Atmion (Pirani method to 10^{3} Torr, technique Bayard-Alpert to 10^{-11} Torr). This is followed by filling the investigated gas (argon, helium or nitrogen) through inlet valve. To monitor the test gas pressure in the discharge chamber in the range of Torr units the Baratron capacitance gauge type is used (limit of 10 Torr, accuracy 0.2% of the maximum measured pressure).

Further, high voltage is applied to the electrodes, and when an ignition of the discharge occurs the Spellman changing form of the discharge by adjusting current that is fixed by camera. At that the time, voltage and current are registered, which builds the current-voltage characteristic.

Registration of the spectral characteristics of the glow discharge in the test gas along the discharge column depends on the spatial resolution of the optical circuit perpendicular to the observation axis and is determined by pixel CCD line height (1 mm). Along observation axis the resolution collimating lens with *f*=75 mm and 15 mm diameter placed on the double focal distance from the axis of the discharge gap was ~ 3.2 mm. Changing the depth of field along the axis of observation can change by the collimating lens diaphragm.

Spectrograph with CCD line used in experiments was calibrated using the SI-10 tungsten lamp. Figure 5 shows dependence calibration of the spectrograph sensitivity with filter ZHS17 and without it in relative units. ZHS17 filter is used to exclude the effect of radiation influence from the spectral range 374-500 nm from the second-order diffraction grating for spectrum in the range of 750-1000 nm.

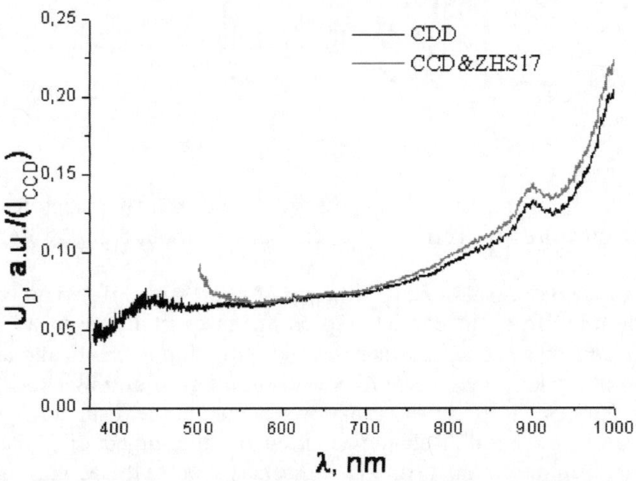

Figure 5. The curve of the spectral sensitivity of the spectrograph with a CCD receiver without filter ZHS17 and with it.

For the measurement of the temporal behavior of the discharge emission photomultiplier tube (PMT) Hamamatsu R1547 is used. Before the PMT the adjustable slit was set. To measure the time evolution of the discharge emission PMT with slit are on the spectrograph position. The time resolution of the PMT was better than 20 ns with the load resistance of 50 ohms.

Discharge's current and voltage, the signals from the photomultiplier and the Baratron gauge are fed through analog-to-digital converter L-Card E20-10 (10 MHz, 14 bits) to the computer. Special programs are used to build CVC, the current and voltage waveforms in time, as well as spectra

3. Experimental results

Test experiments at the NGD facility are carried out in air, nitrogen, argon and helium at pressures of 0.2 - 10 Torr. Initially, the discharge circuit has no ballast resistance. Analysis of the time behavior of discharge's current and voltage under these conditions indicates a quasi-periodic nature of the discharge. The oscillation frequency varies from a few to thousands of Hertz, and depends on the maximum current value at the power supply and the pressure and type of gas. Figure 6 shows a typical waveform view of current and voltage of the discharge in nitrogen at a pressure of 1 Torr at quasi-periodic mode. From these waveforms current-voltage characteristics (CVC) of discharge are constructed of discharge. Examples of CVC of the discharge in argon and air at different initial pressures for the quasi-periodic mode are shown in figure 7. The behavior of the current and voltage for discharges in all the investigated gases without ballast resistance is similar.

Examples of discharge forms in quasi-periodic mode in argon and helium are shown in figure 8. As the pressure in the discharge chamber increases the discharge is usually compressed to the electrode axis. Changes the shape of the discharge in nitrogen, air and argon at change of pressure and discharge current are similar in nature. A typical feature of the discharge in helium is bending the edges of the discharge near the cathode to the anode with increasing pressure.

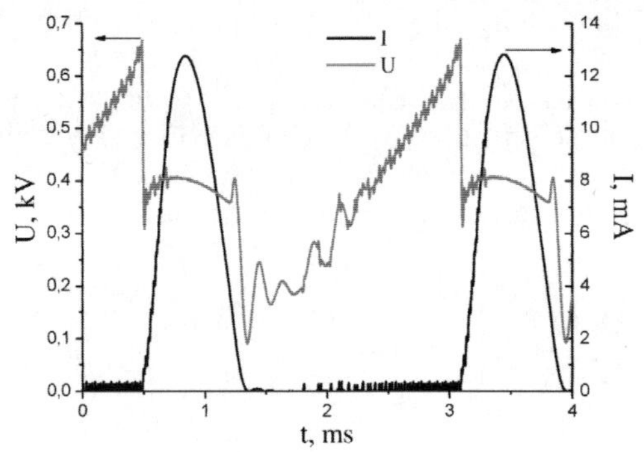

Figure 6. Current and voltage oscillogram without ballast resistance in air at a pressure of 1 Torr

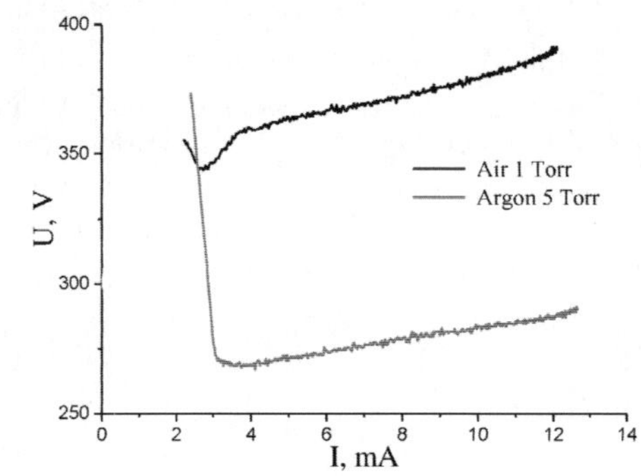

Figure 7. Current-voltage characteristics of the discharge without ballast for argon and air.

Evolution in time of discharge development in argon at a pressure of 1 Torr without ballast is shown in figure 9. The emission from the cathode region increases in a time of no more than 0.1 μs and has duration of 0.5 μs. During the next ten microseconds emission is practically absent, and then it increases in a time of microseconds to a constant level of duration of 500-800 μs. The development character could mean that the first the streamer breakdown occurs at the discharge gap, and thereafter the Townsend one. Study of the nature of discharge occurrence is the subject of further researches.

By increasing the ballast of more than 10 kOm discharge current becomes stationary, which corresponds to the theoretical calculated data [1]. The shapes of the discharge becomes more stable and of all areas typical for glow discharge are clearly visible: aston dark space, the cathode glow cathode dark space, negative glow, Faraday dark space, positive column glow and anode dark space.

Preliminary results of the study the spectral composition of a glow discharge in argon and nitrogen are shown in figures 10-13. The spectra were obtained from the brightest part of the discharge. Figure 11 illustrates the suppression of emission from a second order grating using ZHS17 filter. Emission intensities are given in relative units, taking into account the spectral sensitivity of the spectrograph with a CCD receiver, but the intensity ratio in each experiment corresponds to a real ratio.

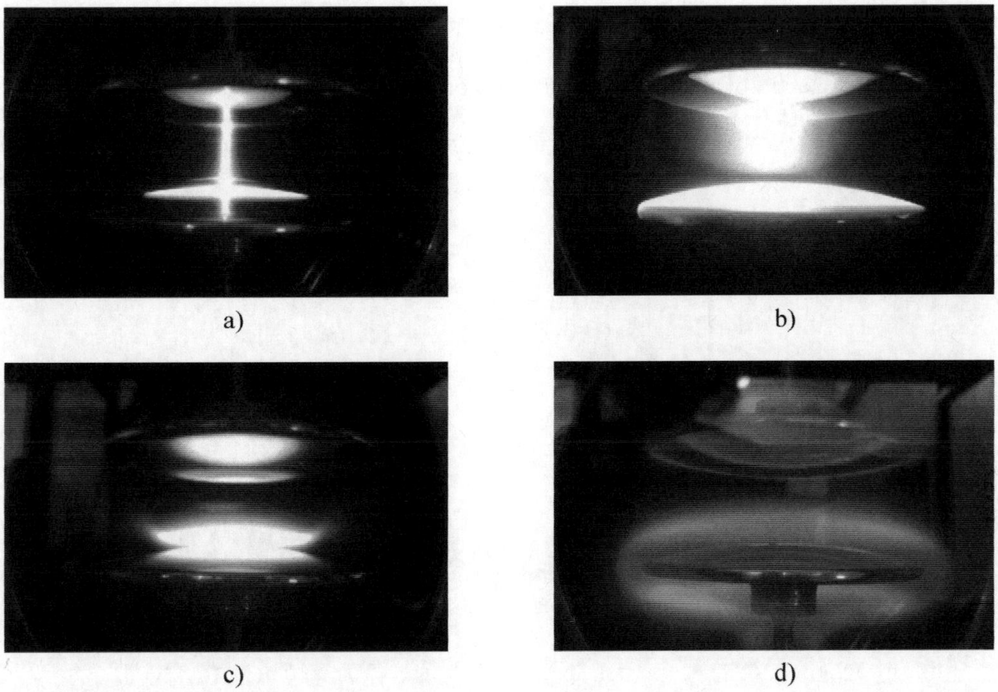

Figure 8. Photos of discharge in Ar at pressure 5 Torr (a); 1 Torr (b) and He at 5 Torr (c); 1 Torr (d).

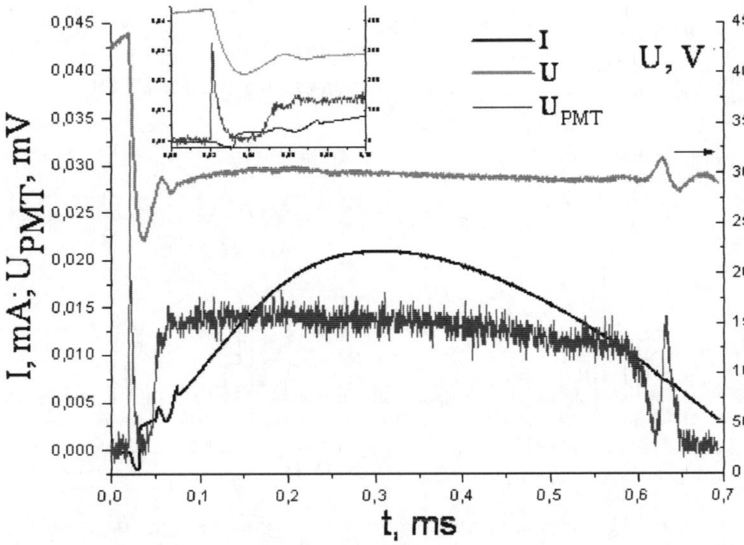

Figure 9. Current, voltage and emission waveforms registered in the PMT at 5 mm above the cathode without ballast resistance in argon at a pressure of 1 Torr.

At low pressures and small discharge currents the line H_α is observed in the emission spectra of the hydrogen, both for nitrogen and for argon, due to hydrogen-containing impurities in the cylinder. The discharge emission spectra of nitrogen identified vibronic transitions molecular nitrogen (C–B) and (B–A) and a positive nitrogen ion (B–X).

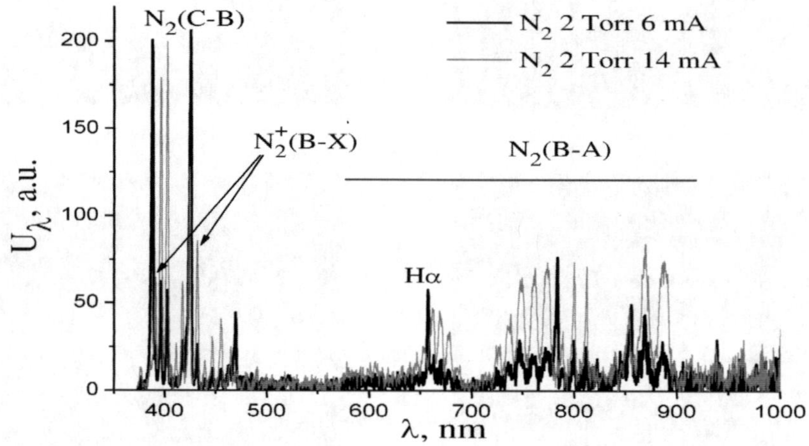

Figure 10. The emission spectra in nitrogen (2 Torr; 6 mA, 14 mA).

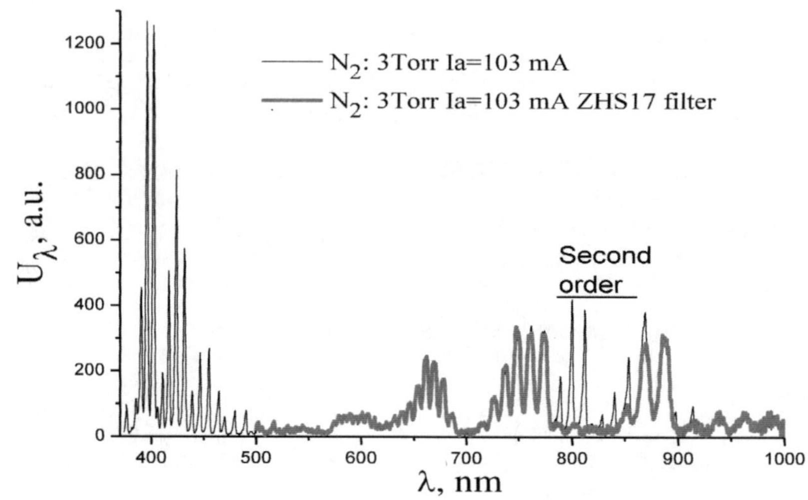

Figure 11. The emission spectra in nitrogen (3 Торр, 103 мА).

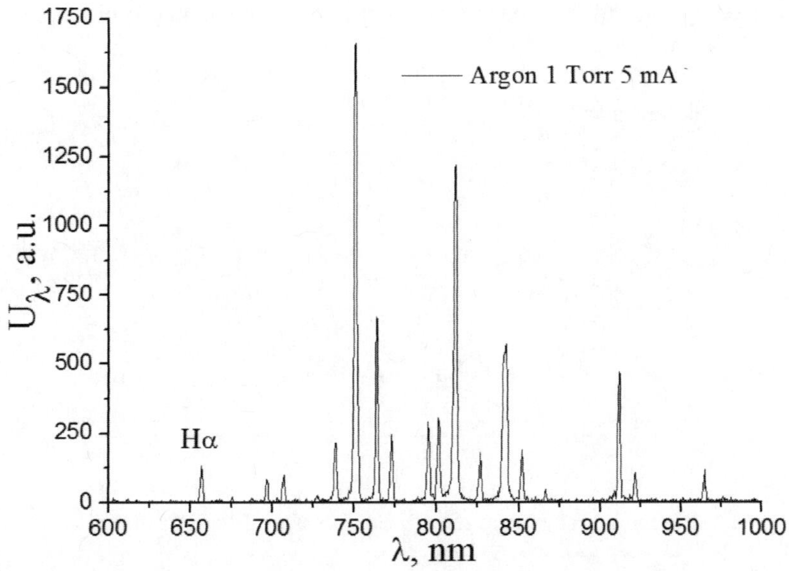

Figure 12. The emission spectra in argon (1 Torr, 5 mA).

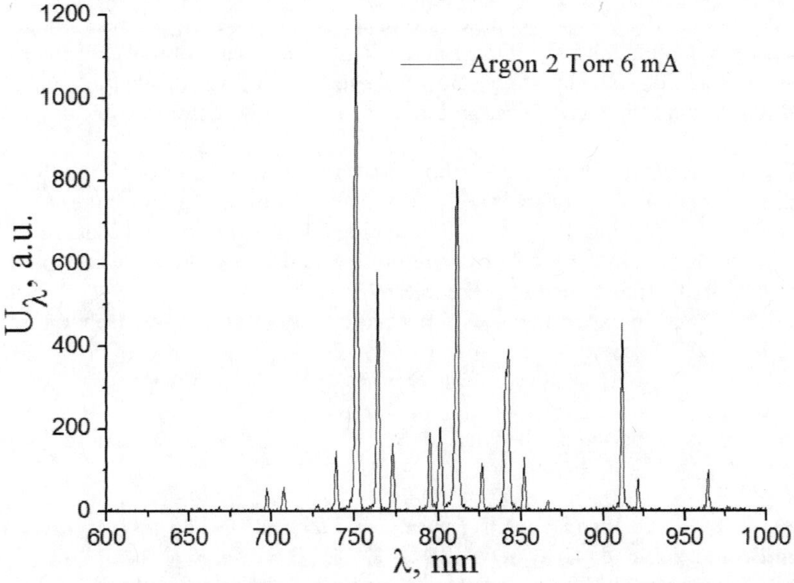

Figure 13. The emission spectra in argon (2 Torr, 6 mA).

The emission spectra of the discharge in argon identified all the main line transitions of singly-ionized atom. The average CVC (3÷6 Torr) in Ar is shown in figure 14.

After analyzing data on the spectra of the corresponding to atomic line positions the full convergence of Ar gas were obtained. In the analysis of the emission spectra on the N2 molecular emission lines were found positive with the transitions of the first group of B–A, C–B of the second group of positive and negative of the first group of B–X. The energy values of atomic transition lines in test gases obtained by spectral analysis will be used in the calculation of physical models of glow discharge.

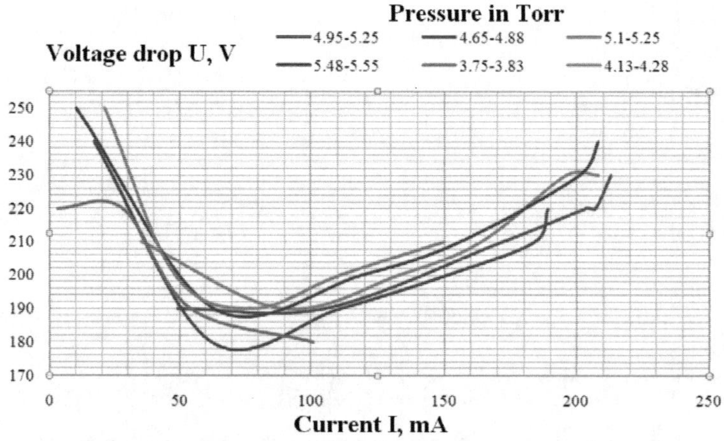

Figure 14. The averaged CVC at pressure 3÷6 Torr in Ar.

4. Conclusion

The experiments carried out at the NGD facility of a glow discharge at a constant pressure allowed to establish the existence of the glow discharge normal regimes. Discharges were obtained in various gases (nitrogen, argon, helium, and air). Quasi-periodic glow discharge modes are investigated. In the future it is planned to continue to study the characteristics of the discharge with simultaneous recording of current-voltage characteristics, pressure and emission spectra in absolute units in different parts of the discharge.

It is planned to investigate the nature of the initial stage of occurrence of discharge in its time evolution. Adding a small amount of hydrogen (less than 1%) in the future will allow determine the concentration of electrons using the Stark broadening. The experimental data on the electrical parameters of the discharge distribution of concentration and temperature in the space between the electrodes will allow in the future verify the theoretical models.

The results will be used in experiments with normal operation of a glow discharge in a high speed gas stream for the GFDST facility.

Acknowledgement
The presented study was supported by the Russian Science Foundation project № 16-11-10275.

References

[1] Surzhikov S T 2013 *Computational Physics of Electric Discharges in Gas Flows* (GmbH, Berlin/Boston:Walter de Gruyter) p 428
[2] Engel A v, Steenbeck M 1932 *Elektrische Gasentladungen, Ihre Physik und Technik* (Berlin Heidelberg GmbH: Springer-Verlag)

Comparison of two-dimensional and quasi -one-dimensional scramjet models by the example of VAG experiment

R K Seleznev [1,2,3]

[1] Dukhov Research Institute of Automatics (VNIIA), Suschevskaya Street 22, Moscow 127055, Russia
[2] Moscow Institute of Physics and Technology (State University), 9 Institutskiy per., Dolgoprudny, Moscow Region, 141701, Russia
[3] A. Ishlinsky Institute for Problems in Mechanics, RAS Vernadsky Street 101(1), Moscow, 119526, Russia

rkseleznev@gmail.com

Abstract. In the paper two-dimensional and quasi-one dimensional models for scramjet combustion chamber are described. Comparison of the results of calculations for the two-dimensional and quasi-one dimensional code by the example of VAG experiment are presented.

1. Introduction
Recently, a study of combustion processes in the ramjet and scramjet is becoming more and more important in connection with attempts to create a hypersonic aircraft. Carrying out physical experiments in this field is connected with a lot of technical difficulties and is too expensive. Special requirements have to meet the fuel and combustion chamber geometry. Therefore, computational and theoretical studies of combustion processes in scramjet begin to play an increasing role. At the first stage of designing the scramjet it is very important to make an accurate prediction of the engine efficiency. In recent years, the quasi-one-dimensional scramjet model is widespread [1–14]. In these models to calculate the chemical concentrations, temperature and pressure fields, the set of ordinary differential equations should be solved, and the area of the combustion chamber is defined as a function of a longitudinal coordinate. However, the quasi-one-dimensional model does not allow to describe a structure of flow fields. There are two-dimensional [15] and three-dimensional [16] models for describing the structure of flows in scramjet chamber.

In [17] the comparison of the results of calculations for the two-dimensional and quasi-one dimensional code has been presented for HyShot-2 experiment. The purpose of this paper is to compares the quasi-one-dimensional and two-dimensional models by the example of numerical simulation of VAG experiment [18].

2. Description of VAG experimental setup
The VAG [18] experiment is similar to T4 at the University of Queensland.[19,20]. In [19] the comparison of data between this two experiments was carried out. The experiment was conducted for a rectangular channel expanding part, as shown in figure 1. The width of the camera was constant and equaled to 94.3 mm. Fuel injection is executed from the gap strut in the center of the chamber. On the top and bottom of the strut there were located 4 holes with a diameter of 2.5 mm, through which

hydrogen is carried out perpendicular to the flow. At the end of the strut there were located 4 holes with a diameter of 2.5 mm, through of hydrogen injected the parallel to the flow. The experiment studied the effect of the strut geometry on the results obtained. The length of the wedge spacers (33.5 mm, 37.6 mm, 39.6 mm, 43.8 mm and 44.8 mm) was varied. In the experiment the pressure distribution on the chamber wall was measured. In this paper only normal hydrogen injection and equivalence ratio φ=0.61 was considered. The following incoming air flow parameters were used for verification:

Incoming flow pressure:	$P = 0.58$ atm;
Incoming flow temperature:	$T = 1258$ K;
Incoming flow Much Number:	$M = 2.44$;
Incoming flow gas mixture:	Air;
Fuel Much Number:	$M = 1$.

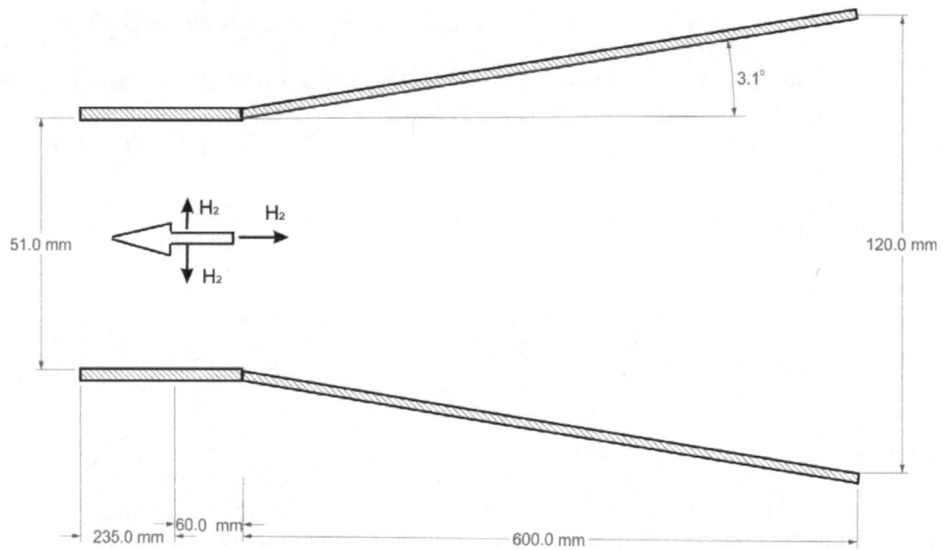

Figure 1. Scheme of the VAG[18] supersonic combustion experiment.

3. Quasi-one-dimensional computational fluid dynamic model

In [17] the quasi-one-dimensional computational fluid dynamic model described in detail.To describe the fluid flow in scramjet the conservation laws for mass, momentum and energy should be specified. The governing equations are based on the following assumptions:1) ideal gas law is fulfilled in the combustor, 2) quasi-one-dimensional flow (all flow variables are functions of the axial distance along the combustor),3) continuous flow (changes in stream properties are continuous functions), 4) steady-state flow. Quasi-one-dimensional approach requires that we state these equations into the set of ordinary differential equations (ODE). The equations solved are:

$$\frac{dU}{dx}=\frac{1}{\alpha}\left\{-\frac{1}{A}\frac{dA}{dx}+\frac{1+\gamma M^2(1-\varepsilon)-\left(h_0/\hat{h}\right)}{\dot{m}}\frac{d\dot{m}}{dx}+\frac{1}{\hat{h}}\left[-\sum_i h_i\frac{dY_i}{dx}+\frac{1}{\dot{m}}\sum_i\left(h_i\frac{d\dot{m}_i}{dx}\right)_{added}\right]-\frac{1}{MW}\frac{dMW}{dx}+\left[\gamma M^2-\frac{c_p(\mathrm{T}_{aw}-\mathrm{T}_w)}{\hat{h}\cdot\mathrm{Pr}^{\frac{3}{2}}A}\right]\frac{2C_f}{D}\right\}. \qquad (1)$$

$$\frac{dT}{dx}=\frac{1}{\hat{c}_p}\left[-\sum_i h_i\frac{dY_i}{dx}+\frac{1}{\dot{m}}\sum_i\left(h_i\frac{d\dot{m}_i}{dx}\right)_{added}-\frac{2C_f c_p\left(T_{AW}-T_W\right)}{\Pr^{\frac{2}{3}}D\cdot A}-\frac{h_0}{\dot{m}}\frac{d\dot{m}}{dx}-U\frac{dU}{dx}\right], \quad (2)$$

$$\frac{d\rho}{dx}=\rho\left[\frac{1}{\dot{m}}\frac{d\dot{m}}{dx}-\frac{1}{U}\frac{dU}{dx}-\frac{1}{A}\frac{dA}{dx}\right], \quad (3)$$

$$\frac{dp}{dx}=-\gamma pM^2\left[\frac{1}{U}\frac{dU}{dx}+\frac{1}{2}\left(\frac{4C_f}{D}\right)+\frac{1}{\dot{m}}\frac{d\dot{m}}{dx}\right], \quad (4)$$

$$\frac{d\left(Y_i\right)}{dx}=\frac{1}{\dot{m}}\frac{d\dot{m}_{i,added}}{dx}+\frac{\dot{\omega}_i MW_i}{\rho U}-\frac{Y_i}{\dot{m}}\frac{d\dot{m}}{dx}, \quad (5)$$

where $\alpha=1/U\left(1-\gamma M^2+U^2/\hat{h}\right)$; $\hat{h}=\hat{c}_p T$; $\hat{c}_p=\tilde{c}_p-\frac{1}{\dot{m}}\left\{\sum_i\dot{m}_i\left(c_{pi}+\tilde{c}_{pi}T\right)_{added}\right\}$; $\tilde{c}_p=c_p+\sum_i T\tilde{c}_{p_i}Y_i$;

$T_{aw}=T\left[1+\Pr^{\frac{1}{3}}\left[\left(\gamma-1\right)/2\right]M^2\right]$; $D=\frac{4A}{P_W}$; $MW=\frac{1}{\sum_i\frac{Y_i}{MW_i}}$; $h_0=h+\frac{U^2}{2}$; $\varepsilon=U_f/U_\infty$ is the ratio of injection velocity in the downstream direction to a flow (under perpendicular injection $\varepsilon=0$); A -is the sectional area being a function of the coordinate x. The skin friction is found using the Eckert Reference Temperature method. $C_f=\frac{0.664\sqrt{C^*}}{\sqrt{\mathrm{Re}}}$; $C^*=\left(\frac{T^*}{T}\right)^{1/2}\left(\frac{1+R/T}{T^*/T+R/T}\right)$; $R=111K$; $T^*=0.5T_W+0.28T+0.22T_{aw}$.The method is valid for estimating frictional drag at subsonic to hypersonic speed and is suitable for aircraft design. The method assumes that the vehicle surface is at the adiabatic wall temperature T_{aw}.

We use the supersonic mixing model [2]. Hydrogen was considered as a fuel. Global kinetic mechanism suggested by Westbrook [21] was used. The set of ODE was solved numerically using the Runge-Kutta method and generalized Newton method [22].

4. Two-dimensional computational fluid dynamic model

For two-dimensional calculations we used the NERAT-2D computer code [15]. NERAT-2D realizes the time-relaxation method. At each time step the following groups of governing equations were integrated successively: the Navier – Stokes and continuity equations, the equations of mass conservation of chemical species, the equation of energy conservation. These equations are formulated in the following form:

$$\frac{\partial\rho}{\partial t}+\mathrm{div}\left(\rho\mathbf{V}\right)=0, \quad (6)$$

$$\frac{\partial\rho u}{\partial t}+\mathrm{div}\left(\rho u\mathbf{V}\right)=-\frac{\partial p}{\partial x}-\frac{2}{3}\frac{\partial}{\partial x}\left(\mu\,\mathrm{div}\mathbf{V}\right)+\frac{1}{r}\frac{\partial}{\partial r}\left[r\mu\left(\frac{\partial \mathrm{v}}{\partial x}+\frac{\partial u}{\partial r}\right)\right]+2\frac{\partial}{\partial x}\left(\mu\frac{\partial u}{\partial x}\right), \quad (7)$$

$$\frac{\partial\rho \mathrm{v}}{\partial t}+\mathrm{div}\left(\rho \mathrm{v}\mathbf{V}\right)=-\frac{\partial p}{\partial r}-\frac{2}{3}\frac{\partial}{\partial r}\left(\mu\,\mathrm{div}\mathbf{V}\right)+\frac{\partial}{\partial x}\left[\mu\left(\frac{\partial \mathrm{v}}{\partial x}+\frac{\partial u}{\partial r}\right)\right]+ +2\frac{\partial}{\partial r}\left(\mu\frac{\partial \mathrm{v}}{\partial r}\right)+2\mu\frac{\partial}{\partial r}\left(\frac{\mathrm{v}}{r}\right), \quad (8)$$

$$\frac{\partial\rho_i}{\partial t}+\mathrm{div}\,\rho_i\mathbf{V}=-\mathrm{div}\,\mathbf{J}_i+\dot{w}_i,\quad i=1,2,\ldots,N_s, \quad (9)$$

$$\rho c_p \frac{\partial T}{\partial t} + \rho c_p \mathbf{V}\,\mathrm{grad}\,T = \mathrm{div}(\lambda\,\mathrm{grad}\,T) + \frac{\partial p}{\partial t} + \mathbf{V}\,\mathrm{grad}\,p + \\ + \Phi_\mu - \sum_{i=1}^{N_s} h_i \dot{w}_i + \sum_{i=1}^{N_s} \rho c_{p,i} D_i (\mathrm{grad}\,Y_i \cdot \mathrm{grad}\,T), \tag{10}$$

where $\Phi_\mu = \mu\left[2\left(\frac{\mathrm{v}}{r}\right)^2 + 2\left(\frac{\partial \mathrm{v}}{\partial r}\right)^2 + 2\left(\frac{\partial \mathrm{u}}{\partial x}\right)^2 + \left(\frac{\partial \mathrm{v}}{\partial x} + \frac{\partial u}{\partial r}\right)^2 - \frac{2}{3}\left(\frac{\partial u}{\partial x} + \frac{\partial \mathrm{v}}{\partial r} + \frac{\mathrm{v}}{r}\right)^2\right]$ is the dissipative function; ; μ, λ are the viscosity and heat conductivity coefficients, c_p is the specific heat capacity of gas mixture; $c_p = \sum_i^{N_s} Y_i c_{p,i}$; Y_i is the mass fraction of species i; $c_{p,i}, h_i$ are the specific heat capacity at constant pressure and specific enthalpy of species i; $\dot{w}_i$ is the reaction rate for species i; D_i is the effective diffusion coefficient of species i; $\rho_i, \mathbf{J}_i$ are the density and mass diffusion flux for species i; $\mathbf{J}_i = -\rho D_i\,\mathrm{grad}\,Y_i$; N_s is the number of species. We used the kinetic mechanism proposed by Evans and Schexnayder [23].

5. Results

Comparison of the pressure distribution calculated by two-dimensional model (blue line) and quasi-one-dimensional model (red line) with experimental data (green triangles) are shown in figure 2. It should be noted that the pressure increase appears to be a more significant under two-dimensional modeling. This can be explained by the fact that the flow in the channel is actually three-dimensional, and therefore it is necessary and very important to consider the fuel injection features for a correct description of the flow and combustion in the channel. One can also see a good agreement between the quasi one-dimensional simulation results and experimental data.

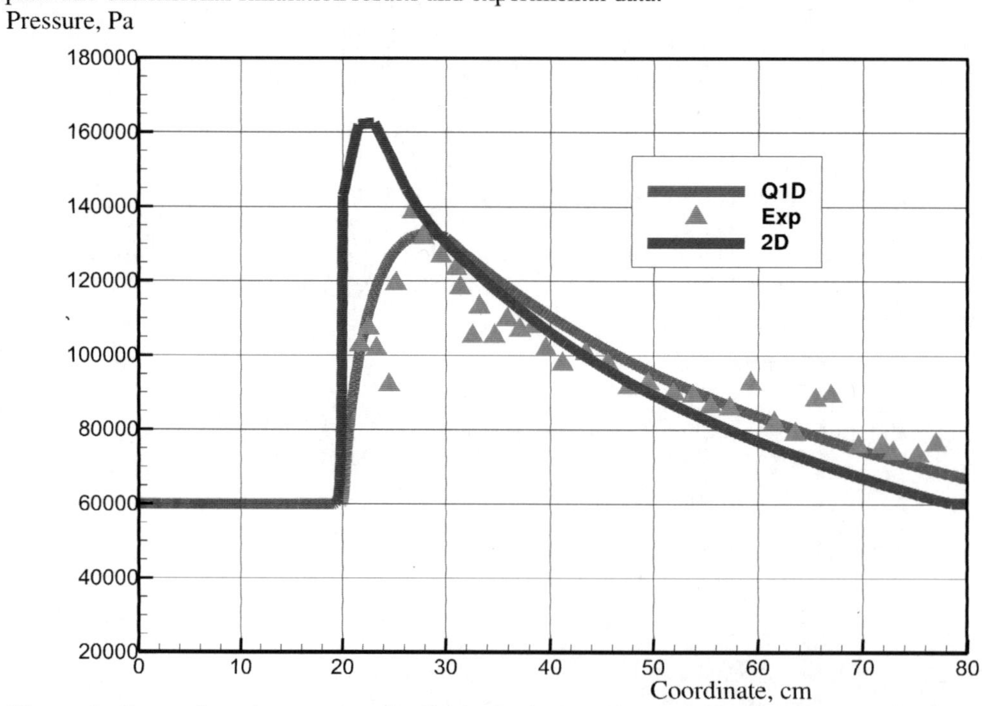

Figure 2. Comparison the pressure distribution calculated by two-dimensional model (blue line) and quasi-one-dimensional model (red line) with experimental data (green triangles).

6. Conclusion

Comparison of the pressure distribution calculated by two-dimensional model and quasi-one-dimensional model with experimental data for VAG experimental setup is presented. This work presents a continuation of our efforts on the verification and validation of numerical methods and computational codes for calculation of various hypersonic vehicles and energetic devices [22-31].

Acknowledgments

Authors are grateful to the prof. S.T. Surzhikov and staff of Laboratory of Radiative Gas Dynamics of Institute for Problems in Mechanics for useful discussions of the problems considered in the study.

The presented study was supported by the Russian Science Foundation project № 16-11-10275.

References

[1] Brien T F, Starkey R P and Lewis M J 2001 Quasi-One-Dimensional High-Speed Engine Model with Finite-Rate Chemistry *J. Propuls. Power* **17** pp 1366–1375

[2] Birzer C, Doolan C 2009 Quasi-one-dimensional model of hydrogen-fueled scramjet combustors *J. Propuls. power* **25** pp 1220–5

[3] Tourani C 2011 Computational simulation of scramjet combustors – a comparison between quasi-one dimensional and 2-D numerical simulations *17th AIAA international space planes and hypersonic systems and technologies conference*

[4] Doolan C, Boyce R 2008 A Quasi-One-Dimensional Mixing and Combustion Code for Trajectory Optimisation and Design Studies *AIAA Paper-2008-2603*

[5] Zabaikin VA 2003 Quality of a high-enthalpy flow upon electric-arc heating of air in a facility for investigating supersonic combustion *Combust. Explos. Shock Waves* **39** pp 23–30

[6] Zhang D, Feng Y, Zhang S, Qin J, Cheng K, Bao W and Yu D 2016 Quasi-One-Dimensional Model of Scramjet Combustor Coupled with Regenerative Cooling *J. Propuls. Power* pp 1–11

[7] Sforza P M 2012 Quasi-One-Dimensional Flow Equations *Theory Aerosp. Propuls. pp* 35–53

[8] Zhao Y, Shen Q and Guan F 2016 Quasi-One-Dimensional Analysis of Supersonic Combustor Performance *52nd AIAA/SAE/ASEE Jt. Propuls. Conf.* pp 1–8

[9] Koo H, Donde P and Raman V 2011 A quadrature-based LES/transported probability density function approach for modeling supersonic combustion *Proc. Combust. Inst.* **33** pp 2203–10

[10] Seleznev R K, Surzhikov S T 2015 Quasi-One-Dimensional and Two-Dimensional Numerical Simulation of Scramjet Combustors *51th AIAA/ASME/SAE/ASEE Joint Propulsion Conference,. AIAA Paper-2015-4166* pp 1-28

[11] Birzer C and Doolan C J 2007 Quasi-One-Dimensional Modeling of Hydrogen Fuelled Scramjet Combustors *AIAA Paper-2007-4314* pp 1-13

[12] Frezzotti M L and Nasuti F 2015 Response Function Modeling in the Study of Longitudinal Combustion Instability by a Quasi-1D Eulerian Solver pp 1–13

[13] Vanyai T, Bricalli M, Brieschenk S and Boyce R R 2014 A Quasi-One-Dimensional Investigation Into the Effect of Combustion Processes on Scramjet Performance *19th AIAA International Space Planes and Hypersonic Systems and Technologies Conference* pp 1–13

[14] Tian L, Chen L, Chen Q, Li F and Chang X 2014 Quasi-One-Dimensional Multimodes Analysis for Dual-Mode Scramjet *J. Propuls. Power* **30** pp 1559–67

[15] Surzhikov S T, Seleznev R K, Tretjakov P K, Zabaykin V A 2014 Unsteady Thermo-Gasdynamic Processes in Scramjet Combustion Chamber with Periodical Input of Cold Air, *50th AIAA/ASME/SAE/ASEE Joint Propulsion Conference, AIAA Paper-2014-3917* pp 1-25

[16] Engblom W, Frate F and Nelson C C 2005 Progress in Validation of Wind-US for Ramjet/Scramjet Combustion *AIAA Paper-2005-1000* pp 1-18

[17] Seleznev R K, Surzhikov S T, Shang J A 2016 Quasi-One-Dimensional Analysis of Hydrogen-Fueled Scramjet Combustors AIAA 2016-4569, *52nd AIAA/ASME/SAE/ASEE Joint*

Propulsion Conference, *AIAA Paper-2016-4569* pp 1–27

[18] Masuya G, Komuro T, Murakamit A, Murayama M and Katsura O 1995 Ignition and Combustion Performance of Scramjet Combustors with Fuel Injection Struts *J. Propuls. Power* **11** pp 301–307

[19] Boyce R R, Paull A, Stalker R J, Wendt M, Chinzei N and Miyajima H 2000 Comparison of Supersonic Combustion Between Impulse and Vitiation-Heated Facilities *J. Propuls. Power* **16** pp 709–717

[20] Stalker R J Shock Tunnels for Real Gas Hypersonics

[21] Ó Conaire M, Curran H J, Simmie J M, Pitz W J and Westbrook C K 2004 A comprehensive modeling study of hydrogen oxidation *Int. J. Chem. Kinet.* **36** pp 603–22

[22] Seleznev R K, Surzhikov S T 2013 A Generalized Newton Method for Differential Equation of Chemical Kinetics *44th AIAA Thermophysics Conference*. *AIAA Paper-2013-3009* pp 1–17

[23] Evans J and Schexnayder C 1980 Influence of Chemical Kinetics and Unmixedness on Burning in Supersonic Hydrogen Flames *AIAA J.* **18** pp 188–193

[24] Surzhikov S T, Zheleznyakova A L, Shang J S, Rivir R B 2013 Simulating gasdynamic interaction and radiative heating within scramjets with hydrocarbon fuels *44th AIAA Thermophysics Conference*

[25] Surzhikov S T 2011 Radiative-Convective Heat Transfer of a Spherically Shaped Space Vehicle in Carbon Dioxide *High Temperature* **49**(1) pp 92–107

[26] Surzhikov S T 2012 Coupled radiative gasdynamic interaction and non-equilibrium dissociation for large-scale returned space vehicles Chemical Physics **398** pp 56–63

[27] Shang J S, Surzhikov S T 2012 Nonequilibrium radiative hypersonic flow simulation *Progress in Aerospace Sciences* **53** pp 46–65

[28] Surzhikov S T. 2012 Radiative-Collisional Models in Non-Equilibrium Aerothemodynamics of Entry Probes *Journal of Heat* Transfer **134** pp 031002-1 – 031002-12

[29] Shang J S, Surzhikov S T 2010 Simulating Nonequilibrium Flow for Ablative Earth Entry *Journal of Spacecraft and Rockets* **47**(5) pp 806-815

[30] Shang J S, Surzhikov S T 2011 Simulating Stardust Earth Reentry with Radiation Heat Transfer *Journal of Spacecraft and Rockets* **48**(3) pp 385-396

[31] Surzhikov S T, Shang J S 2012 Coupled Radiation-Gasdynamic Model for Stardust Earth Entry Simulation *Journal of Spacecraft and Rockets* **49**(5) pp 875-888

Stabilization of solid fuel combustion in a ramjet engine

S A Rashkovskiy[1,2], S E Yakush[1,2] and A A Baranov[1,2]

[1] Institute for Problems in Mechanics of the Russian Academy of Sciences, Ave. Vernadskogo 101 Bldg 1, Moscow, 119526, Russia
[2] Bauman Moscow State Technical University, Baumanskaya 2-ya 5, Moscow, 105005, Russia
E-mail: yakush@ipmnet.ru

Abstract. Combustion stability problem for solid fuel ramjet engine is considered. Stabilization is achieved due to the presence of a wider section in the combustion channel near the air inlet. In this part, playing the role of flame holder, a recirculation zone develops which increases the residence time of combustion products in contrast to the straight part of the channel. In this work, numerical modelling of solid fuel ramjet engine is carried out for PMMA as a fuel. A simple model for gasification rate is used in order to describe the solid fuel decomposition under the incident heat flux. Reacting multicomponent gas equations are solved by a low-dissipation numerical scheme suitable for a wide range of Mach numbers, with Eddy Break-Up model for turbulent combustion. Simulation results are presented for subsonic conditions in the ramjet inlet.

1. Introduction

Solid fuel ramjet engines (SFRE) are considered to be promising propulsion systems of unmanned aerial vehicles. A feature of the workflow in these engines is that the used solid fuels contain a minimum quantity of oxidizing components and incapable to self-sustained combustion. Solid fuel combustion occurs in a forced mode: due to heat supplied from the high-speed flow in the grain channel, the gasification (decomposition) of solid fuel occurs; the products of gasification enter the flow and burn out there, providing the thrust.

Solid fuel ramjet engines are being developed and tested in different countries for over 40 years [1-4]. Despite the high expectations associated with these engines, they have not yet received wide distribution. This is mainly due to instability of their operation and with a very narrow range of operating parameters in which stable combustion of solid fuel is possible. A lack of understanding of the processes occurring in SFRE, and, above all, of the mechanism of instability of the solid fuel forced combustion in the high-speed flow prevent to the wide use of SFRE.

In order to stabilize the solid fuel combustion in SFRE, the solid fuel grains with a profiled channel are used (figure 1). For this purpose, a ledge is formed in the inlet part of the grain channel that acts as a flame holder. It was established experimentally that steady combustion of solid fuel in SFRE is possible only at certain shapes and sizes of the flame holder, but the detailed mechanism of stabilization of combustion is not completely clear; moreover there are no methods for selection of the appropriate size and shape of the flame holder.

Qualitatively, the mechanism of stabilization of combustion in SFRE can be explained as follows. When air, entering the ramjet combustor, flows around the ledge in the grain channel, the recirculation zone occurs, in which the mixing of air with the products of decomposition of the solid fuel, combustion of the mixture and the formation of the high temperature gas mixture occur. The resulting high-temperature gas mixture flows out of flame holder into the grain channel and provides the gasification of solid fuel and combustion of the products of gasification in the channel. In order to provide stable combustion of whole grain, the energy of high-temperature gas mixture in the flame holder should be large enough. This is provided by the flame holder sizes and as long as possible residence time of the particles of the gas mixture, which is also determined by the sizes and shape of the flame holder.

Recently, a number of papers appeared in which attempts were made to investigate numerically the processes in SFRE. In particular, the influence of the shape and the sizes of the flame holder on the flow structure was studied numerically in [5], where the influence of the ratio of the flame holder length to its depth and the angle of slope of the wall at the exit of the flame holder were analyzed. A steady-state gas-dynamic model and steady-state combustion model of solid fuel was used in [5]; it does not allow simulating the occurrence and development of SFRE operation instability. In [6], self-ignition of solid fuel in a ramjet was investigated, but stability of combustion of solid fuels and connection of the stability with the parameters of the flame holder were not considered.

In this paper, we propose a non-stationary model of solid fuel combustion in SFRE and present some screening results of numerical simulation for several inlet parameters of a solid fuel ramjet combustor.

2. Mathematical model

2.1. Solid fuel

Polymethylmethacrylate (PMMA) was taken as the solid fuel because it is quite widespread in current experimental studies [7]. While gasification of PMMA, as a complex polymer, produces whole spectrum of gas components, in this study we take a simple model that PMMA, upon thermal decomposition, gives rise just to gaseous methylmethacrylate (MMA) represented by chemical formula $C_5H_8O_2$.

The mass rate of gasification corresponding to the incident heat flux q_s is described by a simple model taking into account the heating of solid material from its initial temperature T_0 to the boiling temperature T_s which is assumed to be maintained on the surface of decomposing material, as well as the gasification heat ΔH_G:

$$\dot{m} = \frac{q_s}{c_s(T_s - T_0) + \Delta H_G} \tag{1}$$

Here, c_s is the specific heat capacity of solid fuel. The heat flux q_s can generally include the radiation part, but here it is not taken into account due to small sizes of combustor considered. The following properties of PMMA are chosen: density $\rho_s = 1.18$ g/cm^3, boiling temperature $T_s = 200$ °C, specific heat capacity $c_s = 1500$ J/(kg K), heat of gasification $Q_s = 1591$ kJ/kg.

Combustion of MMA in the gas phase is described by a single irreversible gross reaction [7]

$$C_5H_8O_2 + 6O_2 \rightarrow 5CO_2 + 4H_2O + \Delta H_c \tag{2}$$

with heat of combustion per unit mass of fuel $\Delta H_c = 25.6$ MJ/kg.

2.2. Reacting gas mixture

The multicomponent gas mixture is described by the Reynolds-averaged Navier-Stokes equations (RANS) in the fully compressible formulation. Five mixture components are considered: $C_5H_8O_2$, O_2, CO_2, H_2O, N_2. Each component is described by the ideal gas equation of state, the thermal properties (specific heat, internal energy, enthalpy) as functions of temperature are described by polynomials. The temperature of gas mixture was determined from its total energy and composition by solving a corresponding nonlinear equation.

Turbulence is described by the standard $k-\varepsilon$ model, the gas-phase combustion rate is described by the Eddy Break-up (EBU) model [8] which assumes that reaction rate is controlled by the rate of turbulent mixing of fuel and oxidizer. Namely, the rate of combustion reaction (2) is calculated from the following formula [8]:

$$\dot{w} = \rho A \frac{\varepsilon}{k} \min\left(Y_{\mathrm{C3H8O2}}, \frac{Y_{\mathrm{O2}}}{s}, B\frac{Y_{\mathrm{CO2}} + Y_{\mathrm{H2O}}}{1+s} \right) \tag{3}$$

where $A=4$, $B=0.5$ are model constants, ρ is the gas density, k and ε are the turbulent kinetic energy and its dissipation rate, Y_i are mass fractions of corresponding species, $s = 6 \cdot 32/100 = 1.92$ is the mass stoichiometric coefficient of oxygen (i.e., mass of oxygen consumed per unit mass of fuel) in reaction (2).

On the channel boundaries, injection of gaseous fuel occurs due to solid fuel decomposition. The injection mass flux was taken equal to that of solid fuel decomposition (see (1)), i.e., possible incomplete gasification due to charring is not taken into account. The density of gas produced on the walls was determined for the equation of state for pure fuel, assuming the local gas pressure; the temperature of the injected gas was taken equal to the boiling temperature T_s for PMMA. As a result, the linear local injection velocity was obtained and used as the boundary condition for the normal velocity component on the wall. For the tangential component of gas velocity, standard wall functions were applied which implies the turbulent log-law velocity profile. The heat flux from the gas phase onto the fuel surface, q_s, required in (1) for calculation of gasification rate, was also determined from wall functions for the temperature.

3. Numerical implementation

The governing equations (4) are solved in the axisymmetric framework by an explicit finite-volume high-resolution scheme on a uniform Cartesian grid. The "inviscid" fluxes are approximated by a scheme HR-SLAU2 [9] belonging to the AUSM family of numerical schemes. In order to calculate the cell face flux from the state variables defined in the cell centers, the "left" and "right" states on the boundary are first obtained by second-order monotonic interpolation with MINMOD flux limiter, after which the fluxes are found by solving the Riemann problem. An important feature of HR-SLAU2 scheme is that it falls into the category of "all-speed" numerical schemes owing to special control of numerical dissipation depending on the local Mach number. This is a must for internal flow problems where high-speed (including supersonic) zones can coexist with low-speed (subsonic) zones.

Viscosity, diffusion, and heat conduction terms are approximated by standard second-order central difference scheme. Chemical reactions are taken into account at a separate substep, with Strang splitting to maintain the second-order approximation in time: at each time step chemistry is advanced by half-step, then the gas dynamics solver is called to advance the solution by full time step, after which the chemistry solver is called again to integrate the kinetic equations by another half-step.

The solid fuel surface, generally, does not coincide with cell boundaries of Cartesian grid. Also, the shape of the surface can change with time due to fuel burnout. In this work, we apply the "embedded" sharp interface approach [10] in which the internal boundary is described by a level-set of a distance function. In the cells located away from the boundary, normal approximations are used, however, in the near-boundary cells operators are modified to take into account that cells are cut by the surface.

Implementation of wall functions in $k-\varepsilon$ turbulence model bears difficulties because embedded boundaries can cut cells randomly, in contrast to body-fitted grids where the size of near-wall cell can be controlled. To overcome this difficulty, wall functions were implemented not for the cut cells, but for one-dimensional grids extended from the surface into the flow along a normal to the surface, with subsequent interpolation of boundary layer variables onto the Cartesian mesh [11].

4. Geometry, initial and boundary conditions

Numerical simulations were carried out for the geometry which resembles closely the one used in the experiments [1, 2]. The geometry is shown in figure 1, with the main geometrical parameters listed in table 1.

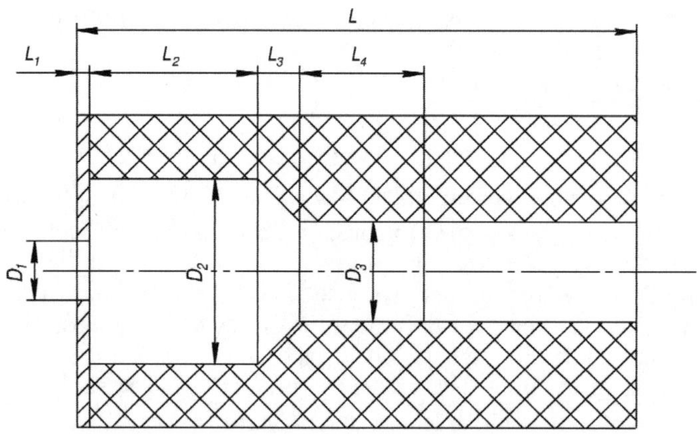

Figure 1. Geometry of solid fuel combustor: air in injected through the inlet of diameter D_1 into the main channel (of diameter D_3) via a flow expansion chamber (of diameter D_2) serving as flame holder due to development of recirculation zone.

Table 1. Combustion chamber measurements.

Parameter	Value (mm)
D_1	10
D_2	30
D_3	15
L	100
L_1	5
L_2	45
L_3	5

The inlet conditions corresponded to air with normal oxygen contents (21% vol.) supplied at a given linear velocity v_{in} , temperature T_{in} and pressure P_{in}. At the initial instant, the channel was filled with air at the same pressure and temperature as in the inlet. The initial axial velocity was set to 50% of the inlet velocity v_{in} in order to provide some gas flow through the channel. Since the cross-section area of the channel varies with distance, the initial flowrate did not correspond to steady-state conditions, rather, some time was required for the system to adjust the initial flow.

Due to strong feedback between the surface reaction and gas flow, solver instability could develop during this initial stage. Therefore, the initial period of 1 ms was simulated with the surface reaction (1) calculated from an artificial constant heat flux $q_s = 1$ kW/m^2, which was sufficient to initialize the gasification, but did not lead to development of high temperature in the combustor. After this period, the heat flux was gradually ramped up to its actual local values, and simulations were carried on without any further intervention.

Note that in the adopted here combustion model (EBU) chemical kinetics of gas-phase reaction (2) is not taken into, and ignition occurs as soon as the fuel is mixed with oxidizer. Further improvement of the combustion model will be performed in the future, so that ignition or extinction of combustion will be addressed. Here, we are interested mostly in the features of reacting flow in an already started SFRJ.

Numerical simulations were carried out in four cases listed in table 2. Results obtained are presented in the next section.

Table 2. Parameters of simulations.

Case	Pressure P_{in} (atm)	Velocity v_{in} (m/s)	Temperature T_{in} (K)
1	1	100	500
2	1	200	500
3	2	100	500
4	2	200	500

5. Results

In figure 2a-d, the temperature, concentrations (of fuel and oxidizer), and reaction rate spatial distributions obtained in Case 1 (see table 2) are plotted. In figure 2a, streamlines demonstrating the flow field developing in the combustion chamber are also shown. One can see that a recirculation zone develops in the flame holder due to abrupt change in the cross-section area.

In the recirculation zone, the gas consists of a mixture of fuel (MMA) emitted at the solid surfaces, and hot combustion products. The reaction zone revealed by the distribution of volumetric heat release rate (figure 2b) separates the fuel-containing zones near the solid surfaces (figure 2c) from the oxygen-rich core flow (figure 2d). Thus, combustion in the solid-fuelled ramjet is of diffusion type.

In figure 3a-d, similar results are shown in Case 2, with the injection velocity increased twice. In figures 4a-d and 5a-d, distributions are shown in Cases 3 and 4 (table 2) for higher pressure in the combustor. Increase in the pressure or inlet velocity results in the increase in the mass inflow of air (i.e., oxidizer) into the combustion chamber. However, this also increases the turbulence level and heat release rate, so that, generally, the distributions of flow parameters in the solid-fuelled ramjet presented in figures 2–4 are featured by similarity.

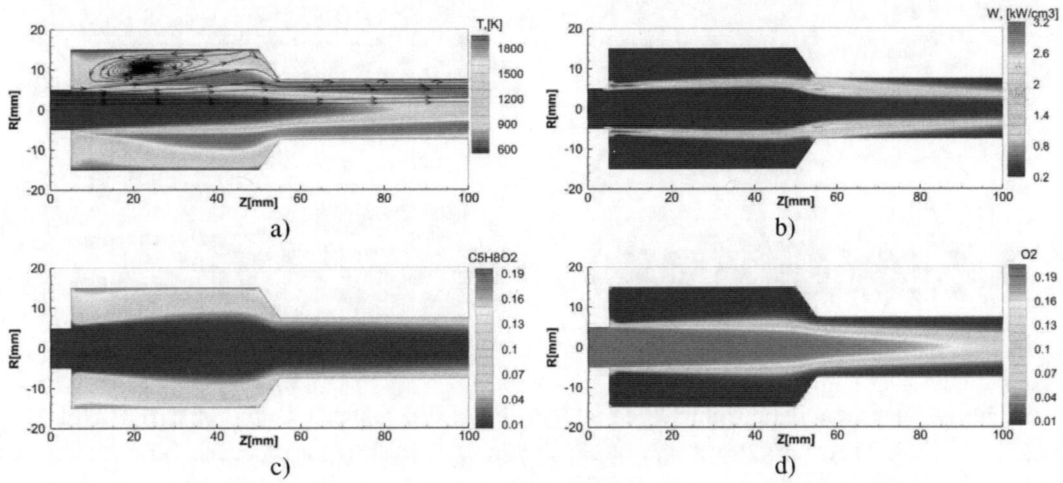

Figure 2. Temperature (a), volumetric heat release rate (b), volume fractions of $C_5H_8O_2$ (c), and O_2 (d) for $P_{in} = 1$ atm, $v_{in} = 100$ m/s, $T_{in} = 500$ K (Case 1 in table 2).

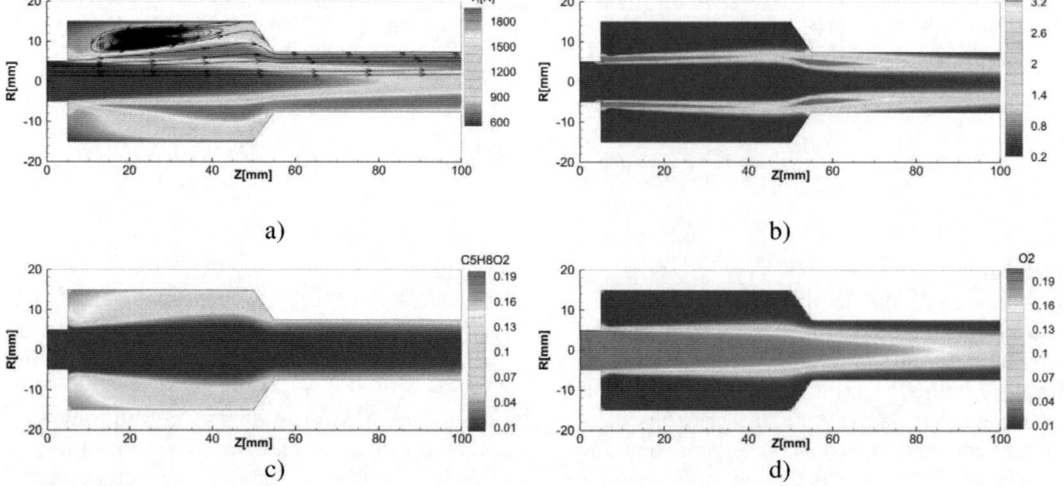

Figure 3. Temperature (a), volumetric heat release rate (b), volume fractions of $C_5H_8O_2$ (c), and O_2 (d) for $P_{in} = 1$ atm, $v_{in} = 200$ m/s, $T_{in} = 500$ K (Case 2 in table 2).

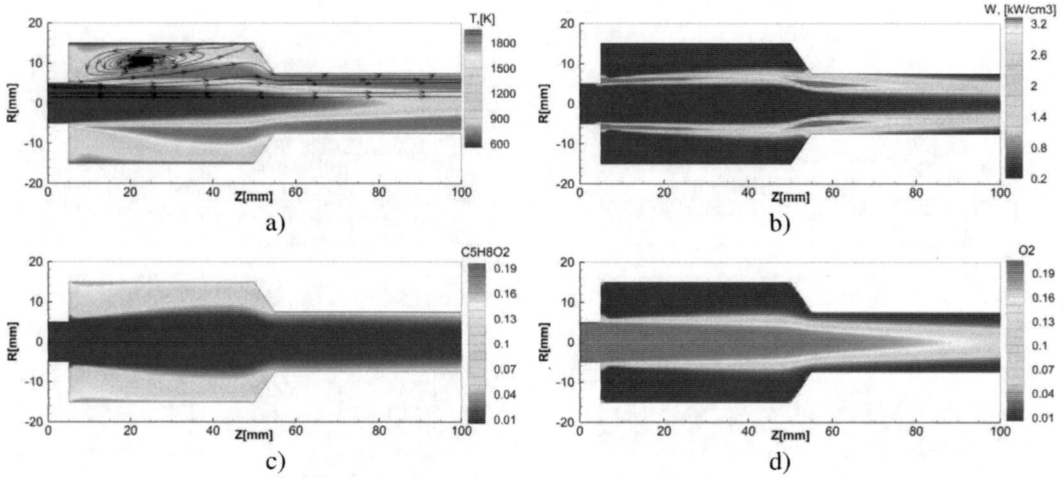

Figure 4. Temperature (a), volumetric heat release rate (b), volume fractions of $C_5H_8O_2$ (c), and O_2 (d) for $P_{in} = 2$ atm, $v_{in} = 100$ m/s, $T_{in} = 500$ K (Case 3 in table 2).

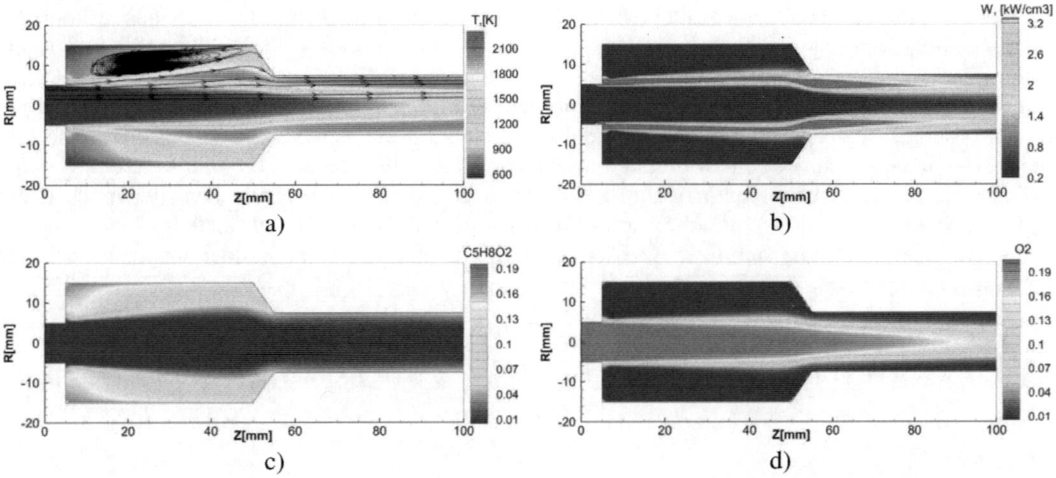

Figure 5. Temperature (a), volumetric heat release rate (b), volume fractions of $C_5H_8O_2$ (c), and O_2 (d) for $P_{in} = 2$ atm, $v_{in} = 200$ m/s, $T_{in} = 500$ K (Case 4 in table 2).

Compare now the time histories of temperature and total combustion power obtained in the four cases listed in table 2. In figure 6, corresponding results are shown, with the curves numbered in accordance with table 2. One can see that initial stage is featured by rapid rise in the combustion power, followed by gradual decrease to steady-state levels. The maximum temperatures are not too sensitive to the inlet parameters, while the combustion powers differ according to the mass flow of oxygen through the ramjet.

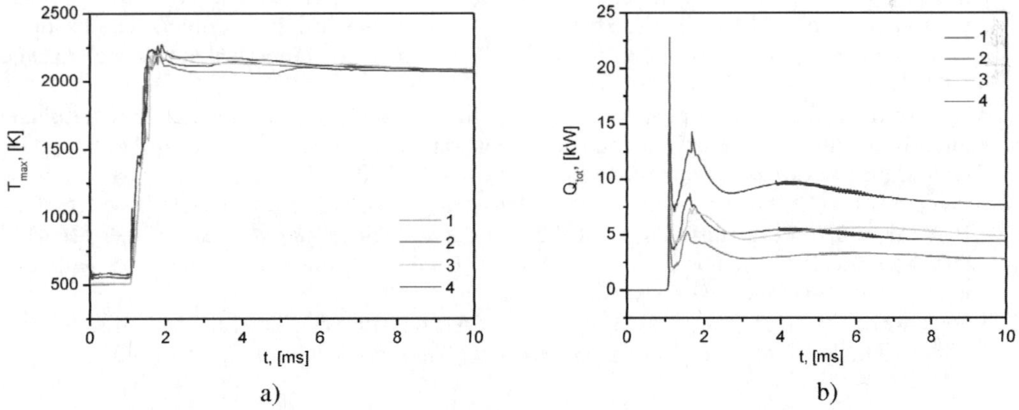

Figure 6. Time histories of the maximum temperature in the combustor (a) and total combustion power (b). (Curves correspond to cases 1–4 in table 2).

6. Conclusions

The results presented in this work demonstrate the capability of the mathematical model and software developed to simulate internal flows in the combustion chamber of a solid fuel ramjet. A more challenging problem which will be tackled in the follow-up work is to run coupled simulations of combustion and fuel burnout, i.e., consider flow with variable shape of the channel. The approach of "embedded" internal boundaries adopted in the numerical implementation provides a tool for describing the shape change without the need to rebuild the mesh each time the solid fuel shape changes.

Another challenging issue is to tackle the problem of ramjet ignition and extinction (flame blow-off). In order to treat this problem, the model of turbulent combustion must be enhanced, so that kinetics of chemical reaction (either global, or detailed) is taken into account.

The main research directions in which this work is going to evolve are to consider much wider range of operating conditions, including subsonic and supersonic ones. The ultimate goal is to find the conditions under which the stabilizer is efficient, study the effect of flame holder shape and perform parameters optimization.

Acknowledgment
This work was supported by Russian Foundation for Basic Research, Grant RFFI 16-29-01084 ofi_m.

References
[1] Zvuloni R, Gany A and Levy Y 1989 Geometric effects on the combustion in solid fuel ramjets *J. Propulsion* **5** 32–7
[2] Ben-Yakar A, Natan B and Gany A 1998 Investigation of a solid fuel scramjet combustor *J. Propulsion and Power* **14** 447–55
[3] Cohen-Zur A and Natan B 1998 Experimental investigation of a supersonic combustion solid fuel ramjet *Journal of Propulsion and Power* **14** 880–9
[4] Wang L, Wu Z, Chi H, Liu C, Tao H and Wang Q 2014 Numerical and experimental study on the solid-fuel scramjet combustor *Journal of Propulsion and Power* **31** 685–93
[5] Pei X and Hou L 2014 Numerical investigation on cavity structure of solid-fuel scramjet combustor *Acta Astronautica* **105** 463–75
[6] Chi H, Wei Z, Wang L, Li B and Wu Z 2015 Numerical investigation of self-ignition characteristics of solid-fuel scramjet combustor *Journal of Propulsion and Power* **31** 1019–32
[7] Novozhilov V, Joseph P, Ishiko K, Shimada T, Wang H and Liu J 2011 Polymer combustion as a basis for hybrid propulsion: A comprehensive review and new numerical approaches *Energies* **4** 1779–839
[8] Magnussen B F and Hjertager B H 1976 On the mathematical modelling of turbulent combustion with special emphasis on soot formation and combustion *Sixteenth Symp. (Int.) on Combustion* The Combustion Institute Pittsburgh PA 711–729
[9] Kitamura K and Hashimoto A 2016 Reduced dissipation AUSM-family fluxes: HR-SLAU2 and HR-AUSM+-up for high resolution unsteady flow simulations *Computers and Fluids* **126** 41–57
[10] Hu X Y, Khoo B C, Adams N A, and Huang F L 2006 A conservative interface method for compressible flows *Journal of Computational Physics* **219** 553–78
[11] Berger M and Aftosmis M J 2012 Progress towards a cartesian cut-cell method for viscous compressible flow *50th AIAA Aerospace Sciences Meeting (Nashville)* Jan. 9-12 1–17

IOP Conf. Series: Journal of Physics: Conf. Series **815** (2017) 012009 doi:10.1088/1742-6596/815/1/012009

On the influence of state-resolved rates of Zeldovich reactions on shock heated air flow parameters

O Kunova and E Nagnibeda

Saint Petersburg State University, 7/9 Universitetskaya nab., St. Petersburg 199034, Russia

E-mail: kunova.olga@gmail.com

Abstract. In this paper, the impact of non-equilibrium rates of Zeldovich reactions N_2+O↔NO+N, O_2+N↔NO+O on shock heated air flows parameters are studied on the basis of state-to-state approach. The comparison of computed macroscopic parameters using different state-specific rate coefficients models for exchange reactions is presented. The influence of NO molecules vibrational excitation on total relaxation process behind a shock wave is shown.

1. Introduction

The important task for investigations of hypersonic flows behind the shock waves appearing near space crafts, in nozzles, in high enthalpy facilities are the accurate description of the non-equilibrium chemistry and vibrational energy relaxation and the prediction of gas parameters. Until the recent time upon modeling of five-component air mixture $N_2/O_2/NO/N/O$ flows within the frame of state-to-state approach the main problem was the lack of models and data for the state-dependent rate coefficients of Zeldovich reactions

$$\mathrm{N}_2(i) + \mathrm{O} \leftrightarrow \mathrm{NO}(i') + \mathrm{N}, \quad (1)$$

$$\mathrm{O}_2(i) + \mathrm{N} \leftrightarrow \mathrm{NO}(i') + \mathrm{O}, \quad (2)$$

that take into account a vibrational state of product molecules NO. Therefore, the assumption that nitrogen oxide molecules remain at the ground vibrational state whereas only the vibrational state of the reactant molecules (N_2 or O_2) is taking into account was often accepted. So the number of works [1, 2, 3, 4, 5] focused on the study of different air flows and did not consider the process of vibrational relaxation of NO molecules. For such assumption there are enough models from which we can distinguish:

- Model [6] based on the theory of reactive hard spheres.
- Given in [7] expression for the state-specific rate coefficients is the approximation of numerical solutions of dynamic problems for reactive molecular collisions.
- The expressions for state-resolved rate coefficients of bimolecular exchange reactions suggested by Bose and Candler [8, 9] on the basis of quasiclassical trajectory calculations.

Until recently the only model that describes the state-depended rates of exchange chemical reactions in consideration of vibrational levels of both included molecules was model proposed by Starikovskii [10] and based on the model of vibronic terms [11]. Recently, the STELLAR database [12] has been compiled on the basis of results of Lino da Silva *et al.* The database includes the rate coefficients of all possible processes in the vibrational non-equilibrium air. There are present the Zeldovich reaction state-specific rate coefficients for a wide range of temperatures. These data were obtained on the basis of expressions of [8, 9] by expanding the model to a full range of molecule vibrational energies and the extrapolation of rate coefficients for temperature range from 100 K to 10^5 K.

In the present paper we consider the flow of five-component air mixture in the relaxation zone behind a shock wave taking into account the full kinetic mechanism. The influence of the models [10] and [12] for Zeldovich reaction rates, exchange chemical reactions themselves and vibrational relaxation of NO molecules on the gas flow parameters behind a shock wave is studied.

2. State-to-state description of shock heated air flows

The complete scheme of kinetic processes in five-component air mixture $N_2(i)/O_2(i)/NO(i)/N/O$ in the relaxation zone behind shock waves includes

- $VT(TV)$ exchanges between vibrational and translational energies

$$A_{ci} + \mathrm{M} \leftrightarrow A_{ci'} + \mathrm{M}, \tag{3}$$

- VV vibration energy exchanges at the collisions of the same molecular species

$$A_{ci} + A_{ck} \leftrightarrow A_{ci'} + A_{ck'}, \tag{4}$$

- VV' exchanges between different species

$$A_{ci} + A_{dk} \leftrightarrow A_{ci'} + A_{dk'}, \tag{5}$$

- dissociation and recombination reactions

$$A_{ci} + \mathrm{M} \leftrightarrow A_{c'} + A_{f'} + \mathrm{M}, \tag{6}$$

- and Zeldovich exchange reactions (1) and (2) in general form

$$A_{ci} + A_{dk} \leftrightarrow A_{c'i'} + A_{d'k'}. \tag{7}$$

Here A_{ci} and A_{dk} indicate the different species of molecules c and d, the symbol "prime" denotes the new species after reaction, i, k and i', k' are vibrational states of molecules before and after the collision, M is a partner in a collision (it can be a particle of an either mixture species).

In the present paper we consider the temperature range corresponding the conditions when the influence of electronic excitation, ionization and radiation remains weaker than that of vibration and chemical relaxation. Therefore in our work these processes are not taken into account.

2.1. Governing equations

The non-equilibrium air flow under the considered conditions is described by the equations for non-equilibrium kinetics coupled to the conservation equations for impulse and total energy.

The system of equations for the considered reacting air mixture includes the kinetic equations for the vibrational level populations of molecular species and number densities of atoms complemented by conservation equations for impulse and total energy. In the zero-order (Euler) approximation of stationary one-dimentional inviscid flow of the considered mixture we obtain:

$$\frac{d(vn_{ci})}{dx} = R_{ci}^{VT} + R_{ci}^{VV} + R_{ci}^{VV'} + R_{ci}^{DR} + R_{ci}^{ZR}, \tag{8}$$

$$i = 0, 1, \dots l_c - 1, \quad c = \mathrm{N_2}, \mathrm{O_2}, \mathrm{NO},$$

$$\frac{d(vn_c)}{dx} = R_c^{DR} + R_c^{ZR}, \quad c = \mathrm{N}, \mathrm{O}, \tag{9}$$

$$\rho v \frac{dv}{dx} + \frac{dp}{dx} = 0, \tag{10}$$

$$v \frac{dE}{dx} + (p + E) \frac{dv}{dx} = 0. \tag{11}$$

Here ρ is the gas density, p is the pressure, v is the velocity, E is the total energy per unit volume presented as a sum of translational, rotational, vibrational and formation energies:

$$E = \frac{3}{2} n k_B T + E_{rot} + \sum_{c,i} \varepsilon_i^c n_{ci} + \sum_c \varepsilon_c n_c, \tag{12}$$

n is the number mixture density, k_B is the Boltzmann constant, ε_i^c is the vibrational energy of a molecule c species at the i-th vibrational level, E_{rot} is the rotational mixture energy, ε_c is the formation energy of a particle c species, n_c, n_{ci} are the number density and vibrational population of i-th level for species c, l_c are the numbers of vibrational levels i of molecules ($l_{\mathrm{N_2}} = 61$, $l_{\mathrm{O_2}} = 46$ and $l_{\mathrm{NO}} = 48$).

The terms R_{ci}^{VT}, R_{ci}^{VV}, $R_{ci}^{VV'}$ describe vibrational energy transitions (3)–(5):

$$R_{ci}^{VT} = \sum_{\mathrm{M}} \sum_{i'=0}^{l_c-1} n_{\mathrm{M}} \left(n_{ci'} k_{c,i'i}^{\mathrm{M}} - n_{ci} k_{c,ii'}^{\mathrm{M}} \right), \quad c = \mathrm{N_2}, \mathrm{O_2}, \mathrm{NO}, \tag{13}$$

$$R_{ci}^{VV} = \sum_{k=0}^{l_c-1} \sum_{i'=0}^{l_c-1} \sum_{k'=0}^{l_c-1} \left(n_{ci'} n_{ck'} k_{c,i'i}^{c,k'k} - n_{ci} n_{ck} k_{c,ii'}^{c,kk'} \right), \tag{14}$$

$$R_{ci}^{VV'} = \sum_d \sum_{k=0}^{l_d-1} \sum_{i'=0}^{l_c-1} \sum_{k'=0}^{l_d-1} \left(n_{ci'} n_{dk'} k_{c,i'i}^{d,k'k} - n_{ci} n_{dk} k_{c,ii'}^{d,kk'} \right), \tag{15}$$

$$d = \mathrm{N_2}, \mathrm{O_2}, \mathrm{NO}, \quad d \neq c,$$

where $k_{c,ii'}^{\mathrm{M}}$, $k_{c,ii'}^{c,kk'}$ and $k_{c,ii'}^{d,kk'}$ are the state-depended rate coefficients for $VT(TV)$, VV and VV' transitions, respectively.

The term R_{ci}^{DR} describes dissociation and recombination (6):

$$R_{ci}^{DR} = \sum_{\mathrm{M}} n_{\mathrm{M}} (n_{c'} n_{f'} k_{rec,ci}^{\mathrm{M}} - n_{ci} k_{ci,diss}^{\mathrm{M}}). \tag{16}$$

Here $k_{ci,diss}^{\mathrm{M}}$ is the dissociation rate coefficient in result of a collision with a particle M, $k_{rec,ci}^{\mathrm{M}}$ is the recombination rate coefficient.

The source term R_{ci}^{ZR} describes Zeldovich exchange reaction (7):

$$R_{ci}^{ZR} = \sum_{i'=0}^{l_{c'}} (n_{c'i'} n_{d'k'} k_{c'i',ci}^{d'k',dk} - n_{ci} n_{dk} k_{ci,c'i'}^{dk,d'k'}), \tag{17}$$

where $k_{ci,c'i'}^{\mathrm{dk},\mathrm{d'k'}}$ is the rate coefficient of the forward reaction, the coefficient $k_{c'i',ci}^{\mathrm{d'k'},\mathrm{dk}}$ corresponds to the backward reaction.

2.2. State-resolved rate coefficients

The relaxation terms in equations (13)–(17) contain state-specific rate coefficients for energy transitions and chemical reactions. So the important problem for modeling of multicomponent reacting mixture flows is the choice of adequate models for the state-dependent rate coefficients of kinetic processes.

In present work the rate coefficients for vibrational energy transitions are calculated using the forced harmonic oscillator model [13]. For dissociation rate coefficients the Treanor-Marrone model [14] modified in [15] for the state-to-state approximation is used.

The state-specific rate coefficients of exchange reactions (1) and (2) are taken from the STELLAR database [12] and computing with using formula proposed in [10]. The state-specific rate coefficients $k_{N_2 i,O}^{NOi',N}$ and $k_{O_2 i,N}^{NOi',O}$ of Zeldovich reactions in dependence of a product molecule vibrational state i' for the considered models is shown in figure 1 for $T = 10,000$ K.

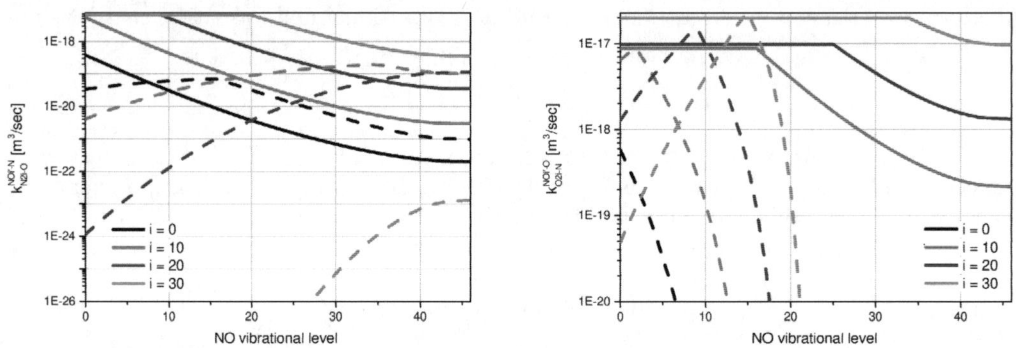

Figure 1. The rate coefficients $k_{N_2 i,O}^{NOi',N}$ (left) and $k_{O_2 i,N}^{NOi',O}$ (right) of reactions (1) and (2) as functions of the NO vibrational level i' for temperature $T = 10,000$ K and different vibrational state i of N_2 and O_2 molecules. STELLAR database: ——; model [10]: – – –.

Calculations show absolutely different behavior of the reaction rate coefficients with rising of NO vibrational level number. The data obtained from the generalization of quasi trajectory calculations [8, 9] take into account the reduction of the probability of forward reactions on the upper levels of the product molecules. At the same time the analytical model shows similar behaviour of reaction rate coefficients for $i < 10$ of N_2 molecules for the reaction (1), and does not describe this decrease for probabilities of reactions from higher levels of reagent molecules. In the case of reaction (2) model [10] gives a rapid non-monotonic variation of reaction rates for $i > 0$, meanwhile the maximum of $k_{O_2 i,N}^{NOi',O}$ increases with the number i and shifts towards increasing i'.

3. Results and discussion

Below we present the results of numerical solution of equations (8)–(11) for the following conditions in the free stream: $T_\infty = 271$ K, $p_\infty = 100$ Pa, $M_\infty = 15$, the mixture composition includes 79% of N_2 and 21% of O_2 molecules. The distributions over vibrational states in the free stream are assumed to be the Boltzmann ones with the temperature T_∞.

Calculations have been made with using the STELLAR database (SD) and Starikovskii model [10] for rate coefficients of reactions (1) and (2). Also we consider the case without NO excitation assuming that NO molecules form and remain on the zero-th vibrational level.

Figure 2 presents the comparison of the air temperature values as functions of the distance from a shock front. The gas temperature behind a shock decreases due to the vibrational and

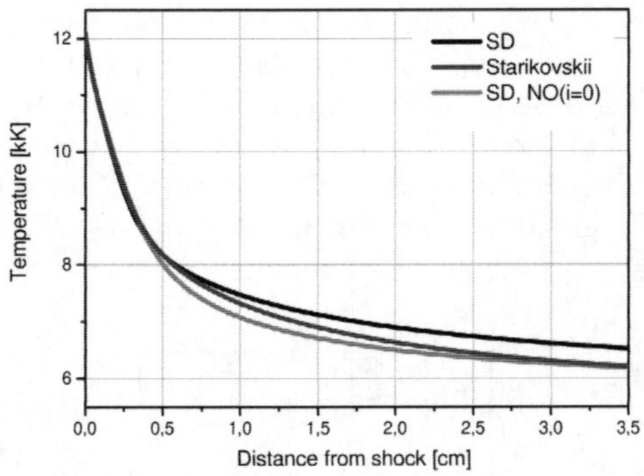

Figure 2. The temperature T as a function of x.

chemical relaxation. At the beginning of the relaxation zone the different models give practically the same temperature values, with x rising these values diverge a little and converge again approaching to the thermal and chemical equilibrium. The maximum temperature discrepancy obtained for three cases does not exceed 6 %.

The air mixture composition is presented in figure 3. Due to the dissociation reactions the number of nitrogen and oxygen molecules reduces with distance from the shock, while the number of free atoms increases. The non-monotonous variation of NO molar fractions is

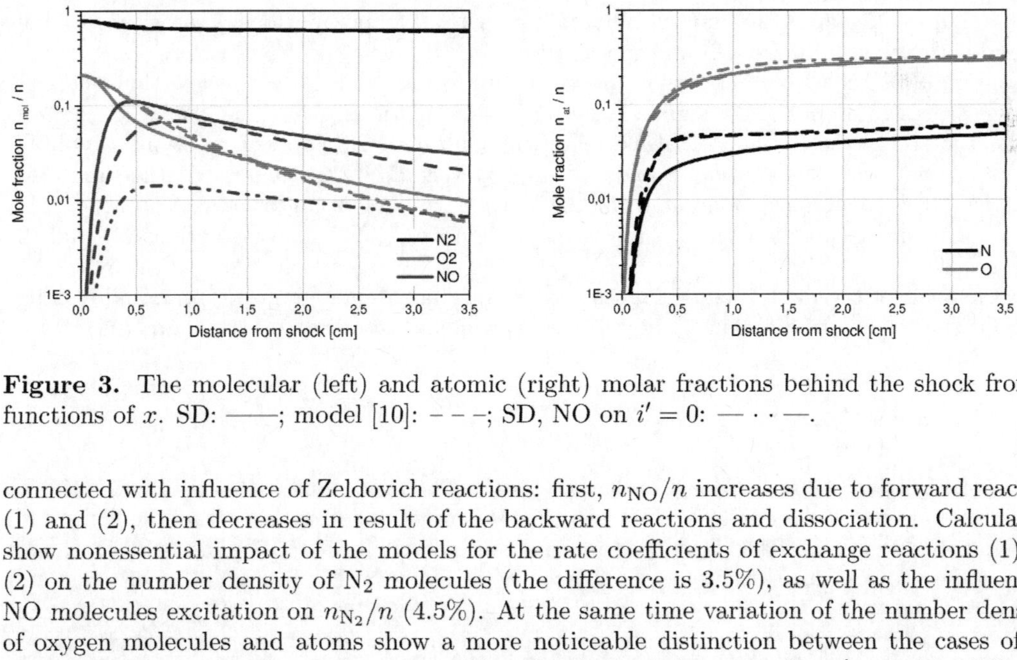

Figure 3. The molecular (left) and atomic (right) molar fractions behind the shock front as functions of x. SD: ——; model [10]: – – –; SD, NO on $i' = 0$: — · · —.

connected with influence of Zeldovich reactions: first, n_{NO}/n increases due to forward reactions (1) and (2), then decreases in result of the backward reactions and dissociation. Calculations show nonessential impact of the models for the rate coefficients of exchange reactions (1) and (2) on the number density of N_2 molecules (the difference is 3.5%), as well as the influence of NO molecules excitation on n_{N_2}/n (4.5%). At the same time variation of the number densities of oxygen molecules and atoms show a more noticeable distinction between the cases of two Zeldovich reaction models with excited NO molecules and the case when $i' = 0$. The difference reaches 15% and 50% for atoms and molecules, respectively. The number density of nitrogen atoms, obtained using model [10] and in case where $i' = 0$, is two-times higher than the number density n_N/n for the case of using the STELLAR database. The greatest impact of a choice of the reaction rate models is found on the number of nitrogen oxide molecules. The value n_{NO}/n

obtained using the database [12] exceed in 4.5 times the data calculated with the use [10] and in 16 times the values found neglecting NO vibrational excitation.

Figure 4 represents the vibrational distributions of air molecules in four positions behind the shock. For molecules of nitrogen and oxygen the excitation process can be seen in the figure, also the deactivation of the lower vibrational levels is noticeable. We can see that all three considered cases give the same values of the N_2 and O_2 level populations for low and intermediate vibrational energies with rising of distance from a shock front. The main difference

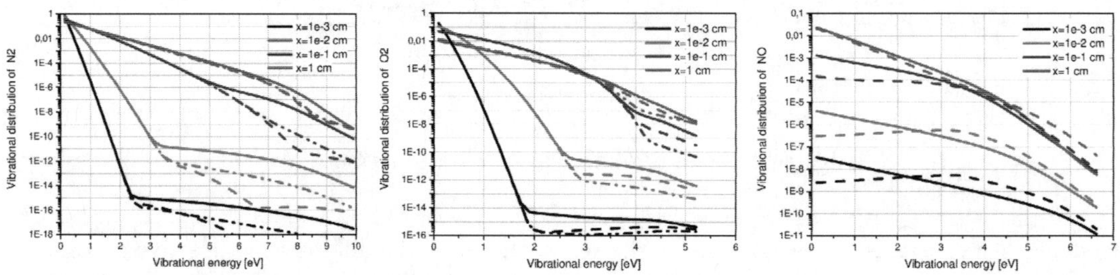

Figure 4. The vibrational distributions $n_{N_2 i}/n$ (left), $n_{O_2 i}/n$ (center) and n_{NOi}/n (right) for different values of x. SD: ——; model [10]: – – –; SD, NO on $i = 0$: — · · —.

between the populations of molecules is found on the upper vibrational levels. The distinctions in the state-resolved rate coefficients of reactions (1) and (2) provide difference in number densities in 4 orders of magnitude for the nitrogen molecules and 1.5 for the oxygen molecules. In the case $i' = 0$ upper O_2 levels are underpopulated, upper N_2 level populations lie between the ones obtained with using database [12] and model [10].

Variation of NO level populations is the most interesting. Calculations show that VT, VV and VV' processes do not affect significantly on NO populations, while the chemical reactions play a major role. Using the STELLAR database for rates of Zeldovich reactions we obtain more populated low levels and less populated upper levels of NO molecules. At the same time the model [10] predicts more populated middle levels in the beginning of relaxation zone. This behavior corresponds to the form of the state-specific rate coefficients of reactions (1) and (2) (see figure. 1).

On the basis of the models considered above we calculated changing of vibrational energies of various species due to different kinetic processes which is described by the source terms:

$$\Omega_c^r = \sum_{i=0}^{l_c - 1} R_i^r \varepsilon_i^c, \quad c = N_2, O_2, NO, \quad r = VT, VV, VV', DR, ZR.$$

Figure 5 presents the vibrational energy source terms. The results made it possible to estimate the vibrational energy gain or loss accounting for different processes as well as the sum of all these contributions. Calculations showed insignificant contribution of VV and VV' exchanges for considered conditions, so they are not presented in figure 5. For molecules of nitrogen and oxygen the process of TV-excitation at the beginning of the relaxation zone plays a dominant role, leading to an increase of the middle and upper vibrational level populations of N_2 and O_2 molecules and a decrease of translational temperature. Then with x rising the second phase begins where the chemical reactions prevail: for N_2 molecules it is Zeldovich exchange reaction (1), the dissociation process is not active; for O_2 molecules the greatest loss of vibrational energy is due to the dissociation reaction. We can see that accumulation of a large number of free atoms N and O switches Zeldovich reactions (1) and (2) in the opposite direction.

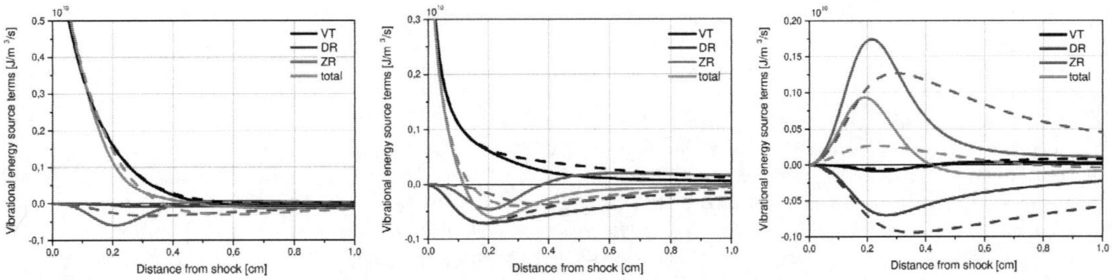

Figure 5. Vibrational energy source terms for N_2 (left), O_2 (center) and NO (right) as functions of x. SD: ——; model [10]: – – –.

In the case of NO molecules (figure 5 (right)) one can notice a delay of all chemical processes in the excitation region in the absence of a sufficient number of free atoms to start the formation of NO molecules. Then on the short distance the forward Zeldovich reactions are the main contributors to NO vibrational energy. With x rising the contributions of exchange reactions and dissociation reaction become of the same order of magnitude and the sum of all energetic contributions tends to zero. As noted above vibrational energy transfers have a small effect on variation of NO number density and their contribution to vibrational energy is about 1.5 order of magnitude lower than including of chemical reactions.

Comparing the Ω_c^{VT} and Ω_c^{DR} obtained using two models for exchange reactions a qualitatively similar behavior may be noted. For model [10] contribution of direct Zeldovich reactions to the vibrational energy of all molecular species is dominant.

4. Conclusion

In the present paper non-equilibrium flows of the reacting five-component air mixture in the relaxation zone behind shock waves are studied using two models for state-resolved rate coefficients of Zeldovich reactions. The results were compared with those obtained solving governing equations with an incomplete scheme of kinetic processes (under the assumption that NO molecules form and remain at the ground vibrational state).

It has been shown that a choice of description of exchange reaction rates and accounting excitation of NO molecules have a substantial effect on the number density of NO molecules. The essential difference between the vibrational level populations of NO molecules derived with using Zeldovich reactions rates from STELLAR database and model proposed by Starikovskii was shown. Neglecting vibrational excitation of NO molecules results in noticeable underestimated NO number densities. This may be important for higher temperatures, when NO molecules may be an essential source of radiation.

Acknowledgments

This study is supported by Saint Petersburg State University (grants No. 6.50.2522.2013 and No. 6.37.206.2016) and Russian Fund of Basic Research (project No. 15-01-02373).

References

[1] Capitelli M, Armenise I and Gorse C 1997 *J. Thermophys. Heat Transfer* **11** 570–578
[2] Colonna G, Tuttafesta M, Capitelli M and Giordano D 1998 *AIAA paper* 98–2951
[3] Colonna G, Tuttafesta M, Capitelli M and Giordano D 2000 *J. Thermophys. Heat Transfer* **14** 455–456
[4] Armenise I and Capitelli M 2005 *Plasma Sources Science and Technology* **14** S9
[5] Kunova O and Nagnibeda E 2014 *Chem. Phys.* **441** 66–76

[6] Warnatz J, Riedel U and Schmidt R 1992 *Advanced in Hypersonic Flows, Modeling Hypersonic Flows* vol 2 (Boston: Birkhäuser)
[7] Polak L, Goldenberg M and Levitskii A 1984 *Numerical Methods in Chemical Kinetics* (Moscow: Nauka) .(in Russian)
[8] Bose D and Candler G V 1996 *J. Chem. Phys.* **104** 2825–2833
[9] Bose D and Candler G V 1997 *J. Chem. Phys.* **107** 6136–6145
[10] Starikovskii A 2003 *Physical-Chemical Kinetics in Gas Dynamics. Electronic journal* `http://chemphys.edu.ru/issues/2003-1/articles/59/` (in Russian)
[11] Rusanov V and Fridman A 1984 *Physics of chemically active plasma* (Moscow: Nauka) (in Russian)
[12] da Silva M L Stellar database `http://esther.ist.utl.pt/pages/stellar.html`
[13] Adamovich I V, Macheret S O, Rich J W and Treanor C E 1998 *J. Thermophys. Heat Transfer* **12** 57–65
[14] Marrone P V and Treanor C E 1963 *Phys. Fluids* **6** 1215–1221
[15] Nagnibeda E and Kustova E 2009 *Non-equilibrium reacting gas flows: kinetic theory of transport and relaxation processes* (Springer)

IOP Conf. Series: Journal of Physics: Conf. Series **815** (2017) 012010 doi:10.1088/1742-6596/815/1/012010

Mathematical simulation of kinetic processes in moving irradiated by neutrons gas medium containing uranium nanoparticles

I V Alexeeva, A P Budnik, A V Sipachev and M N Slyunyaev

JSC "State Scientific Center of the Russian Federation –Institute for Physics and Power Engineering" named after A.I. Leypunsky., 1 Bondarenko Sq., Obninsk, Kaluga Oblast 249033, Russia

budnik@ippe.ru

Abstract. The theoretical model and program complex for mathematical simulation of processes of transformation the nuclear energy into optical radiation energy was developed. The model includes the equations of gas dynamics, as well as the equations describing the kinetic processes in the non-equilibrium plasma excited by uranium fission fragments.
The kinetic processes in the moving irradiated by neutrons argon-xenon gas medium containing uranium nanoparticles was investigated.
The space-time evolution of this medium in nonuniform changing over time neutron field was studied. The space-time evolution of the gas parameters (temperature, density, velocity, pressure), as well as the distribution of the concentration of uranium nanoparticles under different initial velocities of the gas and the size of the nanoparticles was calculated.
The amplifying properties of a laser-active space-nonuniform nuclear-excited moving argon–xenon medium, containing uranium nanoparticles and irradiated by neutrons, was studied.

1. Introduction

There has been 40 years since it was proposed to use finely divided uranium-containing particles dispersed in an active gas medium to convert nuclear energy to optical radiation energy [1].

As compared to traditional techniques for heterogeneous nuclear pumping of active gas media, the use of finely divided uranium-containing particles may lead to the share of energy carried out by fission fragments from condensed phase into the gas medium to increase tenfold and more. This creates opportunities for improving the efficiency of nuclear energy conversion to optical radiation energy.

Dispersal and absorption of laser radiation (LR) by an active medium with a content of finely divided uranium-containing particles is a major factor that hampers the generation of laser radiation in this medium.

It has been proposed recently to use laser-active gas media irradiated by neutrons and containing nanoclusters of uranium compounds [2].

In the beginning of computational and theoretical studies, the number of models and complexes of programs for mathematical modeling of kinetic processes in argon-xenon plasma irradiated by neutrons and containing uranium nanoparticles were designed [2-6]. Using developed models and programs it was demonstrated that it was possible to amplify laser radiation in a nuclear-excited argon–xenon dusty gas plasma [2, 3, 6].

It was further shown by mathematical modeling methods that during generation of laser radiation (LR) in an argon–xenon gas medium irradiated by neutrons and containing uranium nanoparticles the conversion of the uranium fission fragment kinetic energy to LR energy was an order of magnitude as efficient as the conversion of this energy during heterogeneous pumping [7-9].

This makes it possible to expect that a method and devices will be developed with a high efficiency of direct conversion of the fission fragment kinetic energy to the energy of coherent optical radiation.

In [7-9] however, only an immovable homogeneous dusty medium was considered. To prevent uranium nanoparticles from depositing in gas, it appears to be reasonable to blow down this medium. Apart from that, a dusty medium may start to move in the process of irradiation when the heating of gas by fission fragments is non-uniform.

Mathematical modeling of kinetic processes in a laser-active element (LAEL) in a moving argon–xenon gas medium containing uranium nanoparticles, during steady motion was developed in [10-12].

This paper deals with mathematical modeling of kinetic processes in argon-xenon plasma irradiated by neutrons and containing uranium nanoparticles, during unsteady motion of the medium, caused by pumping of this medium through LAEL, and heating of it by neutron-induced spatiotemporal inhomogeneous field of uranium fission fragments as well.

The purpose of the study is to determine the effects of the active medium movement and the spatial non-uniformity on the LR amplification process in a LAEL.

2. The main equations of the model

Let's investigate spatiotemporal evolution of moving into LAEL argon-xenon plasma containing uranium nanoparticles. LAEL is a cylindrical tube of radius R and length Z.

Due to high penetrating ability of neutrons at near to the perpendicular direction of incidence of neutron radiation on LAEL axis can confine ourselves to the study of axially symmetric motion of the gas in it.

We use to describe the spatiotemporal evolution of gas continuity equation environment, the Navier-Stokes equations, the energy balance equation, as well as the transfer of nanoparticles based on convection and diffusion, which in a cylindrical coordinate system are:

$$\frac{\partial \rho}{\partial t}+div\left(\rho\vec{V}\right)=0, \tag{1}$$

$$\frac{\partial \rho u}{\partial t}+div\left(\rho u\vec{V}\right)=-\frac{\partial p_t}{\partial z}+S_u, \tag{2}$$

$$\frac{\partial \rho v}{\partial t}+div\left(\rho v\vec{V}\right)=-\frac{\partial p_t}{\partial r}+S_v, \tag{3}$$

$$\rho c_p\frac{\partial T}{\partial t}+\rho c_p\vec{V}gradT=div\left(\lambda_t gradT\right)+\frac{dp_t}{dt}+Q, \tag{4}$$

where

$$S_u=-\frac{2}{3}\frac{\partial}{\partial z}\left(\mu div\vec{V}\right)+\frac{1}{r}\frac{\partial}{\partial r}\left[r\mu\left(\frac{\partial u}{\partial r}+\frac{\partial v}{\partial z}\right)\right]+2\frac{\partial}{\partial z}\left(\mu\frac{\partial u}{\partial z}\right), \tag{5}$$

$$S_v=-\frac{2}{3}\frac{\partial}{\partial r}\left(\mu div\vec{V}\right)+\frac{\partial}{\partial z}\left[\mu\left(\frac{\partial u}{\partial r}+\frac{\partial v}{\partial z}\right)\right]+2\frac{\partial}{\partial r}\left(\mu\frac{\partial v}{\partial r}\right)+2\mu\frac{\partial}{\partial r}\left(\frac{v}{r}\right), \tag{6}$$

$$\frac{\partial N}{\partial t}=D_f\left(\frac{1}{r}\frac{\partial}{\partial r}r\frac{\partial}{\partial r}N+\frac{\partial^2}{\partial z^2}N\right)-\frac{1}{r}\frac{\partial}{\partial r}rj_r-\frac{\partial}{\partial z}j_z. \tag{7}$$

Here z, r – cylindrical (axial and radial) coordinate, ρ, c_p, T – density, specific heat at constant pressure and temperature; u, v – axial and radial components of velocity $\vec{V}$; p_t – gas pressure; μ – dynamic viscosity and λ_t – gas thermal conductivity; Q – the volumetric power density of fission fragments in the gas, for heating the heavy gas particles (atoms, ions), N – concentration of uranium nanoparticles, D_f – diffusion coefficient of uranium nanoparticles, $\vec{j}$ – the density of the dust particle flow equal to

$$\vec{j} = \vec{v}_p N, \tag{8}$$

containing $\vec{v}_p$ – the dust particle movement velocity, which may differ from the argon–xenon gas medium movement velocity.

Next we confine ourselves to spherical nanoparticles with radius r_p which is less than a few tens of nanometers, which in dense gas (~ 1 atm pressure) moving along with the gas. Then we can assume that $\vec{v}_p$ is the same as the velocity of the gas.

We shall determine the diffusion coefficient D_f using the approximation proposed in [13]:

$$D_f = \frac{k_B T\left(1 + 3{,}12\text{Kn}\right)}{6\pi r_p \mu}, \tag{9}$$

where k_B – Boltzmann constant; Kn – Knudsen number.

3. The power density of fission fragments calculation in LAEL

In LAEL with gas containing nanoparticles of the energy from uranium fission fragments at the coordinates r, z can be defined as the divergence of the energy flux density transported by fission fragments, followed by integration over the volume of LAEL:

$$Q(r,z) = \sum_{i=1}^{n_f} \int_0^{2\pi} d\varphi \int_0^{z_m} dz_1 \int_0^{R} r_1 dr_1 \frac{S_0(r_1,z_1)}{\pi l^2} \left|\frac{\partial T_i(l')}{\partial l'}\right| \frac{\rho(r_1,z_1)}{\rho_0}, \tag{10}$$

containing z_m - the maximum of z, with which uranium fission fragments could reach destination point of calculations, n_f - the number of uranium fission fragments per fission act, $S_0(r_1,z_1)$ - the fission speed in unit volume at point with r_1, z_1, $T_i(l')$ - the energy of fission fragment i, which passed effective distance $l' = \int_0^l \frac{\rho}{\rho_0} dl$, $l = \sqrt{r^2 + r_1^2 - 2rr_1\cos\varphi + z^2}$ - real distance, which fission fragment passed in the gas, r_1 - the distance from the axis of LAEL to fission fragments' source, φ - the angle between directions from the axis of LAEL to the source of fission fragments source on a given point in the cross section, ρ_0 - the normal density of the gas for which the specified dependency $T_i(l)$.

Since the characteristic length in the longitudinal heterogeneity in LAEL, usually much larger than its radius, then, considering the subsonic gas motion may be presented in the form of two components

$$p_t = p_0 + p, \tag{11}$$

where p_0 – the average LAEL's cross-section pressure depends only on the coordinates z, p – component of the pressure, which depends both on the coordinates z, and r, and we can assume that

$$p_0 \gg |p|. \tag{12}$$

Neglecting in (4) p as compared to p_0 and integrating (4) over the cross section of LAEL from z_1 to z_2, taking into account (1) it can be obtained for p_0 the following equation

$$\frac{3V_0}{2}\frac{\partial p_0}{\partial t} = \int_{V_0} QdV - \int_{S_0} \lambda_t gradTd\vec{S} - \int_{z=z_1} udS - \int_{z=z_2} udS, \tag{13}$$

where V_0 - the volume by which the integration is, the S_0 - surface bounding this volume. The last two integrals in (13) are taken in cross-section of LAEL at a fixed z.

Taking into account (11), (12), and that the average pressure varies only along the coordinate z, Navier-Stokes equations can be rewritten

$$\frac{\partial \rho u}{\partial t} + div\left(\rho u\vec{V}\right) = -\frac{\partial p_0}{\partial z} - \frac{\partial p}{\partial z} + S_u, \tag{14}$$

$$\frac{\partial \rho v}{\partial t} + div\left(\rho v\vec{V}\right) = -\frac{\partial p}{\partial r} + S_v. \tag{15}$$

Boundary conditions

$$z=0,\ u=u_{\max}(1-r^2/R^2),\ v=0,\ T=T_{z0},\ p_t=p_0,\ N(z,R,t)=N_0(R); \tag{16}$$

$$r=0,\ \frac{\partial u}{\partial r}=0,\ v=0,\ \frac{\partial T}{\partial r}=0,\ \frac{\partial p_t}{\partial r}=0; \tag{17}$$

$$r=R,\ u=v=0,\ T=T_R(r,t),\ \frac{\partial p_t}{\partial r}=0,\ \ N(z,r,t)=0; \tag{18}$$

$$z=Z,\ \frac{\partial u}{\partial z}=\frac{\partial v}{\partial z}=0,\ \frac{\partial T}{\partial z}=0,\ \frac{\partial p_t}{\partial z}=0,\ \frac{\partial N(0,r,t)}{\partial z}=N_0 \tag{19}$$

To solve the system of equations (1) - (19) the method of finite differences was used. These equations were approximated using five-point finite-difference scheme. The solution method is described in detail in [14-16].

4. Results of spatiotemporal argon-xenon gas medium containing uranium nanoparticles evolution modeling

Figures 1–4 show typical results of the space-time evolution of the velocity distribution in the projection of LAEL its temperature axis, pressure and concentration of uranium nanoparticles. The distribution of the power in LAEL axis at different times are shown on figure 5.

Results of calculations are shown for the particles with a radius in $r_p=5$ nm and $p=0.5$atm initial gas pressure and the following values of the variables: $Z=300$ cm, $R=12$ cm, $u_{max}=2$ m/s, the maximum concentration of uranium particles on LAEL axis was assumed to be $N_0 = 10^{12}$ cm^{-3}. The dependence of the thermal neutron flux density J, causing fission of uranium nuclei, given as

$$J = J_0 \exp(1-(t/\tau_0)-((z-Z/2)/(Z/6))^2)(t/\tau_0), \tag{20}$$

where $\tau_0 = 150$ ms.

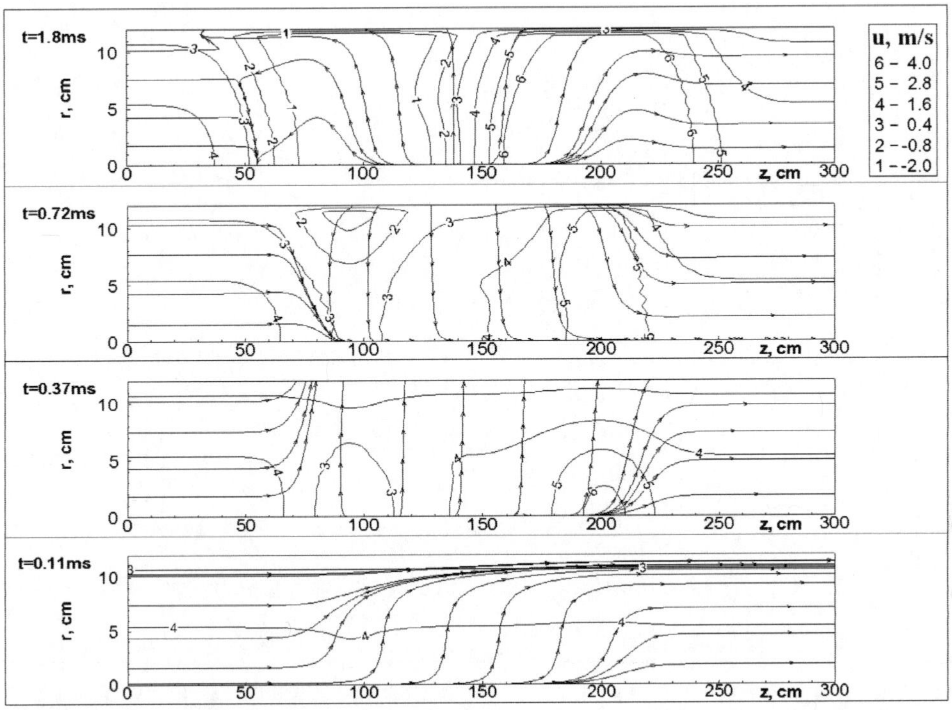

Figure 1. The distribution of the speed projection u to LAEL's axis at different times.

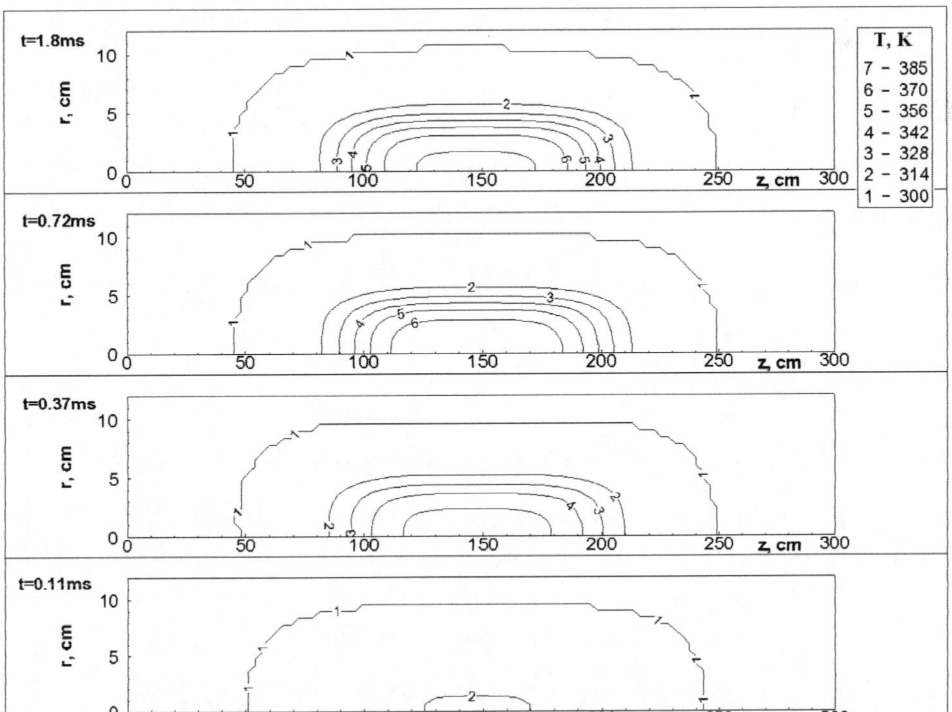

Figure 2. The distribution of the temperature in LAEL at different times.

IOP Conf. Series: Journal of Physics: Conf. Series **815** (2017) 012010 doi:10.1088/1742-6596/815/1/012010

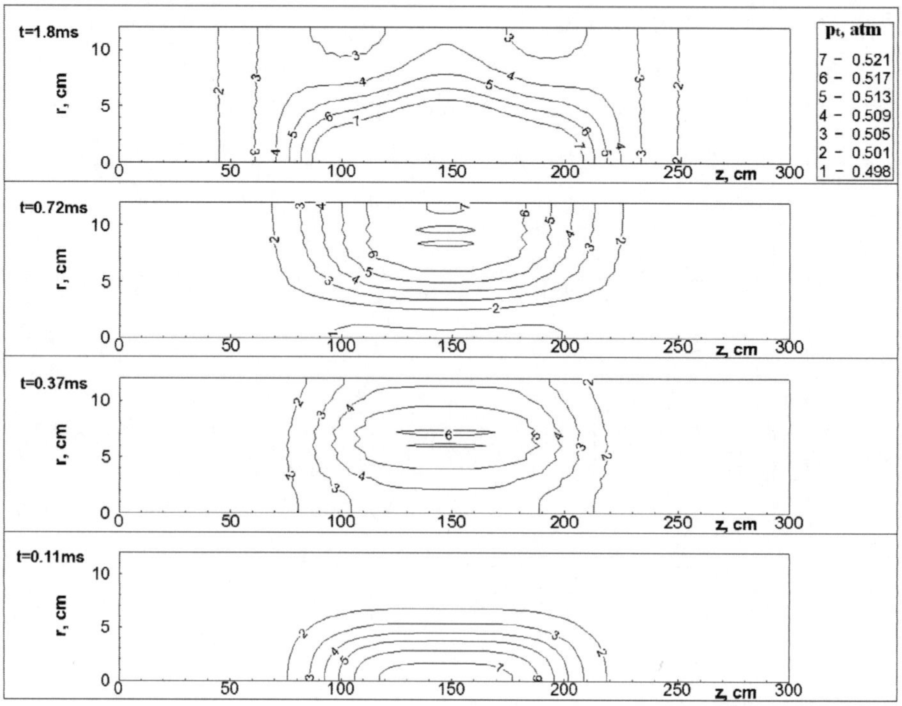

Figure 3. The distribution of the pressure in LAEL at different times.

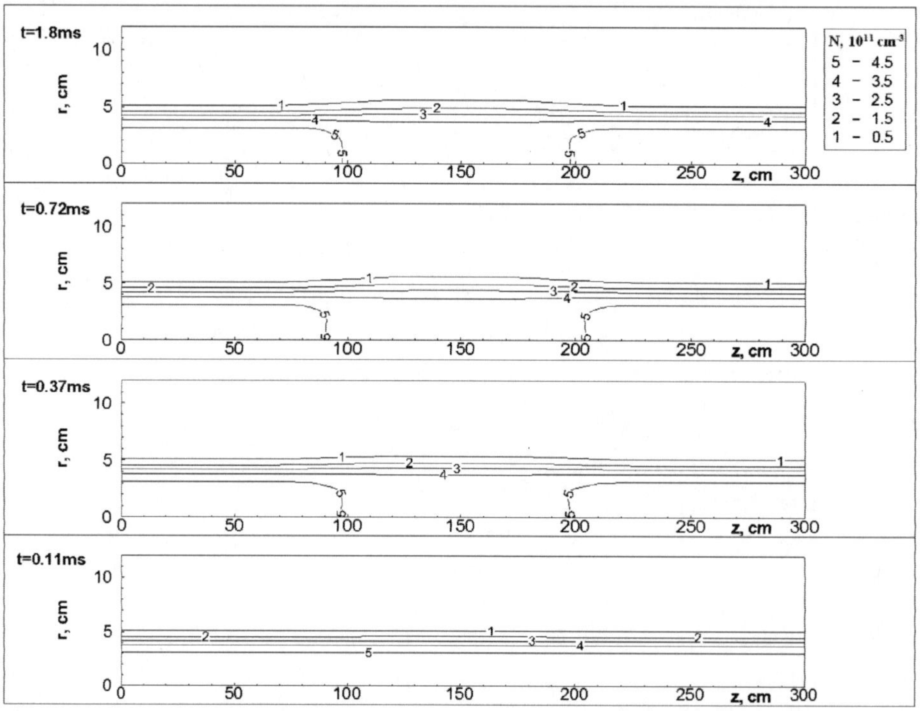

Figure 4. The distribution of the concentration of nanoparticles in LAEL at different times.

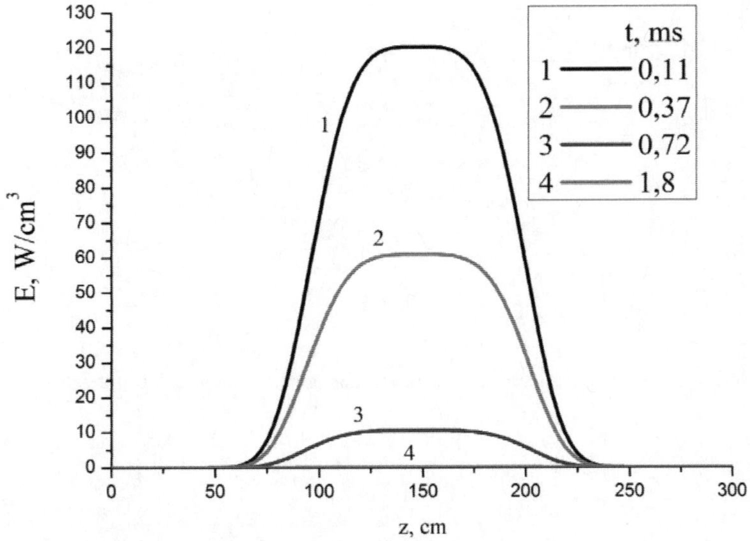

Figure 5. The distribution of the power in LAEL axis at different times.

5. A model of kinetic processes in a dusty argon–xenon laser-active medium excited by fission fragments

A model of kinetic processes in an argon–xenon laser-active medium with a monodispersed dusty component excited by fission fragments was developed in [2, 3]. This model was used in the present study to investigate the kinetic processes in the plasma created by neutron-induced uranium fission fragments.

The kinetic model of an argon–xenon medium with a monodispersed dusty component in the gas component took into account the atomic (Ar^+, Xe^+)and homonuclear molecular (Ar_2^+, Xe_2^+) ions of argon and xenon, the heteronuclear ion $ArXe^+$ and the molecule $ArXe$, argon and xenon molecules in excited states, as well as argon and xenon excimers (Figures 6, 7).

Levels 6s, eight 5d sublevels and six 6p sublevels were considered separately for the excited xenon atom, and the sublevels of the states and 7s were combined to form a single level Xe(p s) (Figure 6). All other xenon states were combined to form a single state designated Xe*.

The sequence of the major kinetic processes leading to the formation of an inverse population in the argon–xenon medium is as follows. While interacting with the argon–xenon medium, fission fragments lose energy, primarily for the formation of atomic ions and excited argon atoms. Further, the collision of atomic ions Ar^+ and excited argon atoms with argon and xenon atoms leads to the formation of both excited xenon atoms and the atomic xenon ions Xe^+, as well as of the molecular homonuclear ions Ar_2^+, Xe_2^+ and the heteronuclear ions.

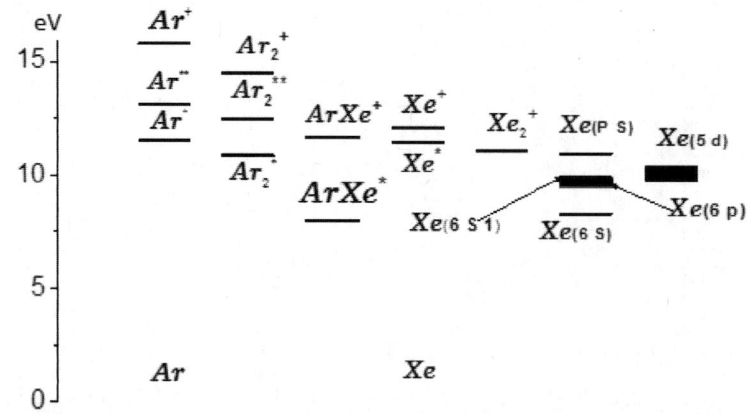

Figure 6. Energy diagram of states considered in the kinetic model.

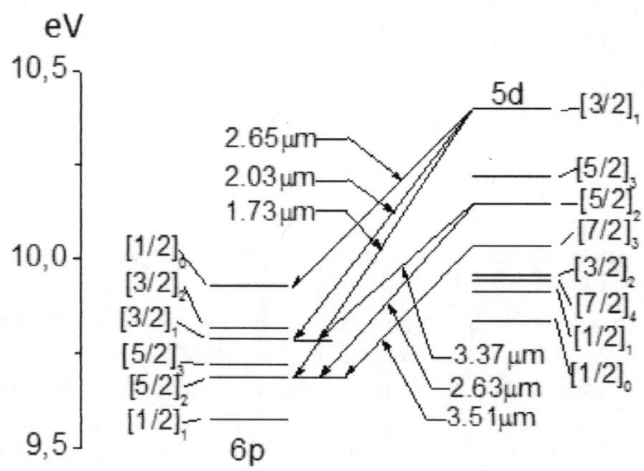

Figure 7. Energy diagram of the xenon atom states considered in the kinetic model at the transition between which it is possible to achieve the generation of laser radiation.

Check the upper excited states of xenon atoms (denoted as PS) procomes as a result of dissociative recombination of molecular ions $ArXe^+$, Xe_2^+. In the future as a result of collisions with argon atoms going population of the upper levels for the transitions with a wavelength of 1.73 micron, 2.03 micron and 2.65 micron.

The up and lower laser levels are populated through multiple channels - due to quenching of excitation in the collision ions with the plasma electrons and by radiation as a result of stimulated and spontaneous emission. These processes lead to the population of the lower laser levels, which are extinguished in collisions with argon atoms and electrons, settling with a low-lying excited states of xenon atoms. These processes lead to the population of the lower laser levels, which are extinguished in collisions with argon atoms and electrons, settling with a low-lying excited states of xenon atoms.

In the model with a monodisperse dust component of the processes of interaction of electrons and ions with the charged nanoparticles are described as follows "plasma chemical" reactions:

$$e + D \rightarrow D(-), \tag{21}$$

$$e + D(n-) \rightarrow D((n+1)-), \tag{22}$$

$$I(+) + D(n-) \rightarrow D((n-1)-), \tag{23}$$

$$I(+) + D \rightarrow D(+). \tag{24}$$

Here we have introduced the notation e – electron, D, $D(+)$, $D(n-)$ – respectively, are electrically neutral, positively and negatively charged nanoparticles 5 nm in radius, n - charge of the nanoparticles in units of electron charge, $I(+)$ - any positively charged atomic or molecular ion of the gas mixture.

Altogether, 57 components were considered and 434 reactions in the argon–xenon medium were taken into account in the model.

6. LR linear amplification by medium coefficient calculation

An important characteristic of the laser-active medium is a gain of the linear laser light. Using the model kinetic processes in a xenon-argon plasma containing uranium nanoparticles may define unsaturated linear laser gain.

In mathematical modeling of the microscopic kinetics of the following circumstances may be used. In typical conditions, the nuclear pump power density of the energy input into the laser-active medium is changed slowly over time (the characteristic time changes $\tau \geq 10$ ms). Diffusion of plasma components can be neglected, because of the characteristic lifetime of most of the major components of argon-xenon medium diffusion length is much smaller than the characteristic spatial scales of energy irregularities. This allows the steady-state approximation of the local kinetics - that is, assume that the concentration of various components of the plasma at a given point are determined by the instantaneous values of the temperature of the heavy particles, gas pressure and power density energy deposition in the local approximation.

In operation, the values were calculated quasistationary linear gain laser radiation at a wavelength of 1.73 microns at a concentration of uranium in the range of nanoparticles $0.125 \cdot 10^{11}$ cm^{-3} to 10^{12} cm^{-3}, LAEL gas pressure in the range of 0.5 atm to 0.6 atm and the specific energy input power ranging from 23 W/cm^3 to 277 W/cm^3.

The table 1 shows typical results, whether linear IS gain calculation gaseous medium quasistationary values at a wavelength of 1.73 microns, depending on the concentration of uranium in the nanoparticulate active laser-xenon, argon-gas atmosphere at a specific energy input power of 230 W/cm^{3}at different gas pressures in LAEL.

The gain of the laser gas medium is designed to disregard the weakening of nanoparticles radiation. There's also shows the calculated data on the dependence coefficients attenuation β laser attenuation uranium nanoparticles (the particle radius - 5 nm) on their concentration for a wavelength of 1.73 microns [2,3,5,6] and the value of the total gain medium, calculated by formula:

$$k = \alpha - \beta. \tag{25}$$

To calculate the total gain medium LAEL containing uranium nanoparticles over the entire range of possible values, the method of linear interpolation based on the calculated parameters of the gain medium.

The changing of intensity I of the laser radiation propagating parallel to the cylinder axis, can be described by the following equation:

$$\frac{dI}{dz} = k(z, r, t) I \tag{26}$$

We define the LR intensity by gas medium containing uranium nanoparticles gain as follows:

$$K(z,r,t) = I / I_0, \tag{27}$$

where I_0 - the intensity of the radiation entering the laser active medium.
Then $K(z,r,t)$ can be written as:

$$K(z,r,t) = \exp\left(\int_0^z k\left(z_1, r, t\right)dz_1\right). \tag{28}$$

Table 1. Linear gains α of the gaseous medium, the attenuation β of the radiation to depending on the concentration of nanoparticles N and gas pressure and the full gain of the medium k at a temperature T = 300 K, and the specific energy input power of 230 W/cm^3.

P, atm	N, 10^{10}cm^{-3}	α, 10^{-2}cm^{-1}	β, 10^{-3}cm^{-1}	k, 10^{-2}cm^{-1}
0.5	100	4.4	2.1	4.2
	50	4.5	1.09	4.4
	25	4.5	0.5	4.5
	12.5	4.6	0.2	4.5
	10	4.6	0.2	4.6
	5	4.6	0.109	4.6
	2.5	4.6	0.05	4.6
	1.25	4.6	0.02	4.6
0.6	100	4.3	2.1	4.1
	50	4.4	1.09	4.3
	25	4.4	0.5	4.4
	12.5	4.5	0.2	4.4
	10	4.5	0.2	4.5
	5	4.5	0.109	4.5
	2.5	4.5	0.05	4.5
	1.25	4.5	0.02	4.5

In the quasi-stationary approximation the local kinetics using data calculations spatio-temporal evolution of environmental parameters (temperature distribution, pressure, concentration of uranium nanoparticles, specific power input energy of fission fragments) and neglecting the refraction of the laser, we can estimate the gain of the laser intensity considered laser-active medium.

Figure 8 shows the results of radial dependence of the intensity $K(z,r,t)$ LR gain at different times from the outlet LAEL.

Conclusion

A two-dimensional non-uniform axial-symmetric model was developed to describe the space-temporal evolution of the concentration of uranium nanoparticles in LAEL. . Also the finite difference method for numerical solution of the model equations was designed.

The mathematical simulation of kinetic processes during direct conversion of the kinetic energy of the uranium fission fragments into the laser energy in the dust LAEL containing uranium nanoparticles was carried out.

The motion of the dusty nuclear-exited plasma induced by forced blowing of gas medium in LAEL and by the nonuniform energy release in the fission fragments' medium irradiated by unsteady space-nonuniform neutron field was investigated.

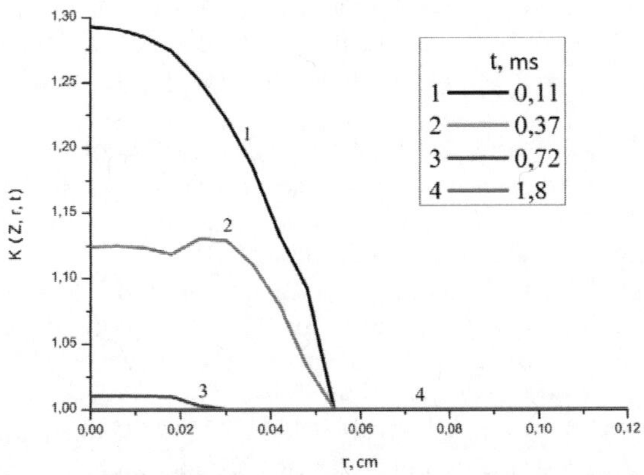

Figure 8. Radial dependence of the LR intensity amplification coefficient at different times.From these results of calculations of the gain of laser intensity, it follows that the medium length of 1 m provides a significant amplification of laser radiation in a single pass.

The space-time evolution of the gas parameters (temperature, density, velocity, pressure), as well as the distribution of the concentration of uranium nanoparticles under different initial velocities of the gas and the size of the nanoparticles was calculated.

Using the results of these calculations in the local microscopic plasma chemical kinetics quasi-stationary approximation it possible to determine the most important characteristics of the active medium - quasi-stationary value of the linear gain of LR at a wavelength of 1,73 microns.

The amplifying properties of a laser-active space-nonuniform nuclear-excited moving argon–xenon medium, containing uranium nanoparticles and irradiated by neutrons allow you to use such medium not only in nuclear pumped lasers but also in optical quantum amplifiers with nuclear pumping.

References

[1] Miley G H, McArhur D, DeYuong R and Prelas M 1989 Fission reactor pumped laser: History and prospects: *Proceedings Conferences 50 Years with nuclear fission* Washington 25–28 April Pub American Nuclear Society pp 333-342

[2] Budnik A P, Kosarev V A and Lunev V P 2009 *in Proc of the IV Int Confer on Physics of Nuclear Pumped Laser and Pulse Reactor* (IPPE, Obninsk,) **1** pp 177-184

[3] Budnik A P, Kosarev V A and Lunev V P 2008 Mathematical simulation of kinetic processes in gas of argon-xenon plasma containing nanoclusters of chemical compounds of uranium (*Obninsk:Preprint IPPE 3141*) p 23 (in Russian)

[4] Budnik A P, Deputatova L V, Fortov V E, Lunev V P and Vladimirov V I 2012 Simulation of kinetic processes, optical and neutron properties of the nuclear-excited uranium dusty plasma of the argon-xenon gas mixture *Ukrainian Journal of Physics* **56** 12 pp 1260-1264

[5] Budnik A P, Deputatova L V, Fortov V E, Kosarev V A, Rykov V A and Vladimirov V I 2009 Simulation of kinetic processes in the nuclear-excited helium non-ideal dusty plasma

Contribution to Plasma Physics 10 pp 765-768

[6] Budnik A P and Lunev V P 2011 The theoretical investigation of neutron and optical properties of argon-xenon gas media containing nanoclusters of the uranium and its chemical compounds by the Monte-Carlo method *Physical-Chemical Kinetics in Gas Dynamics* **11** http://chemphys.edu.ru/issues/2011-11/articles/167/ (in Russian)

[7] Budnik A P and Sipachev A V 2012 Mathematical simulation of kinetic processes in argon-xenon nuclear-exited plasma, containing nanoclusters of uranium *Physical-Chemical Kinetics in Gas Dynamics* **13** (3) http://chemphys.edu.ru/issues/2012-13-3/articles/358/ (in Russian)

[8] Budnik A P and Sipachev A V 2013 Mathematical simulation of kinetic processes in generation of laser radiation in argon-xenon active gas media, containing nanoparticles of uranium *Physical-Chemical Kinetics in Gas Dynamics* **14** (2) (in Russian) http://chemphys.edu.ru/issues/2013-14-2/articles/381/

[9] Budnik A P, Sipachev A V 2014 Mathematical modeling of kinetic processes with generation of laser radiation in helium-argon-xenon dusty plasma containing uranium nanoparticles Physical-Chemical Kinetics in Gas Dynamics **15** (5) (in Russian) http://chemphys.edu.ru/issues/2014-15-5/articles/256/

[10] Budnik A P, Sipachev A V and Slyunyayev M N 2014 Mathematical modeling of kinetic processes in moving argon-xenon dusty plasma containing uranium nanoparticles *Physical-Chemical Kinetics in Gas Dynamics* **15** (6) (in Russian) http://chemphys.edu.ru/issues/2014-15-6/articles/264/

[11] Slyunyayev M N, Budnik A P and Sipachev A V 2015 Simulation of direct conversion of the kinetic energy of the fission fragments of uranium in the laser energy in an argon-xenon dusty plasmas with nanoparticles of uranium *Proceedings of the universitie, Nuclear power* 2 pp 71-80 (in Russian)

[12] Slyunyaev M N, Budnik A P and Sipachev A V 2015 Modeling of direct conversion of the uranium fission product kinetic energy to laser radiation energy in an argon–xenon dusty plasma with uranium nanoparticles *Nuclear Energy and Technology* 1 pp 272-276 (in Russian)

[13] Smirnov B M 1990 Aerosols in gas and plasma (Moscow: IVTAN) (in Russian)

[14] Alexeeva I V, Budnik A P and Sipachev A V 2010 The nonequilibrium radiative plasma dynamics into active media of nuclear-pumped optical amplifiers *Physical-Chemical Kinetics in Gas Dynamics* **9** http://chemphyseduru/issues/2010-9/articles/128/

[15] Alexeeva I V, Budnik A P and Shipachev A V 2004 Mathematical modelling of energy conversion of fission fragment energy into coherent radiation in an inhomogeneous argon-xenon nuclear-excited plasma of a gas moving optical amplifiers with nuclear pumping (Obninsk:Preprint No 3019 IPPE) p 28

[16] Alexeeva I V, Budnik A P and Sipachev A V 2006 Mathematical simulation of kinetic processes in inhomogeneous argon-xenon gas plasma optical quantum amplifiers with direct pumping by fission fragments *Proc. of the Int. conference "VIII Kharitonov readings on problems of physics of high energy densities*", March 21-24, Sarov pp 362-368

State-to-state models of vibrational relaxation in Direct Simulation Monte Carlo (DSMC)

G P Oblapenko[1], **A V Kashkovsky**[2] **and Ye A Bondar**[1,2]

[1] Saint-Petersburg State University, 7/9 Universitetskaya nab., St. Petersburg, 199034 Russia
[2] Institute of Theoretical and Applied Mechanics, 4/1 Institutskaya ul., 630090 Russia

E-mail: kunstmord@kunstmord.com

Abstract. In the present work, the application of state-to-state models of vibrational energy exchanges to the Direct Simulation Monte Carlo (DSMC) is considered. A state-to-state model for VT transitions of vibrational energy in nitrogen and oxygen, based on the application of the inverse Laplace transform to results of quasiclassical trajectory calculations (QCT) of vibrational energy transitions, along with the Forced Harmonic Oscillator (FHO) state-to-state model is implemented in DSMC code and applied to flows around blunt bodies. Comparisons are made with the widely used Larsen–Borgnakke model and the influence of multi-quantum VT transitions is assessed.

1. Introduction

Direct Simulation Monte Carlo (DSMC) [1] methods are widely used for modeling of non-equilibrium high-temperature flows. Modeling of high-temperature flows requires accounting for excitation and relaxation of internal energy modes and chemical reactions, which is essential for correct computation of flow parameters, such as temperature profiles and heat fluxes.

The Larsen–Borgnakke model [2] is commonly used for modeling of rotational and vibrational non-equilibrium in Direct Simulation Monte Carlo methods. It has several advantages, those being 1) it satisfies the detailed balance principle 2) it recovers energy relaxation rates as given by the Jeans and Landau–Teller equations [3]. However, the model is not strictly microscopic, since it requires computation of temperature-dependent collision numbers because rotational and vibrational relaxation times are non constant and strongly depend on the gas temperature [4, 5]; also, the Millikan–White formula [5] (even with Park's correction [6]), commonly used to compute the vibrational collision number, has been shown to poorly approximate VT relaxation times at high temperatures [7, 8, 9]. Therefore, given that the use of state-to-state vibrational transition models is feasible from a computational point of view, it is of interest to assess their influence on various flows.

Various semiclassical models exist for computation of VV and VT transition probabilities, such as the Schwartz-Slavsky-Herzfeld (SSH) model [10], "vibrationally inelastic cross section" (VICS) model [11], and the Forced Harmonic Oscillator model [12]. More accurate models of inelastic processes are based on quasiclassical trajectory calculations (QCT). Performing molecular dynamics simulations using an accurate interaction potential, cross sections of inelastic processes can be recovered, which can then be implemented in the DSMC method. Inelastic cross-sections for collisions of molecular nitrogen with nitrogen atoms were studied in [13, 14, 15], while inelastic cross-sections for collisions of molecular oxygen with oxygen atoms were discussed in [16, 13, 17, 18]. In works [19, 20], the molecular dynamics simulations are implemented within the DSMC method itself, which removes the need for the approximation of QCT results, but leads to an increase in computational cost.

However, while there exist works concerned with the application of accurate state-to-state models of vibrational transitions (either semiclassical or based on QCT calculations), most of are concerned only with spatially homogeneous vibrational relaxation, and do not study the effects of the choice of vibrational relaxation models on non-equilibrium flows around real bodies. Only a few works have investigated 2-dimensional flows with state-to-state vibrational relaxation models [21, 22, 15]. In all of these works, a significant influence of the vibrational transition model on the temperature profiles was found. It should be noted that only in [21] VV exchanges were accounted for. While VV exchanges do not influence the translational and vibrational temperatures, they can significantly affect vibrational level populations [21, 23].

Therefore, the influence of state-to-state models on 2-dimensional bodies warrants further research. The aim of this work is to 1) apply both semiclassical vibrational transition models and QCT-based models to DSMC modeling of various high-temperature 2-dimensional flows 2) compare results obtained using these models to results computed using the Larsen–Borgnakke model 3) study the role of multi-quantum VT transitions and how they affect macroscopic flow parameters.

2. State-to-state vibrational relaxation models

We consider binary flows of nitrogen (N_2/N) oxygen (O_2/O) with the following VT transitions:

$$N_2(i) + N \to N_2(i') + N, \quad N_2(i) + N_2 \to N_2(i') + N_2, \tag{1}$$

$$O_2(i) + O \to O_2(i') + O, \quad O_2(i) + O_2 \to O_2(i') + O_2, \tag{2}$$

where i denotes the initial vibrational levels, and i', denotes the post-collisional vibrational levels.

The following models for VV and VT processes were considered:

1. The Forced Harmonic Oscillator (FHO) model [12]. It is a semiclassical model for VV and VT processes, which allows for multi-quantum VV and VT transitions and accounts for vibrational anharmonicity. It provides good agreement with experimental and QCT data.
2. A model based on the application of the Inverse Laplace Transform to QCT calculations of rate coefficients of VV and VT processes [24, 25] (further referred to as the QCT-ILT model). The QCT calculations of rate coefficients used where taken from works by Billing et. al [26, 27] and the Phys4Entry database [28]. The model gives expressions for cross-sections of VV exchanges and VT transitions (for collisions of molecules with atoms).

It should be noted that the model developed in [24, 25] provides expressions only for monoquantum VT transitions in nitrogen flows, while allowing for multi-quantum VT transitions in oxygen flows. In strongly non-equilibrium flows, especially those of oxygen, the effect of multi-quantum transitions on the flow parameters can be significant [29, 21, 30]. The application of the FHO model and other semiclassical models to modeling nitrogen flows in DSMC has been discussed in [21, 31, 22].

While the QCT-ILT model gives expressions for multi-quantum VT transitions for collisions of oxygen molecules with oxygen atoms, for $|i - i'| > 5$ the model fails to provide physically meaningful results, and thus, the maximum change in the vibrational level considered when performing calculations in oxygen using the model was taken to be 5. The QCT-ILT model also does not provide VT transition probabilities for collisions of molecules with molecules, and, therefore, the FHO model was used to compute them, this combined model is referred to as the combined FHO and QCT ILT-model.

To compute the probabilities of reverse processes, the microscopic detailed balance principle [32] was utilized:

$$P_{rev}(g') = \frac{g^2}{g'^2} P_f(g) \frac{\sigma_{el}(g)}{\sigma_{el}(g')}, \tag{3}$$

where P_{rev} is the probability of a reverse process, P_f is the probability of a forward process, g is the magnitude of the relative velocity of the colliding particles, g' is the post-collisional magnitude of the relative velocity of the colliding particles, $\sigma_{el}(g)$ is the elastic cross-section.

3. Numerical results

All simulations where performed in the SMILE++ [33] software package. Probabilities of dissociation reactions were modeled using the Vibrationally Favoured Dissociation (VFD) model described in [34]. The vibrational temperature T_v in a cell was calculated using the assumption that the vibrational energies of the molecules obey a Boltzmann distribution, and, therefore, the vibrational temperature can be found from the equation

$$\frac{1}{Z_{vibr}(T_v)} \sum_i \varepsilon_i \exp\left(-\frac{\varepsilon_i}{kT_v}\right) = E_{vibr}, \tag{4}$$

where E_{vibr} is the computed vibrational energy in a computational cell, Z_{vibr} is the vibrational partition function, and ε_i is the vibrational energy of vibrational level i of the molecules being considered.

3.1. Model validation: vibrational relaxation of a spatially homogeneous gas

The FHO and QCT-ILT models were used to study vibrational relaxation of a spatially homogeneous gas. The initial conditions were taken to be as follows: the translational temperature T was in equilibrium with the rotational temperature T_r and was taken to be equal to 15000K, while the vibrational temperature T_v was taken to be equal to 300K. The molar fraction of the atoms was taken to be 0.5, while the numeric density was equal to 10^20 $1/\text{m}^3$. Computations done with the FHO model did not account for transitions with $|i - i'| > 5$ due to their low probability and the possibility of obtaining a total inelastic process probability larger than 1, which can lead to unphysical results.

Figure 1 shows the temporal evolution of the translational and vibrational temperatures in a spatially homogeneous gas computed using different vibrational transition models. It can be seen for both cases that all state-to-state models lead to lower vibrational relaxation rates than the Larsen–Borgnakke model. In nitrogen, the combined FHO and QCT-ILT model gives slower vibrational relaxation than the FHO model, however, this is partially due to the absence of expressions for multi-quantum transitions in nitrogen in the QCT-ILT model. In oxygen, the combined FHO and QCT-ILT model gives slightly faster relaxation rates than the FHO model.

Let us consider the effect of multi-quantum transitions on vibrational relaxation rates. Figure 2 shows the temporal evolution of the translational and vibrational temperatures in a spatially homogeneous gas computed using different vibrational transition models with different restrictions on the maximum change in the vibrational level. In nitrogen (figure 2(a)) it can be seen that the effect of multi-quantum transitions when performing computations using the FHO model is small. Computation results using the combined FHO and QCT-ILT model for different number of allowed vibrational energy transitions for molecule-molecule collisions (1- and 5-quantum transitions) are extremely close. Therefore, it can be concluded that multi-quantum VT transitions in nitrogen have a noticeable effect on the vibrational relaxation rate

IOP Conf. Series: Journal of Physics: Conf. Series **815** (2017) 012011 doi:10.1088/1742-6596/815/1/012011

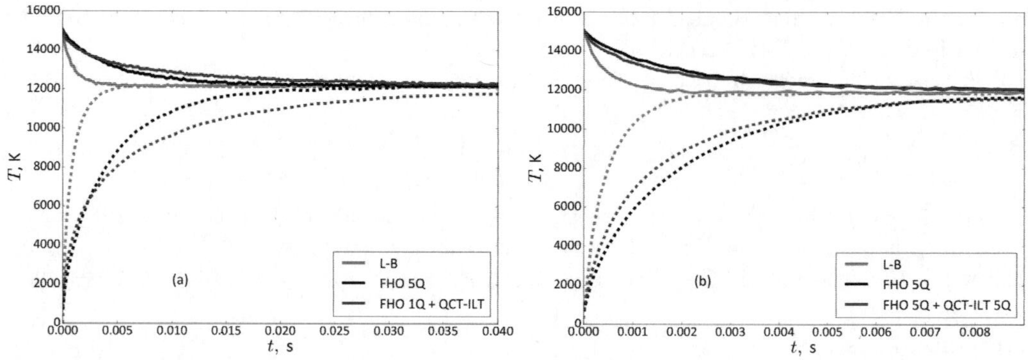

Figure 1. Values of translational (solid lines) and vibrational temperatures (dotted lines) as functions of time in N_2/N (a) and O_2/O (b) mixtures computed various models: Larsen–Borgnakke model (L-B), FHO model with up to 5-quantum transitions (FHO 5Q) and the combined FHO and QCT-ILT model with only one-quantum transitions in N_2/N (FHO 1Q + QCT-ILT) and up to 5-quantum transitions in O_2/O (FHO 5Q + QCT-ILT 5Q).

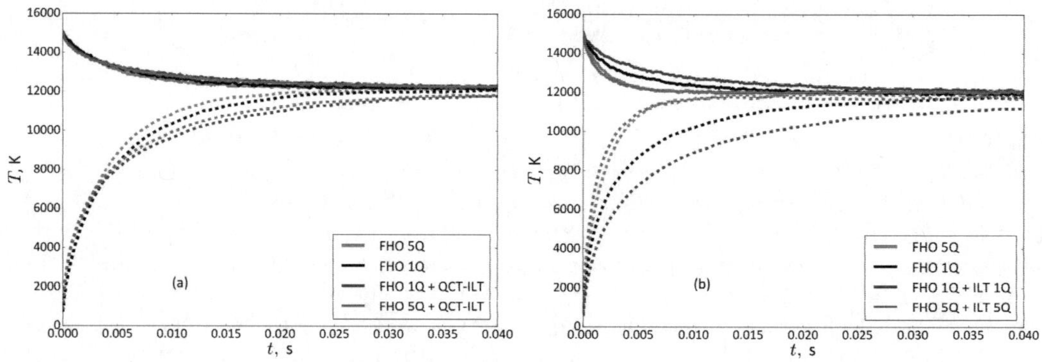

Figure 2. Values of translational (solid lines) and vibrational temperatures (dotted lines) as functions of time in N_2/N (a) and O_2/O (b) mixtures computed various models: FHO model with up to 5-quantum transitions (FHO 5Q), FHO model with monoquantum transitions (FHO 1Q), the combined FHO and QCT-ILT model with only one-quantum transitions in N_2/N (FHO 1Q + QCT-ILT) and O_2/O (FHO 1Q + QCT-ILT 1Q) and up to 5-quantum transitions in O_2/O (FHO 5Q + QCT-ILT 5Q), and the combined FHO and QCT-ILT model with up to 5-quantum transitions for molecule-molecule collisions (computed using the FHO model) and only one-quantum transitions for molecule-atom collisions (computed using the QCT-ILT model) in N_2/N (FHO 5Q + QCT-ILT).

only in molecule-atom collisions. From these results, it follows that in nitrogen flows, accounting for only monoquantum transitions does not significantly affect the relaxation rate.

It can be seen that in oxygen (figure 2(b)), multi-quantum transitions play a significantly more important role. Not including them in the simulations leads to a noticeable decrease in the vibrational relaxation rates. When only monoquantum transitions are accounted for, the combined FHO and QCT-ILT model gives slower vibrational relaxation rates than the FHO model, while when multi-quantum transitions are taken into consideration, the situation is reversed, as can be seen from figure 1(a).

3.2. *Flow around a cylinder*

Let us now consider the influence of the vibrational relaxation models on a 2-dimensional flow around a cylinder. The parameters of the flow are: the free stream velocity was taken to be equal to 7.5 km/s, and the free stream temperature was assumed to be 188K. The density of the free stream flow was taken to be 3.43e-6 kg/m^3. The molar fraction of the atoms in the free stream was equal to 0.5. The radius of the cylinder was taken to be equal to 0.8 m. A wall temperature of 500K was used.

We consider the following vibrational relaxation models:

1. In nitrogen flows, we consider the Larsen–Borgnakke model, the FHO model with monoquantum and 5-quantum transitions, and the QCT-ILT model with monoquantum transitions for molecule-atom collisions and 5-quantum transitions for molecule-molecule collisions.
2. In oxygen flows, we consider the Larsen–Borgnakke model, the FHO model with 5-quantum transitions, and the QCT-ILT model with monoquantum and 5-quantum transitions.

Therefore, we study the effect of multi-quantum transitions in the QCT-ILT model for oxygen flows and provide comparisons with the Larsen–Borgnakke and FHO models. In nitrogen flows, since the QCT-ILT model does not allow for multi-quantum transitions, we consider the influence of the multi-quantum transitions on relaxation rates when performing flow simulations using the FHO model.

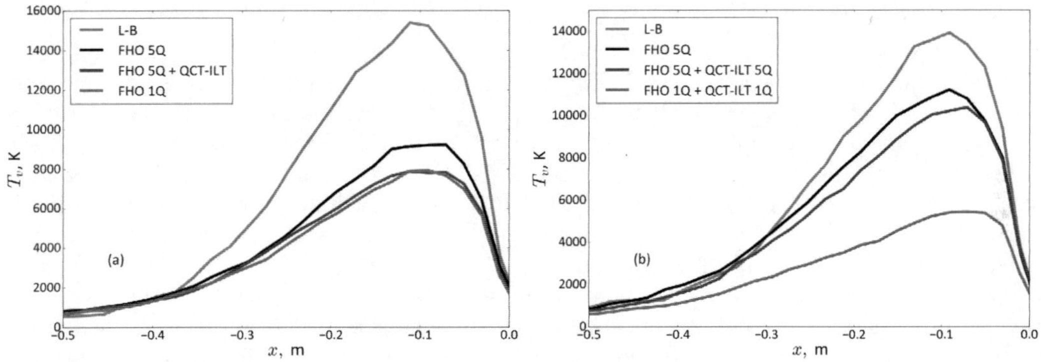

Figure 3. Values of vibrational temperature along the stagnation line as a function of distance from the cylinder in N_2/N (a) and O_2/O (b) mixtures computed various models: Larsen–Borgnakke model (L-B), FHO model with up to 5-quantum transitions (FHO 5Q), FHO model with monoquantum transitions (FHO 1Q), the combined FHO and QCT-ILT model with only one-quantum transitions in O_2/O (FHO 1Q + QCT-ILT 1Q) and up to 5-quantum transitions in O_2/O (FHO 5Q + QCT-ILT 5Q), and the combined FHO and QCT-ILT model with up to 5-quantum transitions for molecule-molecule collisions (computed using the FHO model) and only one-quantum transitions for molecule-atom collisions (computed using the QCT-ILT model) in N_2/N (FHO 5Q + QCT-ILT).

Figure 3 shows the values of vibrational temperature along the stagnation line as a function of distance from the cylinder computed using various models. We see that both for nitrogen and oxygen flows, the Larsen–Borgnakke model gives higher values of the vibrational temperature than the various state-to-state models. In nitrogen (figure 3(a)), all state-to-state models give similar results with significantly lower values of vibrational temperature than the Larsen–Borgnakke model, the difference reaching up to 8000K. Accounting for multi-quantum transitions when using the FHO model leads to a slight increase in the peak value of vibrational temperature. However, accounting for multi-quantum transitions only for molecule-molecule collisions does not have any noticeable effect on the vibrational temperature profile — both the FHO model with only monoquantum transitions (FHO 1Q) and the combined FHO QCT-ILT model with up to 5-quantum transitions for molecule-molecule collisions and only one-quantum transitions for molecule-atom collisions provide virtually identical results. The lack of difference between the FHO and QCT-ILT model computations is not surprising, since the QCT-ILT cross-sections were based on trajectory calculations given in [26, 27], with which the FHO model is in good agreement [12].

In oxygen (figure 3(b)), the difference between the Larsen–Borgnakke model and the state-to-state models is not as pronounced as in nitrogen: the difference between the peak vibrational temperatures for models which account for multi-quantum transitions (FHO 5Q, FHO 5Q + QCT-ILT 5Q) is less than 4000K. We see that multi-quantum transitions have a significant influence on the vibrational temperatures. Accounting for them leads to nearly a 100% (as compared to the model with only one-quantum VT transitions) increase in vibrational temperature. The FHO and combined FHO QCT-ILT models, when multi-quantum transitions are included in the simulation, give noticeably similar results. The FHO model gives a slightly larger value of the peak vibrational temperature.

4. Conclusion

The application of state-to-state VT transitions models to DSMC modeling of rarefied gas flows is studied. A model based on semiclassical collision theory (FHO model) and a model based on the application of the inverse Laplace transform to quasiclassical trajectory calculations (QCT-ILT model) are both implemented in DSMC. Validation of the models is done by considering the problem of vibrational relaxation of a spatially homogeneous gas. It is shown that all state-to-state models give lower vibrational relaxation rates than the commonly used Larsen–Borgnakke model. In nitrogen, the effect of multi-quantum transitions is shown to be negligible, and the difference between the QCT-ILT and FHO models is small. In oxygen, accounting for multi-quantum transitions leads to a noticeably faster relaxation rate compared to state-to-state models with only monoquantum transitions.

The models are applied to a two-dimensional flow around a cylinder, and the vibrational temperature along the stagnation line is studied. In nitrogen flows, vibrational temperature computed using the state-to-state models is twice as low as the vibrational temperature computed the Larsen–Borgnakke model. Again, multi-quantum VT transitions do not affect the vibrational temperature significantly. In oxygen flows, the state-to-state models give values of the vibrational temperature that are slightly lower than those predicted by the Larsen–Borgnakke model. The QCT-ILT model gives smaller values of the vibrational temperature than the FHO model. Accounting only for monoquantum transitions leads to a significant underestimation of the vibrational temperature.

Acknowledgments

This study is supported by the Russian Science Foundation (Project No. 15-19-30016).

References

[1] Bird G A 1994 *Molecular Gas Dynamics and the Direct Simulation of Gas Flows* (Oxford, England, UK: Clarendon)
[2] Borgnakke C and Larsen P 1975 *Journal of Computational Physics* **18** 405–420
[3] Landau L and Teller E 1936 *Phys. Z. Sowjetunion* **10** 34–43
[4] Parker J 1959 *Phys. Fluids* **2** 449
[5] Millikan R and White D 1963 *J. Chem. Phys.* **39** 3209–3213

[6] Park C 1990 *Nonequilibrium Hypersonic Aerothermodynamics* (New York, Chichester, Brisbane, Toronto, Singapore: J.Wiley and Sons)
[7] Ibraguimova L B, Sergievskaya A L, Levashov V Y, Shatalov O P, Tunik Y V and Zabelinskii I E 2013 *J. Chem. Phys.* **139** 034317
[8] Andrienko D A and Boyd I D 2015 *Phys. Fluids* **27** 116101
[9] Kustova E and Oblapenko G 2016 *Phys. Rev. E* **93** 033127
[10] Schwartz R, Slawsky Z and Herzfeld K 1952 *J. Chem. Phys.* **20** 1591
[11] Koura K 1998 *Computers & Mathematics with Applications* **35** 139–154
[12] Adamovich I, Macheret S, Rich J and Treanor C 1998 *J. Thermophys. Heat Transfer* **12** 57–65
[13] Esposito F and Capitelli M 2005 Detailed cross section calculations of atom-molecule energy transfer processes and dissociation for hydrogen, nitrogen and oxygen *ATOMIC AND MOLECULAR DATA AND THEIR APPLICATIONS: Joint Meeting of 14th Internat. Toki Conf. on Plasma Physics and Controlled Nuclear Fusion (ITC14); and 4th Internat. Conf. on Atomic and Molecular Data and Their Applications (ICAMDATA2004)* vol 771 (AIP Publishing) pp 246–254
[14] Jaffe R, Schwenke D and Chaban G 2009 *AIAA paper* **1569**
[15] Kim J and Boyd I 2014 *Phys. Fluids* **26** 012006
[16] Esposito F and Capitelli M 2002 *Chem. Phys. Lett.* **364** 180–187
[17] Andrienko D, Neitzel K and Boyd I 2016 Vibrational relaxation and dissociationin O_2-O mixtures *Proc. 46th AIAA Thermophysics Conference, June 2016* AIAA Paper 2016-4021
[18] Kulakhmetov M, Gallis M and Alexeenko A 2016 *J. Chem. Phys.* **144** 174302
[19] Valentini P, Norman P, Zhang C and Schwartzentruber T E 2014 *Phys. Fluids* **26** 056103
[20] Valentini P, Schwartzentruber T E, Bender J D, Nompelis I and Candler G V 2015 *Phys. Fluids* **27** 086102
[21] Gimelshein S F, Boyd I D, Sun Q and Ivanov M S 33rd Thermophysics Conference, Norfolk, 1999 DSMC modeling of vibration-translation energy transfer in hypersonic rarefied flows vol 99-3451
[22] Li Z, Sohn I, Levin D , Mareschal M and Santos A S 43rd Thermophysics Conference, New Orleans, 2012 State specific vibrational relaxation and dissociation models for nitrogen in shock wave regions (*AIP Conference Proceedings* vol 1501) p 653
[23] YeA B, SF G, AN M and MS I 2014 *Numerical Methods and Programming* **15** 549–559
[24] Baikov B S, Bayalina D K and Kustova E V 2016 *Vestnik of Saint Petersburg University. Series 1. Mathematics. Mechanics. Astronomy* **3 (61)** 389–397
[25] Baikov B S, Bayalina D K, Kustova E V and Oblapenko G P 2016 *AIP Conference Proceedings* **1786** 090005 URL `http://scitation.aip.org/content/aip/proceeding/aipcp/10.1063/1.4967611`
[26] Billing G and Fisher E 1979 *Chem. Phys.* **43** 395–401
[27] Billing G and Kolesnick R 1992 *Chem. Phys. Lett.* **200** 382–386
[28] Planetary entry integrated models http://phys4entrydb.ba.imip.cnr.it/Phys4EntryDB/
[29] Kang S and Kunc J 1994 Energy dependence of vibration-translation transitions in collisions of neutral particles *18th, Aerospace Ground Testing Conference, 18th, Colorado Springs, CO* vol 94-2403
[30] Mishin L and Kustova E 2016 *Vestnik of Saint Petersburg University. Series 1. Mathematics. Mechanics. Astronomy* **3 (61)** 300–308
[31] Boyd I and Josyula E 2011 *Phys. Fluids* **23** 057101
[32] Nagnibeda E and Kustova E 2003 *Kinetic Theory of Transport and Relaxation Processes in Nonequilibrium Reacting Gas Flows* (Saint Petersburg: Saint Petersburg University Press)
[33] Ivanov M, Kashkovsky A, Vashchenkov P, Bondar Y, Levin D, Wysong I and Garcia A 2011 Parallel object-oriented software system for DSMC modeling of high-altitude aerothermodynamic problems *Rarefied Gas Dynamics* (*AIP Conference Proceedings* vol 1333) ed Levin D (Melville, NY: American Institute of Physics) p 211
[34] Haas B L and Boyd I D 1993 *Phys. Fluids A* **5** 478–485

SPH simulation of boron carbide under shock compression with different failure models

S A Dyachkov[1,2], **A N Parshikov**[1], **V V Zhakhovsky**[1]
[1]Dukhov Research Institute of Automatics, ROSATOM, Moscow, 127055, Russia
[2]Moscow Institute of Physics and Technolgy, Dolgoprudny, 141700, Russia
E-mail: `serj.dyachkov@gmail.com`

Abstract. Boron carbide behavior under shock compression is analyzed by comparison of SPH simulations with two sets of plate-impact experiments where the used samples were manufactured using different technology. Various boron carbide failure models are applied to determine the relevant mechanical properties which are responsible for the velocity profiles obtained from VISAR measurements. We show that in order to fit the experimental profiles the strength of the failed state of material should be different in those experiments. This strength is demonstrated to be a unique function of pressure irrespective of the utilized failure models.

1. Introduction

Boron carbide response to high-strain rate shock compression leading to large deformations is of present-day interest. Due to its unique strength properties boron carbide has numerous applications. Nevertheless, under shock loading it is involved into a process of failure [1] what significantly reduces its strength as demonstrated in plate-impact experiments [2].

The comprehensive models for ceramics under shock loading including boron carbide [1, 3] provide all necessary data for simulations of experiments [2]. However, it was found [1] that boron carbide samples behave as different materials in different sets of experiments. For instance, the data [4] cannot be reproduced with provided model parameters and the previous model [5] should be used. The reason for that may be due to different manufacturing technology for boron carbide samples.

In this research we provide a new understanding of manufacturing dependent boron carbide characteristics which define its behavior under shock compression. The analysis is based on the three dimensional SPH simulations of wave profiles in accordance with two different sets of plate-impact experiments [2, 4]. We use the strength model for boron carbide failure based on Johnson and Holmquist models [1,5] and a more simple approach with yield strength temperature dependence.

2. Equation of state for boron carbide under hydrostatic stress

Complex structure of boron carbide causes difficulties in its equation of state definition. In the range of parameters of our interest there is a question about structural phase transition leading to discontinuous volume change. Experimental observation of boron carbide under hydrostatic stress in diamond anvil cells does not demonstrate such kind of behavior [6]. That fact was also proved by DFT calculations [7]. However, at non-hydrostatic stress the question about phase transition is still arguable, but relevant information about it is unavailable.

In our simulations we use a Mie-Gruneisen form for the equation of state:

$$P = P_0 + \Gamma\rho\,(e - e_0)\,, \tag{1}$$

$$P_0 = \frac{\rho_0 c_0^2}{n}\,(\delta^n - 1)\,, \tag{2}$$

$$e_0 = \frac{c_0^2}{n}\left(\frac{\delta^{n-1} - 1}{n-1} + \frac{1}{\delta} - 1\right), \tag{3}$$

$$\delta = \frac{\rho}{\rho_0}, \tag{4}$$

$$\Gamma = 1.625\left(\frac{\rho_0}{\rho} - 1\right) + 0.54, \tag{5}$$

with parameters fitted according to hydrostatic compression of boron carbide in quantum molecular dynamics simulation [7]. The parameters ρ_0, c_0, n are given in the Table 1. In contrast to [1] we do not use any kind of phase transition in this equation of state.

Table 1. B_4C properties.

Material characteristic	Value
Initial density ρ_0, kg/m^3	2550
Sound velocity c_0, km/s	10
EoS exponent n	3
Posson ratio ν	0.162
Bulk modulus K, GPa	255
Shear modulus G, GPa	216
Hugoniot Elastic Limit (HEL), GPa	18.2
HEL strength σ_{HEL}, GPa	14.0
HEL pressure P_{HEL}, GPa	8.8
HEL volumetric strain μ	0.035
Intact strength constant σ_i^0, GPa	6.7
Intact strength constant $\sigma_i^{\max}$, GPa	14.3
Intact strength constant P_i, GPa	2.3
Failure strength constant σ_f^0, GPa	1.0/5.0
Failure strength constant $\sigma_f^{\max}$, GPa	2.0/8.5
Failure strength constant P_f, GPa	2.0/5.0
Hydrostatic tensile strength T, GPa	0.27
Strain rate constant C	0.01
Bulking factor β	1.0

3. Boron carbide failure models

Johnson and Holmquist [1, 5] consider the strength models for boron carbide failure under compression. The idea is that failure begins as soon as von Mises equivalent stress σ_e reaches intact strength limit σ_i. Numerous plate impact experiments with ceramics demostrate hydrostatic pressure dependency for the intact strength. Below the curve $\sigma_i(P)$ the damage

IOP Conf. Series: Journal of Physics: Conf. Series **815** (2017) 012012 doi:10.1088/1742-6596/815/1/012012

zone is located. Boron carbide has significantly lower strength limit $\sigma_f(P)$ for the failed state, and during failure process the damaged material strength limit σ_d reduces from σ_i to σ_f:

$$\sigma_d = \sigma_i - D(\sigma_i - \sigma_f) \tag{6}$$

where D is the damage factor which indicates an amount of accumulated damage starting from 0 (intact) to 1 (failed).

Recent JHB model [1, 3] uses similar form for the intact ($\lambda = i$) and failed ($\lambda = f$) stresses:

$$\begin{cases} \sigma_\lambda = \sigma_\lambda^0 + (\sigma_\lambda^{\max} - \sigma_\lambda^0)\{1.0 - \exp[-\alpha_\lambda(P - P_\lambda)]\}, & P > P_\lambda, \\ \sigma_\lambda = \sigma_\lambda^0 P/(P_\lambda + T), & P < P_\lambda, \end{cases} \tag{7}$$

$$\alpha_\lambda = \sigma_\lambda^0/[(\sigma_\lambda^{\max} - \sigma_\lambda^0)(P_\lambda + T)] \tag{8}$$

with a difference that hydrostatic tensile strength $T = 0$ for the failed state. The constants σ_λ^0, $\sigma_\lambda^{\max}$, P_λ are defined in accordance with experimental data. It should be noted, that strain rate dependency is also included to the intact and failed curves as follows:

$$\sigma_\lambda(\dot\varepsilon^*) = \sigma_\lambda(1.0 + C \ln \dot\varepsilon^*) \tag{9}$$

where $\dot\varepsilon^* = \dot\varepsilon/\dot\varepsilon_0$ – dimensionless quivalent strain rate for $\dot\varepsilon_0 = 1.0\,s^{-1}$, C – constant. In three dimensions equivalent strain can be found as:

$$\dot\varepsilon = \sqrt{\frac{2}{9}\left[(\dot\varepsilon_{xx} - \dot\varepsilon_{yy})^2 + (\dot\varepsilon_{yy} - \dot\varepsilon_{zz})^2 + (\dot\varepsilon_{xx} - \dot\varepsilon_{zz})^2 + \frac{3}{2}(\dot\gamma_{xy}^2 + \dot\gamma_{yz}^2 + \dot\gamma_{xz}^2)\right]}. \tag{10}$$

If process of failure take place, damage D is accumulated according to the kinetic equation:

$$\frac{\mathrm{d}D}{\mathrm{d}t} = \frac{\dot\varepsilon_p}{\varepsilon_p^f}, \tag{11}$$

where $\dot\varepsilon_p$ – plastic strain rate, ε_p^f – total plastic strain to failure. Plastic strain for a cycle of integration $\Delta\varepsilon_p$ can be estimated [8]

$$\dot\varepsilon_p \Delta t = \Delta\varepsilon_p = \frac{\hat\sigma_e^t - \sigma_e^t}{3G}, \tag{12}$$

where $\hat\sigma_e^t$ is equivalent stress calculated at the time step t, but not corrected:

$$\hat\sigma_e = \sqrt{\frac{1}{2}\left[(\hat S_{xx} - \hat S_{yy})^2 + (\hat S_{yy} - \hat S_{zz})^2 + (\hat S_{xx} - \hat S_{zz})^2 + 6(\hat S_{xy}^2 + \hat S_{xz}^2 + \hat S_{yz}^2)\right]}. \tag{13}$$

The resulting stress limit for the damaged material σ_d may be found as (6) using current damage factor D^t. Equivalent stress σ_e must be below σ_d, so if necessary ($\hat\sigma_e > \sigma_d$), we apply the correction to stresses:

$$S_{\alpha\beta} = \frac{\sigma_d}{\hat\sigma_e}\hat S_{\alpha\beta}, \quad \sigma_e = \sigma_d. \tag{14}$$

Using corrected equivalent stress σ_e one can find the plastic strain increment $\Delta\varepsilon_p$.

In the original Johnson-Holmquist model [1, 5] the amount of plastic strain to failure ε_p^f is independent of other model parameters what provides additional degree of freedom. But for instant failure (D changes from 0 to 1 in one step) one can notice that plastic strain to failure is

$$\Delta\varepsilon_p = \frac{\sigma_i - \sigma_f}{3G} = \varepsilon_p^f. \tag{15}$$

For the model to be consistent this expression should be used throughout whole fragmentation process. Thus one can define damage at the next step $t + \Delta t$ as follows:

$$D^{t+\Delta t} = D^t + \frac{\hat{\sigma}_e^t - \sigma_e^t}{\sigma_i - \sigma_f}. \quad (16)$$

Additional pressure ΔP is then added due to bulking of damaged material [5]

$$P_{\mathbf{total}} = P_{\mathbf{EoS}} + \Delta P. \quad (17)$$

It can be estimated, for example, by evaluating elastic energy loss ΔU during a cycle of integration:

$$\Delta U = \frac{(\sigma_e^2)_t}{6G} - \frac{(\sigma_e^2)_{t+\Delta t}}{6G}, \quad (18)$$

and then expressed as:

$$\Delta P^{t+\Delta t} = -K\mu + \sqrt{(K\mu + \Delta P^t) + 2\beta K \Delta U}, \quad (19)$$

where K – bulk modulus, $\mu = \rho/\rho_0 - 1$, β – bulking factor. Further we will refer to this model as JH-based model.

The damage parameter D can be also evaluated as a function of state without kinetic equation. For example, one can add yield strength Y dependency of specific thermal energy e_T what similarly to previous model reduces yield strength from intact Y_i to failed Y_f. For the damage parameter D the following equation is used:

$$D = e_T/e_D, \quad (20)$$

where e_D is the thermal energy of totally failed material. Then the yield strength for damaged material Y_d is defined similar to (6):

$$Y_d = Y_i - D(Y_i - Y_f). \quad (21)$$

Yield strength for the failed material can be taken in the form (8). Further we will refer to this model as Y-based model.

4. Simulation of plate impact experiments

Validation of the presented boron carbide model is performed by comparison with Grady et al. [4] and Vogler et al. [2] plate impact experiments. Simulation setup is shown in the figure 1: an impactor with backer foam hits a boron carbide target producing shock wave which propagates to a lithium fluoride window. Experimental measurements of velocity profiles at the B_4C/LiF interface provided by a VISAR. In SPH simulations the particles at this interface are tracked and their velocities are averaged for a profile.

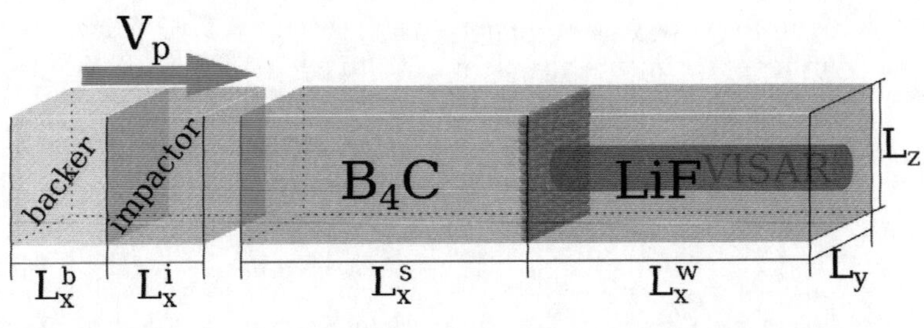

Figure 1. Simulation setup for boron carbide plate impact experiments. Impactor with backer material obtain initial velocity V_p in direction to B_4C target. After collision the shock wave propagates to the interface between boron carbide and lithium fluoride where the velocity of B_4C surface is measured as an average of SPH particle velocities.

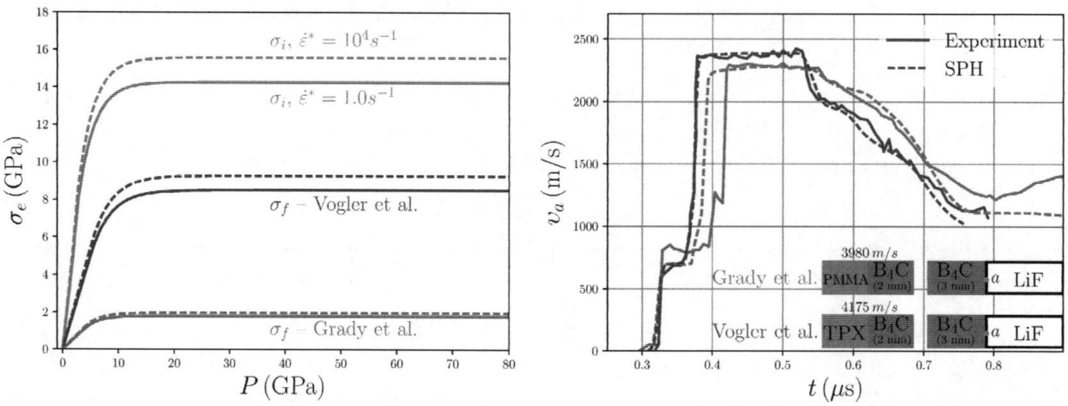

Figure 2. Influence of boron carbide strength at the failed state on wave profiles. Both experiments BC10 and BC-X have similar initial setup, but profiles are rather distinct. SPH simulation with JH-based model and different dependency $\sigma_f(P)$ is shown to be in agreement with each of both experiments.

Each simulated sample has sizes $L_y \times L_z = 0.1 \times 0.1$ mm, and L_x in accordance with specific experiment. Periodic boundary conditions are applied for y and z axes. SPH particles are packed in a dense liquid-like mesh in order to avoid regular grid packing effects. For y and z axes there are an average of 17.5 particles and several thousands for x axis depending on initial data. The SPH kernel Wendland-C^2 [9] and the appropriate Riemann solver [10] are used for our contact modification of SPH [11].

The equation of state for boron carbide provides new bulk modulus $K = \rho_0 c_0^2 = 255$ GPa, so that one have to adjust HEL and other strength properties given in [1]. Using the same HEL volumetric strain $\mu = 0.035$ and Possion ratio we get new values for G and σ_i (see Table 1). For the Y-based model we use $Y_i = 24.3$ GPa, $e_D = 1.547 \times 10^5$ J/kg, and the same parameters for material at the failed state: $Y_f^0 = \sigma_f^0$, $Y_f^{\max} = \sigma_f^{\max}$.

As a result of our simulations we noticed that velocity profiles in different sets of experiments are very sensitive to failed material strength σ_f in both models. Figure 2 demonstrates that two experimental profiles with similar initial setup (BC10 from Grady et al. and BC-X from Vogler et al.) should be simulated with a different strength at the failed state. The reduction of σ_f leads both to elongation of the elastic precursor and smaller amplitude of elastic release what corresponds experimental data. So that in the Table 1 there are two sets of values for $\sigma_f^0, \sigma_f^{\max}, P_f$: first is found for Grady et al. experiments, second is for Vogler et al.

Our simulations of other experiments also demonstrate good agreement (Figure 3). Different boron carbide failure kinetics implemented in JH-based and Y-based models does not appear to affect velocity profile as much as different failed strength σ_f. This property seems to be dependent on boron carbide production technology and can be adjusted by fitting specific velocity profiles.

5. Conclusion

In this research we considered two boron carbide failure models: Johnson-Holmquist based strength model with kinetic equation for damage accumulation and non-kinetic model with thermal dependency for yield strength. Using both models with the same equation of state we performed simulations via contact SPH method for three dimensional samples with periodical boundary conditions and achieved good agreement with the experiments [2, 4]. As it was mentioned earlier [1], different manufactured flavors of boron carbide provide different velocity profiles for plate-impact experiments. We found that the main characteristic that guides boron carbide behavior under shock compression is material strength σ_f at the failed state. It was demonstrated that the strength σ_f is a unique function of material and applied pressure irrespective of the utilized failure models, and it can be adjusted for a specific flavor of boron carbide.

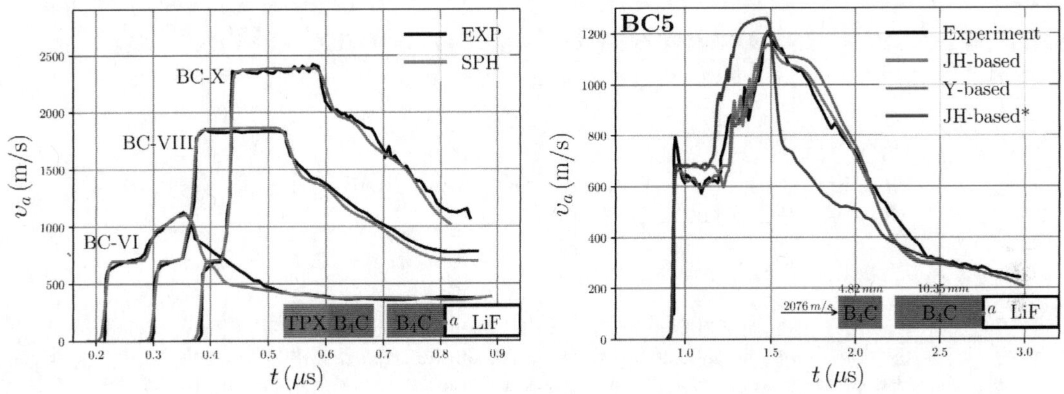

Figure 3. Experimental and simulated by SPH boron carbide wave profiles. On the left three experiments are simulated using JH-based model with σ_f fitted to Vogler et al. data. On the right experiment BC5 by Grady et al. is simulated using two different models: JH-based and Y-based which are in good agreement. However, using σ_f based on Vogler et al. data* the resulting profile become rather distinct.

Acknowledgements
This work is supported by the Russian Science Foundation grant 14-19-01599.

References

[1] Holmquist T J and Johnson G R 2006 *Journal of Applied Physics* **100** 093525
[2] Vogler T, Reinhart W and LC C 2004 *Journal of Applied Physics* **95** 4173–4183
[3] Johnson G R, Holmquist T J and Beissel S R 2003 *Journal of Applied Physics* **94** 1639–1646
[4] Grady D and Moody R 1996 Shock compression profiles in ceramics Tech. rep. Sandia Report SAND96-0551
[5] Johnson G R and Holmquist T J 1999 *Journal of Applied Physics* **85** 8060
[6] Dera P, Manghnani M H, Hushur A, Hu Y and Tkachev S 2014 *Journal of Solid State Chemistry* **215** 85–93
[7] Korotaev P, Pokatashkin P and Yanilkin A 2016 *Modelling and Simulation in Materials Science and Engineering* **24** 015004
[8] Wilkins M L 1999 *Computer Simulation of Dynamic Phenomena* (Springer-Verlag Berlin Heidelberg)
[9] Wendland H 1995 *Advances in computational Mathematics* **4** 389–396
[10] Dukowicz J K 1985 *Journal of Computational Physics* **61** 119–137
[11] Parshikov A N and Medin S A 2002 *Journal of computational physics* **180** 358–382

The Model of Masstransfer Intensification in Channels Using Nanowiresets inside and Nanostructures on the Wall

A A Markov

A. Ishlinsky Institute for Problems in Mechanics, RAS Vernadsky Street 101(1), Moscow, 119526,Russia

E-mail: markov.ipm@yandex.ru

Abstract. The gas and liquid mass transfer intensification is studied using nanowires placed in the flow as well as nanostractures organized on the channel wall. The results of the study can be applied to a drag reduction of the gas and liquid transport in channels. The model of drag reduction is developed using the set of submicronwires placed into the flow. On the wires surface the gas slippage in Knudsen layer is observed. The slippage results in a mass transfer intensification. The sets of wires placed near the channel axis of symmetry as well as placed close to the channel wall are analyzed for both types of cannel surface that are smooth or have the submicron roughness. The slip boundary condition on the surface of a submicron wire and the subsequent averaging procedure over the set of wires are applied to find the intensity of gas slippage as a function of number density of wires distribution as well as of wires shape. The results of numerical simulation of gas transport under a given pressure drop in channels demonstrate the drag reduction up to 300 percent.

1. Introduction

The experimental study of convective and diffusive masstransfer in nanotubes [1, 2] demonstrated the intensification compared to tubes of micron diameter. The numerous application of nanowires are found in [3, 4]. The recent technology allow to obtain the sets of nanowires which diameter is one meter and the length is up to 10m. Let us note that the free pass length of gas molecules in the Knudsen layer close the wire surface can lie in the interval of 10 to 70 that causes the significance of the slippage effect. The problem of drag reduction for liquid transport was studied in [5–7] where numerous references can be found, however, a gas transport was not considered. Reducing the drag of marine vehicles via gas lubrication has long been practiced by injecting gas bubbles or creating a cavitation gas pocket [8] and reached a significant drag reduction (~ 95%) at high Reynolds number flow [9,10]. Since the gas film (or bubbles) does not stay on the solid surface by nature, however, these methods should continue supplying the gas with additional energy, overshadowing the benefit of drag reduction and limiting applications. The prospect of retaining the gas without energy input has been driving the recent explosive interest in superhydrophobic (SHPO) surfaces – a rough surface of a hydrophobic material. Surface roughness generally increases the skin-friction drag in turbulent boundary layer (TBL) flows [11] except for very few specific conditions [12]. However, if the hydrophobic roughness of the SHPO surface retains microscale air pockets and thus maintains a plastron, the resulting slip flow may bring an appreciable drag reduction. Recent establishment in the slips and drag reductions obtained on engineered SHPO surfaces in laminar flows [13–15] have heightened the anticipation that someday an appreciable reduction can be reliably obtained in TBL flows as well [16,17]. Experimental data [10] demonstrates the bigger skin drag reduction for the smaller size of pitch on ultrahydrophobic surface. This result motivated us to consider submicron structure at channel surface filled with a gas. In the present paper the model for gas transport intensification using sets of nanowires is suggested. Close to the wire surface the Knudsen layer causes the gas slippage that intensifies the masstransfer. The numerical simulation of unsteady axially-symmetric flow in a tube for various sets of wires placed in the center of the flow as well as close to the wall surface are analyzed. The slip boundary conditions [18] were applied. The models of gas slippage in Knudsen layers [19, 20] appears to be in good agreement with experimental data. The results of drag reduction using the nanowire sets were reported in [21]. The application of submicron cavities on the channel surface are discussed in [22].

2. The gas flow with outer velocity $U = U_{ex}$ in microchannel over a nanowire

We consider firstly the analytical solution for outer flow problem over a nanowire in the microchannel as follows

$$\frac{1}{\mathrm{R}}\frac{\partial}{\partial R}\left(\mu_g R\frac{\partial U}{\partial R}\right)=\mathrm{P_X},\quad R_0 < R < R_1 .$$

$$-\mu_g U^{/}(R_0) = B_u U(R_0) \text{ and } U(R_1) = U_{ex} .$$

Where the dimensional variables are used. We rewrite the microscale problem for dimensionless variables as follows

$$r = R / R_1 \;,\; u = U / U_{ex}, \text{ and } \mathrm{Re}_{loc} = \frac{U_{ex} R_1 \rho_g}{\mu_g}$$

$$\frac{1}{\mathrm{r}}\frac{\partial}{\partial \mathrm{r}}\left(\frac{1}{\mathrm{Re}_{loc}} r\frac{\partial u}{\partial \mathrm{r}}\right)=\mathrm{p_x},\quad r_0 < r < 1,\quad \frac{-1}{\mathrm{Re}_{loc}}u^{/}(r_0) = b_u u(r_0),\quad \text{and } u(1) = 1 .$$

Let $u_{\max}$ be the maximal gas velocity. Our numerical computation of unsteady Quette flow shows that maximal velocity for steady state takes place on the wire surface. Denoting

$$k = \frac{1}{2}\left(1 + \frac{1}{u_{\max}}\right) \text{ and } u(r_*) = \frac{1}{2}\left(1 + u_{\max}\right).$$

We arrive at $\mathrm{p_x} = -\frac{4}{\mathrm{Re}_{loc}} u_{\max}\psi$, where $\psi = (1-k)/r_*^2$.

3. Gas flow in macrochannel with a set of nanowires

Consider now the gas flow in a macro channel with placed inside of it a set of submicronwires. Let V_{meso} be the mesovolume whose diameter is much less than the diameter of macrochannel, however, the diameter is much more as compared to diameter of the wire. In subsequent computation the mesovolume is chosen as computational grid finitevolume as usual. Using the average of micro pressure gradient over the mesovolume, we arrive at

$$-J_{slip}^{macro} = \frac{1}{V_{meso}}\int_S \mathrm{p_x} dS = \frac{1}{V_{meso}}\sum_j \int_{S_j} \mathrm{p_x} dS = \frac{1}{V_{meso}}\sum_j S_j \hat{\mathrm{p}}_{\mathrm{jx}},$$

where summation over all wire surfaces is assumed. Denoting by n the number density of the wire set, we obtain the macroslippage intensity as follows

$A_u = \dfrac{4nS^{mic}}{V_{meso}\,\mathrm{Re}_{loc}}$ and $S^{mic} = 2\pi r^{mic} L^{mic}$; where r^{mic} and L^{mic} are wire radius and length respectively.

In common case of 3–d gas flow the slippage vector is $\vec{J}_{slip}^{macro} = A_u \mathbf{u}$. The momentum conservation equation reads

$$\frac{\partial \chi \rho \mathbf{u}}{\partial t} + \nabla \cdot (\chi \rho \mathbf{u}\mathbf{u}) + \mathrm{Ma}^{-2} \nabla p = \mathrm{Re}^{-1} \nabla \cdot \boldsymbol{\tau} + \vec{J}_{slip}^{macro}$$

$$\boldsymbol{\tau} = \mu \left[\nabla \mathbf{u} + (\nabla \mathbf{u})^T - \frac{2}{3} (\nabla \cdot \mathbf{u}) \mathbf{I} \right]$$

4. Gasflow in a channel which surface retains microscale air pockets

We consider a gas flow in a channel with submicron pockets filled with gas. The averaging procedure over the mesovolume that is similar to the written above, allows us to represent the boundary slip condition in the form

$$\mu_g \frac{\partial \hat{U}}{\partial R} = \hat{B}_u \hat{U} \text{ , where } \hat{B}_u = \frac{S_b n}{S \cdot R_b}, \text{ or } \hat{B}_u = \frac{n}{r_W}, \; S_b = 2\pi R_b L, \; S = 2\pi r_W L .$$

The values R_b, L_b denote radius and length of air pocket respectively and $\hat{U}$ is the mean flow velocity in the mesovolume.

5. The results of masstransfer simulation in the channel with a nanowire ensemble inside

We consider the nanowire set of radius r_C and length l_s placed on the channel axis at $x = x_s$. The slippage intensity is defined using the functions

$$f_x(x) = \begin{cases} 1, & x_s < x < l_s \\ 0, & l_s \le x \le L \end{cases} \text{ and } f_{r0}(r) = \begin{cases} 1, & 0 < r < r_C \\ 0, & r_C \le r \le 1 \end{cases}, \qquad (1)$$

$$\vec{J}_{slip}^{macro} = A_u f_{r0}(r) f_x(x) \mathbf{u} \sqrt{p\rho}$$

Along with we consider two nanowire sets of radius r_C and length l_s placed on the channel axis at $x = x_s$ and $x = x_s + 2l_s$. The slippage intensity is defined using the functions

$$f_{1x}(x) = \begin{cases} 1, & x_s < x < l_s \\ 0, & (l_s \le x \le L) \cap (0 \le x \le l_s) \end{cases} \text{ and } f_{2x}(x) = \begin{cases} 1, & x_s + 2l_s < x < 3l_s \\ 0, & (3l_s \le x \le L) \cap (0 \le x \le x_s + 2l_s) \end{cases} \qquad (2)$$

$$\vec{J}_{slip}^{macro} = A_u f_{1x}(x) f_{2x}(x) \mathbf{u} \sqrt{p\rho}$$

The results of simulation refer to $x_s = 0.05,\ l_s = 0.5,\ L = 3$.

We obtained the more intensive masstransfer for one nanowire ensemble model (1) compared to two nanowire ensembles model (2).The reason is the gap between nanowires sets (2).

Our computation shows that the bigger value of the Reynolds number the bigger maximal value of the massflow for the same instant of time and more intensive masstransfer.

IOP Conf. Series: Journal of Physics: Conf. Series **815** (2017) 012013 doi:10.1088/1742-6596/815/1/012013

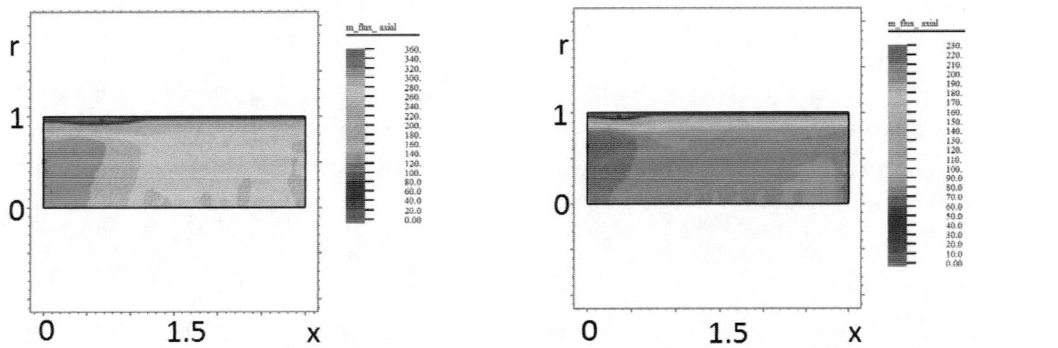

Figure 1. Comparison one nanowire ensemble Ls=1 and two ones of Ls=0.5 each. Mass flow in the channel with a set of two nanowire ensembles of radius $r_C = 0.1$ placed in the tube centre is presented. The massflow distribution is shown at time instant $t = 0.166$ and $\mathrm{Re} = 1$ for one shim of length $l_s = 1$ at $x_s = 0.05$ see (1) (left) and two shims of length $l_s = 0.5$ at $x_s = 0.05$, $x_s = 1.05$ see (2) (right).

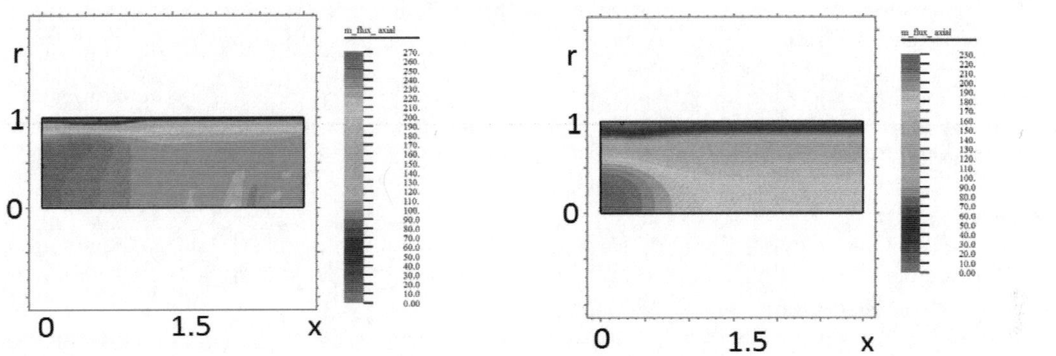

Figure 2. Massflow in the channel with two sets of nanowires of radius $r_C = 0.1$ placed in the tube centre is presented. The massflow distribution is shown at time instant $t = 0.15$ for $\mathrm{Re} = 1$ (left) and $\mathrm{Re} = 0.1$ (right).

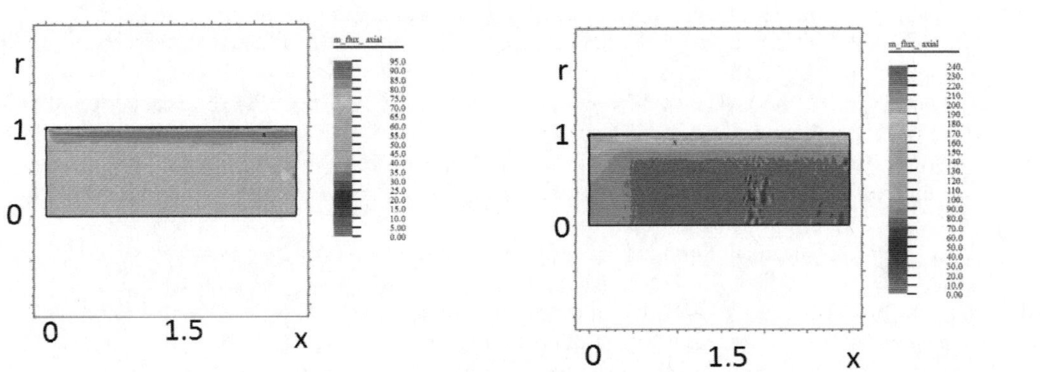

Figure 3. Massflow of gas in the channel with a set of nanowires of radius $r_C = 0.1$ attached to the wall is presented. The massflow distribution is shown at time instant $t = 0.086$ for $\mathrm{Re} = 10$ and $A_u = 150$ (left) and $A_u = 750$ (right).

The transfer intensification coefficient is defined as $K = Q / Q_0$, $Q = 2\pi \int_0^1 u(t,x,r)\rho r \mathrm{d}r$ .The results of numerical simulation using the sets of nanowires of radius r_C as well as using gasslippage on the channel surface due to organized submicron air pockets on the channel wall are as follows. The value Q_0 refers to the integral mass flow when no slippage occurs. In the columns, 1, 2 the results of $Q(L,t)$ computation refer to the flow with a set of wires attached to the channel wall.

$A_u = 0.001, B_u = 1$, $r_C = 0.1$, $K = 2.93$ $A_u = 750, B_u = 1$, $r_C = 0.1$, $K = 3.95$

6. Conclusion

Microanalysis of flow near nanowire and subsequent averaging masstransfer over the set of nanowires placed in the flow, enables to obtain the value of slippage intensity, which depends on number density of wires as well as on the wires shape and size. The numerical simulation of the unsteady gas mass transfer in the channel flow with axial symmetry has been carried out in a wide range of similarity parameters. The results show significant masstransfer intensification (up to 300 percent) due to the set of nanowires placed in the center of channel as well as for wires attached to the channel surface. The organized surface structures on the channel wall provide the gasslippage in the Knudsen layer near the wall. The remarkable masstransfer intensification has been found using the submicron pockets filled with gas on the channel wall.

Acknowledgements

The financial support of this research by the Russian Foundation for Basic Research grant no. 14-08-00664 is gratefully acknowledged.

References

[1] Karnidakis G A and Aluru N *2005* Microflows and nanoflows. Fundamentals and Simulation (Foreword by Chih-Ming Ho) *Interdisciplinary Applied Math. 29. Springer Science+Business Media, Inc.,*p 817

[2] Holt J K, Park H G, Wang Y S M, Artyukhin A B, Grigoropoulos C P, Noy A and Bakajin O 2006 Fast mass transport through sub-2-nanometer carbon nanotubes *Science* **312** pp 1034–1037

[3] Aktas E O, Ozgur H D, Enes K and Mehmet 2011 *Nature Materials.* **10** pp 494–501

[4] Filippov G A , Saltanov G A and Kukushkin A N. Hydrodynamics and heatmasstransfer in the presence of polimer molecular wires 1981 (in Russian) *Moscow Energoatomizdat* p 184

[5] Ceccio S L 2010 Friction drag reduction of external flows with bubble and gas injection *Annu. Rev.Fluid Mech.* **42** pp 183–203

[6] Lay K A, Ryo Y, Simo M, Perlin M and Ceccio S L 2010 Partial cavity drag reduction at high Reynolds numbers *J. Ship Res.* **54** pp 109–119

[7] Hyungmin P, Guangyi S and Chang-Jin K. 2014 Superhydrophobic turbulent drag reduction as a function of surface grating parameters *Journal of Fluid Mechanics* Vol. 747, May 2014, pp 722-734

[8] Walsh M J 1982 Turbulent boundary layer drag reduction usin griblets *AIAAPaper* 0169

[9] Jimenez J 2004 Turbulent flows over rough walls *Annu. Rev. Fluid Mech.* **36** pp 173–196

[10] Jung Y C and Bhushan B 2010 Biomimetic structures for fluid drag reduction in laminar and turbulent flows *J. Phys: Condens. Matter* **22** 035104

[11] Ou J, Perot B and Rothstein J P 2004 Laminar drag reduction in microchannels using ultrahydrophobic surfaces *Phys. Fluids* **16** pp 4635–4643

[12] Choi C H and Kim, C J 2006 Large slip of aqueous liquid flow over a nanoengineered superhydrophobic surface *Phys. Rev. Lett.* **96** 066001

[13] Rothstein J P 2010 Slip on superhydrophobic surfaces *Annu. Rev. Fluid Mech.* **42** pp 89–109

[14] Samaha M A, Tafreshi H V and Gad-el-Hak M 2012 b Superhydrophobic surfaces: from the

lotus leaf to the submarine *C.R. Mec.* **340** pp 18–34
[15] Bocquet L and Lauga E 2011 A smooth future? *Nature Mater.* **10** p 334
[16] Lee C, Choi C H, and Kim C J 2008 Structured surfaces for a giant liquid slip *Phys. Rev. Lett.* **101**, 064501
[17] Lee C, Choi C H and C-J Kim 2008 *Phys. Rev. Lett.* **101**, 064501
[18] Lee A and Kim H Y 2014 Does liquid slippage within a rough channel always increase the flow rate? *Phys. Fluids*, **26**, 072002
[19] Markov A A 2014 Jump-Slip simulation technique for combustion in submicron tubes and sub-micron pores *Computers and Fluids* **99 C** , pp. 83-92
[20] Markov A A, Hobosyan M A and K S Martirosyan K S 2015 Ferrite Synthesis Simulation via Carbon Combustion using Slip, Temperature, and Concentration Gas Species Jump at Pore Surface *Physical-Chemical Kinetics in Gas Dynamics* **16** (1) http://chemphys.edu.ru/issues/2015-16-1/articles/506/
[21] Markov A A, Hobosyan M A and Martirosyan K S 2015 Simulation of heat and masstransfer in pores as applied to synthesis of magnesium–zinc and nickel–zinc ferrite nanoparticles, *Nanomechanics Science and Technology: An International Journal* **6**(3), pp. 1–14
[22] Markov A A 2016 On heat- and masstransfer in channels with setsof nanowires. *Proc. XIInternConf. NPNJ* pp. 95-97. ISBN 978-5-4316-0300-6
[23] Aleksin V A and Markov A A 2015 Intensification of mass transfer in laminar and turbulent channelflows by applying submicron cavities on the channel wall surface *Nanomechanics Science and Technology: An International Journal* **6**(4), pp. 319-334

Diffusion of a passive impurity in a random velocity field

E V Teodorovich
A. Ishlinsky Institute for Problems in Mechanics, RAS Vernadsky Street 101(1), Moscow, 119526,Russia

E-mail: teodor@ipmnet.ru

Abstract. The method of characteristic functional is used for statistical description of the impurity concentration distribution in the random velocity field. A closed set of equations for a mean concentration response to a change of the external source density and for a pair correlation of the concentration fluctuations is obtained.

1. Statement of the problem

The processes of a diffusion in a random velocity field are described by the equation

$$L^{(0)}C(r,t)+u_{\alpha}(r,t)\partial_{\alpha}C(r,t)-\rho(r,t)\equiv\phi[C,\rho,u_{\alpha}]=0 \tag{1.1}$$

Here $C(r,t)$ is the impurity concentration at the space-time point $\{r,t\}$, $L^{(0)}=\partial_t-\kappa\Delta$; κ is the molecular diffusivity; $\rho(r,t)$ is the density of a impurity sources, and $u_{\alpha}(r,t)$ is the random velocity field given statistically and described by a normal distribution with a given pair correlation function

$$\langle u_{\alpha}(r_1,t_1)\cdot u_{\beta}(r_2,t_2)\rangle=D_{\alpha\beta}(r_1,t_1;r_2,t_2)$$

Next, let us introduce a digital notation for the space-time point $\{r_1,t_1\}$ according to the condition $\{r_1,t_1\}\equiv 1$, $\mathrm{d}1=\mathrm{d}r_1\cdot dt_1$ and we will imply an integrating over repeated digital notation and a summing over the components of vector functions (the Einstein rule).

In this notation equation (1.1) takes the form

$$L^{(0)}(1,2)C(2)+V_{\alpha}(1|2,3)u_{\alpha}(2)C(3)-\rho(1)=\phi[C,\rho,u_{\alpha}]=0 \tag{1.2}$$

$$L^{(0)}(1,2)\equiv\left[\partial_t^{(1)}-\kappa\Delta^{(1)}\right]\delta(1-2),\quad V_{\alpha}(1|2,3)\equiv-\delta(1-2)\partial^{(3)}\delta(1-3)$$

2. Characteristic functional and its representation

The concentration $C(1)$ is a random field and the problem of its statistical description in terms of the statistical moments of various orders in various space-time points arises. To do this, one has to know the density distribution of a probability to find a given realization of the concentration field at a given source density $P[C,\rho]$ that can be defined as a mean value of the δ – functional, which is a functional analog of the δ – function generalized to the case of field description.

$$P[C,\rho]=\left\langle\delta\left[C(1)-\tilde{C}(1)\right]\right\rangle \tag{2.1}$$

here $\tilde{C}(1)$ is the realization of the field concentration at the point 1, which is a solution to the equation $\phi[\tilde{C},\rho,u_\alpha]=0$ at given realizations of the velocity field u_α and the source density ρ, the averaging is carried out over the realization of concentration field, which is reduced to averaging over the velocity field realizations u_α.

Using a functional analog of the formula well-known in the theory of generalized functions

$$\delta[\phi(x)] = \delta(x-x_0)\,|\,\phi'(x)\,|,$$

where x_0 is the solution to the equation $\varphi(x)=0$, one gets

$$\delta\left[C(1)-\tilde{C}(1)\right] = \delta\{\phi[C,\rho,u]\}\frac{\delta\phi[C,\rho,u]}{\delta C(1)} \tag{2.2}$$

The calculation of the Jacobian of functional transformation $\delta\phi[C,\rho,u_\alpha]/\delta C$ shows that the result turns out to be independent of the concentration due to a linear dependence of $\phi[C,\rho,u_\alpha]$ on C as well as on the random quantity u_α. Since, the Jacobian in equation (2.2) may be included in a redefinition of the probabilistic measure and hence one can take the value of Jacobian in equation (2.2) be equal to unity.

The characteristic functional Φ is a functional Fourier-transform of the probability density $P[C,\rho]$

$$\Phi[\eta,\rho] = \int d[C]P[C,\rho]\exp\{i\eta(1)\cdot C(1)\} = \int d[C]\langle\delta\{\phi[C,\rho,u_\alpha]\}\rangle\exp\{i\eta(1)\cdot C(1)\} \tag{2.3}$$

Next we apply the formula of the δ-functional representation, which is a functional analog of the δ-function representation in the form of the Fourier-series expansion

$$\delta\{\phi[C,\rho,u_\alpha]\} = \int d[\lambda]\exp\{i\lambda(1)\phi[C(1),\rho(1),u_\alpha(1)]\} \tag{2.4}$$

As the result we obtain a representation of the characteristic functional $\Phi[\eta,\rho]$ in the form of a functional integral over two fields λ and C.

$$\Phi[\eta,\rho] = \int d[\lambda]\cdot d[C]\exp\{i[\eta(1)C(1)+\lambda(1)L^{(0)}(1,2)C(2)-\lambda(1)\rho(1)]\}\times$$
$$\times\left\langle\exp\{i\lambda(1)V_\alpha(1|2,3)u_\alpha(2)C(3)\}\right\rangle \tag{2.5}$$

Using the formula for calculating mean value in the case of a central normal distribution

$$\left\langle\exp\{i\lambda(1)V_\alpha(1|2,3)C(3)u_\alpha(2)\}\right\rangle =$$
$$= \exp\left\{-\frac{1}{2}\lambda(1)V_\alpha(1|2,3)C(3)\lambda(1')V_\beta(1'|2',3')C(3')D_{\alpha\beta}(2,2')\right\} \tag{2.6}$$

we obtain

$$\Phi[\eta,\rho] = \int d[\lambda]\cdot d[C]\exp\{i[\eta(1)C(1)+\lambda(1)L^{(0)}(1,2)C(2)-\lambda(1)\rho(1)]- \\ -1/2\cdot A(1,1';3,3')\lambda(1)\lambda(1')C(3)C(3')\} \tag{2.7}$$

$$A(1,1';3,3') = V_\alpha(1|2,3)V_\beta(1'|2',3')D_{\alpha\beta}(2,2')$$

From the invariance of functional integral in equation (2.7) under a shift of functional variable of integrating $\lambda \to \lambda + \delta\lambda$ it follows the equation in functional derivatives for the characteristic functional $\Phi[\eta,\rho]$

$$\left\{ L^{(0)}(1,2)\frac{\delta}{i\delta\eta(2)} - \rho(1) + A(1,1';3,3')\frac{\delta}{i\delta\eta(3)}\frac{\delta}{\delta\rho(1')}\frac{\delta}{i\delta\eta(3')}\right\}\Phi[\eta,\rho] = 0 \qquad (2.8)$$

From equation (2.8) one can obtain the equation for the logarithm of characteristic functional

$$L^{(0)}(1,2)\frac{\delta \ln\Phi}{i\delta\eta(1)} - \rho(1) +$$

$$+A(1,1';3,3')\left\{\frac{\delta \ln\Phi}{\delta\rho(1')}\frac{\delta \ln\Phi}{i\delta\eta(3)}\frac{\delta \ln\Phi}{i\delta\eta(3')} + \frac{\delta \ln\Phi}{\delta\rho(1')}\frac{\delta^2 \ln\Phi}{i\delta\eta(3)i\delta\eta(3')} + \right.$$

$$\left. + \frac{\delta \ln\Phi}{i\delta\eta(3)}\frac{\delta^2 \ln\Phi}{\delta\rho(1')i\delta\eta(3')} + \frac{\delta \ln\Phi}{i\delta\eta(3')}\frac{\delta^2 \ln\Phi}{\delta\rho(1')i\delta\eta(3)} + \frac{\delta^3 \ln\Phi}{i\delta\rho(1')i\delta\eta(3)i\delta\eta(3')}\right\} = 0 \qquad (2.9)$$

Now we perform a transition to new functional variables by putting

$$\hat{\eta}(1) = \frac{\delta \ln\Phi}{i\delta\eta(1)}, \quad \hat{\rho}(1) = \frac{\delta \ln\Phi}{i\delta\rho(1)} \qquad (2.10)$$

This transition can be carried out with the help of functional Legendre transformation by the transition to new characteristic functional

$$\Psi[\hat{\eta},\hat{\rho}] = -\ln\Phi[\eta,\rho] + i\eta\hat{\eta} + i\rho\hat{\rho} \qquad (2.11)$$

In this case

$$\frac{\delta\Psi}{i\delta\hat{\eta}(1)} = \eta(1), \quad \frac{\delta\Psi}{i\delta\hat{\rho}(1)} = \rho(1) \qquad (2.12)$$

In new variables the equation for characteristic functional $\Psi[\hat{\eta},\hat{\rho}]$ can be written in the form

$$L^{(0)}(1,2)\hat{\eta}(2) - \frac{\delta\Psi}{i\delta\hat{\rho}(1)} + A(1,1';3,3')\times$$

$$\times\left\{ i\hat{\rho}(1')\hat{\eta}(3)\hat{\eta}(3') + i\hat{\rho}(1')\frac{\delta\hat{\eta}(3)}{i\delta\eta(3')} + \hat{\eta}(3)\frac{\delta\hat{\eta}(3')}{\delta\rho(1')} + \hat{\eta}(3')\frac{\delta\hat{\eta}(3)}{\delta\rho(1')} + \frac{\delta}{\delta\rho(1')}\frac{\delta\hat{\eta}(3)}{i\delta\eta(3')}\right\} = 0 \qquad (2.13)$$

3. Meaning of various functional derivatives

Unlike a traditional approach, when equations are used for their solving, in the functional formulation of a statistical description of stochastic system, the equations for characteristic functional are used for finding relations between various statistical characteristics of the system such as correlation functions and functions of mean response to external actions as well as for constructing various approximations on the basis of the analysis of the equations rather then by using some hypothesis of phenomenological type.

In the context of aforesaid one needs to clarify a meaning of functional derivatives of functionals $\Phi[\eta,\rho]$ and $\Psi[\hat{\eta},\hat{\rho}]$ with respect to fields η,ρ and $\hat{\eta},\hat{\rho}$. Note that, according to the definition of the characteristic functional, when obtaining relations for statistical characteristics of the system under consideration one has to take the limit $\eta \to 0$ after carrying out a functional differentiation of characteristic functional.

Below we will consider how the functional derivatives of the characteristic functionals $\Phi[\eta,\rho]$ and $\Psi[\hat{\eta},\hat{\rho}]$ with respect to fields η,ρ and $\hat{\eta},\hat{\rho}$ relate to statistical characteristics of the system.

By definition

$$\frac{\delta\Phi[\eta,\rho]}{i\delta\eta(1)} = \hat{\eta}(1)\Phi[\eta,\rho]|_{\eta\to 0} \to \langle C(1)\rangle \quad (\Phi[0,\rho]=1) \tag{3.1}$$

here $\langle C(1)\rangle$ is the mean concentration at the space-time point 1 at given source density ρ. And further

$$\frac{\delta \ln\Phi[\eta,\rho]}{i\delta\rho(1)} = \hat{\rho}(1)|_{\eta\to 0} \to 0 \tag{3.2}$$

The quantity $\hat{\rho}(1)$ describes the density distribution of fluctuations of the effective source that is not identical with initial ("bare") distribution of source density $\rho(1)$. The matter is that the turbulent velocity field produces a random flux of impurity which divergence is regarded as a density of effective source added to initial one. It should be particularly emphasized that the concept of effective source density arose as a consequence of statistical description of a stochastic system within the framework of the functional formalism rather then as a certain assumption of phenomenological type.

Higher functional derivatives of the characteristic functional $\Phi[\eta,\rho]$ with respect to the field η in the limit $\eta \to 0$ describe statistic moments of high order whereas the derivatives of the characteristic functional logarithm describe so called irreducible (cumulant) means of a statistical system.

In particular

$$\frac{\delta^2\Phi[\eta,\rho]}{i\delta\eta(1)\cdot i\delta\eta(2)}|_{\eta\to 0} \to \langle C(1)\cdot C(2)\rangle$$

$$\frac{\delta^2 \ln\Phi[\eta,\rho]}{i\delta\eta(1)\cdot i\delta\eta(2)}|_{\eta\to 0} \to \langle C(1)\cdot C(2)\rangle - \langle C(1)\rangle\cdot\langle C(2)\rangle = B(1,2) \tag{3.3}$$

$B(1,2)$ is the irreducible pair mean of the concentration fluctuations at the points 1 and 2.

$$\frac{\delta^2 \ln\Phi[\eta,\rho]}{i\delta\eta(1)\cdot\delta\rho(2)} = \frac{\delta\hat{\eta}(1)}{\delta\rho(2)}|_{\eta\to 0} \to \frac{\delta\langle C(1)\rangle}{\delta\rho(2)} = G(1,2) \tag{3.4}$$

$G(1,2)$ is the function of mean response of the concentration field at the space–time point 1 to the action of instantaneous source localized at the point 2 (Green's function).

4. Diagram technique

The powerful method of further analysis is an application of the Feynman diagram technique that enables to get additional information concerning various statistic characteristics of the system and relations between them. The method of Feynman diagram, firstly proposed in quantum electrodynamics for the analysis of perturbation theory series, consists in putting some graphic symbols in correspondence to various terms of the perturbation series. This method turns out to be very fruitful in many branches of mathematics and mathematical physics for obtaining within the framework of perturbation theory the information about the properties of a total perturbation theory series or its infinite subsequence in the case when a knowledge of some first terms of the series appears to be insufficient, and for understanding the phenomenon one needs to operate with partly or totally summarized series to which so called skeleton diagrams are assigned. The method of characteristic functional just corresponds to a description in terms of summarized perturbation theory series. In the statistical theory of turbulence the diagram methods were introduced in [1] and in the theory of turbulent diffusion this has been done in [2] (see, also [3]).

In a "simple" perturbation theory the solution is constructed by treating the convective term in equation (1.1) as small expansion parameter, and as the result one obtains the solution in the form of power series in the velocity. Next term-by term averaging gives a mean value of concentration. In this approach the elements of diagram technique are follows (see figure 1):

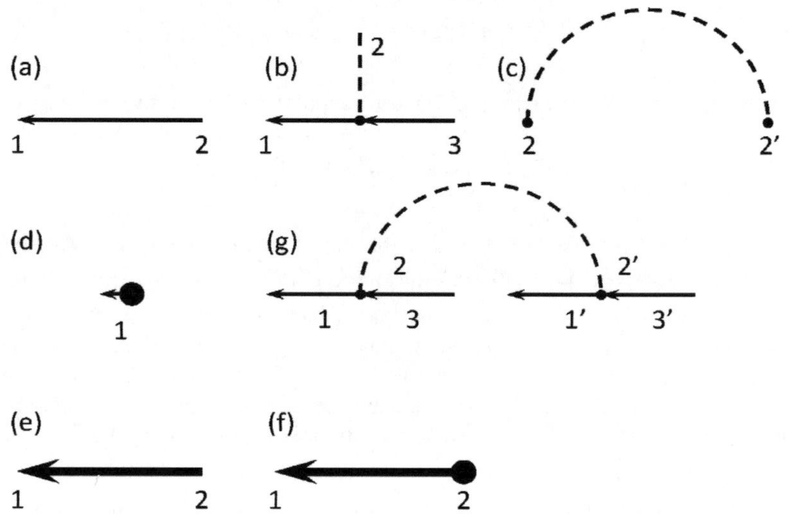

Figure 1. Diagram representation of various elements in “simple” perturbation theory..

(a) Green function $G^{(0)}(1,2)$ of the operator $L^{(0)}(1,2)$ ("bare" Green's function) depicted by a thin line connecting the points 1 and 2 with an arrow incoming into the point 1;

(b) vertex $V_{\alpha}(1|2,3)$ depicted by a point with outgoing arrow 1, incoming arrow 3 and incoming dashed line 2;

(c) pair correlation function (a variance) of the velocity field $D_{\alpha\beta}(2,2')$ depicted by a dashed line connecting the points 2 and 2';

(d) source density $\rho(1)$ depicted by a fat point 1 with outgoing arrow;

(e) Green function defined by equation (4) is depicted by a fat line connecting the points 1 and 2 with an arrow incoming into the point 1.

Diagram representation of mean concentration $<C(1)>=G(1,2)\cdot\rho(2)$ and those for the quantity $A(1,1';3,3)$ defined by equation (2.7) are shown in figure.1, diagrams (f) and (g).

More detail presentation of building the solution within the framework of a "simple" perturbation theory with using the diagram technique one can find in [3].

When studying the high-order statistical moments of the concentration field an application of a "simple" perturbation theory becomes very complex since one has to multiply infinite perturbation-theory series and next to carry out tern-by term averaging of the result obtained. In this context the use of the functional formulation of the problem turns out to be more suitable since in this formalism the averaging procedure is performed from the very beginning and one deals with summarized perturbation-theory series. Within the framework of a functional formulation of the problem the diagram technique is constructed in terms of so called skeleton diagrams corresponding to summarized diagrams of a "simple" perturbation theory (See, [4]). The technique of skeleton diagrams is constructed by following way (see figure 2).

To the characteristic functionals Φ, $\ln\Phi$ and Ψ a light circle, light circle with cross inside and dark circle are assigned respectively.

The actions of operators $\delta/i\delta\eta(1)$ and $\delta/\delta\rho(1)$ to the characteristic functional $\Phi[\eta,\rho]$ are putted into correspondence a line connecting the light circle and the point 1 with incoming or outgoing arrow (see figure 3, diagrams (a) and (b)).

According to these convention the action of various products of two operators $\delta / i\delta\eta(1)$ and $\delta / \delta\rho(1)$ to the characteristic functional $\Phi[\eta,\rho]$ and its logarithm defined by the formulas

(c) $\delta^2\Phi / [i\delta\eta(1)\cdot i\delta\eta(2)] = \langle C(1)\cdot C(2)\rangle$,

(d) $\delta^2 \ln\Phi / [i\delta\eta(1)\cdot i\delta\eta(2)] = \langle C(1)\cdot C(2)\rangle - \langle C(1)\rangle\cdot\langle C(2)\rangle = B(1,2)$

(e) $\delta^2\Phi / [i\delta\eta(1)\cdot \delta\rho(2)] = G(1,2)$

are depicted in figure 2, diagrams (c), (d) and (e).

The representations of the quantities $B(1,2)$ and $G(1,2)$ defined by equations (3.3) and (3.4) are shown in figure 2, diagrams (d) and (e). Note that under the action of the above pointed operators to the logarithm of characteristic functional the diagram that contains unconnected parts is absent, and this corresponds to an account for only irreducible means (cumulants).

We will depict the action of the operators $\delta / \delta\hat{\eta}(1)$ and $\delta / i\delta\hat{\rho}(1)$ to the characteristic functional $\Psi[\hat{\eta},\hat{\rho}]$ by an insert of incoming and outgoing arrows at the points 1 of dark circle. The characteristic functional $\Psi[\hat{\eta},\hat{\rho}]$ corresponds to a description of so called strongly connected diagrams, i.e. those that cannot become unconnected under breaking one line (one-particle irreducible diagrams).

The analysis of the perturbation theory series with exploiting the method of Feynman diagrams shows that in the limit $\eta \to 0$ all diagrams or parts of diagram in which the number of incoming arrows exceeds the number of outgoing arrows disappear. This means that the functional derivatives of the functional $\ln\Phi[\eta,\rho]$ with respect to only fields ρ equal to zero, as well the derivatives of the functional $\Psi[\hat{\eta},\hat{\rho}]$ with respect to only fields $\hat{\eta}$.

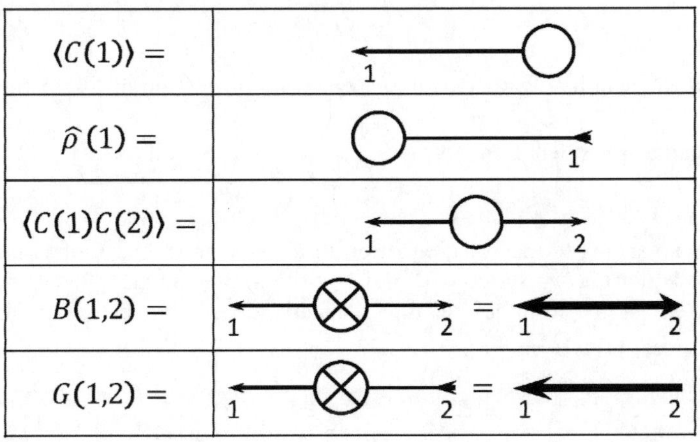

Figure 2. Diagram representation of the action of functional derivatives to characteristic functional.

5. Equations for statistical characteristics of a concentration field

To find equations for the quantities $B(1,2)$ and $G(1,2)$ one has to know some mixed derivatives. Next we consider mixed derivatives

$$\frac{\delta^2\Psi}{\delta\eta(1)\cdot i\delta\hat{\eta}(2)}=\frac{\delta\eta(2)}{\delta\eta(1)}=\delta(1-2)=\frac{\delta\hat{\eta}(1')}{\delta\eta(1)}\cdot\frac{\delta^2\Psi}{\delta\hat{\eta}(1')\cdot i\delta\hat{\eta}(2)}\cdot\frac{\delta\hat{\rho}(1')}{\delta\eta(1)}\cdot\frac{\delta^2\Psi}{\delta\hat{\rho}(1')\cdot i\delta\hat{\eta}(2)} \tag{5.1}$$

As $\eta\to 0$ the first term in the right-hand side of equation (5.1) disappears (two incoming arrows), from what it follows

$$G(1,1')\cdot\delta^2\Psi\delta\hat{\rho}(1')\cdot i\delta\hat{\eta}(2)\equiv G(1,1')G^{-1}(1',2)=\delta(1-2)$$

$$G^{-1}(1,2)=\frac{\delta^2\Psi}{\delta\hat{\rho}(1)\cdot i\delta\hat{\eta}(2)} \tag{5.2}$$

here $G^{-1}(1,2)$ is the reverse Green function.

The other mixed derivative

$$\frac{\delta^2\Psi}{\delta\eta(1)\cdot\delta\hat{\rho}(2)}=i\frac{\delta\rho(2)}{\delta\eta(1)}=0$$

$$\frac{\delta^2\Psi}{\delta\eta(1)\cdot\delta\hat{\rho}(2)}=\left\{\frac{\delta\hat{\eta}(1')}{\delta\eta(1)}\cdot\frac{\delta}{\delta\hat{\eta}(1')}+\frac{\delta\hat{\rho}(1')}{\delta\eta(1)}\cdot\frac{\delta}{\delta\hat{\rho}(1')}\right\}\frac{\delta\Psi}{\delta\hat{\rho}(2)}$$

Using equations (3.4) and (5.2) in the limit $\eta\to 0$ one obtains

$$\left.\frac{\delta^2\Psi}{\delta\eta(1)\cdot\delta\hat{\rho}(2)}\right|_{\eta\to 0}=-B(1,1')G^{-1}(2,1')+G(1,1')R(1',2)=0 \tag{5.3}$$

here

$$R(1,2)=\left.\frac{\delta^2\Psi}{\delta\hat{\rho}(1)\cdot\delta\hat{\rho}(2)}\right|_{\eta\to 0} \tag{5.4}$$

The quantity $R(1,2)$ can be regarded as a pair correlation function of fluctuations of the effective source density.

Equation (5.3) can be rewritten in the form

$$B(1,2)=G(1,1')G(2,2')R(1',2') \tag{5.5}$$

This equation is an analog to the Wyld equation in the statistical theory of turbulence obtained by summing the perturbation theory series with the help of the technique of Feynman's diagrams [1]. Further we will need for the formulas relating the derivatives with respect to fields η,ρ and the derivatives with respect to fields $\hat{\eta},\hat{\rho}$.

$$\frac{\delta}{\delta\hat{\eta}(1)}=G^{-1}(1',1)\frac{\delta}{\delta\rho(1')},\quad \frac{\delta}{\delta\hat{\rho}(1)}=G^{-1}(1,1')\frac{\delta}{\delta\eta(1')}+R(1,1')\frac{\delta}{i\delta\rho(1')} \tag{5.6}$$

and the reverse formulas

$$\frac{\delta}{i\delta\eta(1)}=B(1,1')\frac{\delta}{\delta\hat{\eta}(1')}+G(1,1')\frac{\delta}{i\delta\hat{\rho}(1')},\quad \frac{\delta}{\delta\rho(1)}=G(1',1)\frac{\delta}{\delta\hat{\eta}(1')} \tag{5.6a}$$

With the aim to get the equations for Green's function, we act to Eq. (2.13) by the operator $\delta/\delta\hat{\eta}(4)$, this gives

$$L^{(0)}(1,2)\delta(1-4)-\frac{\delta\Phi}{\delta\rho(1')\cdot i\delta\hat{\eta}(4)}+$$

$$+A(1,1';3,3')\left\{\frac{\delta\hat{\eta}(3')}{\delta\rho(1')}\delta(3-4)+\frac{\delta\hat{\eta}(3)}{\delta\rho(1')}\delta(3'-4)+\frac{\delta}{\delta\hat{\eta}(4)}\cdot\frac{\delta^3\ln\Phi}{\delta\rho(1')\cdot i\delta\eta(3)\cdot i\delta\eta(3')}\right\}=0 \quad (5.7)$$

In equation (5.7) the terms being proportional to $\hat{\rho}$ and $\hat{\eta}$ are omitted since they include diagrams in which the number of incoming arrows exceeds the number of outgoing arrows that disappear in the limit as $\eta\to 0$.

After performing the limiting transition $\eta\to 0$ in equation(5.7) and usingequations (3.9), (5.2) and (5.6) one gets the expression for inverse Green's function

$$G^{-1}(1,4)=L^{(0)}(1,4)+A(1,1';4,3')G(3',1')+A(1,1';3,4)G(3,1')+$$

$$+A(1,1';3,3')\frac{\delta^4\ln\Phi}{\delta\rho(1')i\delta\eta(3)i\delta\eta(3')i\delta\rho(4')}G^{-1}(4',4) \quad (5.8)$$

The second term in the left-hand side of equation (5.8) contains Green's function at equal times. If in the formula for $V(1|2,3)$ to introduce an infinitely small time of delay in the concentration response to a velocity change, this term vanishes due the causality condition of Green's function (in the Feynman diagram technique a closed loop containing Green's function gives zero contribution). The last term in the right-hand side of equation (5.8) corresponds to an account for "one-particle irreducible diagrams" and we assume the contribution from these diagrams can be neglected (one-loop approximation).

$$G^{-1}(1,2)=L^{(0)}(1,2)-\Sigma(1,2),$$
$$\Sigma(1,2)=-A(1,1';3,2)G(3,1') \quad (5.9)$$

Equation (5.9) may be rewritten in the form of Dyson's equation in the quantum-field theory

$$G(1,2)=G^{(0)}(1,2)+G^{(0)}(1,1')\Sigma(1',2')G(2',2) \quad (5.10)$$

The quantity $\Sigma(1,2)$ is an analog to the "self–energy operator" in the quantum–field theory.

Similarly, under the action of the operator $\delta/\delta\hat{\rho}(4)$ to equation (2.13) in the limit as $\eta\to 0$ one finds

$$R(1,4)+A(1,4;3,3')B(3,3')+A(1,1';3,3')\{\ \frac{\delta^4\ln\Phi}{i\delta\rho(1')i\delta\eta(3)i\delta\eta(3')\delta\eta(4')}G(4',4)-$$

$$-\frac{\delta^4\ln\Phi}{\delta\rho(4')i\delta\rho(1')i\delta\eta(3)i\delta\eta(3')}R(4',4)\ \}=0 \quad (5.11)$$

In the "one-loop approximation" we obtain

$$R(1,2)=-A(1,2;3,3')B(3,3') \quad (5.12)$$

The diagram equations by Wyld and Dyson are shown in figure 3 diagrams (a) and (c), and the quantities $R(1,2)$ and $\Sigma(1,2)$ depicted by a rectangular with two outgoing arrows and by one incoming and one outgoing arrows are presented in figure 3, diagrams (b) and (d). It should be emphasized that the Wyld equation (5.5) and the Dyson equation (5.10) are exact ones since under their obtaining no additional assumptions or any approximation were used. The approximation is defined by a choice of the form for Σ and R. The set of equations (5.9) and (5.12) in combination with equations by Wyld (5.5) and Dyson (5.10) is a closed one and this enables to find the statistic characteristics of the system, namely, the pair correlation function of the concentration fluctuations and the function of mean response of the concentration to instant change of the source density.

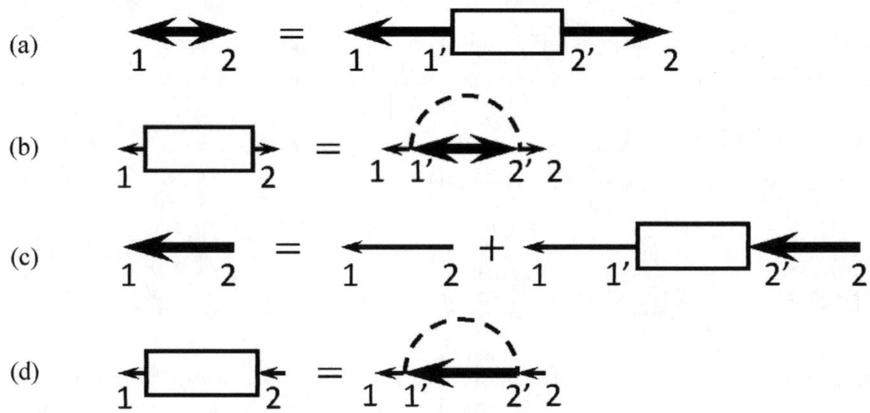

Figure 3. Diagram equations by Wyld and Dyson (a) and (c) and diagram representations for R(1,2) and Σ(1,2).

It is a certain surprise that the equations by Wyld and Dyson obtained in present paper turn out to be identical with those in the statistical theory of turbulence [5].

Note that the equation for Green's function appears to be independent of the statistic characteristics of the concentration field such as $B(1,2)$ and $R(1,2)$, whereas these quantities are related to each other by the Wyld equation into which the Green function enters. Therefore, one must firstly solve the equation for Green's function and then use this solution for obtaining the equation for statistical characteristics of the concentration field.

The author thanks Prof. Vladimir Frost for fruitful discussions.

References

[1] Wyld H W 1961 Formulation of the theory of turbulence in an incompressible fluid *Ann. Phys.* **14** 2 pp 143–165

[2] Phythian R and Curtis W D 1978 The effective long-time diffusivity for a passive scalar in a Gaussian model fluid flow *J. Fluid Mech.* **89** 2 pp 241–250

[3] Teodorovich E V 1991 Turbulent diffusion of a passive impurity *J. Appl. Math. Mech. (PMM)* **55** 2 pp 220–226

[4] Teodorovich E V 1994 Diagram equations of the theory of fully developed turbulence *Theor. Math. Phys.* **101** 1 pp 1177–1183

[5] Teodorovich E V 2013 On one possibility of closuring the chain of equations for statistical moments in turbulence theory 2013 *J. Mod. Phys.* **4** 1 pp 56–63

Numerical solution of the Navier-Stokes equations by discontinuous Galerkin method

M M Krasnov[1], P A Kuchugov[1,2], M E Ladonkina[1,2], A E Lutsky[1], V F Tishkin[1,2]

[1]Keldysh Institute of Applied Mathematics of RAS, 4 Miusskaya Sq., 125047, Moscow, Russia
[2]Lavrentyev Institute of Hydrodynamics of SB RAS, 15 ac. Lavrentyeva Av., 630090, Novosibirsk, Russia

E-mail: ladonkina@imamod.ru

Abstract. Detailed unstructured grids and numerical methods of high accuracy are frequently used in the numerical simulation of gasdynamic flows in areas with complex geometry. Galerkin method with discontinuous basis functions or Discontinuous Galerkin Method (DGM) works well in dealing with such problems. This approach offers a number of advantages inherent to both finite-element and finite-difference approximations. Moreover, the present paper shows that DGM schemes can be viewed as Godunov method extension to piecewise-polynomial functions.

As is known, DGM involves significant computational complexity, and this brings up the question of ensuring the most effective use of all the computational capacity available. In order to speed up the calculations, operator programming method has been applied while creating the computational module. This approach makes possible compact encoding of mathematical formulas and facilitates the porting of programs to parallel architectures, such as NVidia CUDA and Intel Xeon Phi.

With the software package, based on DGM, numerical simulations of supersonic flow past solid bodies has been carried out. The numerical results are in good agreement with the experimental ones.

1. Introduction

The solution of real-world aerothermodynamics problems in areas with complex geometry demands the use of unstructured grids, which often should also consist of mesh elements of different scale, and the application of high accuracy numerical methods. Galerkin method with discontinuous basis functions or Discontinuous Galerkin Method (DGM) [1] works well in dealing with such problems. This approach offers a number of advantages inherent to both finite-element and finite-difference approximations. In particular, it provides an adjusted order of accuracy, is applicable for meshes of arbitrary structure, allows implement boundary conditions of various kind relatively easily and reliably theoretically grounded [1–5].

While using DGM, the approximation of the solution within the cell is in the form of polynomials of degree N with time-dependent coefficients. During the numerical solution it is necessary to correctly calculate flux functions on cell's faces. The approach for calculation of flux function values for diffusion terms in the form of half sum has been proposed in work [6]. In [2] it is shown the necessity of using stabilizing additions for calculation of numerical fluxes while solving heat conductivity equations. In the present paper, it is proposed to use the second method for calculation values of flux functions for diffusion terms [7].

For all its merits, DGM involves significant computational complexity, and this brings up the question of ensuring the most effective use of all the computational capacity available. In the present paper, in order to speed up the calculations, grid-operator approach to the programming has been applied while creating the computational module for solving three-dimensional Navier-Stokes equations based on DGM. This approach makes possible compact encoding of mathematical formulas and facilitates the porting of programs to parallel architectures, such as NVidia CUDA and Intel Xeon Phi. Moreover, the usage of meta-programming features of C++ allows to do a lot of preliminary calculations on the compilation stage.

2. The description of DG method for the Navier-Stokes equations

Let's consider the Navier-Stokes equations, written in the form of first-order equations:

$$\begin{gathered}\partial_t \boldsymbol{U} + \nabla \cdot \boldsymbol{F}(\boldsymbol{U}) - \nabla \cdot \boldsymbol{G}(\boldsymbol{U}, \boldsymbol{\tau}, \boldsymbol{q}) = 0, \\ \boldsymbol{\tau} = \left(\lambda - \frac{2}{3}\mu\right) \hat{I}(\nabla \cdot \boldsymbol{v}) + 2\mu \boldsymbol{S}(\boldsymbol{v}), \quad \boldsymbol{S}(\boldsymbol{v}) = \frac{1}{2}\left(\nabla \boldsymbol{v} + (\nabla \boldsymbol{v})^*\right), \quad \boldsymbol{q} = -k\nabla T,\end{gathered} \tag{1}$$

and supplemented with the appropriate initial and boundary conditions, the type of which depends on the specific objectives and will be specified further. Vector of conservative variables $\boldsymbol{U}$ is given in the following form:

$$\boldsymbol{U} = (\rho, \rho u, \rho v, \rho w, E)^T, \tag{2}$$

where ρ is the density of the fluid, u, v, w, are the components of velocity $\boldsymbol{v}$, p is the pressure, ϵ is the specific internal energy and $E = \rho\left(\epsilon + \dfrac{u^2+v^2+w^2}{2}\right)$, is total energy per unit volume. Perfect gas equation of state will be used to determine the pressure value p

$$p = (\gamma - 1)\rho\epsilon, \tag{3}$$

where γ is the specific heats ratio. For components of flux functions $\boldsymbol{F}(\boldsymbol{U})$ and $\boldsymbol{G}(\boldsymbol{U}, \boldsymbol{\tau}, \boldsymbol{q})$ one can obtain:

$$\begin{aligned} F_x(\boldsymbol{U}) &= \left(\rho u, \rho u^2 + p, \rho uv, \rho uw, (E+p)u\right)^T, \\ F_y(\boldsymbol{U}) &= \left(\rho u, \rho vu, \rho v^2 + p, \rho vw, (E+p)v\right)^T, \\ F_z(\boldsymbol{U}) &= \left(\rho u, \rho wu, \rho wv, \rho w^2 + p, (E+p)w\right)^T, \end{aligned} \tag{4}$$

and

$$\begin{aligned} G_x(\boldsymbol{U}, \boldsymbol{\tau}, \boldsymbol{q}) &= \left(0, \tau_{xx}, \tau_{xy}, \tau_{xz}, u\tau_{xx} + v\tau_{xy} + w\tau_{xz} - q_x\right)^T, \\ G_y(\boldsymbol{U}, \boldsymbol{\tau}, \boldsymbol{q}) &= \left(0, \tau_{yx}, \tau_{yy}, \tau_{yz}, u\tau_{yx} + v\tau_{yy} + w\tau_{yz} - q_y\right)^T, \\ G_z(\boldsymbol{U}, \boldsymbol{\tau}, \boldsymbol{q}) &= \left(0, \tau_{zx}, \tau_{zy}, \tau_{zz}, u\tau_{zx} + v\tau_{zy} + w\tau_{zz} - q_z\right)^T. \end{aligned} \tag{5}$$

To apply DGM, the area Ω, where the solution is sought, will be covered with a tetrahedral mesh T_h. An approximate solution of the set of the equations (1) will be sought in the form of polynomials $P(x)$ of degree N with time-dependent coefficients on each mesh element T_j:

$$\begin{aligned} \boldsymbol{U}_h(\boldsymbol{x}, t) &= \sum_{k=0}^{D_P} \boldsymbol{U}_k(t)\phi_k(\boldsymbol{x}), \\ q_{ih}(\boldsymbol{x}, t) &= \sum_{k=0}^{D_P} q_{ik}(t)\phi_k(\boldsymbol{x}), \\ \tau_{ijh}(\boldsymbol{x}, t) &= \sum_{k=0}^{D_P} \tau_{ijk}(t)\phi_k(\boldsymbol{x}), \end{aligned} \tag{6}$$

where $i, j = x, y, z$, $D_P = C^3_{N+3} - 1$, is the dimension of the space of polynomials, ϕ_k is the corresponding basis function. In the present paper, Taylor basis is being used as the basis functions:

$$\phi_k = \left(\frac{x - x_c}{\Delta x}\right)^\alpha \left(\frac{y - y_c}{\Delta y}\right)^\beta \left(\frac{z - z_c}{\Delta z}\right)^\delta, \tag{7}$$

where $\alpha + \beta + \delta = 0, 1, 2, \dots N$, x_c, y_c, z_c, are the coordinates of the mass center of particular tetrahedron, Δx, Δy and Δz, are the projections of the cell on the coordinate axis x, y and z.

An approximate solution of (1) in DGM is sought as a solution of the following system [1,6]:

$$\begin{aligned}
&\frac{d}{dt}\int_{T_j} \phi_k(\boldsymbol{x}) U_h(\boldsymbol{x}, t) d\Omega + \oint_{\partial T_j} \phi_k(\boldsymbol{x}) h_F\left(U_h^+, U_h^-, n\right) d\sigma - \\
&\quad - \int_{T_j} \left(\frac{\partial \phi_k(\boldsymbol{x})}{\partial x} F_x\left(U_h(\boldsymbol{x}, t)\right) + \frac{\partial \phi_k(\boldsymbol{x})}{\partial y} F_y\left(U_h(\boldsymbol{x}, t)\right) + \frac{\partial \phi_k(\boldsymbol{x})}{\partial z} F_z\left(U_h(\boldsymbol{x}, t)\right)\right) d\Omega - \\
&- \oint_{\partial T_j} \phi_k(\boldsymbol{x}) h_G\left(U_h^+, \tau^+, q^+, U_h^-, \tau^-, q^-, n\right) d\sigma + \int_{T_j} \left(\frac{\partial \phi_k(\boldsymbol{x})}{\partial x} G_x\left(U_h(\boldsymbol{x}, t), \tau(\boldsymbol{x}, t), q(\boldsymbol{x}, t)\right) + \right. \\
&\quad \left. + \frac{\partial \phi_k(\boldsymbol{x})}{\partial y} G_y\left(U_h(\boldsymbol{x}, t), \tau(\boldsymbol{x}, t), q(\boldsymbol{x}, t)\right) + \frac{\partial \phi_k(\boldsymbol{x})}{\partial z} G_z\left(U_h(\boldsymbol{x}, t), \tau(\boldsymbol{x}, t), q(\boldsymbol{x}, t)\right)\right) d\Omega = 0,
\end{aligned} \tag{8}$$

$$\begin{aligned}
\int_{T_j} \boldsymbol{\tau} \cdot\cdot \Phi_k d\Omega = &\oint_{\partial T_j} \left(\lambda - \frac{2}{3}\mu\right)\left(\Phi_k \cdot\cdot \hat{I}\right) h_v\left(v^+, v^-, n\right) d\sigma - \\
&- \int_{T_j} \left(\nabla\left(\left(\lambda - \frac{2}{3}\mu\right)\left(\Phi_k \cdot\cdot \hat{I}\right)\right), v\right) d\Omega + \oint_{\partial T_j} \mu\left[h_S\left(v^+, v^-, n\right) \cdot\cdot \Phi_k\right] d\sigma - \\
&- \int_{T_j} \left(v, \nabla \cdot (\mu \Phi_k)\right) d\Omega, \quad \forall \phi_k(\boldsymbol{x}), k = 0 \dots D_P
\end{aligned} \tag{9}$$

$$\begin{aligned}
\int_{T_j} \phi_l(\boldsymbol{x}) q_h(\boldsymbol{x}, t) d\Omega + &\oint_{\partial T_j} \phi_l(\boldsymbol{x}) h_T\left(T_h^+, T_h^-, k_h^+, k_h^-\right) n d\sigma - \\
&- \int_{T_j} \nabla\left(k(\boldsymbol{x}) \phi_l(\boldsymbol{x})\right) T_h(\boldsymbol{x}, t) d\Omega = 0,
\end{aligned} \tag{10}$$

where $\boldsymbol{U}_h(x, t)$ is the solution vector, $\boldsymbol{n}$ is the outward normal vector to the element boundary ∂T_j, $\boldsymbol{h}_F\left(\boldsymbol{U}_h^+, \boldsymbol{U}_h^-, \boldsymbol{n}\right)$, $\boldsymbol{h}_G\left(\boldsymbol{U}_h^+, \boldsymbol{\tau}^+, \boldsymbol{q}^+, \boldsymbol{U}_h^-, \boldsymbol{\tau}^-, \boldsymbol{q}^-, \boldsymbol{n}\right)$, $\boldsymbol{h}_v\left(\boldsymbol{v}^+, \boldsymbol{v}^-, \boldsymbol{n}\right)$, $\boldsymbol{h}_S\left(\boldsymbol{v}^+, \boldsymbol{v}^-, \boldsymbol{n}\right)$ and $\boldsymbol{h}_T\left(T^+, T^-, \boldsymbol{n}\right)$ are flux functions calculated at element boundary ∂T_j. Quantities denoted by $\boldsymbol{U}_h^-$ are calculated at the boundary ∂T_j of the element T_j using the values within the cell, while ones indicated by $\boldsymbol{U}_h^+$ are calculated using the values of the adjunction cell. Function $\boldsymbol{h}_F\left(\boldsymbol{U}_h^+, \boldsymbol{U}_h^-, \boldsymbol{n}\right)$ is monotone and it satisfies the matching condition:

$$\boldsymbol{h}_F\left(\boldsymbol{U}_h(\boldsymbol{x}, t), \boldsymbol{U}_h(\boldsymbol{x}, t), \boldsymbol{n}\right) = \boldsymbol{F}\left(\boldsymbol{U}_h(\boldsymbol{x}, t)\right). \tag{11}$$

In the present paper, Godunov [8] and Rusanov–Lax–Friedrichs [9,10] fluxes have been used:

$$\begin{aligned}
\boldsymbol{h}_F\left(\boldsymbol{U}_h^+(\boldsymbol{x}, t), \boldsymbol{U}_h^-(\boldsymbol{x}, t), \boldsymbol{n}\right) = &\frac{1}{2}\left(\boldsymbol{F}\left(\boldsymbol{U}_h^+(\boldsymbol{x}, t)\right) + \boldsymbol{F}\left(\boldsymbol{U}_h^-(\boldsymbol{x}, t)\right) - \right. \\
&\left. - B\left(\boldsymbol{U}_h^+(\boldsymbol{x}, t) - \boldsymbol{U}_h^-(\boldsymbol{x}, t)\right)\right), B = \max\left(\left|v^+\right| + c^+, \left|v^-\right| + c^-\right),
\end{aligned} \tag{12}$$

where $\boldsymbol{v}$ is the velocity and c is the speed of sound.

It is necessary to calculate flux functions h_T, h_v and h_S at element boundary in equations (9)–(10). The following expressions are used for this aim:

$$\begin{aligned} h_T\left(T_h^+, k_h^+, T_h^-, k_h^-\right) &= \left(\frac{k^+T^+ + k^-T^-}{2}\right), \\ h_v\left(v^+, v^-, n\right) &= \left(\frac{v^+ + v^-}{2}, n\right), \\ h_S\left(v^+, v^-, n\right) &= \frac{v^+ + v^-}{2} \otimes n + n \otimes \frac{v^+ + v^-}{2}. \end{aligned} \tag{13}$$

It is necessary to calculate the flow values of $h_G\left(U_h^+, \boldsymbol{\tau}^+, \boldsymbol{q}^+, U_h^-, \boldsymbol{\tau}^-, \boldsymbol{q}^-, n\right)$ at the element boundary in (8) using stabilizing additions [7] as for heat conductivity equation [2,3]:

$$\begin{aligned} \hat{q}\left(T_h^+, q^+, T_h^-, q^-, n\right) &= \frac{q^+ + q^-}{2} + C_3\left(k^+T^+ - k^-T^-\right)n, \\ \hat{\boldsymbol{\tau}}\left(U_h^+, \boldsymbol{\tau}^+, U_h^-, \boldsymbol{\tau}^-, n\right) &= \frac{\boldsymbol{\tau}^+ + \boldsymbol{\tau}^-}{2} + C_1\left(\left(\boldsymbol{v}^+ - \boldsymbol{v}^-\right), n\right)\hat{I} + C_2\frac{\left(v^+ - v^-\right) \otimes n + n \otimes \left(v^+ - v^-\right)}{2}, \\ h_G\left(U_h^+, \boldsymbol{\tau}^+, \boldsymbol{q}^+, U_h^-, \boldsymbol{\tau}^-, \boldsymbol{q}^-, n\right) &= G_x\left(U_h^+, U_h^-, \hat{\boldsymbol{\tau}}, \hat{q}\right)n_x + G_y\left(U_h^+, U_h^-, \hat{\boldsymbol{\tau}}, \hat{q}\right)n_y + \\ &+ G_z\left(U_h^+, U_h^-, \hat{\boldsymbol{\tau}}, \hat{q}\right)n_z. \end{aligned} \tag{14}$$

All the integrals appearing in equations (8)–(10) are calculated using numerical Gaussian quadrature formulas, the number of integration points corresponding to the required calculation accuracy [11, 12].

As is known, to ensure the monotonicity of the numerical solution obtained by this method, it is necessary to apply so called slope limiters, especially in the case of strong discontinuities [13–15]. In the present paper the classical Cockburn limiter [1] has been used due to the fact that it is easily implemented in the multidimensional case on grids of arbitrary structure.

The system of ordinary differential equations (8), which determines the changes in time of numerical solution, can be written as

$$\frac{d\boldsymbol{U}^p}{dt} = \boldsymbol{A}^{-1}\boldsymbol{R}\left(\boldsymbol{U}^p\right), \tag{15}$$

where $\boldsymbol{A}$ is the mass matrix, $\boldsymbol{U}^p$ is the global vector of degrees of freedom (the expansion coefficients of all the sought-for functions), k is the equation number in the system (1) and $\boldsymbol{R}\left(\boldsymbol{U}^p\right)$ is the vector of the right hand sides.

Remark. For high order accuracy methods it is necessary to use higher-order scheme in time. In this paper we use the Runge–Kutta scheme of the third order [1].

In the case when viscosity and heat conductivity are vanishing, it is possible to show that equations (8) can be obtained by sequential projection of initial data on the space of piecewise-polynomial functions E_P and by the subsequent solution of Riemann problem on the cell's faces and that these solution can be used as the new initial data at the time moment $t + \Delta t$.

Thus, we consider the problem with the initial data in the form of piecewise-polynomial function $U_p\left(x, 0\right)$. If during some time period Δt_0 the exact solution of this problem is known, then, receiving it at time moment Δt_0, it can be again projected to E_P, and so on. As is known, from works [16, 17] the exact solution with piecewise-polynomial data for sufficiently small values Δt can be obtained with accuracy $O\left(\Delta t\right)$ as the solution of Riemann problem with the data to the left and to the right of the discontinuity. It is known that in the multidimensional case around the vertices and the edges of the cells a more complex problem arises which has no exact solution at the moment. However, due to the finite speed of propagation of perturbations for hyperbolic problems, the area of influence of these data near these points and edges does not exceed $c\Delta t$, where c is the maximum speed of propagation of disturbances. In the limit $\Delta t \to 0$ we obtain the function $U_p(\boldsymbol{x}, t) = \sum_{k=0}^{D_P} U_{kj}(t)\,\phi_{kj}(\boldsymbol{x})$. Coefficients $U_{kj}(t)$ are calculated using the following equations:

$$\frac{d}{dt}U_{kj}(t)\left(\int_{T_j}\phi_{kj}^2(\boldsymbol{x},t)dx\right) - \int_{T_j}\left(\frac{\partial\phi_{kj}(\boldsymbol{x},t)}{\partial x}F_x(U(\boldsymbol{x},t)) + \frac{\partial\phi_{kj}(\boldsymbol{x},t)}{\partial y}F_y(U(\boldsymbol{x},t)) + \right.$$
$$\left. + \frac{\partial\phi_{kj}(\boldsymbol{x},t)}{\partial z}F_z(U(\boldsymbol{x},t)) + \frac{\partial\phi_{kj}(\boldsymbol{x},t)}{\partial t}U(\boldsymbol{x},t)\right)dx +$$
$$+ \oint_{\partial T_j}\left(F_x^*(U(\boldsymbol{x},t))n_x + F_y^*(U(\boldsymbol{x},t))n_y + F_z^*(U(\boldsymbol{x},t))n_z\right)\phi_{kj}(\boldsymbol{x},t)\,d\sigma = 0, \quad (16)$$

where the flow values $F_x^*(U(\boldsymbol{x},t))$, $F_y^*(U(\boldsymbol{x},t))$, $F_z^*(U(\boldsymbol{x},t))$ are obtained as a result of Riemann problem solution with the initial data $U_p^+(\boldsymbol{x},t)$ and $U_p^-(\boldsymbol{x},t)$ on the left and right sides of discontinuity. It should be noted that, in the case when $\phi_{kj}(\boldsymbol{x},t)$ doesn't depend on time t, equations (16) coincide with expressions (7)-(10) if Godunov fluxes [8] are used as flow functions. Thus, discontinuous Galerkin schemes can be regarded as Godunov method extension [4, 5] to piecewise-polynomial functions.

3. Program realization of DG method

As is known, DGM involves significant computational complexity, and this brings up the question of ensuring the most effective use of all the computational capacity available. In order to speed up the calculations, the grid-operator approach to the programming of mathematical physics problems [18] has been applied while creating the computational module. This method is aimed at maximizing parallelization on shared memory and facilitates programs porting to parallel architectures, such as NVidia CUDA and Intel Xeon Phi. It also makes possible compact encoding of mathematical formulas.

In order to use operator method, it is necessary to clearly identify the parts, which may be designed in the form of operators, in general algorithm. In DGM they are volume integration and calculation of fluxes through cell faces and limiting. Their precise mathematical formulation has allowed to implement them relatively easily and very efficiently [19, 20].

Let's illustrate the approach feasibility on an example of volume integrals calculations in equations (8). The expansion coefficients of $\boldsymbol{U}_h$ by basis functions serve to ensure the input of the corresponding procedure. Domain-specific language (DSL) with simple grammar is implemented by means of meta-programming features of C++ (using expression templates), and it allows to input the expressions, which will be later used for computations, as the integration function parameters. Finally, the program appears as follows:

```
I1ev = volume_integral(rho_u, rho_v, rho_w)(U);
I2ev = volume_integral(rho_u * u + p, rho_u * v, rho_u * w)(U);
I3ev = volume_integral(rho_v * u, rho_v * v + p, rho_v * w)(U);
I4ev = volume_integral(rho_w * u, rho_w * v, rho_w * w + p)(U);
I5ev = volume_integral((E + p) * u, (E + p) * v, (E + p) * w)(U);
```

All the other operators are implemented in the same manner. It should be noted here, that according to DGM, surface integrals of flux functions through the particular cell face are calculated twice. However, it doesn't contradict the operator programming method and doesn't significantly degrade the performance.

IOP Conf. Series: Journal of Physics: Conf. Series **815** (2017) 012015 doi:10.1088/1742-6596/815/1/012015

Thus, one can get simply readable and effective numerical code portable to various parallel architectures.

4. Some examples of calculations using DG method

This section illustrates numerical results of calculations of several model problems using numerical method and its software implementation presented above.

4.1. Flow past a triangular prism

One of such examples may be taken in the form of the problem of shock wave interaction with a stationary obstacle, such as a triangular prism. In this case, a complex pattern is formed by secondary shock waves and weak discontinuities, as well as the result of their interaction with each other and with vortices formed in the flow at the base of the prism. A mesh with a total number of about 10^7 cells has been used in calculation. Four basis functions have been used to represent the mesh function in the cell. Numerical domain has the size of $L_x = 0.3$ m, $L_y = 0.15$ m, $L_z = 0.1$ m, the prism apex is located at the distance of $L_V = 0.1$ m from the left border of the domain on symmetry axis $y = L_y/2$, the prism base $b = 0.02$ m and half apex angle is 30°. The background pressure was 0.05 MPa, temperature – 300 °K, Mach number of primary shock wave was 1.3, specific heats ratio $\gamma = 1.4$.

Figure 1 shows the numerical results of the corresponding calculation at different time moments. The comparison of the resulting flow structure with experimental shadowgraphs, see, for example, [21], suggests their good agreement. One of the inconsistencies lies in the lack of the secondary vortices, which may be due to the insufficient resolution of the used numerical grid.

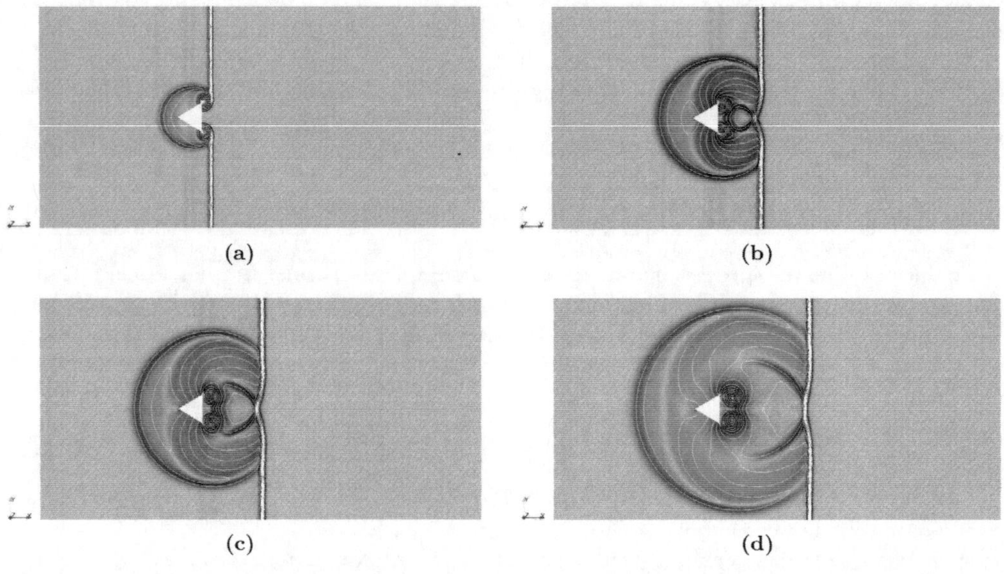

Figure 1. Numerical Schlieren with marked density contours illustrating the problem of the shock wave interaction with a solid triangular prism at time moments t = 53, 102, 130 and 172 μs.

Another exponential problem is the modeling of the supersonic inviscid flow past another triangular prism [22], which this time allows to assess numerical results in a quantitative sense. In this case, oblique bow shock wave is formed and its slope angle β to the axis of symmetry can be calculated analytically. Also a characteristic pattern of shock waves reflected from the boundaries is formed, as well as the result of their interaction. Initial setup parameters for this problem are different, in particular, $L_x = 0.12$ m, $L_y = 0.03$ m, $L_z = 0.01$ m, $L_V = 0.005$ m, the height of the prism is $l = 0.011$ m. The density distribution of the steady flow for half apex angles of the prism of 20° is shown below, on figure 2. The numerical value of the angle β for

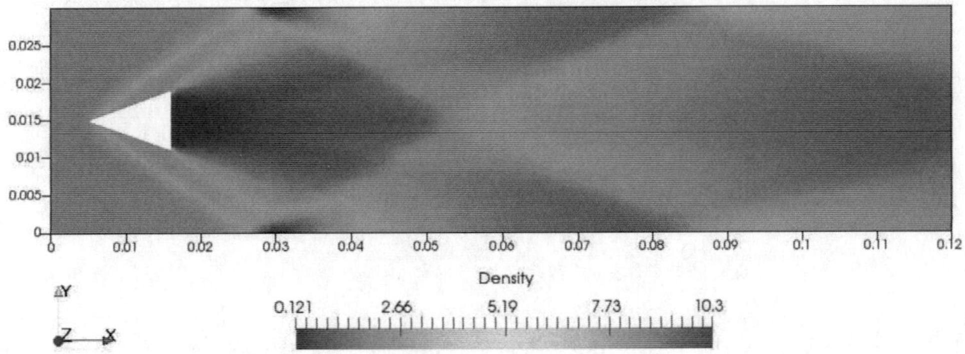

Figure 2. Density distribution in the supersonic flow past a triangular prism with half apex angle of 20°.

the considered option is 34.9° and its theoretical value – 34.58°, indicating a good agreement of the numerical experiment with analytical data within 1.5%.

4.2. Flow in the compression corner

In this section a more complex problem which illustrates the structure of the supersonic separated flow and shock waves interaction with the boundary layer is being considered. According to the experiment described in [23], the flow in the compression corner has been modelled. Geometric configuration of the solid body model with sizes in millimeters is shown on figure 3. It consists of a horizontal plate with a ramp which is inclined at 30°. The following wind tunnel flow

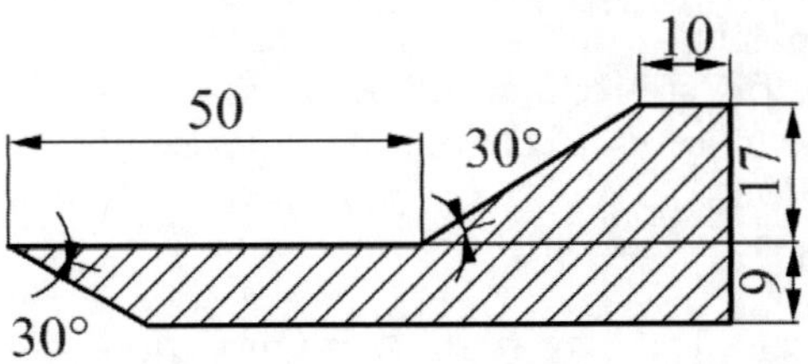

Figure 3. Configuration of the solid body model past which the flow pattern is investigated.

parameters have been used: Mach number of 6.01, pressure $p_{0\infty} = 9.72 \cdot 10^5$ Pa, temperature $T_{0\infty} = 380$ °K and Reynolds number calculated using the model length $L = 50$ mm as a spatial scale $\mathrm{Re}_L = 6 \cdot 10^5$. The calculations have been performed for the angle of attach $\alpha = 0°$ using unstructured tetrahedral mesh with $\sim 10^6$ cells. The viscous and thermal conductivity have been taken into account. Dynamic viscosity coefficient was calculated according to Sutherland's law.

At high flow velocity, shock waves, which have been formed over the separation zone, interact with the shock wave at the reattachment point which leads to the appearance of a complicated flow structure affecting the shear layer above the region of the reverse flow. Numerical experiments have been aimed at reproducing this pattern. As shown on figure 4a,

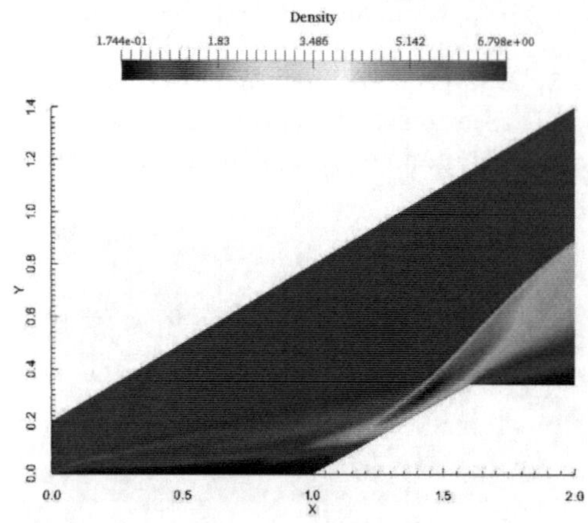

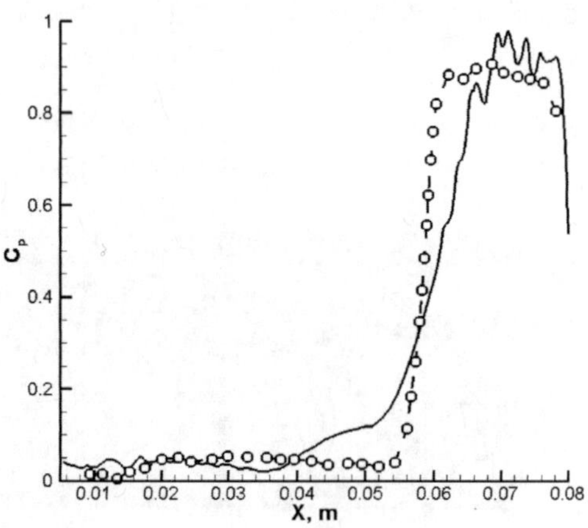

Figure 4. (a) Density distribution of the steady flow; (b) pressure coefficient over the symmetry axis of the solid body model in the numerical calculation (solid line) and in the experiment (dashed line with circles).

all the waves, as well as the separation and reattachment points, can be clearly evidenced together with the shear layer above the region of the reverse flow. The distribution of the pressure coefficient C_p on the wall along the symmetry axis of the model, shown on figure 4b, qualitatively agrees with the numerical and experimental results given in [23]. It should be noted that the boundary layer exceeds the experimental one, but this fact can be attributed to the use of a coarse mesh in calculations. The same reason can be given for some smoothing of C_p profile at the location of the ramp mounting.

5. Conclusions

In this paper numerical simulations of hypersonic flow past various solid bodies have been carried out. The interaction of the considered flow with the obstacle generates a set of shock and rarefaction waves, numerous Mach legs, triple points as well as vortices that develop as time goes and interact with each other. In this case the steady flow is of particular interest. Despite the fact that computations performed have been based on coarse mesh, their results are in good agreement with the experimental and numerical ones and the complex flow pattern is accurately captured. This fact, in turn, confirms the effectiveness of DGM application to solving aerodynamic problems in areas with complex geometry.

The use of the meta-programming of C++ language has allowed to perform part of the calculations at compilation stage, and thus to reduce overall program execution time, while the use of the operator approach has permitted to create an efficient software implementation of the algorithm, which can be easily ported to the modern parallel computing architecture, including graphics accelerators.

Acknowledgments

The research has been performed under the financial support of Russian Science Foundation, grant No. 16-11-10033.

References

[1] Cockburn B 1998 *An introduction to the Discontinuous Galerkin method for convection-dominated problems* vol 1697 (Springer Berlin Heidelberg) chap Advanced Numerical Approximation of Nonlinear Hyperbolic Equations, pp 150–268

[2] Arnold D N, Brezzi F, Cockburn B and Marini L D 2001 *SIAM Journal on Numerical Analysis* **39** pp 1749–1779

[3] Pani A K and Yadav S 2011 *Journal of Scientific Computing* **46** pp 71–99

[4] Ladonkina M E and Tishkin V F 2015 *Doklady Mathematics* **91** pp 189–192

[5] Ladonkina M E and Tishkin V F 2015 *Differential Equations* **51** pp 895–903

[6] Bassi F and Rebay S 1997 *Journal of Computational Physics* **131** pp 267–279

[7] Ladonkina M E, Neklyudova O A, Tishkin V F and Utiralov D I 2014 The no-slip boundary conditions for discontinuous Galerkin method Preprint 32 *Keldysh Institute of Applied Mathematics of RAS*

[8] Godunov S K 1959 *Mat. Sb. (N.S.)* **47(89)** pp 271–306

[9] Rusanov V V 1962 *USSR Computational Mathematics and Mathematical Physics* **1** pp 304–320

[10] Lax P D 1954 *Communications on Pure and Applied Mathematics* **VII** pp 159–193

[11] Mysovskikh I P 1981 *Interpoliatsionnie kubaturnie formuli* (M.:Nauka)

[12] Li B Q 2006 *Discontinuous Finite Elements in Fluid Dynamics and Heat Transfer* (Springer-Verlag London)

[13] Ladonkina M E, Nekludova O A and Tishkin V F 2012 Research of the impact of different limiting functions on the order of solution obtained by RKDG Preprint 34 *Keldysh Institute of Applied Mathematics of RAS*

[14] Ladonkina M E, Neklyudova O A and Tishkin V F 2013 *Mathematical Models and Computer Simulations* **5** pp 346–349

[15] Ladonkina M E, Nekludova O A and Tishkin V F 2013 The high order limiter for RKDG on triangular meshes Preprint 53 *Keldysh Institute of Applied Mathematics of RAS*

[16] Teshukov V M 1980 *Journal of Applied Mechanics and Technical Physics* **21** pp 261–267

[17] Men'shov I S 1990 *USSR Computational Mathematics and Mathematical Physics* **30** pp 54–65

[18] Krasnov M M 2015 *Matem. Mod.* **27** pp 109–120

[19] Krasnov M M and Ladonkina M E 2016 Discontinuous Galerkin method on three-dimensional tetrahedral meshes. the usage of C++ template metaprogramming Preprint 24 *Keldysh Institute of Applied Mathematics of RAS*

[20] Krasnov M M, Ladonkina M E and Tishkin V F 2016 Discontinuous Galerkin method on three-dimensional tetrahedral meshes. the usage of the operator programming method Preprint 23 *Keldysh Institute of Applied Mathematics of RAS*

[21] Chang S M and Chang K S 2000 *Shock Waves* **10** pp 333–343

[22] Chaudhuri A, Hadjadj A and Chinnayya A 2011 *Journal of Computational Physics* **230** pp 1731–1748

[23] Zapryagaev V, Kavun I and Lipatov I 2013 *Progress in Flight Physics* **5** pp 349–362

Supercomputer modeling of flow past hypersonic flight vehicles

M K Ermakov and I A Kryukov

A. Ishlinsky Institute for Problems in Mechanics of the Russian Academy of Sciences, Vernadsky Prospect 101(1), Moscow, 119526, Russia

E-mail: ermakov@ipmnet.ru

Abstract. A software platform for MPI-based parallel solution of the Navier-Stokes (Euler) equations for viscous heat-conductive compressible perfect gas on 3–D unstructured meshes is developed. The discretization and solution of the Navier-Stokes equations are constructed on generalized S.K. Godunov's method and the second order approximation in space and time. Developed software platform allows to carry out effectively flow past hypersonic flight vehicles simulations for the Mach numbers 6 and higher, and numerical meshes with up to 1 billion numerical cells and with up to 128 processors.

1. Introduction

For the X–43 model of hypersonic flight vehicle, that made three demonstration unmanned flights in the NASA Hyper–X program, the supercomputer calculations of a hypersonic flow past the model have been made. Supercomputer codes development of spatial modeling is one of the stages in a creation of hypersonic flight vehicle virtual model [1]. This virtual model has to provide physical description and modeling of a hypersonic direct-flow propulsion jet engine and a spatial flow past the model for flight velocities within Mach numbers $4 \leq M \leq 15$ and heights 20÷50 km. A need of the virtual model for hypersonic flight vehicle development is connected with essential technical difficulties and high material inputs on carrying out tests in ground wind tunnels and flight experiments, a difficult spatial structure of interacting among themselves and with boundary layer shock waves, and also with a difficulty of mathematical model that includes, in perspective, physic-chemical and radiation gas-dynamics processes. The purpose of this initial stage of work consisted in development of an effective gas-dynamic parallel supercomputer code and accomplishment of test calculations of the flow past the provided in publications hypersonic flight vehicle X–43.

2. Virtual model of a hypersonic flight vehicle

Virtual model of a hypersonic flight vehicle proposes to create spatial 3–D object model, to build unstructured numerical mesh around the model, to carry out calculations and to visualize results. A preparation of spatial 3–D model for a hypersonic flight vehicle is made by CAD systems and is described precisely in [2]. A surface of the model is rather complicated; it includes small radius of curves and sharp edges of adjacent surfaces. It is generally connected with large manual works to build regular mesh. An approach based on molecular dynamics [3] allowed avoiding losses due to automatic construction of unstructured meshes. The constructed unstructured mesh is converted into graph that is used for supercomputer modeling.

3. Software platform for parallel calculations

The description of approaches of the organization of calculations for grid problems for parallel calculations is present in [4]. In this paper own approach which doesn't demand reorganization of logic of work and structures of data of the consecutive program is realized. Performance of a technological chain of work on a flow of objects of irregular shape requires carrying out the following preliminary operations:

1) Creation of a link graph between numerical cells.
2) Partitioning set of numerical cells into subareas. This step was carried out with use of a widespread open software package METIS.
3) Formation of the control data determinating calculations for each of numerical subareas.

The scheme of the organization of works in consecutive and parallel options of a code is presented in figure 1. One could see that in a cycle of calculation of a time step, a change come down to splitting calculations for "internal" and for "external" groups of cells, and also to a call of the subroutines (functions for C/C++) providing start of asynchronous compact operations of exchange and expectation of the end of operations. Additional operations are marked in red color. To provide universality of data exchange between areas the structures of sparse matrixes representation are used.

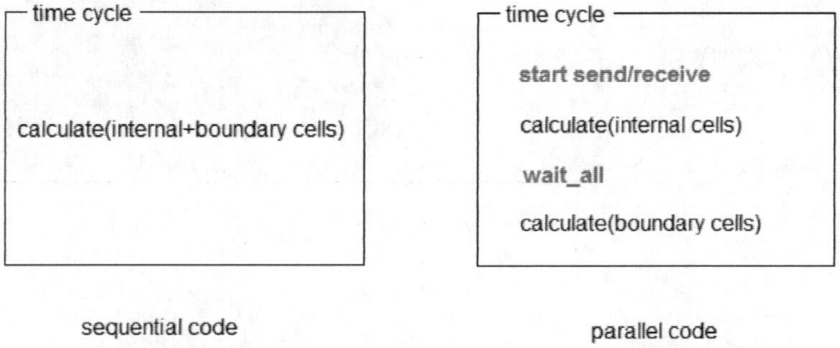

Figure 1. Sequential and parallel schemes of calculation flow.

The similar approach has been used earlier in the software platform INMOST [4].

As the existing tools of settlement grids are limited by value of about 30 million cells, that for receiving more detailed grids is used the subdivision method consisting in splitting each control volume into 8 smaller in such a way that splitting the next volumes are coordinated among themselves on the common side (figure 2). For each face splitting looks in such a way that the middle of the edges of a face connect, forming on each face four new triangles instead of previous one. So, if the middle of the edges of the triangle formed by tops 1, 2 and 3, to denote as 12, 13 and 23, then new triangles are (1, 12, 13), (12, 2, 23), (13, 23, 3) and (12, 13, 23). In a tetrahedron, unlike a boundary triangle, unique subdivision on smaller isn't present any more. Uniquely, however, the new tetrahedrons containing tops 1, 2, 3 and 4 are determined: (1, 12, 13, 14), (2, 12, 23, 24), (3, 13, 23, 34) and (4, 14, 24, 34). Other tetrahedrons can be set, for example, as (12, 34, 13, 14), (12, 34, 14, 24), (12, 34, 23, 13) and (12, 34, 23, 24).

By subdivision of each cell into 8 more small we succeed to build numerical meshes for hypersonic flight vehicle X–43 with about 240 million cells, 700 million cells and 2 billion cells.

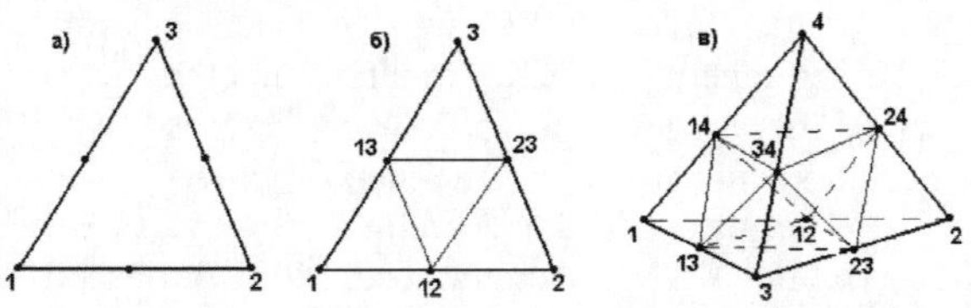

Figure 2. Triangle and tetrahedron partitioning schemes.

4. Region partitioning

The examples of spatial region partitioning for hypersonic flight vehicles X–43 and X–51, for 128 and 64 cuts correspondingly, are presented in figures 3, 4. The partitioning has been made by widespread open software METIS on PC at the preliminary preparation stage for supercomputer calculations. The RAM requirements at partitioning stage do not exceed requirements at other preliminary stages.

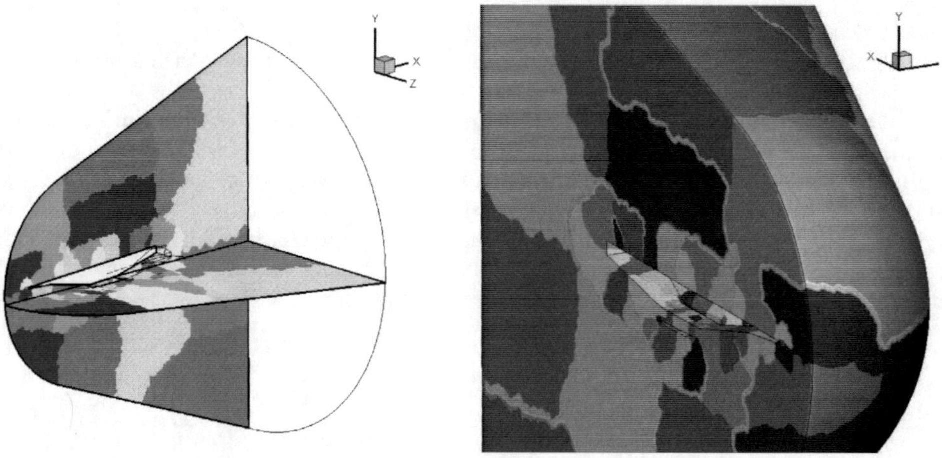

Figure 3. 128-cut partitioning of numerical region around X–43.

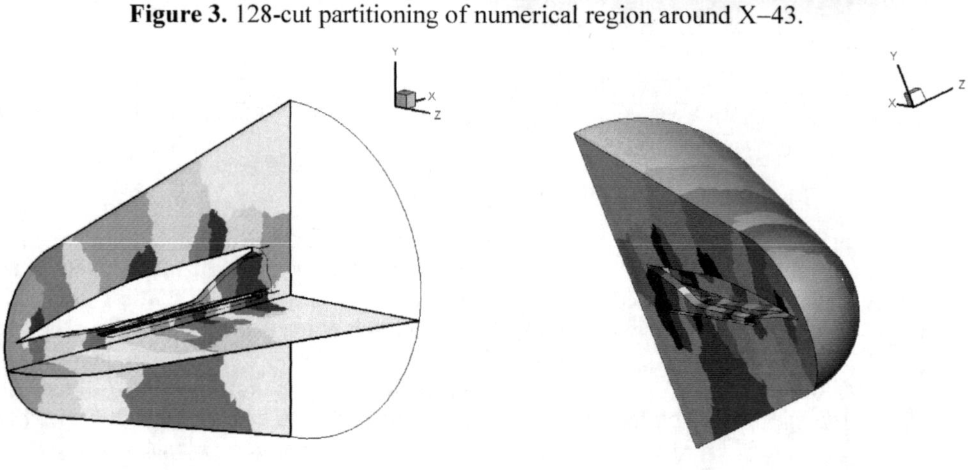

Figure 4. 64-cut partitioning of numerical region around X–51.

5. Supercomputing calculations

The sequential code ug3D was used to process it into parallel version. This code was recently used for a modeling of a hypersonic flow past a model objects in a hypersonic shock tube [5]. The ug3D code models gas flows on the base of the Navier-Stokes and the Euler equations for unstructured spatial meshes on the base of second order generalized S.K. Godunov's method.

The supercomputer calculations carried out for numerical mesh with 240 million cells. On 8 processors the typical calculation time for one time step is about 20 seconds and boundary cells exchange is about 1/20 second. The main additional preliminary processing was held outside parallel system, however it requires essential RAM amounts. The 240 million cells mesh is the maximum mesh for which we were able to check the correct working capacity of software platform. The efficiency of parallelization is above 89 % for up to 128 processor calculations. To build data structures at the preliminary stages for a mesh with 500 million cells a computer with more than 64 GB is needed, and for mesh with 2 billion cells – a computer with at least 256 GB RAM.

The figure 5 presents shock-wave high-resolution structure of the flow of the Mach number 6 past hypersonic flight vehicle X–43, pressure distribution in the symmetry plane, vehicle surface and few sections.

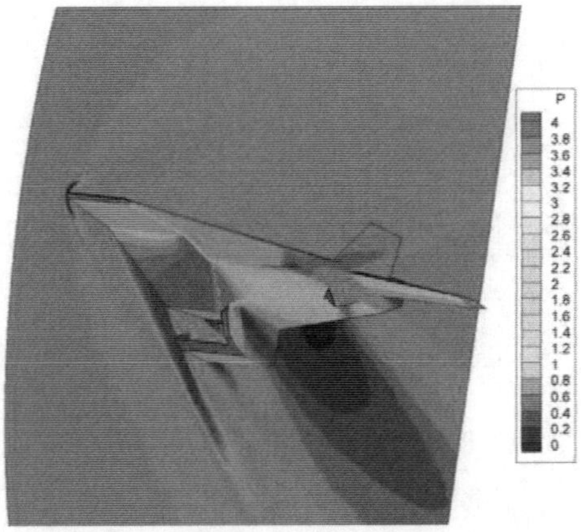

Figure 5. Pressure distribution in the symmetry plane and on the X–43 surface in a flow of the Mach number 6.

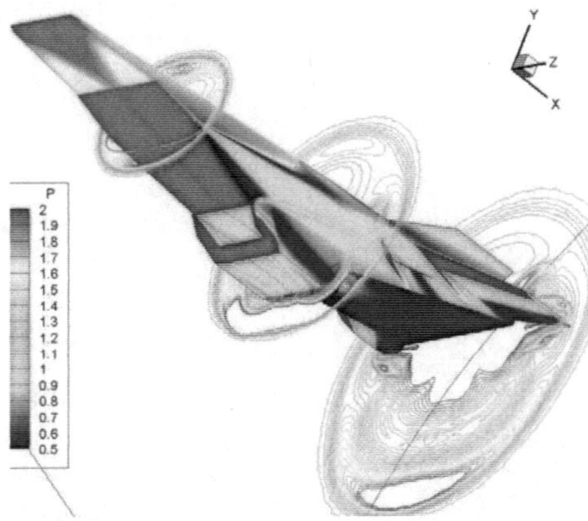

Figure 6. Pressure distribution on the X–43 surface and in some sections in a flow of the Mach number 6.

Combustion in propulsion jet engine has not been considered. The developed supercomputer code is the base for implementation of extended models of flows past hypersonic flight vehicles.

6. Conclusions

The software platform for numerical codes parallelization that does not require to rebuild the data structures and the nucleus of numerical part is described. The platform tools are applied for MPI-based parallelization for the Navier-Stokes (Euler) equations for spatial unstructured mesh calculations. The spatial unstructured meshes for hypersonic flight vehicle X–43 with up to 2 billion cells are built. Parallel code allows to carry out flow past hypersonic flight vehicle calculations with the Mach number 6 and higher and numerical meshes with up to 1 billion numerical cells and with up to 128 processors. The efficiency of parallelization is above 89 % for up to 128 processor calculations. Developed parallel code allows to carry out calculations with higher detailing in boundary layers and sharp gradient regions, to advance to conjugate simulations of integrated configuration and propulsion system, and also to implement optimization in the studied systems. The developed supercomputer code is also the base for implementation of extended models of flows past hypersonic flight vehicles.

Acknowledgments

Authors thank Dr. M.A. Kotov and Dr. A.L. Zheleznyakova for digital surfaces of hypersonic flight vehicles creation and unstructured tetrahedral meshes building for calculations on their basis. The work was performed within the framework of the Program of basic research of the Russian academy of sciences.

References

[1] Zheleznyakova A L and Surzhikov S T 2013 *On the Path of Creation of a Model of Virtual Hypersonic Flight Vehicle. I* (Moscow: A.Ishlinsky Institute for Problems in Mechanics RAS) p 160

[2] Kotov M A 2013 *Physical-Chemical Kinetics in Gas Dynamics* **14** (http://chemphys.edu.ru/issues/2013-14-4/articles/427)

[3] Zheleznyakova A L and Surzhikov S T 2011 *Doklady Physics* **56** pp 385–390
[4] Danilov A A, Terekhov K M, Konshin I N and Vassilevski Yu V 2015 *Supercomputing Frontiers and Innovations* **2** pp 55–66
[5] Kotov M A, Kryukov I A, Ruleva L B, Solodovnikov S I and Surzhikov S T 2014 *AIAA 2014–2657 AIAA Aerodynamic Measurement Technology and Ground Testing Conference*

Turbulent flow over an axisymmetric body with annular cavity

I E Ivanov[2,3], I A Kryukov[1,3], E V Larina,[3] and G S Glushko[1]

[1] A. Ishlinsky Institute for Problems in Mechanics of the Russian Academy of Sciences, Vernadsky Prospect 101(1), Moscow, 119526, Russia
[2] Lomonosov Moscow State University, Moscow, 119991, Russia
[3] Moscow Aviation Institute (National Research University), Moscow, 121993, Russia

E-mail: kryukov@ipmnet.ru

Abstract. The paper concerns the numerical simulation of turbulent flow over axisymmetric body with an annular cavity. The influence of the ratio of the length to the cavity depth on the flow structure in the cavity was numerically studied in the case of supersonic flow. The open or closed type of the flow was obtained depending on the ratio of the length to the cavity depth. Comparison with experimental data [9, 10, 13] shows a good agreement between calculated and experimental results. The phenomenon of gas dynamic hysteresis for the implementation of the open - closed flow type studied experimentally in [10] was numerically reproduced. Also it has been shown that experimental conditions [13] also may cause this type of hysteresis.

1. Introduction

The practical interest in a flow over an axisymmetric body with an annular cavity is caused by a rather wide use of such constructions in aerospace applications. Just as in a rectangular cavity there can be two types of flows depending on the ratio of the annular cavity length to the depth of that: an open flow type (in a short cavity) or a closed flow type (in a long cavity). In the first case a single annular recirculating flow emerges in the cavity, in the second case two separate flows appear in the opposite corners of the cavity. The intermediary state between these two types, where flow may also switch from one to the other, is termed as transitional. In addition the type of cavity flow is affected by the Mach number, the Reynolds number and the nature and state of the approach boundary layer. Many researchers pay extra attention to the possible gas dynamic hysteresis in the flow over cavities related to the reorganization of the separated flow regime. For example, the gas dynamic "open-closed" type hysteresis was experimentally found in the supersonic flow over an axisymmetric body with an annular rectangular cutout on a cylindrical surface [9, 10]. A supersonic flow over an axisymmetric body and the emergence of open and closed cavity flows were experimentally and numerically analyzed in [13, 14]. For comparison some results of numerical simulations for plane and three-dimensional cavity flows can be found in [7] and some experimental results can be found in [8].

2. Mathematical model and numerical method

The unsteady axisymmetric RANS set of equations closed by the two-equations k-ε turbulence model is used as a mathematical model. The version of the k-ε turbulence model used in this article was constructed in [3, 4] and reveals the influence of the non-equilibrium between the generation and dissipation of turbulence and compressibility. This model was tested for a variety of supersonic flows and showed rather satisfactory results [5, 6].

The method described in [1, 2] was used for numerical solving the set of equations mentioned above. A high-resolution Godunov-type scheme based on MUSCL variable reconstruction and slope limiters was used as a basic scheme for approximating the inviscid part of the set of equations. The viscous terms of the set of equations were approximated by an unstructured grid analogue of the central difference method; all source terms were approximated by the explicit-implicit method [1]. The no-slip conditions were specified on the wall, the symmetry conditions were specified in the symmetry line and the non-reflective boundary conditions were specified on the free boundaries.

3. Numerical Simulation of Guvernyuk et al experiment [9, 10]

A supersonic flow with M=3 at zero angle of attack on an axisymmetric body which consists of a sharp cone and a cylinder with an annular cavity was numerically simulated following the experiment described in [9, 10]. The annular cavity has a rectangular section as shown in figure 1. The cone semi-apex angle β=20°, distance between the cone body and the cavity l=30 mm and the cylinder diameter D=45 mm and depth of the cavity h=8 mm are all fixed. The cavity "floor" cylinder diameter can be evaluated as $d=D-2h$ =29 mm. The relative cavity length $s=L/h$ may be varied within the $10<s<15$ range due to the L variation (80 mm$<L<$120 mm).

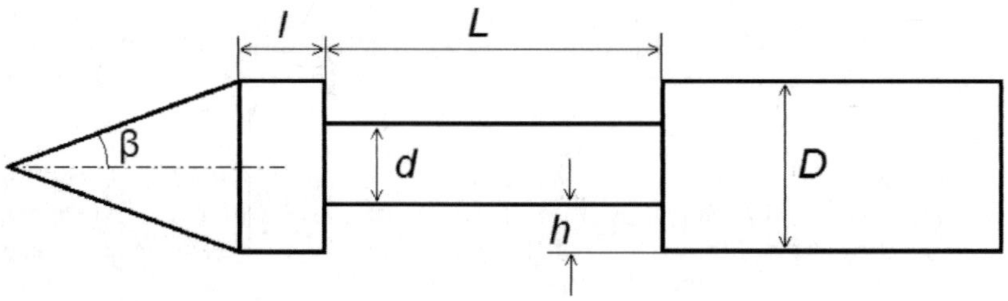

Figure 1. An axisymmetric body with an annular cavity [9, 10]

The computational mesh contains 400×150 cells for regular calculations and 800×300 cells for more accurate ones. The cavity is filled with 160×30 and 320×60 cells respectively.

In numerical calculations presented here as in the experiments [9, 10], the open-flow type was always obtained for small relative cavity length values $s<10$. In this case the flow detaches from the cavity front lip and reattaches to the right vertical cavity wall near the rear edge. A recirculating flow consisted of one large and possibly several small vortexes in the corner regions with a subsonic flow regime is inside the cavity. The recirculating flow region is separated from the external flow by a mixing layer.

A supersonic ($M=3$) turbulent ($Re_\infty=1.7\times10^6$) flow over the axisymmetric body with the cavity was calculated for the relative cavity length range of $s=10\div16.25$ or for the cavity length range of $L=80\div130$ mm (see figure 1). The incident flow parameters are $P_\infty=16770$ Pa and $T_\infty=96.43$°K. Figures from 2 to 4 show calculated results for $L=81$ mm ($s=10.125$).

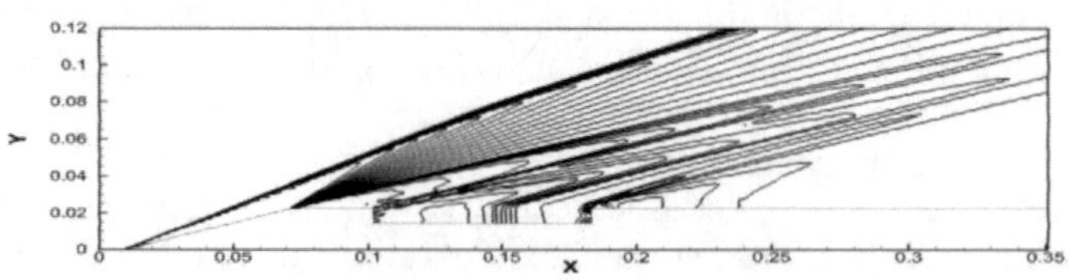

Figure 2. Pressure contour lines for $L=81$ mm.

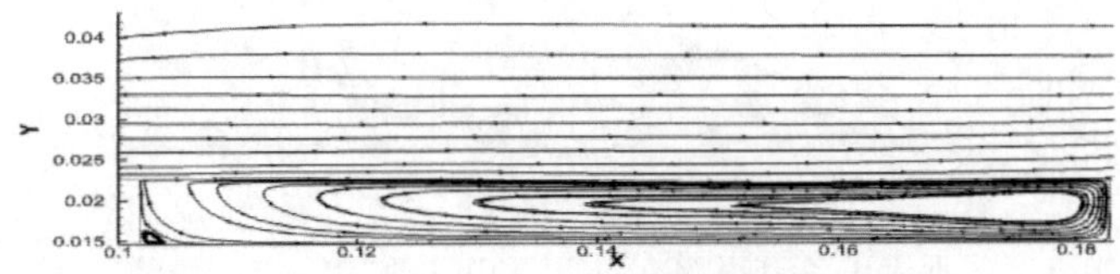

Figure 3. Streamlines inside the cavity for $L = 81$ mm.

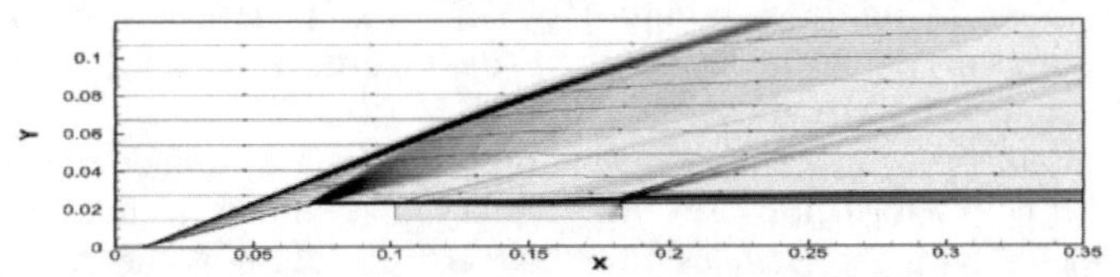

Figure 4. Density gradient and streamlines for $L = 81$ mm.

Figure 2 shows the pressure contour lines for $L = 81$ mm. Whole computational region has a size 0.4×0.2 m. Therefore most of the computational region is shown in the figure. Streamlines inside the cavity and in some vicinity of cavity are shown on figure 3. Figure 4 contains the numerical schlieren image (density gradient contours) and streamlines. From these figures the open-flow type with one large vortex and one small vortex in the forward corner can be easily seen.

When the relative cavity length exceeds a certain critical value $s > s_2$, there is the closed-flow type in the cavity. A closed type flow is characterized by a presence of two non-interacting or weakly interacting recirculating zones; the first recirculating zone is behind the front face of the cavity and the other zone is in vicinity of the rear face. The zones are separated by the external flow along the bottom of the cavity. The external flow is characterized by presence of strong shock waves and expansion waves caused by supersonic flow over the corner recirculating zones.

The flow over the body with annular cavity was calculated for the cavity length L=110 mm ($s = 13.75$). The results are shown in figures 5–7.

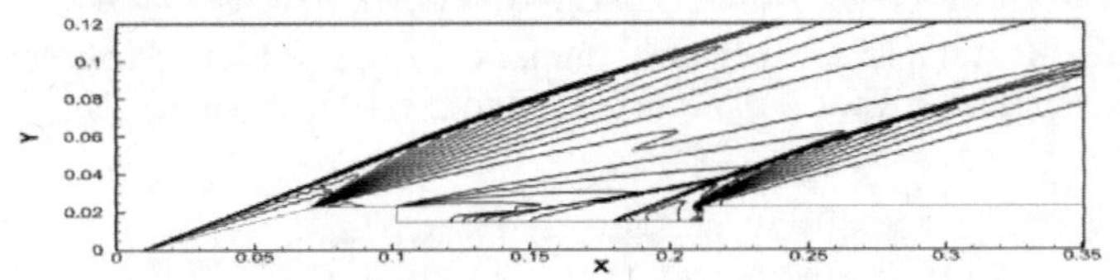

Figure 5. Pressure contour lines for $L = 110$ mm.

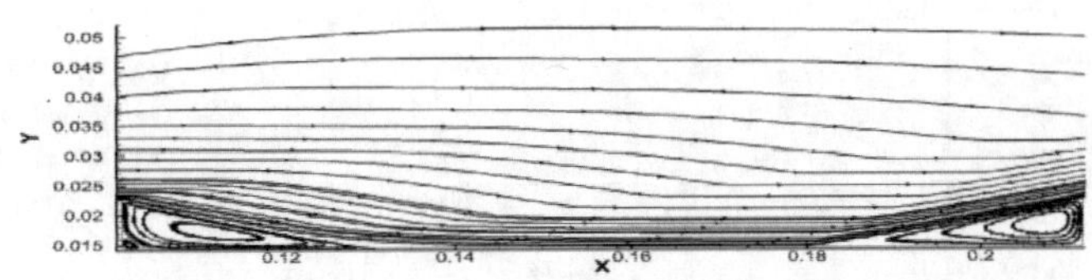

Figure 6. Streamlines inside the cavity for $L = 110$ mm.

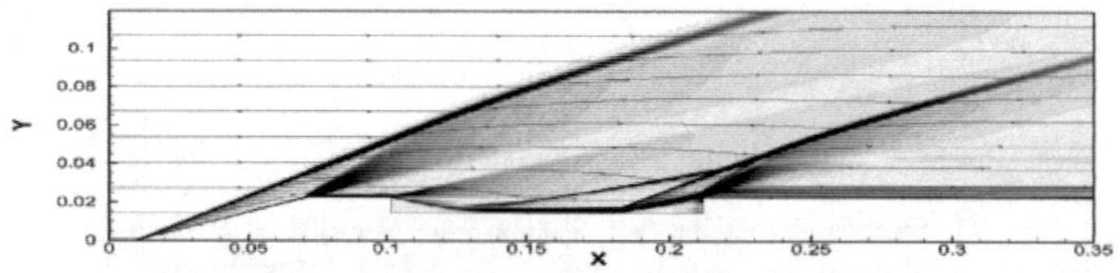

Figure 7. Density gradient and streamlines for $L = 110$ mm.

Figure 5 shows the pressure contour lines for L=110 mm. Streamlines inside the cavity and in some vicinity of cavity are shown in figure 5. Figure 7 contains the numerical schlieren image (density gradient contours) and streamlines. In these figures the closed-flow type with two vortexes in the opposite corners of the cavity can be easily seen.

As mentioned before, the open-flow cavity type is realized for small relative cavity length values, while the flow inside the cavity formed by several vortexes is separated from the external flow by a shear layer that spans the cavity and reattaches onto the rear face at or near the top corner (see figures 3 and 8a). The shear layer is curved towards the external flow. As s increases the mixing layer bends towards the cavity bottom while the vortex flow inside the cavity undergoes deformation: the main vortex looks to take shape of a butterfly, and the external flow gets closer to the cavity bottom (see figures 6 and 8b).

As a result of numerical simulation the critical values of the relative cavity length ($s_1 = 10.2$ ($L = 81.5$ mm) and $s_2 = 13.8$ ($L = 110.5$ mm)) when the flow structure changes from the open to the close type and vice versa were evaluated. Between the s_1 and s_2 values the flow type can be either open or closed depending on the flow development history.

Figure 8. Comparison between numerical (top) and experimental schlieren (bottom) for L=81 mm (left) and L=110 mm (right).

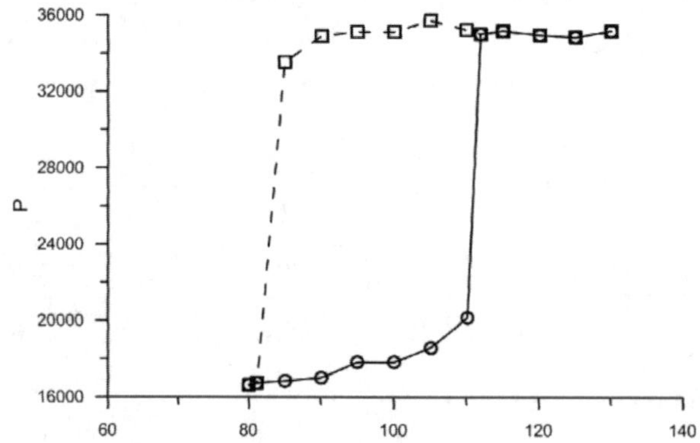

Figure 9. Dependence pressure P on the cavity length L in the rear corner of the cavity.

The presence of a gas dynamic hysteresis is illustrated in figure 9 by a dependence of the pressure, measured near the rear cavity corner, on the cavity length. The solid curve was obtained at an increasing s and the flow type changes from open to closed. The dashed curve was obtained at a decreasing s; in this case the flow type changes from closed to open. Thus, the gas dynamic "open-closed" type hysteresis in the supersonic flow over an axisymmetric body with a rectangular annular cutout on the side surface of a cylindrical part of the body forms while the cutout length changes.

These results satisfactorily (accurate to 0.5 mm by transition points) match the experimental results from [10].

4. Numerical simulation of Mohri and Hillier experiment [13]

In this case the supersonic flow (M=2.2 at a zero angle of attack) over an axisymmetric body consisting of a sharp cone and a cylinder with a annular cutout was numerically simulated. The difference between this case and the previous one is in the fact that here the cavity starts right after the cone. Compare figures 1 and 10.

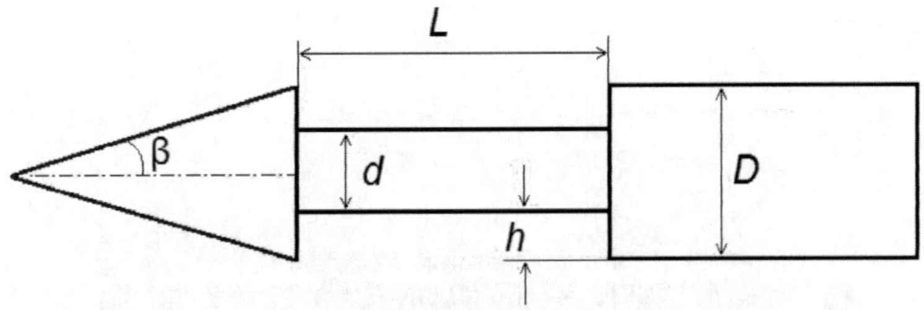

Figure 10. An axisymmetric body with an annular cavity [13].

The semi-apex angle is $\beta = 16°$, the cavity begins just behind the cone, the cavity 'floor' cylinder diameter is d=9 mm, the cylinder diameter is D=15 mm, the cavity depth is h=3 mm, $D = d + 2h$. The relative cavity length $s = L/h$ varies in the $1.67 < s < 14.33$ range due to the L variation (3 mm $< L <$ 41 mm). The incident flow parameters are $P_\infty = 25250$ Pa, $T_\infty = 146.34$°K, $\mathrm{Re}_\infty = 0.87 \times 10^6$.

Figures from 11 to 13 show calculated results for L=37 mm (s=12.33). These figures show the pressure contour lines, streamlines inside the cavity and the numerical schlieren image (density gradient contours). It can be seen in figure 12, the flow in the cavity at L=37 mm has two recirculating zones, the first one is formed by the boundary layer separating from the front cavity lip and reattaching to the cavity 'floor' (the flow behind the backward step). The second recirculating zone is located between the separation point of the boundary layer from the cavity 'floor' and the reattachment point on the cavity back face (the flow over the forward step). These two recirculating zones are separated by the external flow along the cavity 'floor' and don't noticeably influence one another at large cavity length L values.

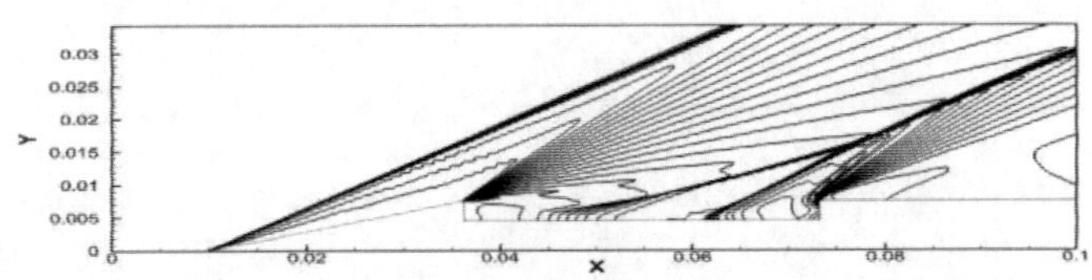

Figure 11. Pressure contour lines for L=37 mm.

For comparison the numerical density contours (upper part of figure) and experimental schlieren image (bottom part of figure) are shown together in a single scale in figure 14. A good agreement in shape and position of all main discontinuities in the flow can be seen.

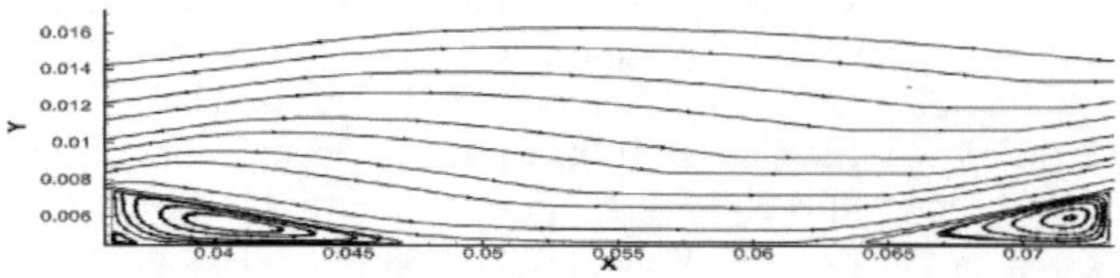

Figure 12. Streamlines inside the cavity for $L = 37$ mm.

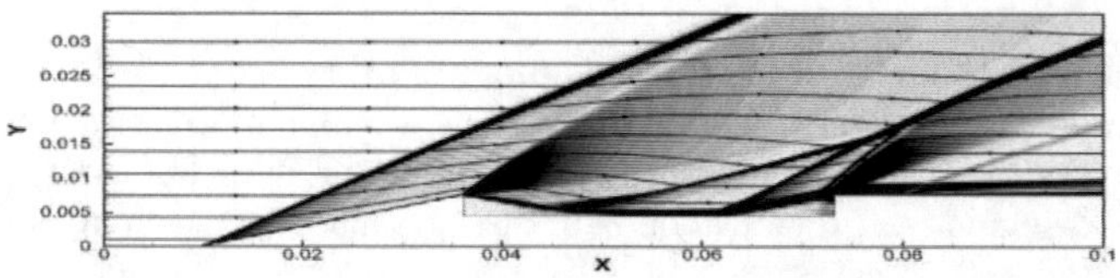

Figure 13. Density gradient and streamlines for $L = 37$ mm.

Figure 14. Comparison between CFD density destribution (top) and experimental schlieren [13] (bottom) for $L = 37$ mm.

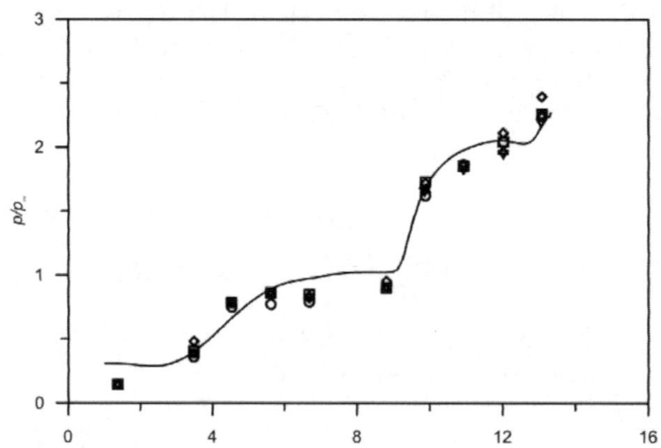

Figure 15. Experimental [13] and computational surface pressure distribution on the cavity floor for $L = 37$mm.

5. Analysis of calculated results and comparison with the experiment

The comparison of calculated and experimental data in figures 14 and 15 shows that the results are matching well in flow structure (see figure 14), in detached zone length, in shock waves' and expansion fans' position. Also the local flow parameters are matching satisfactorily in relative pressure distribution on the cavity floor. The points of flow attachment and separation from the cavity bottom were predicted rather well by the location of pressure increase areas. The pressure level in the external flow near the cavity bottom is 5÷8% higher in the simulation than in the experiment. The largest pressure level difference is registered in the separation zone behind the forward lip of cavity at a maximum 3 mm distance behind the cone: in that area the measured relative pressure is $P/P_\infty < 0.2$, compared to the calculated $P/P_\infty = 0.28 \div 0.3$. This difference can be explained by the following way: in the experiment, as stated in [13], the flow over the conical fore-body of the axisymmetric body is laminar; a tripping disk was inserted behind the cone for flow turbulization. The disk adjoined directly on the front corner of the cavity and had a diameter slightly bigger than the diameter *D*. As a result a turbulent flow mode with a corresponding flow structure (figure 14) was set in the cavity. In the numerical simulation due to the *k-ε* model peculiarities the laminar-turbulent transition happens very closely to the conical fore-body apex and thus the flow over almost the whole conical fore-body is turbulent. The difference in the flow mode in calculation and experiment and also in the turbulence parameters in the cavity can lead to different levels of calculated and experimental pressure in the front separation zone (base pressure).

Experimental and calculated data on supersonic flow over sharp cones was processed and published in the book [11] (in the chapter "Base pressure behind the backward step"). The dependence of base pressure on the Mach number in an incident flow was determined. The relative base pressure takes values $P/P_\infty = 0.31 \div 0.46$ for air flow $M = 2.2$ over a cone with a semi-apex angle of 15°, which slightly exceeds the pressure behind the front face of the cavity in this article's calculation and is a considerably higher than the pressure measured in experiment [13].

The ratio of the mean base pressure to the pressure in the incident flow in the case of a flow with a Mach number $M = 2.2$ and an isentropic index γ is described well by a semi-empirical formula

$$p_b = 0.15 + \left(1 + \frac{\gamma - 1}{2} M^2\right)^{-(\gamma+1)/\gamma}$$

obtained at the Khristianovich Institute of Theoretical and Applied Mechanics (Siberian branch of Russian Academy of Sciences) [12] based on a lot of experimental data in the Mach number range 1.5÷14 [15]. In the case of our calculation for $M = 2.2$ and $\gamma = 1.4$ this formula gives an estimation of base pressure $P/P_\infty = 0.46$, which corresponds to the value shown in [11] at this Mach number.

As a result of the numerical research a gas dynamic "open-closed" flow type hysteresis was also found for the experimental conditions. The hysteresis is close to [9; 10] with critical parameters of relative cavity length $s_1 = 10.5$ and $s_2 = 12.75$. In [13] the gas dynamic hysteresis was not studied experimentally.

6. Conclusion

The supersonic flow over an axisymmetric body with an angular cavity is studied numerically. Two physical experiments for incidental flow Mach numbers $M = 3.0$ and $M = 2.2$ are numerically simulated. The flow has either of an open or of closed type depending on relative cavity length. The hysteresis phenomenon was numerically reproduced for various flow types (open and closed) for both experiments.

Acknowledgement

The work was performed within the framework of the Program of basic research of the Russian academy of sciences and was funded by RFBR grants №16-01-00444a, №16-38-60185, №14–08–01286.

References

[1] Glushko G S, Ivanov I E and Kryukov I A 2010 *Math. Models Comput. Simul.* **2** 407-22

[2] Ivanov I E and Kryukov I A 1996 *Matematicheskoe Modelirovanie* **8** 6 47–55

[3] Kryukov I A 2009 *Vestnik of Moscow Aviation Institute* **16** 2 101-8

[4] Glushko G S, Ivanov I E and Kryukov I A 2010 Turbulence modeling for supersonic jet flows *Physical-Chemical Kinetics in Gas Dynamics* **9** http://chemphys.edu.ru/issues/2010-9/articles/142/

[5] Ivanov I E, Kryukov I A and Larina E V 2013 Numerical simulation of shock wave turbulent boundary layer interaction for compression corner flow *Physical-Chemical Kinetics in Gas Dynamics* **14** http://chemphys.edu.ru/issues/2013-14-4/articles/424/

[6] Ivanov I E, Kryukov I A and Larina E V 2014 Numerical simulation of high-speed viscous intake flow *Physical-Chemical Kinetics in Gas Dynamics* **15** http://chemphys.edu.ru/issues/2014-15-4/articles/240/

[7] Lawson S J and Barakos G N 2011 *Progress in Aerospace Sciences* **47** 186–216

[8] Kotov M A, Ruleva L B, Solodovnikov S I and Surzhikov S T 2014 Investigation of shock wave process of model flow in hypersonic aerodynamic shock tube // *Physical-Chemical Kinetics in Gas Dynamics* **15** http://chemphys.edu.ru/issues/2014-15-3/articles/223/

[9] Guvernyuk S V, Zubkov A F, Simonenko M M, and Shvetz A I 2014 *Fluid Dynamics* **49** 540–6.

[10] Guvernyuk S V, Zubkov A F and Simonenko M M 2014 On the observation of the aerodynamic hysteresis in supersonic flow over a ring cavity on an axisymmetric body *Advances in the Mechanics of Continuous Media, Collected papers at the Int. Conf. timed to the 75th birthday of Academician V. A. Levin* (Irkutsk : Megaprint) pp 163–8

[11] Lavruhin G N and Popovich K F 2009 *Ajerogazodinamika Reaktivnyh Sopel* vol 2 (Moscow: FIZMATLIT) p 312

[12] Zubkov A I, Garanin A F, Safronov V F, Sukhanovskaya L D and Tretyakov P K 2005 Supersonic flow past an axisymmetric body with combustion in separation zones at the body nose and base *Thermophysics and Aeromechanics* **12** 1-12

[13] Mohri K and Hillier R 2011 *Shock Waves* **21** 175-91

[14] Sinha J 2013 *Advances in Aerospace Science and Applications* **3** 83-90

[15] Lutsky A E, Menshov A S and Khankhasaeva Y V 2014 The use of free boundary method for solving the problem of the flow past moving bodies KIAM Preprint № 93

Influence of Energy Input on the Flow Past Hypersonic Aircraft X-43

Ya V Khankhasaeva, V E Borisov and A E Lutsky

Keldysh Institute of Applied Mathematics, Moscow, Russia

hanhyana@mail.ru, narelen@gmail.com, lutsky@kiam.ru

Abstract. This paper deals with a numerical study of the influence of energy sources on the flow past hypersonic aircraft X-43. Flight mode with M = 6 and angle of attack $\alpha = 0°$, $4°$ with energy deposition in areas around various parts of HA was considered. It is shown that energy input in front of the bow of the HA leads to a significant weakening of the bow shock wave and an increase in aerodynamic efficiency of the vehicle. The results of studies on the impact of energy input in the scramjet intake are also presented.

1. Introduction

In recent decades, research on development of hypersonic aircrafts (HA) using scramjets (supersonic combusting ramjet) for prolonged flight in the atmosphere has greatly intensified. Operating principle of scramjet is similar to that of a conventional ramjet, in which air flow entering through the intake is compressed due to the special shape of the aircraft and is decelerated to subsonic speed. In the ramjet for M > 6, deceleration of the incoming air flow to subsonic speeds leads to the values of temperature and pressure that are problematic because of the restrictions imposed by the thermal strength of the engine combustion chamber structural materials. The airflow in a scramjet is only partially decelerated and remains supersonic throughout the entire engine, therefore temperatures within are kept at a tolerable level and the engine can be used at flight speeds of M > 6. The positive results obtained during the testing of unmanned aircraft X-43A (figure 1) in 2004 within the framework of NASA Hyper-X program [1-4] show promise in the use of scramjet at flight speeds of M > 6.

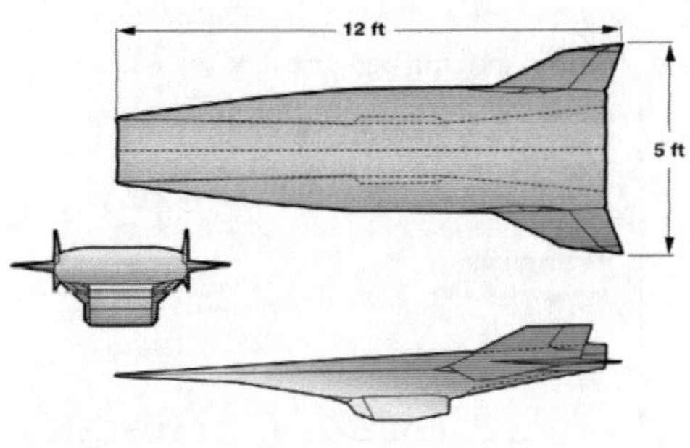

Figure 1. Experimental unmanned hypersonic flying vehicle X–43 [4].

The development of modern aircrafts requires the search and development of new effective tools to manage the characteristics of the gas flow in the vicinity of the aircraft's surface, control the transfer of heat and mass transfer in the boundary layer, reduce surface friction, delay the laminar-turbulent transition, control flow separation, reduce the ignition time and manage the process of combustion of fuel in a ramjet and scramjet engines.

One of the methods for improving the aerodynamic characteristics of prospective aircraft is a controlled action on the oncoming flow. It can be performed in various ways, in particular by using energy input localized in a small closed region. The possibility of remote energy input into a supersonic flow is confirmed in many experiments [5–10]. The high-temperature wake with reduced values of Mach number, total pressure, and velocity is formed behind the energy source, which enables us to vary the flow regime. If the energy source and body are of comparable sizes then the flow around the body is quasi-uniform and drag can be reduced by changing straightforwardly the parameters of the oncoming flow. This oncoming flow requires large energy expense and is impractical. However, energy input even in a relatively small space region can lead to restructure of the bow shock-wave ahead of the body. This method of drag reduction is rather efficient. Power expenses on this energy influence are substantially smaller than the gain in propulsion power from lower drag.

To date, a large amount of numerical research in this field has been performed [11-14]. However, in most of these works considered flow regimes are for moderate Mach numbers. This work is focused on numerical study of the impact of energy input on the flow past hypersonic aircraft X-43 at M = 6, characteristic for long atmospheric flight hypersonic aircrafts.

2. Problem Statement

The results presented below were obtained in the framework of the unsteady Reynolds averaged Navier–Stokes equations (URANS) for a viscous compressible gas with the Spalart–Allmaras turbulence model. For the simulation of energy input URANS system was supplemented with a source term in the conservation-of-energy equation. A detailed description of the mathematical model, numerical methods and program complex used in this work, is given in [15, 16].

The problem considered is the hypersonic flow past hypersonic aircraft X-43. Conditions of the incoming flow were consistent with the height h = 30 km [17]. Calculations were performed for the Mach number M = 6 and angles of attack $\alpha = 0°$ and $4°$. The Reynolds number was $\mathrm{Re} = 2.0 \times 10^6\ \mathrm{m}^{-1}$, pressure and density of the incoming flow $p_\infty = 1200$ Pa и $\rho_\infty = 1.84 \times 10^{-2}$ kg/m^3, respectively.

Due to the symmetry conditions the calculations were carried out for half of the X-43 model. Calculation area was the half of the cone, in the center of which the device was located. A hexagonal block-structured grid with 8 991 802 elements was used. Detailed description of the grid model used can be found in [18].

For $\alpha = 0°$ the energy input location was $x \in [-3.74;-3.70]$, $y \in [0.12;0.13]$, $z \in [0.;0.2]$ (figure 2). The total power of energy input was $Q = 6.8\%N$ where $N = F_x U$ – power necessary to overcome the drag for the undisturbed flow. We assume that energy input is stationary and spatially homogeneous in its area.

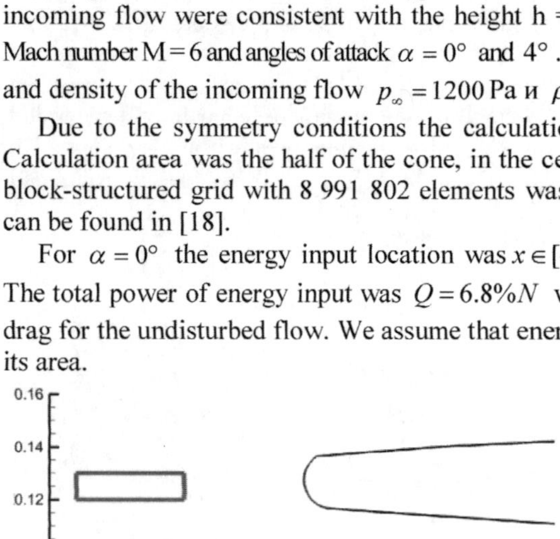

Figure 2. Location of energy input area in front of the nose.

For the purpose of modeling the fuel combustion process in the engine, calculations with the energy input inside the engine for the angle of attack $\alpha = 0°$ were made (figure 3).

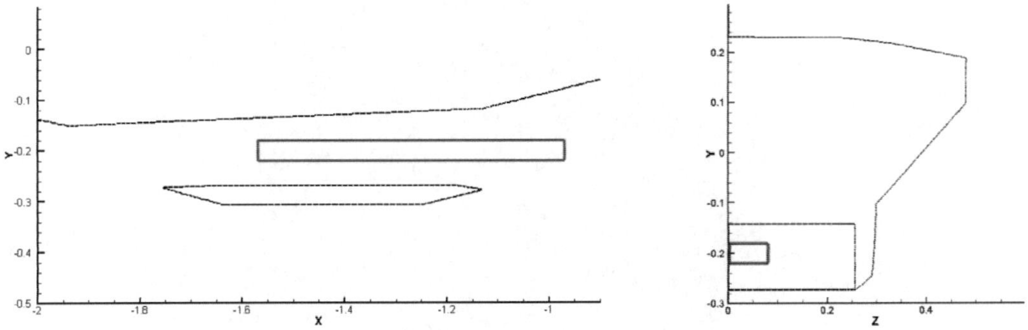

Figure 3. Location of energy input area in the engine.

The amount of energy deposited was determined by the following relationship:

$$q_{\max} = \frac{I_{\infty} f_{O_2} H_U}{L V_{energ}}$$

where I_{∞} [kg/s] – air mass flow through the inlet, $f_{O_2} = 0.2315$ – ratio by mass of oxygen in the air, $H_U = 119.54$ MJ/kg – heat of combustion (hydrogen), $L = 34.5$ – stoichiometric air-to-fuel ratio (hydrogen), $V_{energ} = 0.00192$ m^3 – volume of the area of energy input. Thus $q_{\max}$ is the highest possible (within the framework of this approach) amount of the energy released for this fuel.

3. Numerical simulation results

3.1. The case of undisturbed flow, effect of angle of attack

First, let's consider an undisturbed flow (figures 4 and 5).

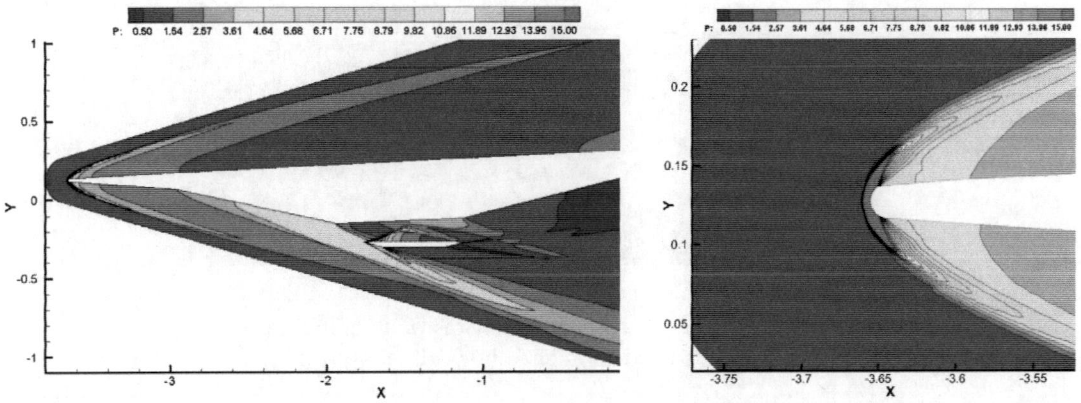

Figure 4. Pressure distribution in the case of undisturbed flow, slice $z = 0.01$, $\alpha = 0°$.

Figure 5. Pressure distribution on the model surface, slice $z = 0.01$, x = $1.65, \alpha = 0°$.

As shown in figures 4 and 5, there is a bow shock and the shock wave from the wedge at the bottom of the fuselage in front of the air intake. X-43 uses the fuselage nose part to form the shock in front of the intake. The airflow in a scramjet is only partially decelerated and remains supersonic throughout the entire engine. The main flow features inside the engine are the multiple reflected shocks (figure 6). The rear surface of the body after the scramjet plays a role of a nozzle and is intended to accelerate a supersonic gas flow coming from the engine.

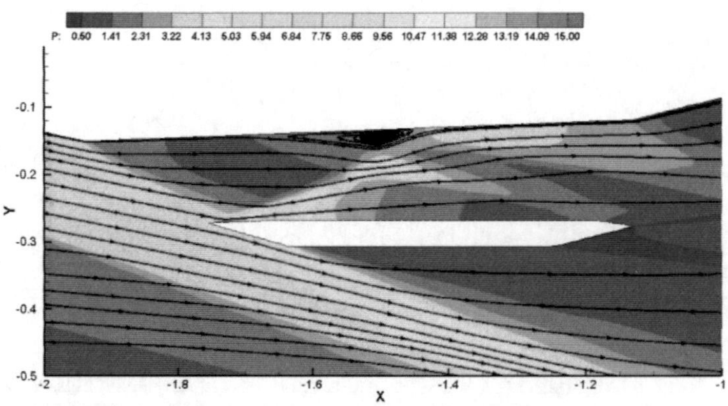

Figure 6. Flow in the engine, pressure distribution in the case of undisturbed flow, slice $z = 0.01$, $\alpha = 0°$.

Let's consider the effect of the angle of attack. Figure 7 shows pressure distribution with the angle of attack $\alpha = 4°$.

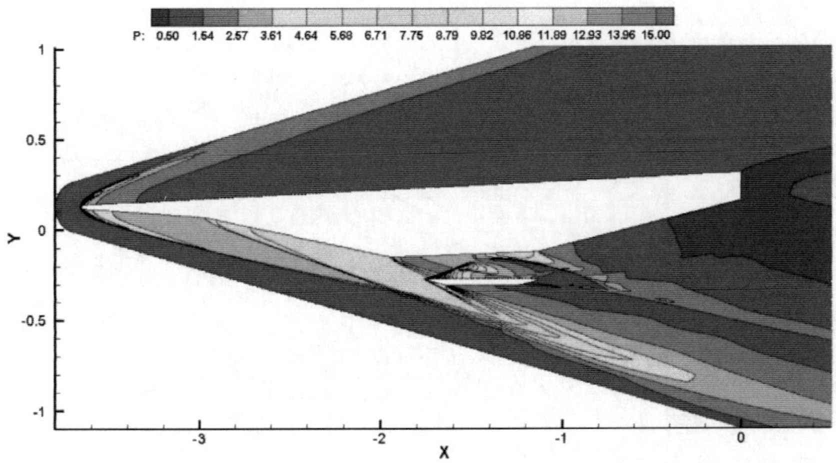

Figure 7. Pressure distribution in the case of undisturbed flow, slice $z = 0.01$, $\alpha = 4°$.

In general, the current structure remains the same. The differences lie in the fact that by increasing the angle of attack rarefaction occurs above the upper surface and pressure increases on the lower surface. The last point, in turn, increases the pressure of the flow entering the engine, see table. 1.

Table 1. Parameters of the flow in the air intake in the case of undisturbed flow, M = 6, $P_{0\infty} = 1578.88$, drag and lift coefficients C_D and C_L, aerodynamic quality K.

α	Air flow (ρu)	$P_{o\ in}$	$P_{o\ out}$	$(P_{o\ in} - P_{o\ out}) / P_{o\ in}$	C_D	C_L	K
0	0.239	422.11	317.88	24.69%	0.01633	0.00407	0.25
4	0.372	610.42	484.01	20.71%	0.02311	0.05441	2.353

Also, due to the angle of attack, the shock wave comes closer to the front edge of the air intake, which allows it to capture a larger part of the flow. It is obvious that an increase in the angle of attack also increases lift.

Comparison of the obtained drag and lift coefficients with experimental data [4] is shown in figure 8.

3.2. Impact of the energy input in front of the nose

Let's consider results obtained for energy input in front of the nose part of the model.

Energy input ahead of the body substantially changes the flow structure (figure 9). Shock waves issue from the energy input region. The front of the bow shock wave is changed by the wake formed behind the energy input region. In the space between the energy input region and the nose there is formed an area with lower pressure in comparison with the oncoming flow. The weakening of the bow shock wave leads to a decrease in drag (table 2) which in turn raises the aerodynamic quality. The angle of the bow shock wave is also changed, so now it comes closer to the front edge of the air intake, which allows it to capture a larger part of the flow. Since energy input reduces the intensity of the bow shock wave, speed and, therefore, stagnation pressure, of the flow entering the air intake is higher than in the case without energy input.

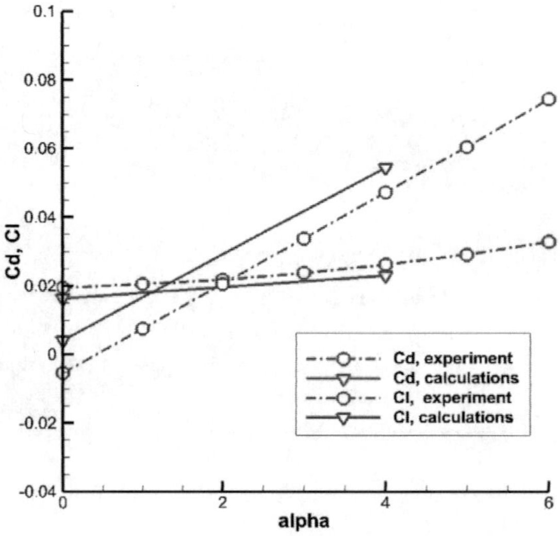

Figure 8. Drag and lift coefficients, comparison with experimental data.

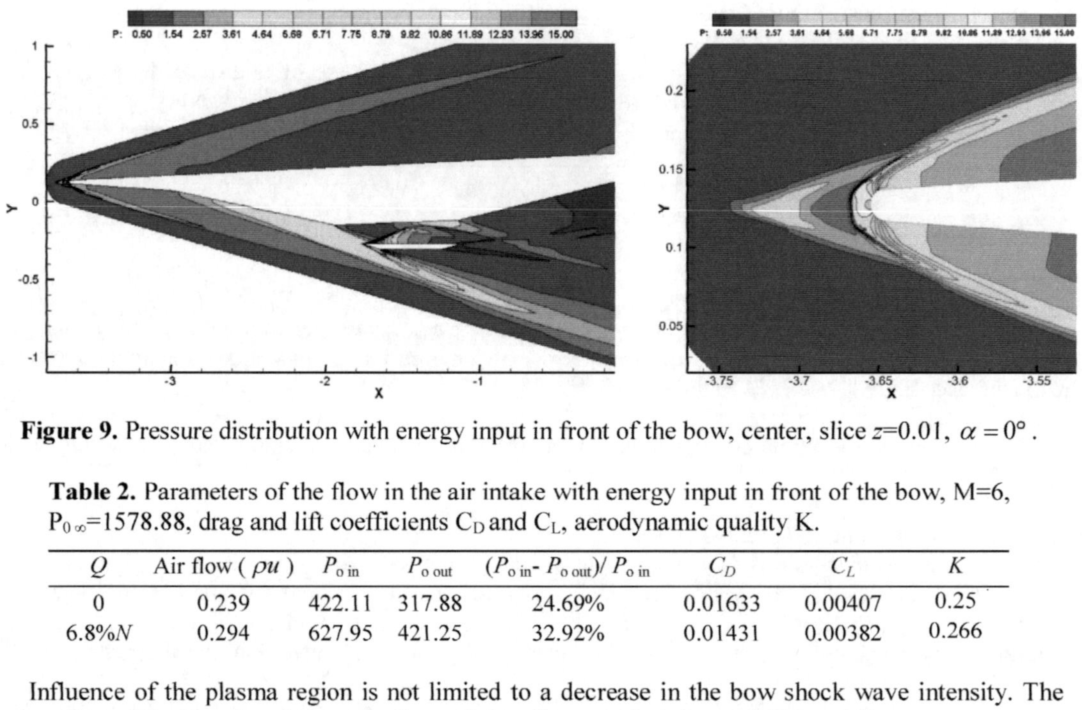

Figure 9. Pressure distribution with energy input in front of the bow, center, slice z=0.01, $\alpha = 0°$.

Table 2. Parameters of the flow in the air intake with energy input in front of the bow, M=6, $P_{0\,\infty}$=1578.88, drag and lift coefficients C_D and C_L, aerodynamic quality K.

Q	Air flow (ρu)	$P_{o\ in}$	$P_{o\ out}$	$(P_{o\ in} - P_{o\ out})/P_{o\ in}$	C_D	C_L	K
0	0.239	422.11	317.88	24.69%	0.01633	0.00407	0.25
6.8%N	0.294	627.95	421.25	32.92%	0.01431	0.00382	0.266

Influence of the plasma region is not limited to a decrease in the bow shock wave intensity. The parameters of the boundary layer near the model surface are also changed (figure 10).

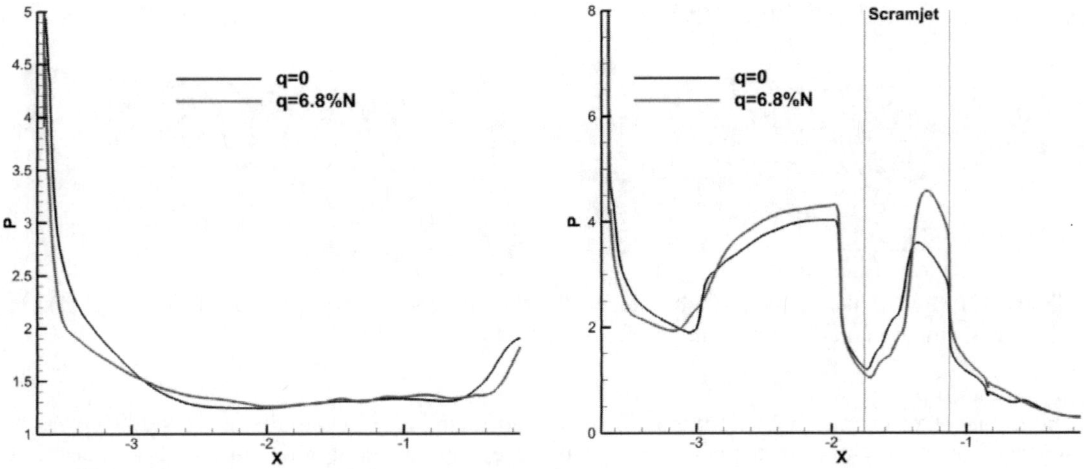

Figure 10. Pressure distribution on the top (left) and bottom (right) surfaces of the model, slice $z = 0.01$, $\alpha = 0°$.

3.3. Impact of the energy input inside the engine

Let's consider results obtained with energy input in the engine for $\alpha = 0°$ and $q = q_{max}$.

As shown on Figures 11, 12 and Table 3, energy input in this way significantly affects the structure of the flow in the air intake. Shock-wave structure moves to the inlet cross-section due to the counter pressure created by energy input, thus increasing the intensity of the incident shock from the bottom wall of the intake. The area of separation located on the upper wall of the air intake significantly increases in size and under the influence of the counter pressure moves to the inlet section. Considering the integral characteristics of the flow in the engine (table. 3), it can be concluded that we get close to off-design operating mode of the air intake. Further research on this topic is scheduled to take place in subsequent papers.

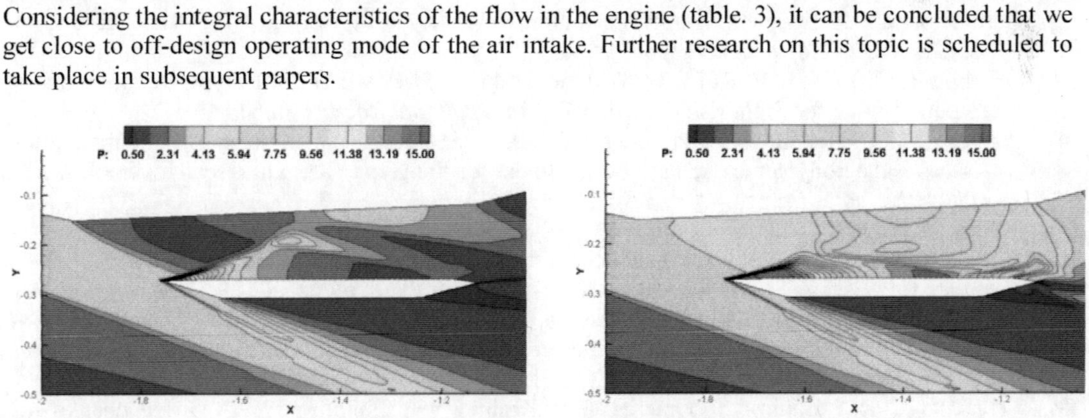

Figure 11. Flow in the engine, pressure distribution in the case of undisturbed flow (left) and with energy input in the engine (right), slice $z = 0.01$, $\alpha = 0°$.

Table 3. Parameters of the flow in the air intake with energy input in the engine, M = 6, $P_{0\infty} = 1578.88$, drag and lift coefficients C_D and C_L.

q	$P_{o\,in}$	$P_{o\,out}$	$(P_{o\,in} - P_{o\,out})/P_{o\,in}$	C_D	C_L
0	422.11	317.88	24.69%	0.01633	0.00407
q_{max}	330.84	130.18	60.65%	0.01457	0.01064

IOP Conf. Series: Journal of Physics: Conf. Series **815** (2017) 012018 doi:10.1088/1742-6596/815/1/012018

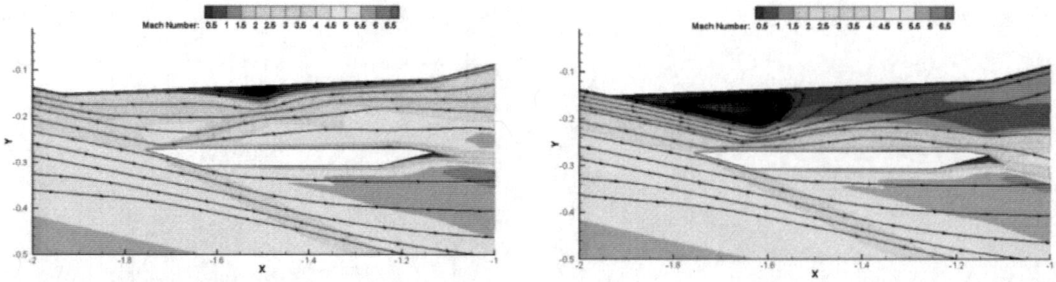

Figure 12. Flow in the engine, Mach number distribution in the case of undisturbed flow (left) and with energy input in the engine (right), slice $z = 0.01$, $\alpha = 0°$.

4. Conclusions

The influence of energy input on the aerodynamic characteristics of the aircraft X-43A for various angles of attack was investigated. It is shown that energy input in front of the nose leads to a decrease in drag. For the purpose of modeling the fuel combustion process in the engine, calculations with the energy input inside the engine were made, off-design operating mode of the air intake was obtained.

Acknowledgments
This work was supported by the Russian Foundation for Basic Research project No. 16-31-00399.

References

[1] Reubush D E, Nguyen L T and Rausch V L 2003 Review of X-43A return to flight activities and current status *12th AIAA International Space Planes and Hypersonic Systems and Technologies, International Space Planes and Hypersonic Systems and Technologies Conf.* (Norfolk, Virginia) *AIAA* 2003-7085

[2] Drummound J P, Bouchez M and McClinton C R 2002 Overview of NATO background on scramjet technology *NATO report*

[3] Huebner L D, Rock K E, Ruf E G, Witte D W and Andrews E H 2001 Hyper-X flight engine ground testing for flight risk reduction *J. Spacecraft and Rockets* **38** pp 844-52

[4] Engelund W C, Holland S D and Cockrell C E et al 1999 Propulsion system airframe integration issues and aerodynamic database development for the Hyper-X flight research vehicle // *XIV ISOABE*

[5] Mishin A I, Klimov A I and Gridin A Y 1992 *Tech.Phys.Lett.* **18** pp 86-92

[6] Fomin V M, Lebedev A V and Ivanchenko A I 1998 *Dokl. Akad. Nauk* **361** pp 58-60

[7] Grachev L P, Esakov I I and Khodataev K V 1999 *Tech. Phys.* **69** pp 14-18

[8] Kolesnichenko Yu F, Brovkin V G, Azarova O A, Grudnitsky V G, Laskov V A and Mashek I 2003 Ch MW energy deposition for aerodynamic application *41st Aerospace Science Meeting and Exhibit AIAA Paper* 2003-361

[9] Tretyakov P K, Garanin A F, Grachev G N, Krainev V L, Ponomarenko A G, Ivanchenko A I and Yakovlev V I 1996 *Dokl. Akad. Nauk* **351**

[10] Leonov S, Bityurin V, Yuriev A, Pirogov S and Zhukov B 2003 Problems in energetic method of drag reduction and flow/flight control *41st Aerospace Science Meeting and Exhibit AIAA Paper* 2003-35

[11] Georgievsky P Yu and Levin V A 1988 *Tech.Phys.Lett.* **14** pp 684-7

[12] Georgievsky P Yu and Levin V A 2003 *Fluid Dyn.* **5** pp 154-67

[13] Azarova O, Knight D and Kolesnichenko Y 2011 Pulsating stochastic flows accompanying microwave filament/supersonic shock layer interaction *Shock Waves* **21** pp 439-50

[14] Lutsky A E and Khankhasaeva Ya V 2015 The 3D flow problem for an aircraft model with active influence on the flow *Yakutian Math.J.* **22** pp 72-8

[15] Borisov V E, Kudryashov I Yu and Lutsky A E 2016 *Keldysh Institute of Appl. Math. Preprint* **2**
[16] Borisov V E and Lutsky A E 2016 *Keldysh Institute of Appl. Math. Preprint* **68**
[17] Zheleznyakova A L and Surzhikov S T 2013 On the Path to Creating a Virtual Model of Hypersonic Aircraft (*Moscow: A. Ishlinsky Institute for Problems in Mechanics*)
[18] Lutsky A E and Severin A V 2016 *Keldysh Institute of Appl. Math. Preprint* **102**
[19] Zheleznyakova A L and Surzhikov S T 2013 *High Temperature* **51** pp 816-29
[20] Prisell E G 2006 The feasibility of the scramjet; an analysis based on first principles *25th International Congress of the Aeronautical Sciences ICAS 2006* pp 1-14

On the concept of the interactive information and simulation system for gas dynamics and multiphysics problems

O Bessonov[1] and P Silvestrov[2]

[1] Institute for Problems in Mechanics of the Russian Academy of Sciences
prosp. Vernadskogo 101-1, Moscow, 119526, Russia
[2] Tactical Missiles Corporation
ul. Ilyicha 7, Korolev, Moscow district, 141080, Russia

E-mail: bess@ipmnet.ru

Abstract. This paper describes the general idea and the first implementation of the Interactive information and simulation system – an integrated environment that combines computational modules for modeling the aerodynamics and aerothermodynamics of re-entry space vehicles with the large collection of different information materials on this topic. The internal organization and the composition of the system are described and illustrated. Examples of the computational and information output are presented. The system has the unified implementation for Windows and Linux operation systems and can be deployed on any modern high-performance personal computer.

1. Introduction

Progress in high tech strongly depends on the development of information and computing technologies. Such technologies are widely used in countries developing aircrafts (US, France, Japan etc.) for predicting their characteristics and for optimization of the parameters at the design stage, as well as for analysis of data at the stages of the ground and flight testing and for prognostic research on the subject. However, the accuracy of computations often doesn't meet modern requirements.

The latest achievements in supercomputer technologies and mathematical modeling of complex dynamic processes for atmospheric re-entry space vehicles make it possible to create computer models, the use of which will significantly reduce the amount of land and flight testing, lower the influence of the associated risks and improve the economic efficiency of finding optimal integral and engineering solution.

Currently there is no single software package for the complete multiphysical investigation of the interaction of a space vehicle with the surrounding gas. However, leading scientific institutions and design centers in the field of aerospace research have many years of experience and possess computer codes for solving different parts of this global problem [1-9].

Thus arises the problem of systematization and structuring of both existing and emerging computer codes into the single software package with a modular architecture that would allows to fully solve problems of modeling different sorts of interactions (mechanical, thermal, chemical) of a vehicle with the surrounding environment along the flight pass.

This integrated software package should include computational modules (solvers) for modeling the aerodynamics (aerothermodynamics) of prototypes of re-entry vehicles and their constructive elements, as well as the following tools and facilities:

- for selection of a prototype model from the bank of prototypes;
- for setting input data and computational parameters;
- for selection, preparation and visualization of electronic surfaces and grids;
- for querying physical and chemical characteristics of substances needed for the simulation;
- for running a computation and controlling its progress;
- for presentation and visualization of all kinds of the results (graphical, moving, textual, interactive);
- for requesting and presentation of information resources on the topic of the simulation.

The goal is to eventually reach the world-class level of research in this area and surpass it in many ways. In particular, the new software package will allow to conduct the following research:

- to determine the aerodynamic and thermal loads on the surface and elements of perspective re-entry vehicles;
- to model thermogasdynamic processes in external and internal flows, taking into account the combustion;
- to model high-temperature flows in the shock layer near a vehicle with the integrated ramjet engine;
- to model interactions of high-temperature gases with the heat shield of a re-entry vehicle;
- to model unsteady flows in an integrated ramjet engine including the process of ignition in the combustion chamber;
- to simulate non-equilibrium chemical and photochemical processes and the formation of clusters and aerosol particles in exhaust jets of the engine;
- to model signatures of re-entry vehicles and ways of suppressing natural and artificial atmospheric disturbances.

The full-featured system of physical and mathematical models of the aerothermodynamics will allow to carry out numerical modeling of external configurations of re-entry vehicles as a whole and their key elements (edges of a body and wings, nose cone, engine) in particular. Simulations of key elements can be performed with higher accuracy to improve the reliability of the integral results.

This system will also provide access to a set of standard test problems formulated on the basis of ground and flight experimental data. These formulations can be used for performing preliminary calculations.

Additionally, a database on gas dynamics and multiphysics will be established in order to assist in performing numerical simulations and accelerate the study of external flows and physical processes in engines. This database will contain information about the physical-chemical properties of gases at high speeds, the kinetics of the combustion of fuel components, the thermal, thermodynamic and spectral optical properties of substances, as well as the catalytic properties of surfaces.

This paper presents the first variant of such integrated environment, the Interactive information and simulation system (IISS). Scientific and information content of the system is based on the experience acquired during the previous research in the fields of radiation gas dynamics, high speed aerothermodynamics and physical / chemical mechanics of non-equilibrium media and multi-temperature thermodynamics (see, for example, [4-9]).

2. Function and composition of the Interactive information and simulation system

Interactive information and simulation system considered in this paper is an integrated environment comprising computational modules with necessary input data, grids, information materials, scientific papers and reports on the themes of computational modules, collections of results of previous computations on these themes, databases on physical and chemical gas dynamics and other constituent parts. All these components are interconnected into a hierarchical / hypertext structure and form a unified information space. The system is organized as an integrated package with rich graphical user interface. It can be deployed on a high-performance computer working under Windows or Linux operating system.

The IISS is driven by the interactive master control program which provides both the internal functionality of the system and the graphical interaction with a user. Graphical interface is implemented as a natural and convenient icon-based and menu-based environment.

The master control program combines different tools and features which allow to provide the following capabilities:

- navigation through hierarchical and hypertext links;
- search and browse of information materials, papers and reports;
- view of graphical and video information, including surfaces and grids;
- query in databases and help subsystems;
- selection of a computational module and creation of a project to work with it;
- preparations for running a computational module (preview and editing of input data, selection and preview of a grid etc.);
- running a computational module, controlling the run, watching the output, stopping the computation;
- viewing textual and graphical results of the computation;
- launching external applications (table editors, graphical packages etc.);
- tuning and customization of the environment;
- user authentication and access control;
- accumulation of new results in the special database (not yet implemented);
- facilities for running computations remotely (not yet implemented).

3. Architecture of the interactive information and simulation system

Architecture of the IISS is based on the idea of separation of the code of the master control program from the application and information content. The graphical master program includes the above tools and features and is independent on the content of the IISS. It is controlled by the set of configuration files in XML format organized in the hierarchical manner. These files describe the structure of the application and information content of the IISS, appearance and layout of graphic elements on the screen, icons with their underlying functions, menus and other characteristics of the graphical interface. There is also another sort of configuration files which describe mechanisms and details of running application programs: nomenclature and types of input and output files, input data and options, grids, computational modules and scripts to run them.

Thus, the set of configuration files and scripts in conjunction with the directory tree containing all information and application data fully describes the whole content of the IISS and all working scenarios. Any modifications or additions of the content or scenarios are performed simply by editing or rebuilding configuration files and don't require to change and recompile the source code of the master control program. This ensures easy system expansion and development, e.g. addition of new prototypes and information materials, complication of operating scenarios, support of new external applications etc. Additionally, isolation of the master control environment from the data allows to implement several variants of the IISS with different application and information content in accordance with the specific needs of customers.

4. Description of the graphical interface and regimes

Graphical master control program of the IISS supports two main styles of the operation – the information regime and the computing regime. After the initialization and user's authentication, the system enters the information regime.

4.1. *Information regime*

In this regime, the screen is split into two fields – the upper one (menu field) with headings and icons corresponding to the objects to be selected, and the lower one (information field) with the descriptions of these objects and action buttons (figures 1 and 2). Clicking on an icon or action button either loads another screen (inferior with respect to the current one) with the similar organization, or performs some action (previewing a grid or a surface, browsing a paper or an information material, viewing a picture or a movie, opening a project for performing a computation etc.). There are also some navigation buttons (return, home etc.). All these control elements provide navigation in the hierarchical information space of the IISS. Owing to them, the user is able to browse information and illustrative materials to become familiar with the subject, to look through the previous computational or experimental results, and finally to create a new project for running a computational module.

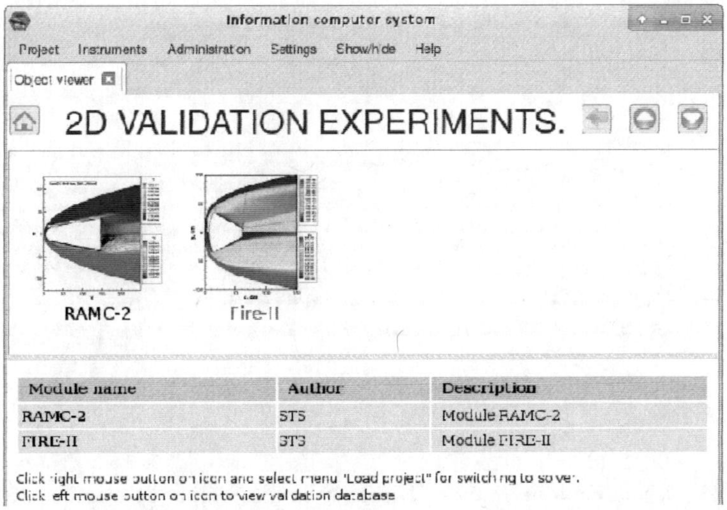

Figure 1. Example of the information regime. Left-click on an icon (RAMC-2 or Fire-II) in the menu field loads the screen with the validation data, right-click switches to the computing regime.

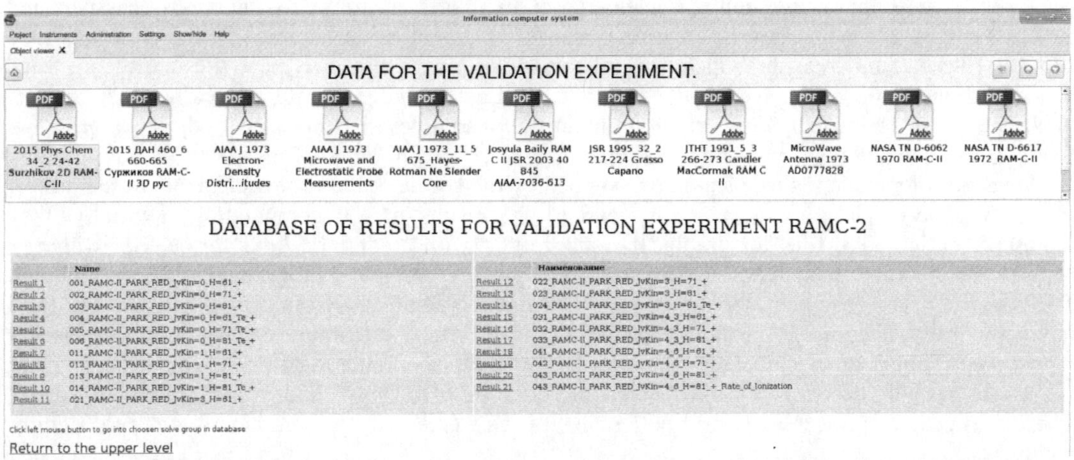

Figure 2. Example of the information regime. Screen with the validation data for RAMC-2 experiment. Icons in the menu field represent information materials on the topic (papers and reports), action buttons in the information field switch to the screens with particular validation results.

The structure and content of the information space is mapped on the similar hierarchical structure of configuration files: each directory in the tree contains the XML file "dirinfo.dif" with the full description of each element or field on the screen (its layout, appearance, caption, associated action, name of an HTML file with the content of a field, link to the inferior directories, hypertext links to other elements etc.).

4.2. Computing regime

After clicking on a correspondent icon on the information screen, the system switches to the computing regime (figure 3). In this regime, a new field appears on the left side of the screen – project description tree (the remaining part of the screen retains the previous structure with the menu and information fields).

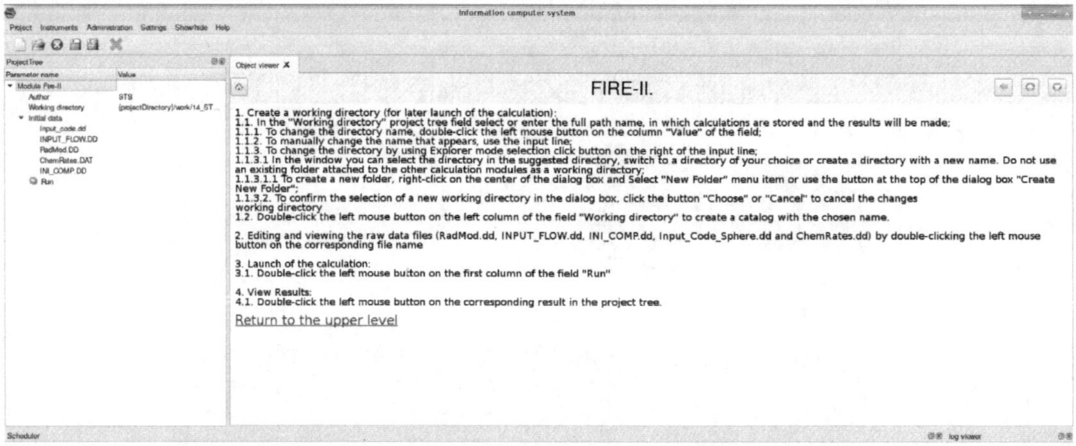

Figure 3. Example of the computing regime. Elements in the project description tree (left) represent the author's abbreviation, working directory, input files and action button for starting a computation. Text in the information field (right) describes the working procedure.

A project description tree contains elements of two sorts – informative captions (passive) and actions (active). Elements may be organized into the hierarchical structure with the ability to hide subtrees (this is similar to the appearance of the file tree structure in typical file managers). Each active element is associated with some function, such as: initialize, create a working directory, set input options or parameters, view or edit an input file, read a description, select and view a grid, and finally start a computation (run a computational module).

After the computation is terminated (or stopped by the special button) new elements of the project tree may appear. These elements correspond to the results of the computation. Typically, they represent configuration files for graphical visualization (with predefined links to the files with raw results). By clicking on such an element, the corresponding graphical program (ParaView, Gnuplot, TecPlot etc.) is launched and outputs the results of the computation (see figure 4), with the ability of interactive manipulation with a picture (as appropriate). After watching these results, the user may repeat preparation actions (e.g. change input data) and run the computational module again.

Project trees are described by the special configuration files "*.prj" and "*.prs" (in XML format). These files contain all necessary information about each element, such as captions, associated actions and links.

The approach of the project description tree is very rich and flexible. It allows to organize complicated scenarios of the work without modifying the source code of the graphical master program. In particular, using the operating systems scripts (command files) allows to build dynamic scenarios with accumulation of new results in the special database, as well as to support multiphase, multiuser, asynchronous (batch) and remote computations.

When the project description tree is open, it remains possible to work with the information field (on the rights side of the screen). After clicking on the correspondent navigation button, the system closes the project description tree and returns to the general information regime.

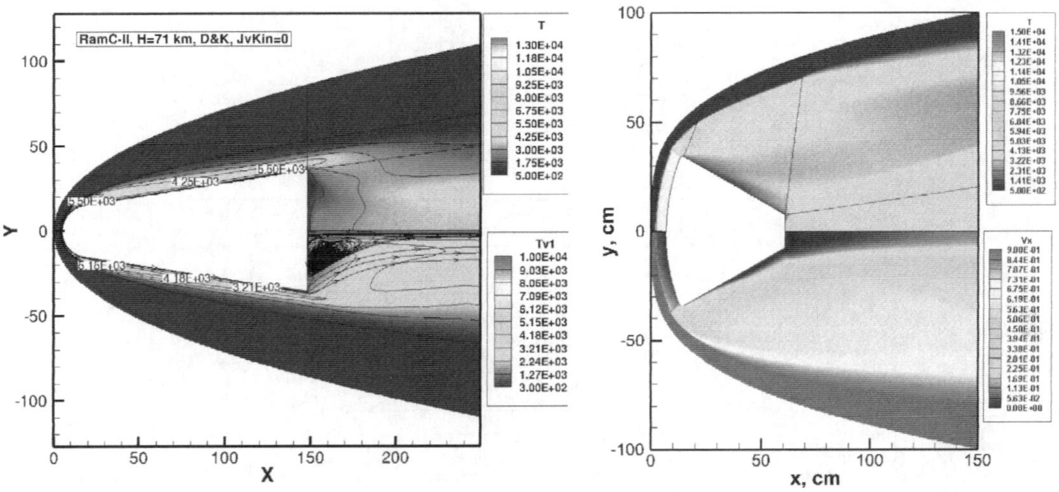

Figure 4. Examples of the computational results for RAMC-2 (left) and Fire-II (right) vehicles in the graphical form as obtained after clicking on the corresponding elements in the project description tree.

4.3. Other regimes

Besides two main regimes, the graphical control program supports several more subsystems:

- application-specific databases driven by the PostgreSQL database engine; they include data on the kinetics of the combustion, on kinetic parameters and transport coefficients, on spectral optical properties, on physical-chemical kinetics, on chemical reactions, on chemical kinetics of high-temperature gases and some others;
- interactive help subsystem; with the future extension of the information system, it will include descriptions of elements of the IISS and working regimes;
- grid manipulation and conversion utilities; currently this subsystem provides facilities for conversion of grids from different formats and for interactive 3D-view;
- administration and access control facilities; they are used to manage users and assign them different levels of privileges;
- system customization tools; they allow to define the location of the system, determine system utilities for dealing with different sorts of files (depending on the operating systems), rebuild configuration files etc.

The diagram on figure 5 illustrates the structure and content of the IISS (in the simplified form). In this example, the user selects FIRE 2 space vehicle as a particular case of the prototype P1, specifies the small grid for this vehicle, previews it and then creates the project description tree in order to run Code 1. Elements of this tree illustrate phases of the further computational session.

5. Implementation of the graphical master control program

The master control program of the IISS is written on C++ language. The graphical interface is implemented using the open-source cross-platform application framework Qt [10, 11], version 5.5.

IOP Conf. Series: Journal of Physics: Conf. Series **815** (2017) 012019 doi:10.1088/1742-6596/815/1/012019

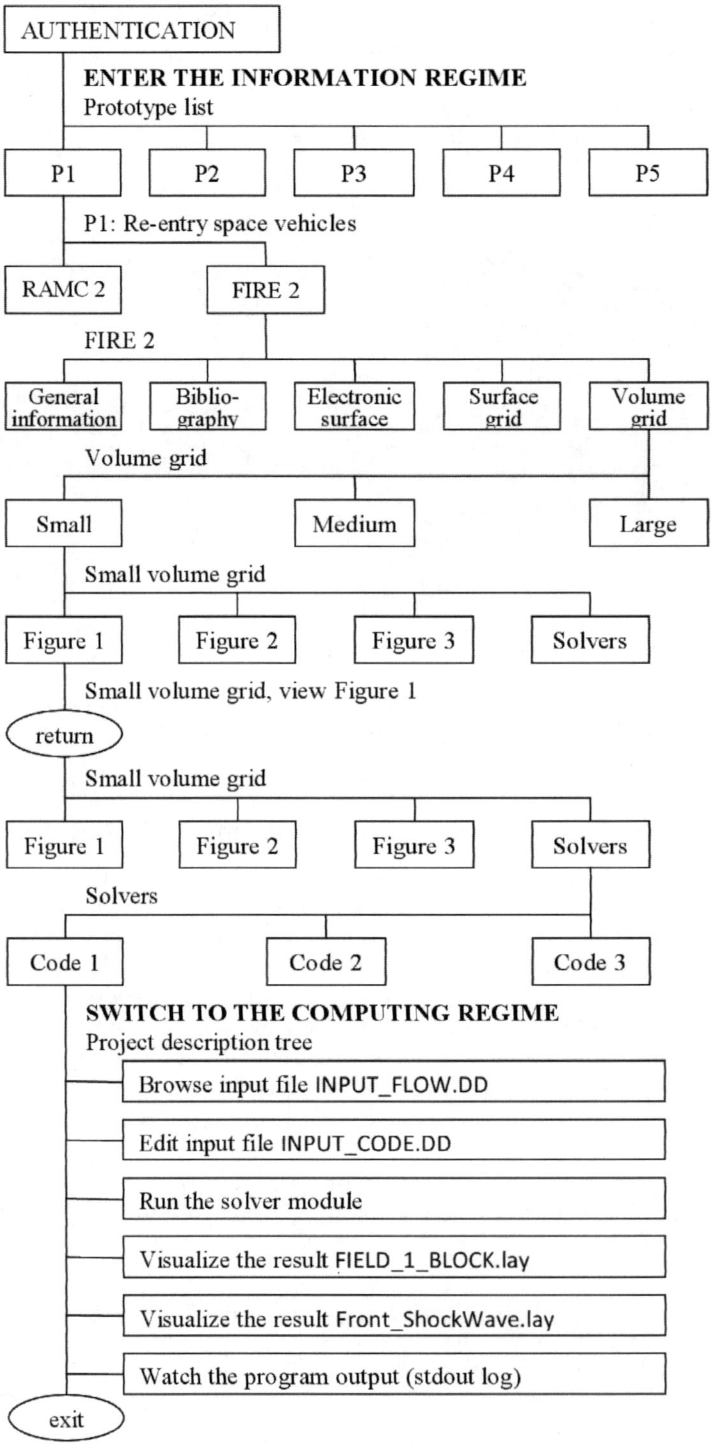

Figure 5. Example of the structure and content of the Interactive information and simulation system.

The program can work on the Microsoft Windows as well as on different variants of Linux. System-dependents parts of the code use "ifdef" and "ifndef" constructs to distinguish between Windows and Linux, thus the source code is unified. Total length of the current variant of the program exceed 5000 lines of code.

Computational modules can be written on any programming languages and compiled by any suitable compilers. Modules are launched indirectly from within command files (scripts). There are two sets of scripts – for Windows (files "*.bat") and for Linux (Bash scripts). If some module needs a run-time library, the required environment can be established inside the corresponding script. Similarly, a complex running procedure can be algorithmized within a script also (e.g. running a parallel job with the use of OpenMP and/or MPI environment). Generally, the script approach allows to release the master control program from doing complicated actions and particular tunings.

6. Conclusion

The presented interactive information and simulation system and its main element, graphical master control program, provide integration of computational modules, grid manipulation tools, information resources and other elements into the unified information and computational environment. The system is being extended continuously, both by adding application modules (solvers) accompanied with corresponding information elements (papers, documents, visual elements, results of computations), and by enhancing the capabilities of the master control program. The resulting computational platform will become the efficient and convenient technological and educational environment for performing computations, visualization and analysis of results for gas dynamics and multiphysics problems.

Acknowledgements

The work was performed within the framework of the Program of basic research of the Russian academy of sciences. The authors wish to thank Dr. H. Ilyasov and Acad. S. Surzhikov for their initial contribution to the project and for useful discussions, and Dr. A. Rybakov for his help and support in performing the work.

References

[1] Candler G V and MacCormack R W 1991 Computation of weakly ionized hypersonic flows in thermochemical nonequilibrium *J. Thermophysics* **5** 266-73

[2] Grasso F and Capano G 1995 Modeling of ionizing hypersonic flows in nonequilibrium *J. Spacecraft Rockets* **32** 217-24

[3] Josyula E and Bailey W F 2003 Governing equations for weakly ionized plasma flowfields of aerospace vehicles *J. Spacecraft Rockets* **40** 845-57

[4] Zheleznyakova A L and Surzhikov S T 2010 Flow field around FIRE II space vehicle under angle of attack *Physical-Chemical Kinetics in Gas Dynamics* **9** 20 (in Russian) http://chemphys.edu.ru/issues/2010-9/articles/139/

[5] Surzhikov S T 2011 Radiative-Convective Heat Transfer of a Spherically Shaped Space Vehicle in Carbon Dioxide *High Temperature* **49**(1) pp 92–107

[6] Surzhikov S T 2012 Coupled radiative gasdynamic interaction and non-equilibrium dissociation for large-scale returned space vehicles *Chemical Physics* **398** 56–63

[7] Shang J S and Surzhikov S T 2012 Nonequilibrium radiative hypersonic flow simulation *Progress in Aerospace Sciences* **53** 46–65

[8] Surzhikov S T 2012 Radiative-Collisional Models in Non-Equilibrium Aerothemodynamics of Entry Probes *Journal of Heat Transfer* **134** 031002-1–031002-12

[9] Surzhikov S T 2015 Two-dimensional numerical analysis of flow ionization in the RAM-C-II flight experiment *Russian J. Physical Chemistry* B **9** 69-86

[10] https://www.qt.io/

[11] Dalheimer M K 2002 *Programming with Qt: Writing Portable GUI Applications on Unix and Win32* (O'Reilly Media)

Mode intermittence in a wake from two cylinders

G V Gembarzhevskii [1,2], A K Lednev [1] and K Yu Osipenko [1,3]
[1] A.Ishlinsky Institute for Problems in Mechanics, Russian Academy of Sciences, Moscow, Russia
[2] Moscow Aviation Institute (National Research University), Moscow, Russia
[3] National Research Univercity Moscow State University of Civil Engineering (NRU MGSU), Moscow, Russia

E-mail: gvgemb@ipmnet.ru

Abstract. Near turbulent intermittent wake from a set contained of two cylinders is investigated in the case of moderate Reynolds number when distinguished wake global modes are competing. This study is performed within the framework of the reduced-order oscillator model where each von Karman vortex street behind a cylinder is modeled by an oscillator of van der Pole type. First distinguished feature of the model is using a nonlinear form in model equations for treating of these streets interaction. The second feature is that the oscillator equations are generalized for explicit dependence of the oscillation frequency on oscillation amplitude. Dependence of the global modes frequencies of oscillation on the model parameters is adduced. An example of attraction domain for an asymmetric wake global mode is studied.

1. Justification

The problem of our interest is in the field of turbulent flows in the electrical constant power industrial lasers, some relevant references to this topic can be found in papers [1 - 4]. We try to adopt the technique of flow control by the electrical discharge in the fast-flow CO2-laser where the glow discharge exists ab initio. A promising result has been achieved for the plasma flow of near wake from two parallel cylinders as is in figure 1 [5]. In that study the wake from cylinders in the case of side-by-side arrangement was studied for the mixture composed of nitrogen and helium (at pressure of tens mbar). The cross section of the discharge chamber was of 55x940 mm, the diameter of the quartz cylinders was of 15 mm and the diameter of the copper water-cooled cathode was of 4 mm. The cylinders touched the cathode in cross arrangement as one can see from figure 1.

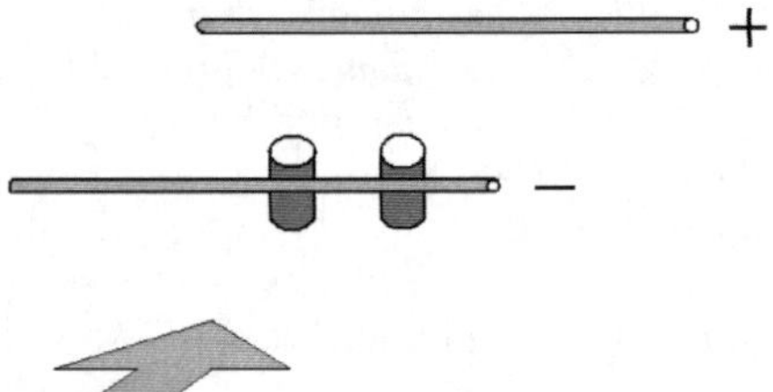

Figure 1. Wake flow in the discharge chamber.

This cylinders wake may be treated as a incompressible turbulent flow since the value of the flow Mach number wasn't higher than 0.15 and the value of Reynolds number defined in terms of cylinder diameter and non-perturb velocity was of Re ~ 1000. Mean velocity was measured by means of Prandtl tube. One-point spectra of the wake velocity were measured in 120 mm beneath the axis of one cylinder by means of a transducer of Siddon-Ribner type. Its signal was processed by a "PALSE 3560 C" complex of Bruel & Kjar. Experimental velocity spectra for the wake at non-self-sustained glow discharge and relevant spectra for the neutral gas flow (that is a reference data) were compared. When the distance between cylinder axes was about of $L/D \approx 2.2$, the velocity spectrum with one peak was observed in the case of gas wake, as is seen from figure 2a [5].

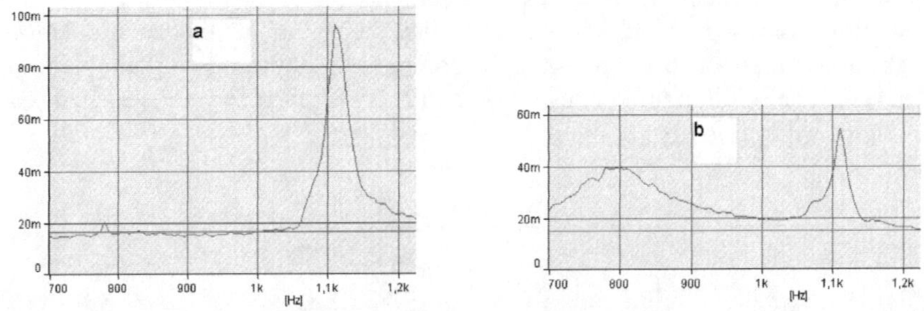

Figure 2. The wake velocity spectra (in arbitrary units): a - for neutral gas flow, b - for discharge plasma wake.

This result was treated as a manifestation of one global hydrodynamic mode existing in the case of gas wake with above mentioned parameters. On the contrary, for a relevant plasma wake flow the two-peak velocity spectrum was observed and interpreted as a property of intermittent flow of two concurrent global modes. One of global modes of the plasma wake was the same as for neutral gas flow (since its oscillation frequency was the same), whereas another new mode had lower frequency as compared to previous one. When current density at cathode tube was of about $2\,\mathrm{A/m}$ (the case in figure 2 b), the dominant part of the energy of coherent oscillations was concentrated in the lower frequency peak. For treatment of such effect one has to use some model of the wake. The previous variant of the model used by us see in [6].

2. Model of wake from two cylinders

Brief sketch of von Karman street formation behind the bluff body is as follows. When fluid pusses cylinder a stagnant zone forms on leeward side of the cylinder where fluid dragged in boundary layer (on one side of the cylinder, principally) accumulates. After half a period this stagnant fluid leaves the cylinder in leeward direction what forms favorable conditions for accumulation of stagnant fluid flowing out boundary layer on opposite side of the cylinder (and with opposite sine of circulation). After second half a period this portion of stagnant fluid takes off the cylinder also and so on. Sequence of rotating in opposite direction moles forms von Karman street behind the cylinder. Accordingly, using some oscillator model for Karman street modeling is natural [7]. Van-der-Pole model is a generic one among oscillator models then we used it as a basis for modeling two cylinders wake. Two von Karman streets constituting the wake of two cylinders are modeled, correspondingly, by two nonlinearly interacting van der Pol oscillators according to the set of equations (1) and (2):

$$\left(1+\Delta\left(\rho^2-4\right)\right)\frac{\mathrm{d}^2X_1}{\mathrm{d}t^2}+X_1-\varepsilon\left[1-X_1^2-\lambda X_2^2-lX_1X_2\right]\frac{\mathrm{d}X_1}{\mathrm{d}t}=h_1 \quad (1)$$

$$\left(1+\Delta\left(r^2-4\right)\right)\frac{\mathrm{d}^2X_2}{\mathrm{d}t^2}+X_2-\varepsilon\left[1-X_2^2-\lambda X_1^2-lX_1X_2\right]\frac{\mathrm{d}X_2}{\mathrm{d}t}=h_2 \quad (2)$$

Here, variables X_i are identified with the transverse velocity component at the characteristic points of wake from each cylinder (behind the region of the wake formation). Variables h_i in the right-hand sides of these equations represent the stochastic processes (a certain analog of the Langevin forces), which describe the emergence of stochasticity of the trace. In contrast to analogous models [8–10], nonlinear interaction between the streets (represented by the oscillators) is taken into account by means of a quadratic form of general view (within the square brackets in the set (1) – (2)). Moreover, the explicit dependence of an oscillation frequency on the oscillator amplitude is taken into account and is described by the expression in the parentheses. Accordingly, the set of equations (1) and (2) contains four parameters $\varepsilon, \Delta, \lambda, l$ while the inheritable parameter of the van der Pole oscillator is assumed to be small $\varepsilon \ll 1$ according to the experimental data.

The Krylov–Bogolyubov method known from the theory of nonlinear oscillations [11, 12] completed by the above mentioned smallness of the parameter ε appears to be suitable for the problem under consideration. The solution to equations (1) and (2) is searched for in the form of oscillations at fundamental frequency with slowly varying amplitude and phase.

$$X_1 = \rho \cos\left(\int_0^t \frac{d\tau}{\sqrt{1+\Delta\left(\rho^2-4\right)}} + \theta \right)$$
$$X_2 = r \cos\left(\int_0^t \frac{d\tau}{\sqrt{1+\Delta\left(r^2-4\right)}} + \psi \right) \tag{3}$$

By performing averaging over fast oscillations, one obtains the “reduced” system of “truncated” equations (4) – (6) for slowly varying amplitudes a phase.

$$\Phi_\rho \frac{d\rho^2}{dt} = \varepsilon \rho^2 \left[1 - \frac{\rho^2 + \lambda r^2 \left(2-\cos(2P)\right) + l \rho r \cos(P)}{4} \right] + h_\rho \tag{4}$$

$$\Phi_r \frac{dr^2}{dt} = \varepsilon r^2 \left[1 - \frac{r^2 + \lambda \rho^2 \left(2-\cos(2P)\right) + l \rho r \cos(P)}{4} \right] + h_r \tag{5}$$

$$\frac{dP}{dt} = \frac{1}{\sqrt{1+\Delta\left(\rho^2-4\right)}} - \frac{1}{\sqrt{1+\Delta\left(r^2-4\right)}} - $$
$$- \frac{\varepsilon}{8} \left[\lambda \left(\frac{r^2}{\langle \Phi_\rho \rangle} + \frac{\rho^2}{\langle \Phi_r \rangle} \right) \sin(2P) + l \rho r \left(\frac{1}{\langle \Phi_\rho \rangle} + \frac{1}{\langle \Phi_r \rangle} \right) \sin(P) \right] + h_P \tag{6}$$

Here, we use the definitions

$$P = \int_0^t \left(\frac{1}{\sqrt{1+\Delta\left(\rho^2-4\right)}} - \frac{1}{\sqrt{1+\Delta\left(r^2-4\right)}} \right) d\tau + \theta - \psi$$
$$\Phi_\rho = 1 + \Delta\left(0{,}5\,\rho^2 - 4\right), \quad \Phi_r = 1 + \Delta\left(0{,}5\,r^2 - 4\right) \tag{7}$$

The dynamic model (4) – (7) makes it possible to obtain global harmonic modes of a wake as being the stable stationary points of system (4) – (6) with zero Langevin forces h_j in the equations. The result of the analysis is that the wake from two cylinders may exist in the form of five different global modes [13]. They are: an asymmetric mode consisted of two different Karman streets, another asymmetric mode with one street fully damped, and three symmetric modes of equal Karman streets which are sin-phase, anti-phase, and variable angle synchronized (in accord to general consideration [14]). In some domains of the model parameters some wake modes intermit. In all there are six different intermittence regimes within the framework of our model [13]. Configuration of all modes and its stability analysis may be done straightforward analytically except for the case of asymmetric mode $AS_{\rho r}$ with unequal oscillation amplitudes of oscillators-streets $\rho \neq r$.

To analyze configuration and stability of the modes it is useful to introduce the collective variables [6] R^2, D, and P by using the transform

$$R^2 = \rho^2 + r^2, \quad D = 0.5\left[1 - 2\rho r / \left(\rho^2 + r^2\right)\right] \tag{8}$$

Than coordinates of asymmetric mode $AS_{\rho r}$ in new variables are solutions to the set of equations (9), (10), and (11) which give stable stationary points to system (4)–(6) with zero Langevin forces

$$\cos^2 P = \frac{3\lambda - 1}{2\lambda} \tag{9}$$

$$R^2 = \frac{4}{1 + l\left(0.5 - D\right)\cos P} \tag{10}$$

$$\frac{1}{\sqrt{1 + \Delta\left[R^2\left(0.5 + N\right) - 4\right]}} - \frac{1}{\sqrt{1 + \Delta\left[R^2\left(0.5 - N\right) - 4\right]}} - \\ -\frac{\varepsilon R^2}{8}\left[\lambda\left(\frac{0.5 - N}{\Phi_\rho} + \frac{0.5 + N}{\Phi_r}\right)\sin\left(2P\right) + l\left(0.5 - D\right)\left(\frac{1}{\Phi_\rho} + \frac{1}{\Phi_r}\right)\sin\left(P\right)\right] = 0 \tag{11}$$

Here

$$N = \sqrt{D\left(1 - D\right)} \tag{12}$$

Accordingly to equation (11) only low values of variable $D << 0.5$ may serve as solution in the case of $\varepsilon \to 0$, than configuration and stability analysis for the mode $AS_{\rho r}$ may be performed by semi-analytical way [13]. The set of equations $(9 - 12)$ may have in principle a four-branch solution as angle P may settle in four quadrants accordingly to equation (9). Really, the set $(9 - 12)$ has solutions in three quadrants only (the second, the third and the fours for the case $l > 0$). Among them we found one branch stable to infinitesimal perturbations in some domain of model parameters. It is branch (13).

$$P = \frac{-l}{|l|}\arccos\left(\frac{-l}{|l|}\sqrt{\frac{3 - 1/\lambda}{2}}\right) \tag{13}$$

The stability domain for branch (13) is presented in figure 3 for intermediate value of parameter $\Delta = 0.125$ (half-analytical evaluation for the case of $\varepsilon \to 0$, [13]). Numerical calculations of model system of "exact" equations $(4) - (6)$ with sufficiently low values of parameter ε and with initial conditions belonging to this domain on figure 3 demonstrate stable focuses or stable cycles. Moreover, stable cycles exist even in essentially wider domain of parameters λ and l, than it is on figure 3.

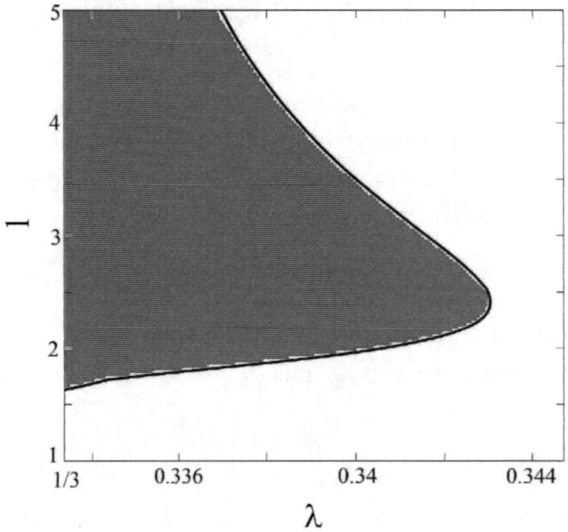

Figure 3. Domain (in red) of existence of stable focuses of wake mode $AS_{\rho r}$ on the plane of model parameters λ and $l>0$ in the case $\Delta = 0.125$; $\varepsilon \to 0$. (The domain right boundary corresponds to criterion of stability $a_1 a_2 - a_3 > 0$ in notations [15]).

3. Distribution of oscillation frequencies of the wake modes

Principle feature of the wake is its oscillation frequency. It is of special interest for comparison with the experimental data. Distribution of the wake modes oscillation frequencies is adduced beneath. The frequencies are in non-dimensional form such that solitary cylinder demonstrates the oscillation frequency being equal to unity.

I. For symmetric mode S_0 of wake consisted of identical Karman streets synchronized sin-phase $\rho^2 = r^2 = 4/(1+\lambda+l)$, $P=0$ and the oscillation frequency is as $\omega^2 = \dfrac{1+\lambda+l}{1+(1-4\Delta)(\lambda+l)}$ The domain of existence and stability of this mode is $\lambda < 1, l > -2\lambda$.

II. Symmetric mode S_π of identical streets synchronized in anti-phase $\rho^2 = r^2 = 4/(1+\lambda-l)$, $P=\pi$ possesses oscillation frequency $\omega^2 = \dfrac{1+\lambda-l}{1+(1-4\Delta)(\lambda-l)}$. The domain of existence and stability of this mode is $\lambda < 1, l < 2\lambda$.

III. Symmetric mode S_φ of streets synchronized at a variable angle differed from 0 or π: $P = \arccos(-l/2\lambda)$; $\rho^2 = r^2 = 4/\left(1+3\lambda - l^2/\lambda\right)$ Its oscillation frequency is $\omega^2 = \dfrac{1+3\lambda-\left(l^2/\lambda\right)}{1+(1-4\Delta)\left(3\lambda-\left(l^2/\lambda\right)\right)}$. This mode exists and is stable in the domain $-(1/3) < \lambda < 0, |l| < -2\lambda$ and $\lambda < -(1/3), \sqrt{\lambda(1+3\lambda)} < |l| < -2\lambda$.

IV. Asymmetric mode AS_{01} with total quenching of Karman street behind one of the cylinders: $\rho^2 = 4,\ r^2 = 0$. Its oscillation frequency is as $\omega^2 = 1$. The domain of existence and stability of this mode is $\lambda > 0.5$.

V. Asymmetric mode $AS_{\rho r}$ of two streets with different amplitudes. Its form and the domain of existence and stability discussed earlier, in section 2. Expression for the mode oscillation frequency in the limit $\varepsilon \to 0$ is $\omega^2 = \dfrac{2-|l|\sqrt{1.5-(0.5/\lambda)}}{1+(1-4\Delta)\left(1-|l|\sqrt{1.5-(0.5/\lambda)}\right)}$.

Information concerning the oscillation frequencies of the modes that may concurrent is summarized in figure 4. These data are represented by level lines of square of oscillation frequency on the model parameters plane λ, l while the value of the third parameter is intermediate: $\Delta = 0.125$. The value of parameter ε is essential only for asymmetric mode $AS_{\rho r}$ data. In this case the parameter ε assumed to be small enough. Aria $0 < l < 8$ is adduced at this figure only as for area $l < 0$ is a mirror image of the opposite one for $l > 0$.

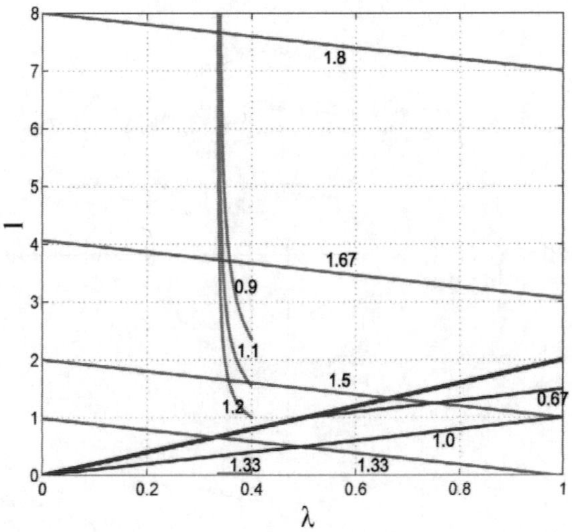

Figure 4. Dependence of square of oscillation frequencies for three wake global modes S_0, S_π and $AS_{\rho r}$ as functions of model parameters λ, l. (Area $l < 0$ is a mirror image of the opposite one: $l > 0$). The frequency dependence is represented by level lines and the values of these constant levels of frequency squares are depicted at corresponding lines. Information for S_0 mode is given in green, for S_π - in blue, and for $AS_{\rho r}$ - in red. Heavy blue line is a boundary of the mode S_π stability domain.

According to preliminary discussion the principle question is as follows. Whether or not the advanced model accommodates the distribution of oscillation frequencies, which has been observed in our experiment? These oscillation frequencies are as in figure 2 whereas the wake frequency for solitary cylinder is of 0.9 kHz. The desired agreement may be achieved if we assume the intermittence of S_0 and S_π wake modes, for example, in the case of following model parameters: $\lambda = 0.65$; $l = 0.99$ and $\Delta = 0.14$. Another opportunity is S_0 and $AS_{\rho r}$ intermittence in the case of the model parameters: $\lambda = 0.344$; $l = 6.41$; $\Delta = 0.1$ and $\varepsilon = 0.1$, for example. Here it should be mentioned that for the last set of parameters, the $AS_{\rho r}$ stationary point is unstable and the trajectory localizes in some area around this point (as it was mentioned above and in [13]).

4. Domain of attraction for asymmetric mode AS_{pr}

Among properties of the wake mode its attraction domain is of principle one. Information concerning the wake modes attraction domains in phase space may be useful in analysis of velocity spectra and other topics. Accordingly, numerical calculation of the attraction area for asymmetric mode AS_{pr} was performed. The results are illustrated by figure 5 for the case $\lambda = 0.34$, $l = 2.5$, $\Delta = 0.125$, and $\varepsilon = 0.1$. One can see that the mode attraction domain has characteristic shape of a roll with central core, and the attraction domain volume is not large as compared to total volume of the phase space. Accordingly, the existence probability of asymmetric mode AS_{pr} have to be low in comparison with existence probability of competing symmetric mode. Precisely this situation (of practically zero intermittence coefficient) one can see in our experimental data for neutral gas wake (figure 2a).

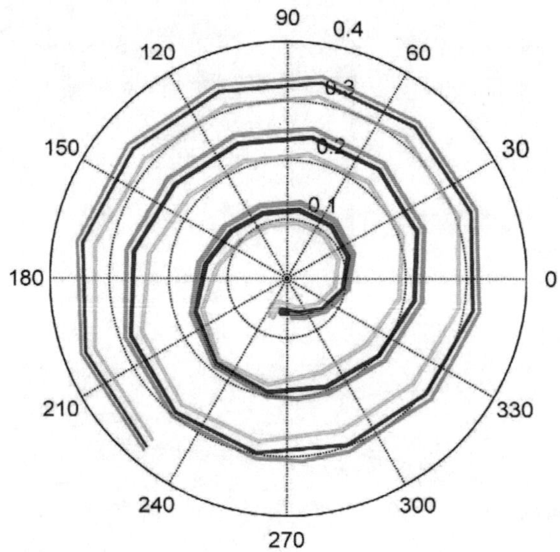

Figure 5. Attraction domain in the phase space R^2, D, P of our model for wake asymmetric mode AS_{pr}. The domain is illustrated by level cuts $R^2 = const$ which colored in green for $R^2 = 8$, blue for $R^2 = 4$, and brown for $R^2 = 2$ (three turns of infinite spiral are illustrated only).

5. Brief resume

If one exclude from consideration mode AS_{01} with total quenching of one Karman street (that is the case of strongly interacting streets [14]), the intermittent wake of two cylinders within the framework of our model may exist in two regimes. These regimes are: the first - intermittence of sin- and anti-phase synchronized von Karman streets (S_0 and S_π modes), and the second - intermittence between streets synchronized at the angle $\sim \pi/2$ (that is AS_{pr} mode) and one symmetric mode (S_0 or S_π).

Acknowledgements

This study is an extension of the work supported by Russian Fund for Basic Researches (grant No 13–01–00742) and may ground the appropriate claim No 17–01–00394.

References

[1] Panchenko V Ya, Zavalov Yu N, Galushkin M G, Grishaev R V, Golubev V S and Dubrov V D

IOP Conf. Series: Journal of Physics: Conf. Series **815** (2017) 012020 doi:10.1088/1742-6596/815/1/012020

2006 The development of turbulence in the active medium of a fast-flow gas-discharge laser *Laser Physics* **16** 1 pp 40–51

[2] Osipov A I, Uvarov A V and Vinnichenko N A 2006 Influence of the initial vibrationally nonequilibrium state of a medium on the structure of von Karman vortex street *Physics of Fluids* **18** 10 105106

[3] Huang H and Wang Y 2010 Kinetic modeling and optimum design of the discharge tube for the CO2 laser with computational fluid dynamics method *Optical Engineering* **49** 11 114201

[4] Raizer Yu P 2001 Gas discharge physics (Springer) p 449

[5] Gembarzhevskii G V 2009 Electric-discharge effect in a plasma wake flow: Pulsation energy redistribution toward lower frequencies *Technical Physics Letters* **35** 3 pp 241–244

[6] Gembarzhevskii G V 2015 An Approach to Plasma Wake Studying *Journal of Modern Physics* **6** 1 pp 46–57

[7] Landa P S 1996 *Nonlinear oscillations and waves in dynamical systems* (Springer) p 544

[8] Peschard I and Le Gal P 1996 Coupled Wakes of Cylinders *Physical Review Letters* **77** 15 pp 3122–3125

[9] Ryabinin A N 1997 Two applications of “fluid oscillator” model *Mathematical Modeling* **9** 7 pp 26–35 (in Russian) (now *Mathematical Models and Computer Simulations*)

[10] Le Gal P, Chauve M P, Lima R and Rezende J 1990 Coupled wakes behind two circular cylinders *Physical Review* A **41** 8 pp 4566–4569

[11] N Bogoliubov and Y A Mitropolsky 1961 *Asymptotical methods in the theory of nonlinear oscillations* (Gordon & Breach) p 537

[12] Nayfeh A H 1993 *Introduction to perturbation techniques* (Wiley-VCH) p 536

[13] Gembarzhevskii G V, Lednev A K and Osipenko K Yu 2015 Simulation of evolution of the two cylinders plasma wake under the electric discharge influence *Technical Physics Letters* **41** 12 p 1132–1135

[14] Pikovsky A, Rozenblum M and Kurths J 2003 Synchronization: *A universal concept in nonlinear science* (Cambridge University Press) p 432

[15] Korn G A and Korn T M 2000 *Mathematical handbook for scientist and engineers* (Dover Publications) p 1152

Natural oscillations of a gas in an elongated combustion chamber

S V Nesterov, L D Akulenko and V G Baydulov

A. Ishlinsky Institute for Problems in Mechanics of the RAS, 101/1, prospect Vernadskogo, Moscow, 119526, Russia

E-mail: baydulov@gmail.com

Abstract. For the analysis of the frequencies and shapes of the natural oscillations of a gas in an elongated rectilinear combustion chamber, this chamber can be treated as a kind of an organ pipe that has the following specific features:

1. the chamber has an inlet and outlet nozzles;
2. a gas mixture burns in the combustion chamber;
3. the combustion materials flow out from the outlet nozzle;
4. the gas flows in such a way that its velocity in the larger part (closer to the outlet nozzle) of the chamber exceeds the speed of sound (Mach number $M>1$). There are only separate domains (one or several), where $M<1$.

The excitation of the natural oscillations of the gas and an increase in the amplitude of such oscillations can lead to instability of the combustion process [1].

1. A simplified model of the natural oscillations of the gas in an elongated chamber

Let the transverse dimensions of the chamber be substantially smaller than its longitudinal dimension. We assume that the chamber is of constant cross section, the length of the chamber is l, and both its ends are open. The gas flows into the chamber at constant velocity U and flows out of it with the same velocity U (figure 1). It is required to find the frequencies and shapes of the natural oscillations of the gas. Using the linearized equations of acoustics for a moving medium [2], we obtain the equations for the sound potential

$$\frac{\partial^2 \Phi}{\partial t^2} + 2U\frac{\partial^2 \Phi}{\partial x \partial t} - \left(c^2 - U^2\right)\frac{\partial^2 \Phi}{\partial x^2} = 0 \tag{1}$$

and the boundary conditions

$$\Phi(0,t) = \Phi(l,t) = 0 \tag{2}$$

Boundary conditions (2) provide zero values for the sound pressure at the left-hand and right-hand boundary points. The variable component of the velocity and the sound pressure are expressed in terms of the potential Φ as follows:

$$u(x,t) = -\frac{\partial \Phi}{\partial x} \equiv -\Phi', \ p(x,t) = \rho\frac{\partial \Phi}{\partial t} \equiv \rho\dot{\Phi} \tag{3}$$

where ρ is the average gas density.

We will seek the solution of the boundary-value problem (1), (2) in the following form ($c^2 > U^2$, c is the speed of sound in the immobile medium)

$$\Phi = \hat{\Phi}(x)e^{i\omega t} \tag{4}$$

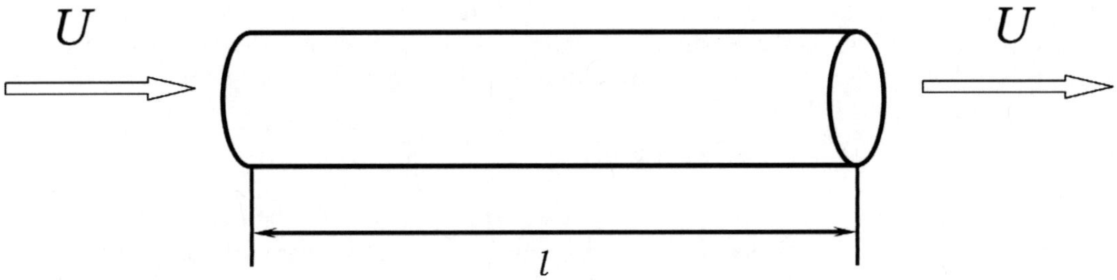

Figure 1. Geometry of the problem.

The conventional calculations yield

$$f_n = \frac{\omega_n}{2\pi} = \frac{n}{2l}c\left(1-M^2\right),\ \varphi_n = \tilde{A}_n \exp\left(\pm i\pi n M\frac{x}{l}\right)\sin\left(\pi n\frac{x}{l}\right) \tag{5}$$

where n is the number of the oscillation mode, v is the velocity of the gas in the pipe, and $M = U/c$ is the Mach number. From the second relation of (3) it follows that the functions φ_n characterize the natural modes of the acoustic pressure.

Comments to the relations (5):

The natural oscillation frequencies and shapes can exist only when $M<1$. For $M<1$, all natural oscillation shapes φ_n are complex-valued. Only for $M=0$, when the steady-state flow is absent, the natural shapes become real. The real natural shapes can be represented as a superposition of the complex-valued natural shapes. This implies that the sound pressure distribution in the case of acoustic oscillations substantially differs from the sound pressure distribution for the case when the flow is absent.

Table 1. Sound velocity in the methane – air mixture.

$v=\frac{1}{3}$	$397.2\,m/s$
$v=\frac{1}{2}$	$380.8\,m/s$
$v=\frac{1}{5}$	$410.3\,m/s$
$v=\frac{4}{5}$	$351.2\,m/s$

Furthermore, the steady-state flow always leads to a decrease in the natural frequencies.

From the above presented considerations it follows that the speed of sound c is a rather important parameter. If the chamber is filled with a gas at a given temperature T, the speed of sound can be defined rather accurately by

$$c = c_0\sqrt{\frac{T}{T_0}} \tag{6}$$

where $T_0 = 273K$, T is the temperature of the gas in kelvins, and c_0 is the speed of sound for $T = T_0$.

Let the chamber be filled with a mixture of two gases and let ν and $1-\nu$ be the mass fractions of the respective components in the mixture. Then, if the speed of sound is known for each of the components, the speed of sound for the mixture can be defined by one of the following expressions [3]:

$$c = (1-\nu)c_1 + \nu c_2,\ c^2 = (1-\nu)c_1^2 + \nu c_2^2 \tag{7}$$

where c is the speed of sound for the mixture, c_1 is the speed of sound for the first component, c_2 is the speed of sound for the second component, and ν is the mass fraction of the second component.

As an example, we present the data for the methane-air mixture: $T = 273K$, $c_1 = 430\,m/s$ (methane), $c_2 = 331.5\,m/s$ (air) (table. 1).

2. The dependence of the natural frequencies on the change in the speed of sound along the combustion chamber

Let c_0 be the speed of sound in the gas mixture at the inlet of the combustion chamber. Due to a number of causes, the speed of sound will change as the distance from the inlet increases. For example, the mix proportion changes, the combustion materials are formed, and the temperature changes. Abstracting from the causes of the change in the speed of sound, we assume that this speed is a known function of the coordinate x measured along the combustion chamber from the inlet nozzle to the outlet nozzle:

$$c = c_0\left(1 + F(z)\right) \tag{8}$$

where c_0 is the speed of sound at the chamber inlet and $z = x/l$ is the dimensionless length.

In addition we assume the inequality $M \ll 1$ for the Mach number. The determination of the natural frequencies and shapes of the oscillations of the gas is reduced to the Sturm-Liouville boundary-value problem

$$\Phi'' + \frac{\lambda}{1+F(z)}\Phi = 0;\ \Phi(0) = \Phi(1) = 0 \tag{9}$$

The natural frequencies ω are related to the eigenvalues λ by $\omega = \sqrt{\lambda}c_0/l$, $f = \omega/2\pi$ (Hz). As an example, we consider the following functions $f(z)$ that characterize the sound speed distributions: 1) $F(z) = z/4$; 2) $F(z) = 8ze^{-5z}$. The results of calculations of natural frequencies for different types of sound speed distribution $F(z)$ are presented in the table 2.

Using the accelerated convergence method [4], we find the eigenvalues λ_n and, thereby, the natural frequencies for three cases: $F(z) = 0$); 1); 2). The calculations for the natural oscillation modes show that the shapes of these oscillations for the sound speed distributions (8) under consideration are close to a sinusoidal form, i.e., they are close to the unperturbed oscillation modes. Using the accelerated convergence method [4], we calculated the natural frequencies up to the sixth digit after the point and then rounded the result. In a similar way, the calculations can be performed for any functions $F(z)$.

Table 2. The dependence of natural oscillation frequencies f_n from the form of sound speed distribution

n	$F(z)=0,\ f_n$	$F(z)=0.25z,\ f_n$	$F(z)=8z\exp(-5z),\ f_n$
1	$0.5\frac{c_0}{l}$	$0.5298\frac{c_0}{l}$	$0.5739\frac{c_0}{l}$
2	$1.0\frac{c_0}{l}$	$1.0591\frac{c_0}{l}$	$1.1490\frac{c_0}{l}$
3	$1.5\frac{c_0}{l}$	$1.5886\frac{c_0}{l}$	$1.7155\frac{c_0}{l}$
4	$2.0\frac{c_0}{l}$	$2.1181\frac{c_0}{l}$	$2.2811\frac{c_0}{l}$
5	$2.5\frac{c_0}{l}$	$2.6486\frac{c_0}{l}$	$2.8470\frac{c_0}{l}$

If the Mach number is known as a function of the coordinate x, the natural oscillation frequencies and modes can be determined by using the accelerated convergence method. However, this method needs a modification, since the equation contains a term with the complex unit. We suggest a simplified expression for the natural frequencies f_n, similar to (5), in accordance with [1]:

$$f_n = \frac{n}{2l}\bar{c}\left(1-\hat{M}^2\right), \tag{10}$$

where $\bar{c}$ is the mean value of the speed of sound, and $\hat{M}$ is the generalized Mach number $\hat{M}=\bar{U}/\bar{c}$; the averaging is performed along the chamber length. The conditions for which expression (10) provides an acceptable accuracy can be obtained by means of the accelerated convergence method.

3. Calculations of the acoustic oscillations of the gas in the combustion chamber with taking into account the design features

A simplified scheme for the combustion chamber that is used in engineering applications is shown in figure 2. The chamber has a length of 1 m. The inlet nozzle has a square cross section of the area $S_0 = 400\ cm^2$. The outlet nozzle has a rectangular cross section of the area $S_1 = 800\, cm^2$. Near the middle of the chamber, there is an elevation of the top surface located between the points $x_1 = 0.48\, m$ and $x_2 = 0.58\, m$. If the length of the chamber is taken as the unit of length, the cross-sectional area can be approximated as follows:

$$S(x) = S_0\left(1+a_1 x\right)\left[1+a_2\left(\text{th}\left(\frac{x-a_3}{b_1}\right)+\text{th}\left(\frac{a_4-x}{b_2}\right)\right)\right] \tag{11}$$

By varying the parameters a_i and b_i, a rather wide class of design features of the chamber can be covered.

At the first stage, we assume that gas flow is absent from the chamber. It is required to determine the natural oscillation frequencies and shapes for a chamber open at both ends. We arrive at the Sturm-Liouville problem

$$\begin{gathered}\frac{d}{dx}\left(S(x)\frac{du}{dx}\right)+\lambda S(x)u=0 \\ \Phi(0)=\Phi(1)=0\end{gathered} \tag{12}$$

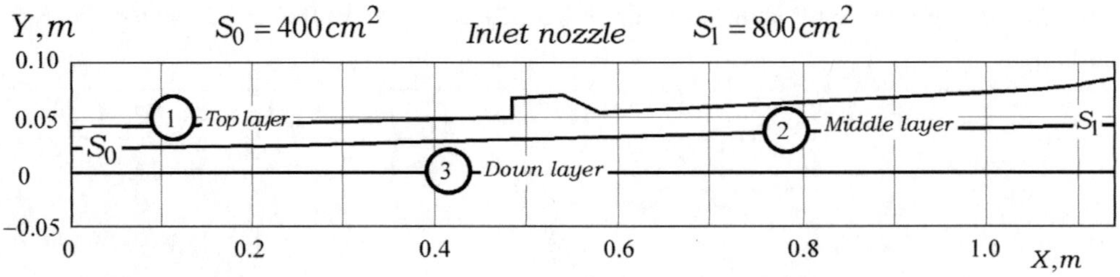

Figure 2. Scheme of the combustion chamber used in technical applications.

After finding the eigenvalues λ_n and the respective eigenfunctions Φ_n, we determine the natural frequencies in the dimensional units:

$$f_n = \sqrt{\lambda_n}\frac{c}{2\pi l}$$

Table 3. Natural oscillation frequencies f_n, Hz, for the chamber without an elevation. $a_1 = 1.875$, $a_2 = 0$.

$f_1 = 0.49339\frac{c}{l}$	$f_1 = 0.5\frac{c}{l}$
$f_2 = 0.99633\frac{c}{l}$	$f_2 = 1.0\frac{c}{l}$
$f_3 = 1.49748\frac{c}{l}$	$f_3 = 1.5\frac{c}{l}$
$f_4 = 1.99809\frac{c}{l}$	$f_4 = 2.0\frac{c}{l}$
$f_5 = 2.49847\frac{c}{l}$	$f_5 = 2.5\frac{c}{l}$

Table 4. Natural oscillation frequencies f_n, Hz, for the chamber with an elevation $a_1 = 1.875$; $a_2 = 0.2$; $a_3 = 0.48$; $a_4 = 0.58$; $b_1 = 0.001$; $b_2 = 0.02$

$f_1 = 0.47608\frac{c}{l} Hz - 2\%$
$f_2 = 1.02134\frac{c}{l} Hz + 2\%$
$f_3 = 1.46111\frac{c}{l} Hz - 2.7\%$
$f_4 = 2.02913\frac{c}{l} Hz + 1.6\%$
$f_5 = 2.47321\frac{c}{l} Hz - 1\%$

Using the accelerated convergence method [4], we found the solutions of the boundary-value problem (12). The computational results are presented in tables 3,4.

The right-hand column presents the natural frequencies for the uniform chamber, $a_1 = 0$. For the first natural frequency, f_1, the difference as compared to the nonuniform chamber makes up 1.3%; for all other frequencies, this difference is less than 1%. The perturbed frequencies virtually coincide with the relevant values $\tilde{f}_n = \frac{n}{2l}c,\ Hz$ for the natural frequencies of the straight pipe, $S(x) = S_0 = \text{const}$.

Table 4 presents the first five natural frequencies for the chamber that has an elevation, as well as the differences of these frequencies from the respective values obtained for the chamber without an elevation. The elevation located as indicated above leads to a decrease in the odd natural frequencies and an increase in the even frequencies.

For comparison, consider the case where there is a cavity on the top wall of the chamber. The results for this case are presented in table 5.

The cavity on the top wall leads to an increase in the odd natural frequencies and a decrease in the even frequencies. The plots of the shapes of the natural oscillations for all cases considered show that these shapes virtually coincide with sinusoids $\sin\frac{\pi n x}{l}$ for the uniform chamber, $S(x) = S_0$, whereas the derivatives of the velocity have essential features near the elevations or cavities, as was mentioned previously [4].

Table 5. Natural oscillation frequencies f_n for the chamber with a cavity $a_1 = 1.875;\ a_2 = -0.2;\ a_3 = 0.48;\ a_4 = 0.58;\ b_1 = 0.001;\ b_2 = 0.02$

$f_1 = 0.51142\frac{c}{l}$: $\delta = +3.7\%$
$f_2 = 0.94839\frac{c}{l}$ -4.9%
$f_3 = 1.53991\frac{c}{l}$ $+2.2\%$
$f_4 = 1.94530\frac{c}{l}$ -2.7%
$f_5 = 2.54078\frac{c}{l}$ $+1.7\%$

For subsonic flows in the chamber, $M < 1$, the natural frequencies for both cases can be calculated according to (5).

4. Chamber through which a supersonic flow is blown

Numerical computations show that for the chamber with positive defect (an elevation on the top wall) a supersonic flow ($M > 1$, air blowing) is observed virtually everywhere. However, $M < 1$ in the neighborhood of the defect for $0.48 \le x \le .58$. In this domain, standing acoustic waves (natural oscillations) may exist. This neighborhood is sketched in figures 3,4.

Table 6. Natural oscillation frequencies f_n for region with defect

$f_n = \frac{(2n+1)}{4l} 634(1-0.25)$	
$n = 0$	$f_1 = 1427\ Hz$
$n = 1$	$f_2 = 4461\ Hz$
$n = 2$	$f_3 = 7435\ Hz$
$n = 3$	$f_4 = 10409\ Hz$

IOP Conf. Series: Journal of Physics: Conf. Series **815** (2017) 012021 doi:10.1088/1742-6596/815/1/012021

While the left-hand boundary is fixed fairly accurately, the right-hand boundary does not coincide with the critical point at which $M=1$. For this reason, we take the point $x=x_0=0.56$ as the left-hand boundary.

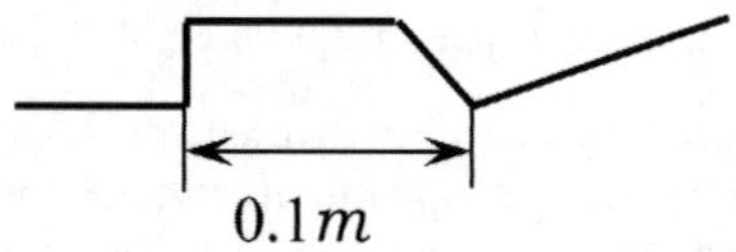

Figure 3. Region of the chamber, satisfying the condition of existence of standing waves.

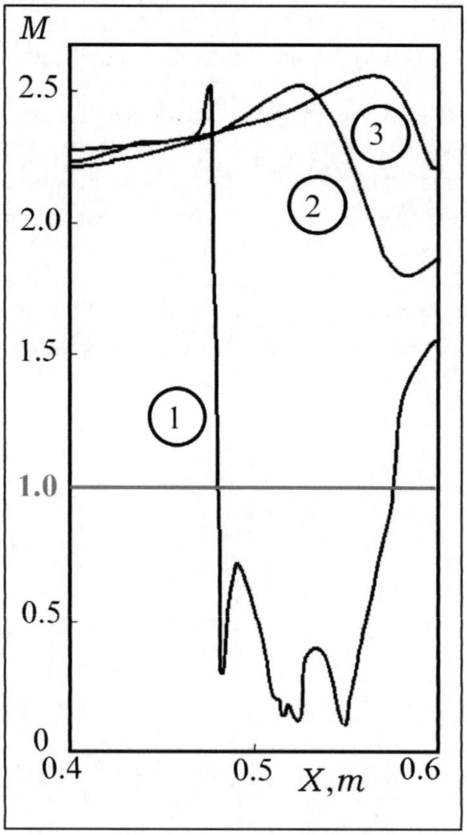

Figure 4. The dependence of the Mach number of the longitudinal coordinate of the Top (1), middle (2) and lower (3) layers of the chamber in the defect area.

The boundary conditions should be taken as follows:

- at the left-hand point of the elevation, the speed of sound is equal to zero;
- at the right-hand point, the pressure is equal to zero.

Then we arrive at the following boundary-value problem for the sound pressure:

$$\frac{d^2u}{dz^2}+\lambda u=0,\ \left.\frac{du}{dz}\right|_{z=0}=0,\ \left.u\right|_{z=z_0}=0$$

$$u=C_1\cos\left[\tfrac{\pi}{2l}(2n+1)z\right],\ \lambda_n=\left[\tfrac{\pi}{2l}(2n+1)\right]^2$$

$$f_n=c^2\frac{\sqrt{\lambda_n}}{2\pi}=\frac{c}{4l}(2n+1);\ l=0.08\ m$$

The mean speed of sound in this region is defined by $c = 331.5 \cdot \sqrt{\frac{1000}{273}} = 634.5\, m/s$, and the average Mach number is $M = 0.5$. Hence, in accordance with (6), we have results presented in table 6.

The standing waves in the domain where $M < 1$ can be excited due to the vortex shedding or due to a thermomechanical interaction, as it is the case for a heated Helmholtz resonator [5]. If there are several regions in which $M < 1$, several types of standing acoustic waves may occur in each of these regions.

References

[1] D T Harrje and F H Reardon (eds) 1972 *Liquid Propellant Rocket Combustion Instability* (Washington: NASA) p 637.
[2] Blokhintsev D I 1981 *Acoustics of a Nonhomogeneous Moving Medium* (Moscow: Fizmatlit) p 206 (in Russian)
[3] Kudryavtsev B B 1956 *Akusticheskij Zhurnal.* **2** 167 (in Russian)
[4] Akulenko L D and Nesterov S V 2005 *High-Precision Methods in Eigenvalue Problems and Their Applications* (Boca Raton: Chapman and Hall/CRC Press) p 255
[5] Akulenko L D and Nesterov S V 2004 *Fluid Dynamics.* **39** 358
[6] Akulenko L D, Baidulov V G and Nesterov S V 2006 *J. Appl. Math. Mech. (PMM).* **70** 504

Numerical simulation of the flow over a hypersonic waverider using the method for splitting into physical processes

D S Yatsukhno

A.Ishlinsky Institute for Problems in Mechanics of the Russian academy of sciences, Vernadsky prospect 101(1), Moscow, 119526, Russia

E-mail: yatsukhno-ds@rambler.ru

Abstract. The aim of the study is to obtain hypersonic waverider aerodynamic characteristics over the angle-of-attack range from 0° to 10° at Mach 4. The simplified virtual model of the waverider was designed. The Navier-Stokes equations solution was performed using computer code into which the method for splitting into physical processes was implemented. The unstructured grid was used in the computational process.

1. Introduction

The hypersonic waveriders is a class of vehicles having the following performance advantages:

- The reusable space vehicles
- An ability to take off and land horizontally like conventional airplane
- High aerodynamic efficiency

The waverider shape is derived from the inviscid conical flow field and the shock wave shape. The upper surface is produced by the axisymmetric method of characteristics. The lower surface is generated by solutions of the Taylor-Maccoll differential equations [1] given as

$$\frac{\gamma-1}{2}\left(1-V_r^2-\left(\frac{dV_r}{d\theta}\right)^2\right)\left(2V_r+\cot(\theta)\frac{dV_r}{d\theta}+\frac{d^2V_r}{d\theta^2}\right)-\frac{dV_r}{d\theta}\left(V_r\frac{dV_r}{d\theta}+\frac{dV_r}{d\theta}\frac{d^2V_r}{d\theta^2}\right)=0 \quad (1)$$

$$V_\theta=\frac{dV_r}{d\theta} \quad (2)$$

Here γ – ratio of specific heats, θ – angle from the conical axis, V_r – radial velocity, V_θ – normal velocity. Shock wave shape and streamlines about cone are depicted in Figure 1. Waveriders have a performance advantage of maximization lift-to-drag ratio compared with other hypersonic vehicles. It is achieved due to a shock wave attached everywhere along the leading edge. A second feature of the waverider is that upper surface is oriented, so that it does not disturb the overcoming flow. The general waverider disadvantage is that the high lift-to-drag ratio can be realized for a specific operation condition with respect to flight altitude, Mach number and shock wave shape [2].

The viscous effects are not considered in the above mentioned family of vehicles designing. However, waveriders have a higher lift-to-drag ratio if viscous effects as well as transition from laminar to turbulent flow are taken into consideration. This class of aircraft is named a viscous-optimized waveriders [3, 4]. Thus, waverider designing is a complex process including geometric modeling using methods of computational fluid dynamics (CFD) and optimization process.

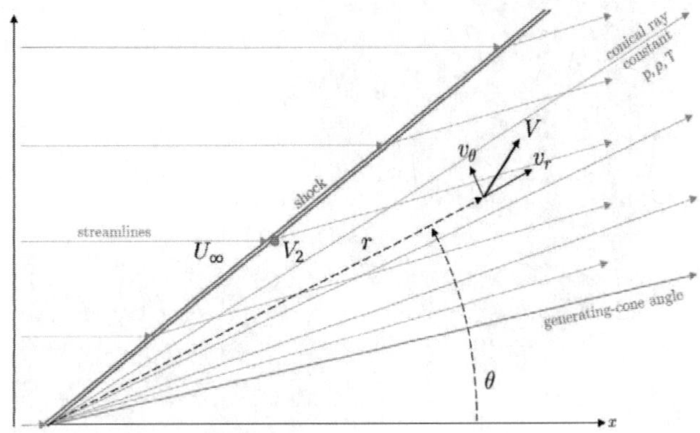

Figure 1. Taylor-Maccoll inviscid cone similarity solution [2].

2. Design of the virtual prototype

The waverider virtual model used for calculations in this study is shown in figure 2. This vehicle is based on viscous-optimized waverider [5] which was designed using Maryland Axisymmetric Waverider Program (MAXWARP) [3, 4]. The model background includes CFD predictions and experimental data obtained from tests in the Unitary Plan Wind Tunnel (UPWT) [6]. The main purpose of this work is to compare current numerical simulation results with mentioned data. It was performed with aid UST3D computer code which has already been validated [7, 8].

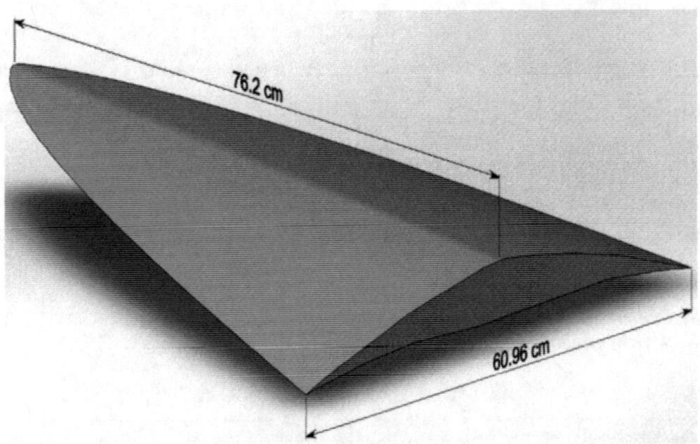

Figure 2. Waverider model designed for free-stream Mach number of 4.

The elements of the singular cone flow method were used for the mentioned waverider designing. The model has a sharp leading edge. The waverider is assumed to have optimal aerodynamic performance at the Reynolds number per foot of $2{\cdot}10^6$.

3. Mesh

The three-dimensional unstructured mesh was generated for the present computational study. It consists of 2262354 finite elements and 393370 nodes. A slip boundary condition is imposed by 64094 fictive elements which were generated along the waverider surfaces. The mesh is shown in figure 3.

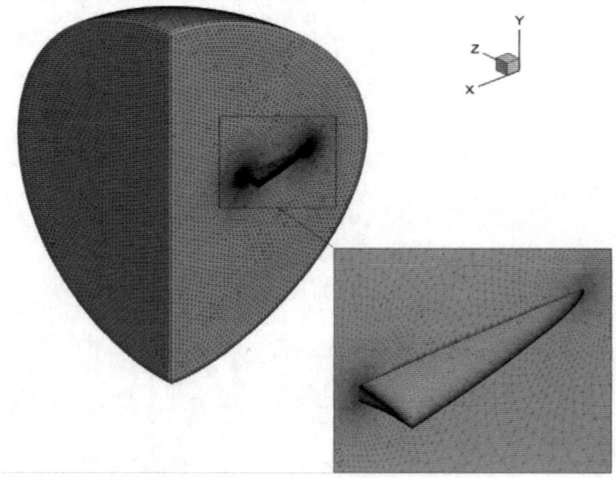

Figure 3. Waverider model mesh.

The most important unstructured mesh feature is the application of the finite volumes method elements [7]. The averaged values of the finite volume derivatives are determined in the framework of this approach. The Navier-Stokes equations are approximated by the following relations

$$\left\langle \frac{\partial f}{\partial \alpha} \right\rangle_{V_i} = \frac{1}{V_i} \int_{V_i} \frac{\partial f}{\partial \alpha} dV = \frac{1}{V_i} \int_{V_i} f\mathbf{i} \cdot d\mathbf{S} \approx \frac{1}{V_i} \sum_{j=1}^{4} S_i^j n_{i,j}^{\alpha} f_i^j \tag{3}$$

$$f = \left(u, \mathrm{v}, w, \mathrm{p}, \mathrm{q}_{\alpha}, \tau_{\alpha\beta}\right); i = 1 \ldots \mathrm{N}; j = 1 \ldots 4; \alpha = x, y, z; \beta = x, y, z$$

Here u, v, w – components of the velocity vector; p – pressure; q_{α} – components of the heat flux vector; $\tau_{\alpha\beta}$ – components of the viscous stress tensor; V_i – i-element volume; S_i^j – i-element j-face surface area; $n_{i,j}^{\alpha}$ – i-element j-face normal component.

4. Numerical simulation

4.1. Problem formulation

The three-dimensional Navier-Stokes equations are used for perfect viscous compressible gas flow describing and presented in the vector form

$$\frac{\partial \mathbf{w}}{\partial t} + \frac{\partial \boldsymbol{F}^x(\mathbf{w})}{\partial x} + \frac{\partial \boldsymbol{F}^y(\mathbf{w})}{\partial y} + \frac{\partial \boldsymbol{F}^z(\mathbf{w})}{\partial z} = \frac{\partial \boldsymbol{G}^x(\mathbf{w})}{\partial x} + \frac{\partial \boldsymbol{G}^y(\mathbf{w})}{\partial y} + \frac{\partial \boldsymbol{G}^z(\mathbf{w})}{\partial z} \tag{4}$$

$$\mathbf{w} = \left(\rho, \rho u, \rho \mathrm{v}, \rho w, \rho E\right)^T$$

$$\boldsymbol{F}^x = \left(\rho u, \rho u^2 + p, \rho u \mathrm{v}, \rho u w, \rho u E + p u\right)^T$$

$$\boldsymbol{F}^y = \left(\rho \mathrm{v}, \rho u \mathrm{v}, \rho \mathrm{v}^2 + p, \rho \mathrm{v} w, \rho \mathrm{v} E + p \mathrm{v}\right)^T$$

$$\boldsymbol{F}^z = \left(\rho w, \rho u w, \rho \mathrm{v} w, \rho w^2 + p, \rho w E + p w\right)^T$$

$$G^x = \left(0, \tau_{xx}, \tau_{yx}, \tau_{zx}, u\tau_{xx} + \mathrm{v}\tau_{yx} + w\tau_{zx} - q_x\right)^T$$

$$G^y = \left(0, \tau_{xy}, \tau_{yy}, \tau_{zy}, u\tau_{xy} + \mathrm{v}\tau_{yy} + w\tau_{zy} - q_y\right)^T$$

$$G^z = \left(0, \tau_{xz}, \tau_{yz}, \tau_{zz}, u\tau_{xz} + \mathrm{v}\tau_{yz} + w\tau_{zz} - q_z\right)^T$$

In the above equations $\mathbf{w}$ – conservative variables vector; F^x, F^y, F^z – projections of the convective flow vector; G^x, G^y, G^z – projections of the viscous flow vector, ρ – density, E – specific total energy. The components of the viscous stress tensor are given as

$$\begin{aligned}
\tau_{xx} &= \mu\left(\frac{4}{3}\frac{\partial u}{\partial x} - \frac{2}{3}\frac{\partial \mathrm{v}}{\partial y} - \frac{2}{3}\frac{\partial w}{\partial z}\right), \ \tau_{xy} = \tau_{yx} = \mu\left(\frac{\partial \mathrm{v}}{\partial x} + \frac{\partial u}{\partial y}\right), \\
\tau_{yy} &= \mu\left(\frac{4}{3}\frac{\partial \mathrm{v}}{\partial y} - \frac{2}{3}\frac{\partial u}{\partial x} - \frac{2}{3}\frac{\partial w}{\partial z}\right), \ \tau_{yz} = \tau_{zy} = \mu\left(\frac{\partial w}{\partial y} + \frac{\partial \mathrm{v}}{\partial z}\right), \\
\tau_{zz} &= \mu\left(\frac{4}{3}\frac{\partial w}{\partial z} - \frac{2}{3}\frac{\partial u}{\partial x} - \frac{2}{3}\frac{\partial \mathrm{v}}{\partial y}\right), \ \tau_{xz} = \tau_{zx} = \mu\left(\frac{\partial u}{\partial z} + \frac{\partial w}{\partial x}\right)
\end{aligned} \tag{5}$$

Here μ – dynamic viscosity. In addition to the Navier-Stokes equations the perfect gas state equation is considered

$$p = (\gamma - 1)\rho U = (\gamma - 1)\rho\left[E - 0.5\left(u^2 + \mathrm{v}^2 + w^2\right)\right] \tag{6}$$

Here U – specific internal gas energy. The thermophysical parameters temperature dependence is not taken into account.

The solution method basis is the original unsteady Navier-Stokes system splitting into physical processes [8]. Each of the time steps calculation fall into three stages. First, the flow parameters reference values are determined without convective effects. The Euler equations are the result of the initial system simplification. Second, mass flows over the cell boundaries are considered. Accordingly, the effects of transition are calculated. And, finally, flow parameters are determined by laws of conservation. The mass, momentum and energy redistribution are accomplished at this stage.

4.2. Boundary conditions

The four types of the boundary conditions were used for the numerical simulation. The free-stream boundary conditions are imposed on the front boundary and are described as

$$\begin{aligned}
\rho_{enter} &= \rho_\infty \\
u_{enter} &= u_\infty \\
\mathrm{v}_{enter} &= \mathrm{v}_\infty \\
w_{enter} &= w_\infty \\
E_{enter} = E_\infty &= \frac{p_\infty}{(\gamma - 1)\rho_\infty} + \frac{u_\infty^2 + \mathrm{v}_\infty^2 + w_\infty^2}{2}
\end{aligned} \tag{7}$$

Here $p_\infty, \rho_\infty, u_\infty, \mathrm{v}_\infty, w_\infty, E_\infty$ – free-stream flow parameters (pressure, density, velocity vector components, specific total energy).

The open boundary conditions were implemented for the gas outflow from computational domain:

$$\left.\frac{\partial \rho}{\partial \boldsymbol{n}}\right|_{exit} = 0$$

$$\left.\frac{\partial u}{\partial \boldsymbol{n}}\right|_{exit} = \left.\frac{\partial \mathrm{v}}{\partial \boldsymbol{n}}\right|_{exit} = \left.\frac{\partial w}{\partial \boldsymbol{n}}\right|_{exit} = 0 \tag{8}$$

$$\left.\frac{\partial E}{\partial \boldsymbol{n}}\right|_{exit} = 0$$

Here $\boldsymbol{n}$ – specific normal vector.

The impermeability boundary conditions were applied in the Euler equations solution process on the solid wall. The normal velocity is equal to zero. The tangent velocity behavior is characterized as continuous. It is described by no-slip boundary conditions:

$$\left.\frac{\partial p}{\partial \boldsymbol{n}}\right|_{wall} = 0 \tag{9}$$

$$\left.u^n\right|_{wall} = \left.\mathrm{v}^n\right|_{wall} = \left.w^n\right|_{wall} = 0$$

The insulated wall boundary condition was realized for the energy equation:

$$\left.\frac{\partial T}{\partial \boldsymbol{n}}\right|_{wall} = 0 \tag{10}$$

The Navier-Stokes equations describe viscous gas model on the wall by a slip boundary conditions. The insulated wall boundary condition was used for the energy equation:

$$\left.\frac{\partial p}{\partial \boldsymbol{n}}\right|_{wall} = 0$$

$$\left.u^n\right|_{wall} = \left.\mathrm{v}^n\right|_{wall} = \left.w^n\right|_{wall} = 0 \tag{11}$$

$$\left.\frac{\partial T}{\partial \boldsymbol{n}}\right|_{wall} = 0$$

The fixed temperature is supported on the body surface:

$$\left.T\right|_{wall} = T_{wall} \tag{12}$$

The symmetry boundary conditions are realized in the symmetry plane. It is the same as the inviscid boundary conditions on the insulated wall:

$$\left.\frac{\partial p}{\partial \boldsymbol{n}}\right|_{symm} = 0$$

$$\left.u^n\right|_{symm} = \left.\mathrm{v}^n\right|_{symm} = \left.w^n\right|_{symm} = 0 \tag{13}$$

$$\left.\frac{\partial T}{\partial \boldsymbol{n}}\right|_{symm} = 0$$

5. Results

The numerical simulation results are presented in figures 5 – 8. The comparison between CFD predictions, experimental data and computational study results was performed at Mach 4 and various angles of attack assuming that base pressure equal to free-stream pressure. The obtained aerodynamic characteristics have the similar type of behavior as experimental data. The lift and drag coefficients increase as angle of attack increases. The reasonable accordance between experimental data and solutions was obtained for the angle-of-attack range from 0° to 4°. The calculation error increases at angles of attack above 4°.

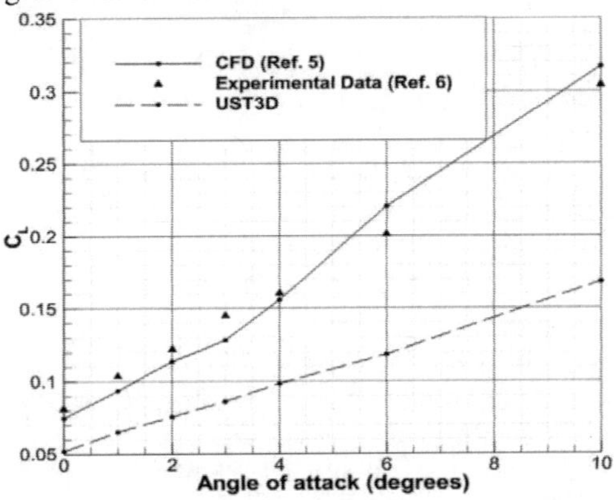

Figure 5. Lift coefficient values for different angles of attack at Mach 4.

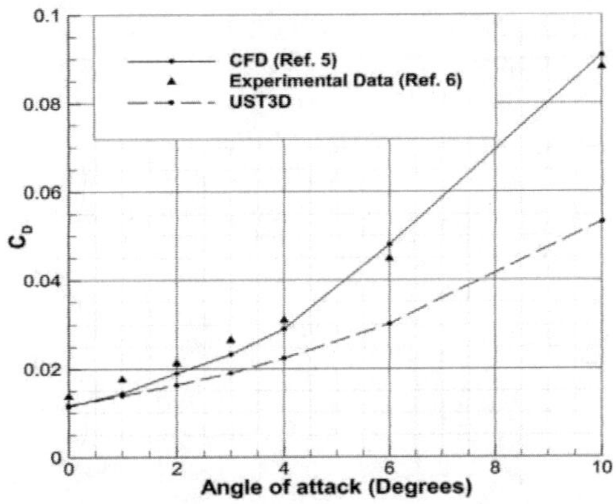

Figure 6. Drag coefficient values for different angles of attack at Mach 4.

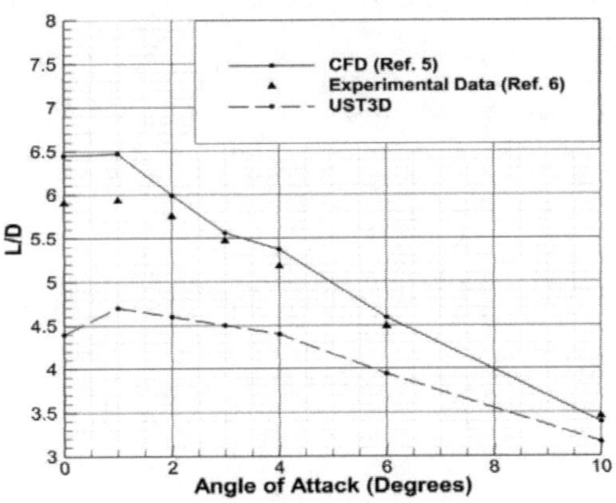

Figure 7. Lift-to-drag ratio values for different angles of attack at Mach 4.

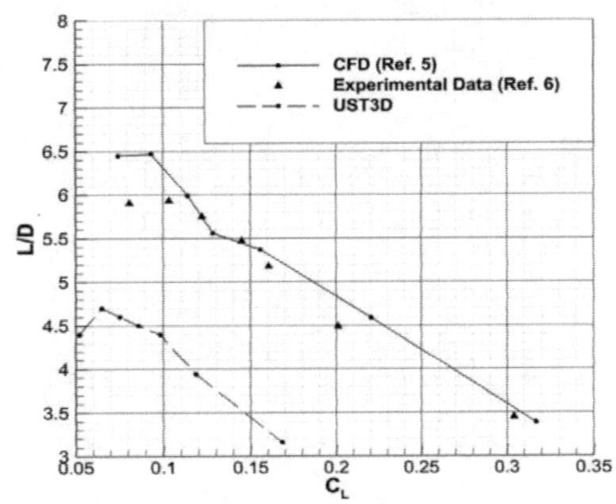

Figure 8. Lift-to-drag ratio as lift coefficient function at Mach 4.

The maximum lift-to-drag ratio occurs at 1° angle of attack. This parameter decreases as angle of attack increase. The aerodynamic performance degrades significantly at high angles of attack due to the drag coefficient increasing faster than lift coefficient growth. The acceptable agreement between the experimental values and UST3D code solutions occurs at high angles of attack.

6. Conclusions

The results obtained with the aid of the UST3D code showed that the aerodynamic parameters can be obtained with reasonable accuracy. It should be noted that the geometric model reproduction precision effects the numerical simulation success significantly. The difference in the definition of the reference area is another factor resulting in the error.

Acknowledgments
The work was performed within the framework of the Program of basic research of the Russian academy of sciences and was supported by the Russian Foundation for Basic Research grant No.16-01-00379. Author thanks Professor S.T. Surzhikov for the help.

References

[1] Maccoll J W 1937 The conical shock wave formed by cone moving at a high speed. *Proc. of the Royal Society of London. Series A, Containing Papers of a Mathematical and Physical Character* vol 159 pp 459–472

[2] Maxwell J R 2016 Hypersonic Waverider Stream Surface Actuation for Variable Design Point Operation *52nd AIAA/SAE/ASEE Joint Propulsion Conference July 25–27 Salt Lake City UT AIAA Paper 16–4706*

[3] Bowcutt K G, Anderson J D 1987 Viscous optimized hypersonic waveriders *AIAA 25th Aerospace Sciences Meeting January 12–15 Reno Nevada AIAA Paper 87–0272*

[4] Corda S, Anderson J D 1988 Viscous optimized hypersonic waveriders designed from axisymmetric flow fields *AIAA 26th Aerospace Sciences Meeting January 11–14 Reno Nevada AIAA Paper 88–0369*

[5] Cockrell Ch E 1993 Interpretation of waverider performance data using computational fluid dynamics *AIAA 24th Fluid Dynamics Conference July 6–9 Orlando FL AIAA Paper 93–2921*

[6] Bauer S X S, Covell P F, Forrest D K, McGrath B E 1990 Preliminary assessment of a Mach 4 and a Mach 6 waverider *Proc. of first international hypersonic waverider symp., University of Maryland, College Park MD* pp 1–24

[7] Zheleznyakova A L and Surzhikov S T 2013 Application of the Method of Splitting by Physical Processes for the Computation of a Hypersonic Flow over an Aircraft Model of Complex Configuration *High Temperature* **51** (6) pp 816–829

[8] Zheleznyakova A L and Surzhikov S T 2014 Calculation of a Hypersonic Flow over Bodies of Complex Configuration on Unstructured Tetrahedral Meshes Using the AUSM Scheme *High temperature* **52** (2) pp 274–284

Validation of computational code UST3D by the example of experimental aerodynamic data

S T Surzhikov

A. Ishlinsky Institute for Problems in Mechanics of the Russian Academy of Sciences, Vernadsky prospekt 101(1), Moscow, 119526, Russia
Dukhov All-Russian Scientific Research Institute of Automatics, Moscow, Russia

E-mail: surg@ipmnet.ru

Abstract. Numerical simulation of the aerodynamic characteristics of the hypersonic vehicles X-33 and X-34 as well as spherically blunted cone is performed using the unstructured meshes.
It is demonstrated that the numerical predictions obtained with the computational code UST3D are in acceptable agreement with the experimental data for approximate parameters of the geometry of the hypersonic vehicles and in excellent agreement with data for blunted cone.

1. Introduction

The author's computational code UST3D (**U**n-**S**tructure **T**etrahedral **3-D**imentional) is intended for calculation of aerodynamic characteristics of hypersonic vehicles of complex geometry on the basis of the Navier-Stokes equations.

This computational code was created with the use of the explicit method of the splitting on physical processes. The simplest variant of the method oriented on gas dynamic problem for structured orthogonal grids has its origin in [1,2]. Recently such a method was applied for unstructured tetrahedral grids [3].

High efficiency of this method was demonstrated by numerical study of aerodynamic characteristics of hypersonic vehicles (HV) which are similar to X–43 and X–51 [3,4]. The mathematical modeling of aerodynamic characteristics was performed with the use of multi-nuclear working stations of the middle performance. The method of parallelization on the base of compiler 'Intel ® Parallel Studio XE 2015' was used.

Development of the splitting method by improvement approach for calculation of a decay of a discontinuity on edges of tetrahedral elements had resulted in decreasing the numerical dissipation [5]. However, unfortunately, the calculation time was increased about in three times. Nevertheless, it was concluded that there are reasons to use both these methods depending on the purpose of a numerical simulation.

It is well known that to achieve high degree of reliability of numerical predictions it is necessary to use validated computer codes. General goal of the present paper is the validation of the computational code UST3D by the examples of rebuilding aerodynamic coefficients of some hypersonic vehicles that were measured in experimental research on wind tunnels.

Experimental data for models of two hypersonic vehicles, X–43 [6] and X–51 [7], and for spherically blunted cone [8] are analyzed and numerically reconstructed in the given paper.

It should be noted that numerical technology of the validation of a computer code includes several significant stages. These are:

- Creation of electronic surface of investigated vehicles,
- Creation of computational grids for surface and volume,
- Numerical rebuilding of aerodynamic characteristics of aircrafts known from experimental measurements.

The electronic images of surfaces HV, investigated in present work, as well as the surface and volume computational meshes used in present work were created in [9], where numerical format of *.neu was used for the grids. Numerical approaches to creation of the computational meshes were based on works [10–12].

The computational code UST3D consists of three parts. The first part is the preprocessor, in which all necessary data for each elementary computational domain are collected. Each elementary computational volume of the tetrahedral is characterized by the following data:

- the unique number of each vertex of each side,
- the direction cosines of each side;
- area of the each side,
- volume,
- numbers of elementary volumes bounded with each of fourth edges,
- characteristics of the volumes bordering with the each surface (there are several kinds of the volumes: the volume inside the computational domain, the fictive volume outside the free surface of the computational domain, the fictive volume inside solid body).

The second part is the main computational solver, which realizes numerical simulation of the Navier-Stokes equations. The third part is the postprocessor, which is intended for graphical representation of obtained results of numerical simulation.

2. Aerodynamic characteristics of the X–33

Experimental data that are taken as the base for the validation are given in [7]. It was proposed that the semi-scale prototype of a rocket-based vehicle X–34 will demonstrate key design and operational aspects of a single-stage-to-orbit (SSTO), reusable launch vehicle (RLV). The X–33 concept had a lifting body shape with two integrated linear aerospike rocket engines and flies a suborbital trajectory to simulate important aero-thermodynamic aspects of ascent and reentry environments for a full-scale RLV.

The X–33 vehicle is a lifting-body delta platform with twin vertical tails, canted fins, and body flaps. The canted fins have a dihedral of 20 deg and a –8.58 deg incidence angle. Reference dimensions and the geometry details are given in [7], where two configurations of the X–33 vehicle were investigated. These are: the F–Loft and Rev–F configurations. The Rev–G configuration has minor modifications to the aft, upper surface of the vehicle.

We used experimental data obtained at two facilities of NASA Langley Research Center (LaRC) Aerothermodynamics Facilities Complex (AFC).

Nominal flow conditions realized in the 20–Inch Mach 6 and the 31–Inch Mach 10 Air Tunnels are presented in table 1.

Table 1. Nominal flow conditions

M	6	10
p_∞, **erg/cm³**	11180.	2397.
T_∞, **K**	62.1	48.9
γ	1.4	1.4
T_{surf}, **K**	300.	300.

Table 2. Reference dimensions for the X–33 model

Dimension	0.007-scale model
S_{ref}, cm^2	73.2
S_{mid}, cm^2	28.2
c.g.$_{ref}$, cm	8.91
X_{nose}, cm	5.64

The results of numerical simulation for the aerodynamic coefficients C_L and C_D and their comparison with available experimental data [7] are shown in figure 1. The coefficients of the lift-body force C_L and of the drag-force C_D data flow coordinate system are calculated as follows:

$$C_L = C_y \cos\alpha - C_x \sin\alpha \ , \ \ C_D = C_x \cos\alpha + C_y \sin\alpha \ , \tag{1}$$

where C_x=C_A and C_y=C_N are axial force coefficient and normal force coefficient; α is the angle of attack.

Aerodynamic coefficients C_x and C_y are calculated as following:

$$C_x = F_x / q_\infty S_{mid} \ , \ \ C_y = F_y / q_\infty S_{mid} \ , \tag{2}$$

where $q_\infty = 0.5\rho_\infty V_\infty^2$,

$$F_x = \iint\limits_s \left[(p - p_\infty)\omega_x^n + \tau\omega_x^\tau \right] ds \ ,$$

$$F_y = \iint\limits_s \left[(p - p_\infty)\omega_y^n + \tau\omega_y^\tau \right] ds \ , \tag{3}$$

p and p_∞ are the pressure near surface and in input flow; τ is the frictional force; that is tangential to elementary surface ds with unit vector τ; and $\omega_x^n, \omega_y^n, \omega_x^\tau \omega_y^\tau$ are the directional cosines:

$$\omega_x^n = \cos\ (\mathbf{i} \cdot \mathbf{n}) \ , \ \omega_y^n = \cos\ (\mathbf{j} \cdot \mathbf{n}) \ ,$$

$$\omega_x^\tau = \cos\ (\mathbf{i} \cdot \boldsymbol{\tau}), \ \omega_y^\tau = \cos\ (\mathbf{j} \cdot \boldsymbol{\tau}), \tag{4}$$

where $\mathbf{i}, \mathbf{j}$ are the unit vectors of the Cartesian coordinate system, $\mathbf{n}, \boldsymbol{\tau}$ are the normal and tangent to the elementary surface.

For recalculation of the coefficients for dimensions of experimental models it is used the following formula

$$C_{L,D}^* = C_{L,D}\, S_{mid} / S_{ref} \ . \tag{5}$$

All calculations were performed for the following set of parameters of the computational mesh: Nel = 3 275 358 is the total number of elemental volumes, *Npoints* = 569 087 is the total number of points, *Nsurf,el* = 72 079 is the number of tetrahedral elements on the streamlined surface, and $N_{fict,el}$ = 93 160 is the number of fictive elements.

The calculations were performed for half-space bounded by the plane z=0. Boundary conditions of the symmetry were used on this surface. The surface and volume meshes are shown in figure 2. Distribution of temperature near the surface is also shown in figure 1a.

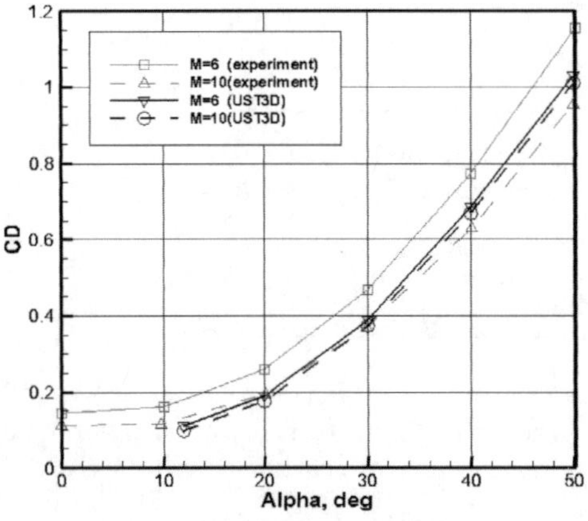

a

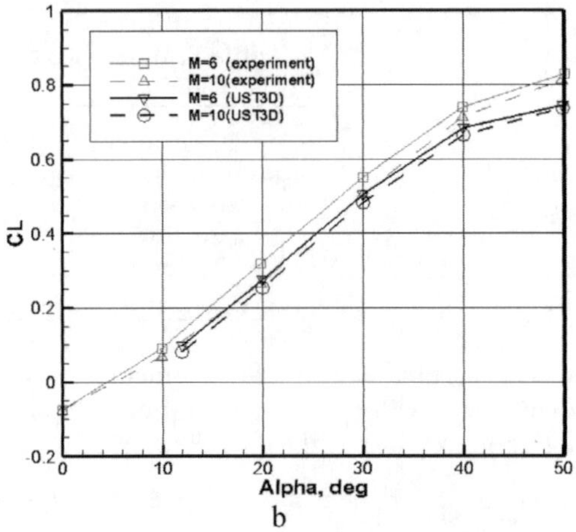

b

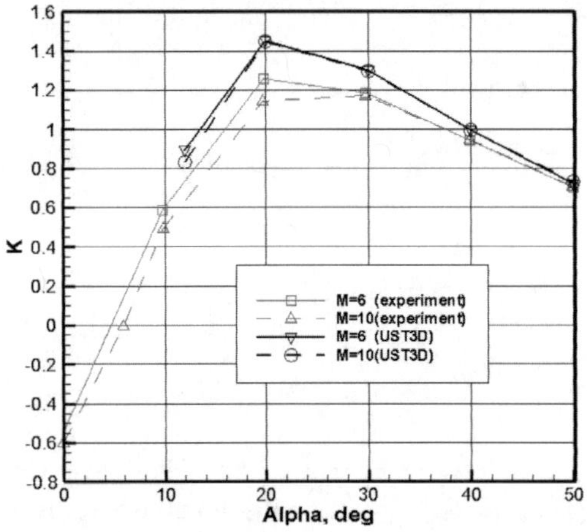

c

Figure 1. Aerodynamic coefficients of drag (a), lifting (b), and lifting-to-drag relation (c) for the model X–33.

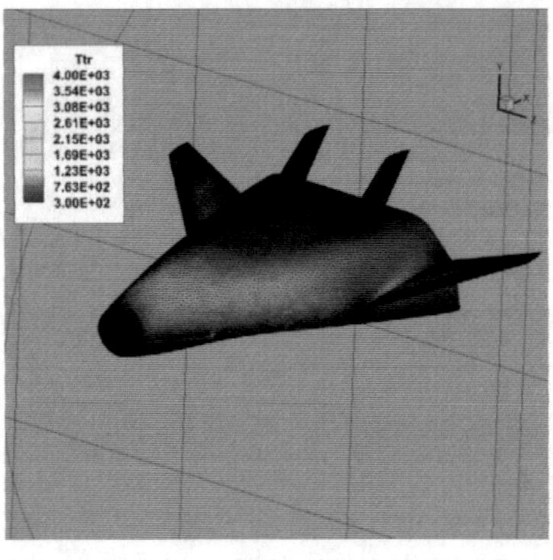

a

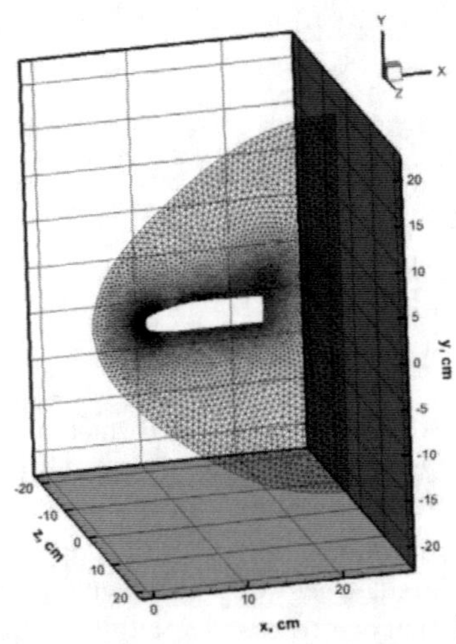

b

Figure 2 (a) Surface and (b) volume computational grid for the model X–33.

Distributions of Mach number (*a*), pressure (*b*), and temperature (in K; *c*) along the surface and in the plane z=0 at Mach numbers M = 10 and M = 6 and the angle of attack of 50 deg are shown in figures 3 and 4. Some disagreement between the experimental and calculation data is explained by the differences between real experimental geometry and the electronic geometry that was created on the basis of some images from scientific papers.

Differences between the real and the electronic geometries does not impedes to investigate the aerodynamic characteristics of the hypersonic vehicles, similar to the X–33, because created surface and volume grids allow to perform an analysis of gas dynamic functions similar to those shown in figures 3 and 4, and to perform cross-verification of various computational codes.

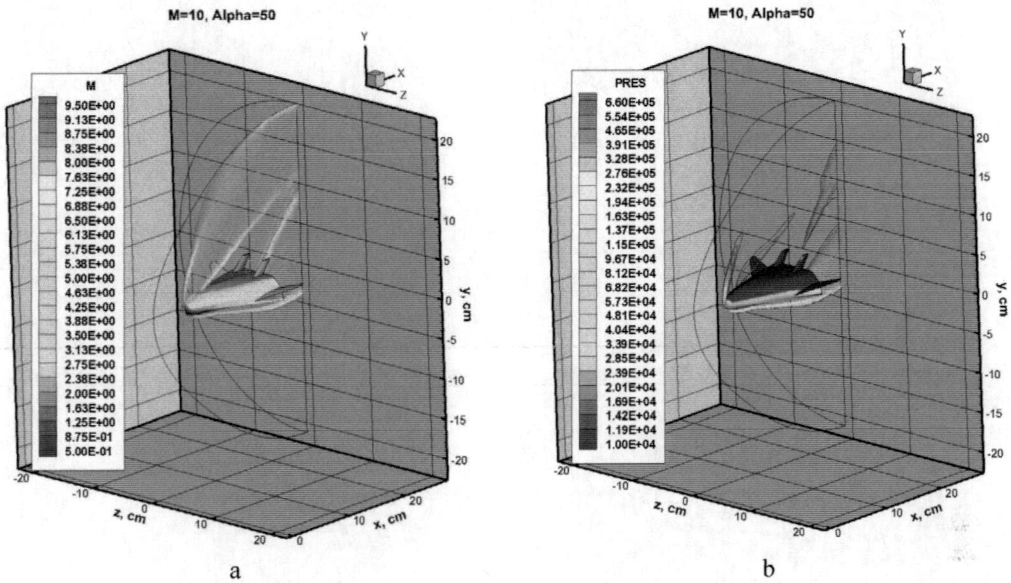

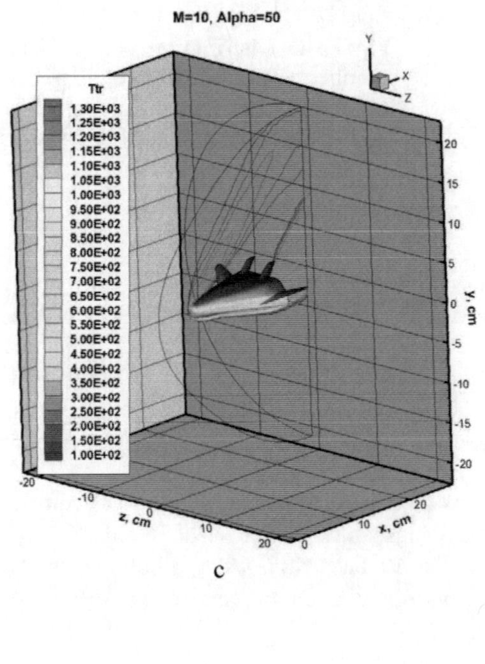

Figure 3. Distributions of (a) Mach number, (b) pressure (in erg/cm^3), and (c) temperature (in K) along surface and in the plane z=0 at M=10 and angle of attack of 50 deg.

IOP Conf. Series: Journal of Physics: Conf. Series **815** (2017) 012023 doi:10.1088/1742-6596/815/1/012023

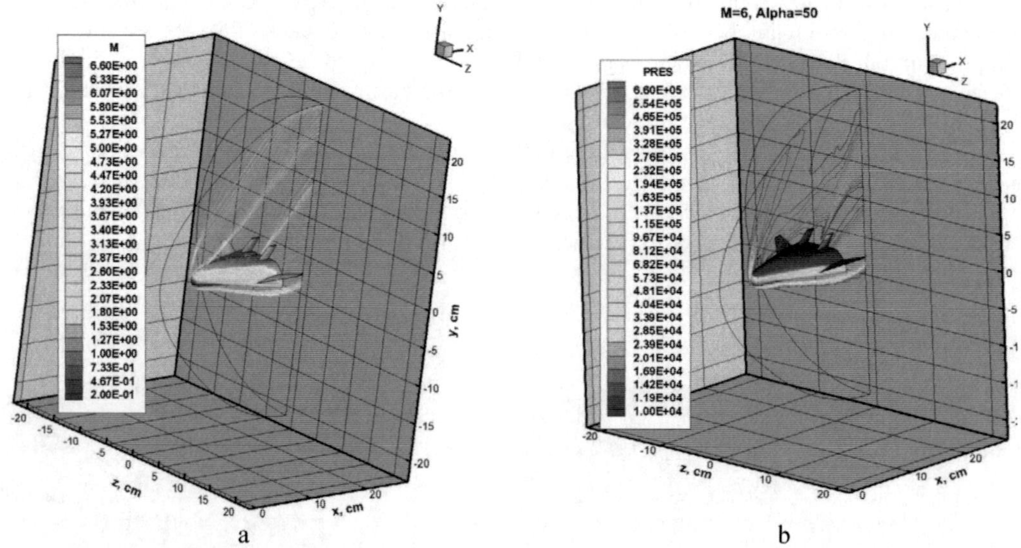

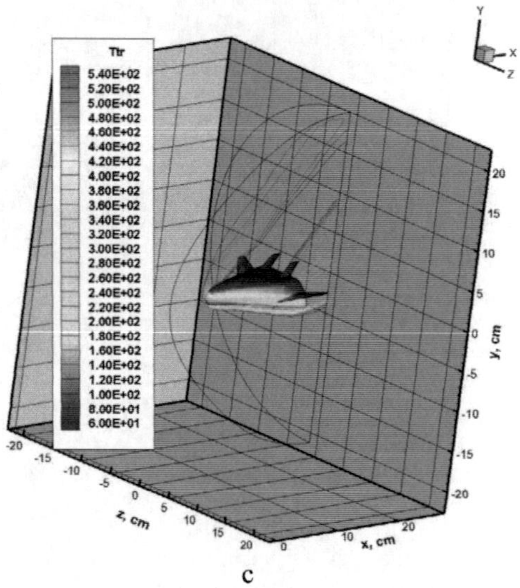

Figure 4. Distributions of (a) Mach number, (b) pressure (in erg/cm^3), and (c) temperature (in K) along surface and in the plane z=0 at M=6 and angle of attack of 50 deg.

3. Aerodynamic characteristics of the X–34

Aerodynamic experimental data for the X-34 are presented in [6]. The model X–34 is a model of an unmanned suborbital vehicle, designed to be used as a flying test bed to demonstrate key vehicle and operational technologies applicable to future reusable launch vehicles (RLVs). It was presumed that the X–34 will be air-launched from a L–1011 carrier aircraft at approximately Mach number M = 0.7 and altitude of ~12 km. An onboard engine will be used for acceleration the vehicle to speeds exceeding M = 7 and altitudes ~76 km. Then an unpowered entry will follow including autonomous landing.

One can note that the X–34 platform is similar to the Space Shuttle Orbiter. Real construction (length) of the X–34 is approximately 16.5 m with the wingspan of 8.54 m. A double-delta wing with sweep angles with 80 and 45 deg is employed. The X–34 has been designed and build by the Orbital Sciences Corporation (OSC). The parameters of trajectory of the X–34 are presented in [6].

The most of the data in [6] were obtained with models representing the latest so called Outer-Mold-Line geometry. The models were built in two scales, namely, 0.018 scale (there were two similar models) and 0.033 scale.

The X–34 reference dimensions, which were used in present paper, are presented in table 3.

Table 3. Reference dimensions for the X–34 model

Dimension	0.018-scale model
S_{ref}, cm^2	107.68
S_{mid}, cm^2	24.0
c.g.$_{ref}$, cm	19.2
X_{nose}, cm	24.83

The aerodynamic data of this paper were obtained with the use of three NASA Langley Research Centre facilities: the Low-Turbulence Pressure Tunnel – LTPT; the Unitary Plan Wind Tunnel – UPWT, and the 20–Inch Mach 6 Air Tunnel. Brief information about these facilities is in [6].

The nominal flow conditions borrowed from [6] and used in this paper are shown in table 4.

Table 4. Nominal flow conditions for the X–34 model

Facility	**M**	**p_∞, erg/cm^3**	**T_∞, K**
LTPT	0.4	850 000	315
	0.8	623 000	288
	0.9	561 000	280
	1.05	215 000	266
UPWT	2	144 000	181
20 – Inch Mach 6	6	2 244	60
	6	4 937	61.7
	6	10 322	61.7
	6	15 259	63.3

All calculations in this paper were carried out for the X–34 0.018-scale model.

All calculations were performed for the following set of parameters of the computational mesh: Nel = 4 916 505 is the total number of elemental volumes, $Npoints$ = 875 947 is the total number of points, $Nsurf,el$ = 160 292 is the number of tetrahedral elements on the streamlined surface, and $N_{fict,el}$ = 221 872 is the number of fictive elements.

The calculations were performed for half-space bounded by the plane z = 0. Boundary conditions of the symmetry were used on this surface. The volume mesh is shown in figure 5.

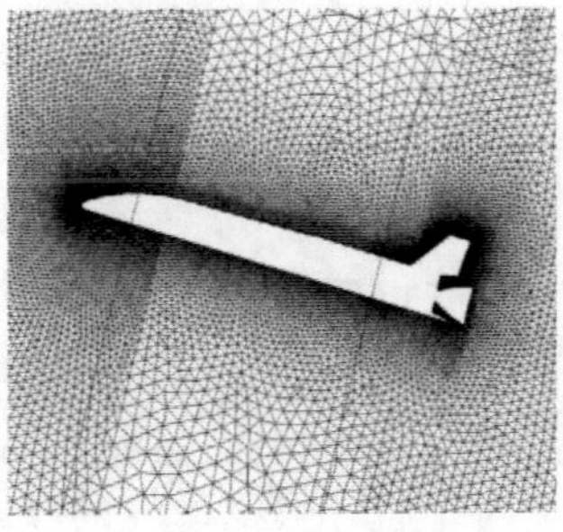

Figure 5. Computational grid on the plane $z = 0$ for the model X–34.

The results of numerical simulation for the aerodynamic coefficients of lifting C_L and drag C_D for the model of X–34 are shown in figure 6. Experimental data obtained in [6] are also shown in this figure.

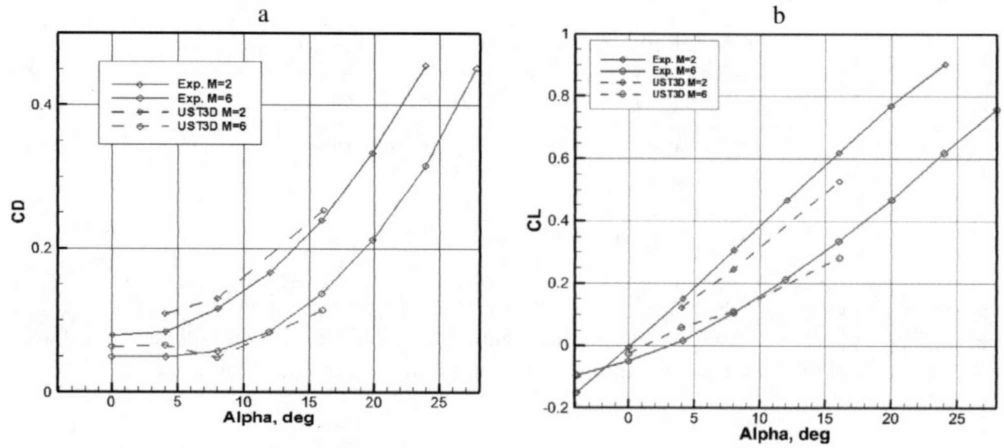

Figure 6. Aerodynamic coefficients of (a) drag and (b) lifting for the model X–34.

Distribution of pressure and Mach number along the surface and on the plane z = 0 are shown in figure 7.

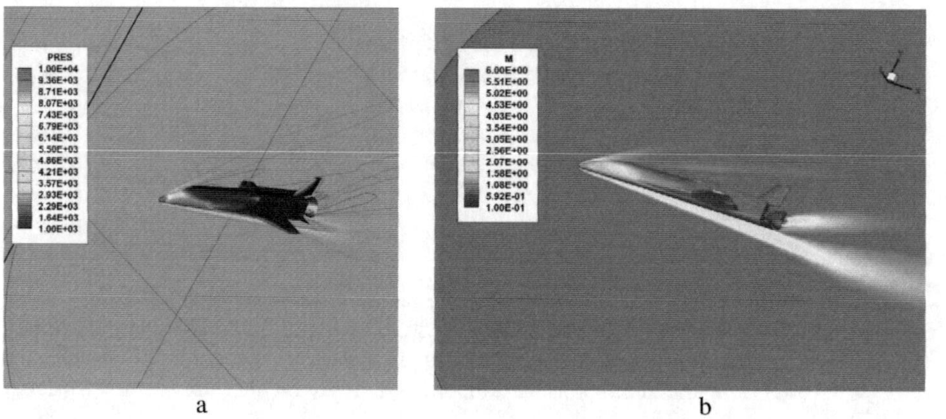

Figure 7. Distributions of (a) pressure (in erg/cm^3) and (b) Mach number for the model X–34 at angle-of-attack of α = 16 deg.

4. Aerodynamic characteristics of the spherically blunted cone

In the present paper the numerical simulation of the aerodynamic coefficients of the spherically blunted cone has been performed to validate the computational code UST3D. It is significant that the geometry of the blunted cone is known precisely.

Detailed experimental investigation of the aerodynamic coefficients of the spherically blunted cone has been performed in [8].Force and moment tests were conducted in this work for sharp and spherical blunted cone configuration having half the cone angle of 9^0 in the Langley 11-inch hypersonic tunnel at Mach number of 6.77, the Reynolds number per meter of 5.315×10^6, and the angle-of-attack range from 0^0 to 180^0.

Calculations of the aerodynamic coefficients are performed with the use of equations (1) – (5). Schematic of the calculational task and a definition of the sign of aerodynamic moment are shown in figure 8.

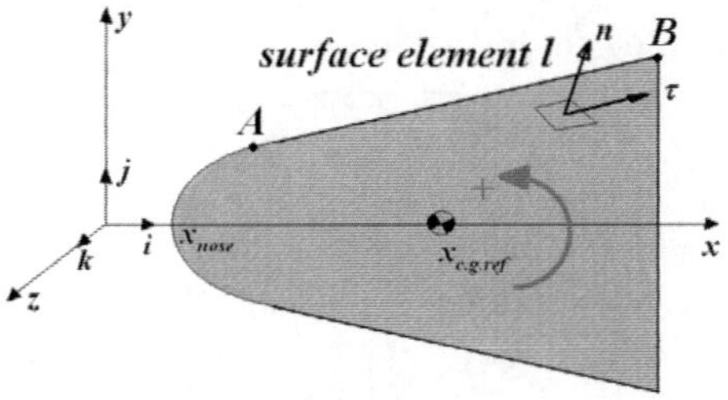

Figure 8. Geometry of experimental model [8].

A pitch moment for elementary surface element is defined as follows:

$$\Delta M_{l,z} = x_l \Delta F_{l,y} - y_l \Delta F_{l,x} \ , \qquad (6)$$

where x_l, y_l are the coordinates of the l-th surface element, which are counted out from the origin of the laboratory frame of reference; $\Delta F_{l,x}$, $\Delta F_{l,y}$ are the projections of the surface force effecting on the l-th surface element.

The elementary moment of the pitch relative to the center of gravity is calculated as follows:

$$\Delta M_{l,z,c.g.} = \Delta M_{l,z} - \left(x_{cg}\Delta F_{l,y} - y_{cg}\Delta F_{l,x}\right) = \Delta F_{l,y}\left(x_l - x_{cg}\right) - \Delta F_{l,x}\left(y_l - y_{cg}\right) , \qquad (7)$$

where x_{cg}, y_{cg} are the coordinates of the center of gravity in the laboratory frame of reference.

It should be noted that the longitudinal coordinate $x_{cg,ref}$ is counted from the critical point of the HV, therefore

$$x_{cg} = x_{nose} + x_{cg,ref} \ , \qquad (8)$$

where x_{nose} is the longitudinal coordinate of the critical point.

It was supposed that

$$y_{cg} = y_{cg,ref} = z_{cg} = z_{cg,ref} = 0 \ .$$

In accordance with (6) the positive moment is counted in counter clock-wise direction.

The coefficient of the pitch moment is calculated by the following formula:

$$m_z = \frac{\sum_{l}^{N_{el.surf}} \Delta M_{l,z,c.g.}}{L_{ref} \cdot q_\infty S_{ref}},$$

here L_{ref} is the reference length.

Initial conditions used in the calculations are presented in table 5.

Table 5. Nominal flow conditions for blunted cone

p_∞, erg/cm^3	188 000
T_∞, K	56.9
γ	1.4
M_∞, K	6.77

Reference dimensions for this model are shown in table 6.

The results of numerical simulation for the aerodynamic coefficients of lifting C_L , drag C_D , and pitch moment for the spherically blunted cone are shown in figures 9. Experimental data obtained in [8] are also shown in this figure.

Table 6. Reference dimensions for blunted cone

R_N ,	cm	0.926
X_A ,	cm	5.000
$Y_A = R_N \cos\alpha$,	cm	0.914
X_B ,	cm	17.12
Y_B,	cm	2.858
x_{nose} ,	cm	4.219
x_{cg},	cm	11.298
$x_{cg,ref}$,	cm	7.079
L_{ref},	cm	5.715
S_{ref},	cm^2	25.65

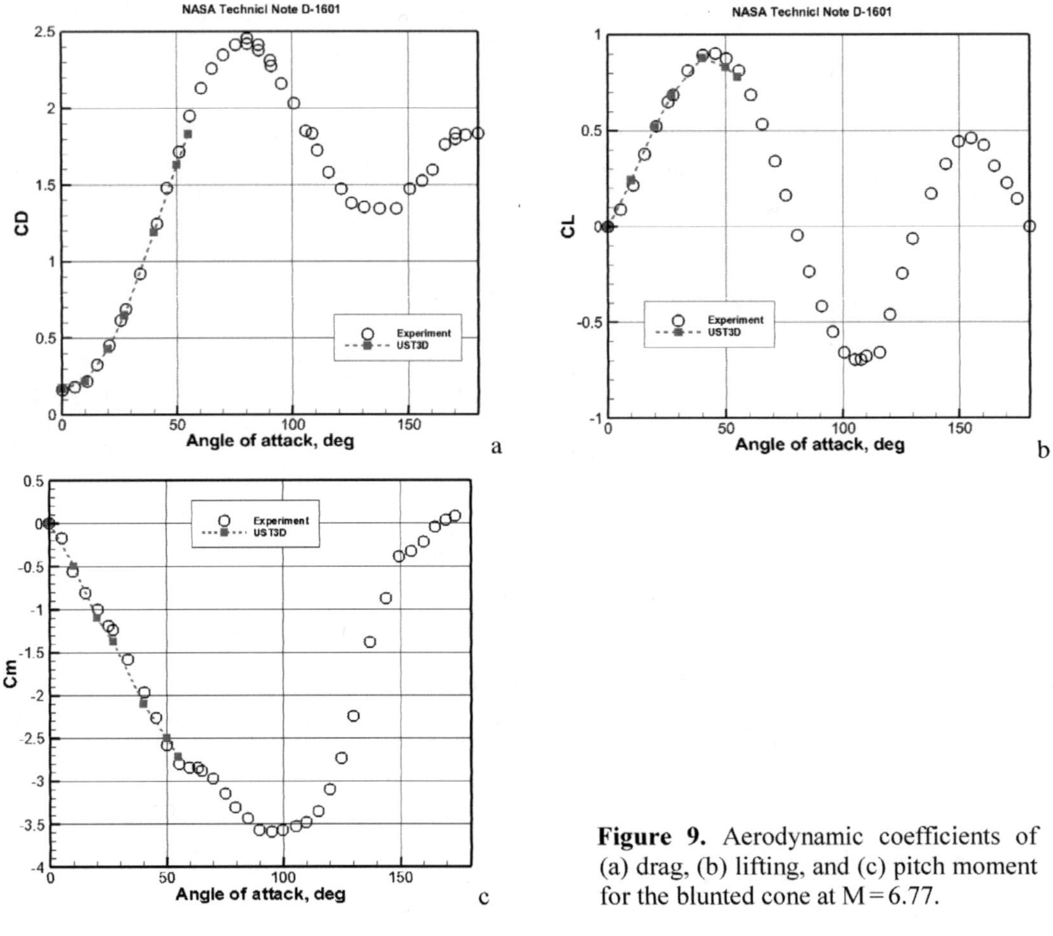

Figure 9. Aerodynamic coefficients of (a) drag, (b) lifting, and (c) pitch moment for the blunted cone at M=6.77.

All calculations were performed for the following parameters of the computational mesh: $Nel = 686\,145$ is the total number of elemental volumes, $Npoints = 140\,889$ is the total number of points, $Nsurf{,}el = 37\,116$ is the number of tetrahedral elements on the streamlined surface, and $N_{fict,el} = 102\,556$ is the number of fictive elements.

Distribution of Mach, pressure and temperature along the surface and on the plane z = 0 are shown in figure 10.

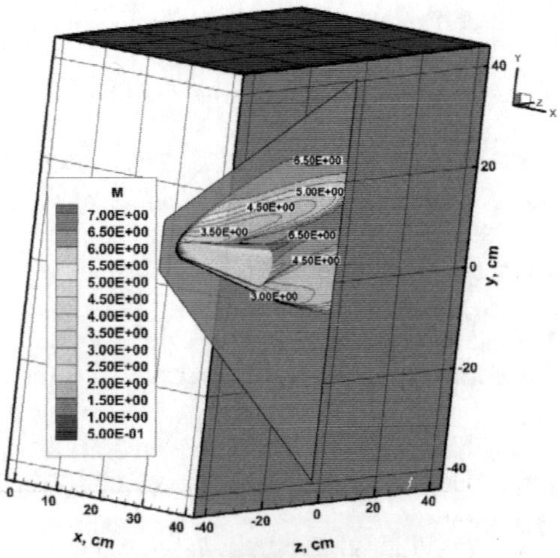

a

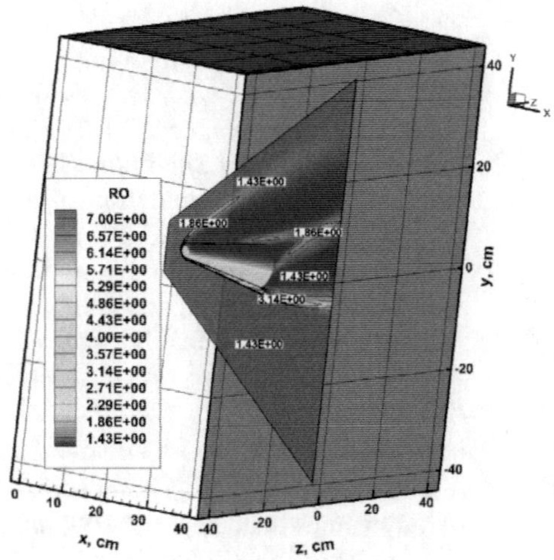

b

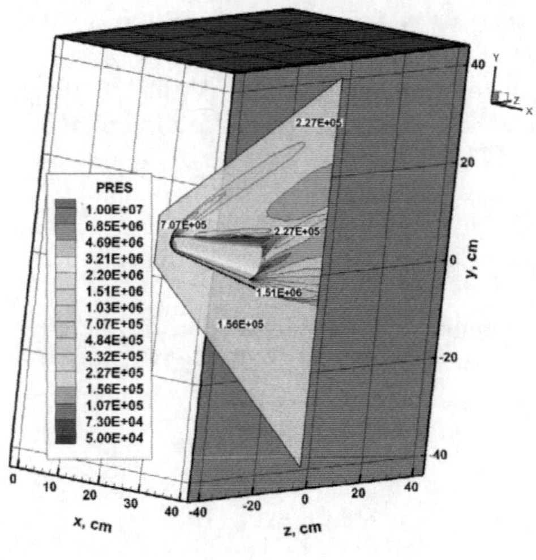

c

Figure 10. Distributions of (a) Mach number, (b) pressure (in erg/cm^3), and (c) temperature (in K) along surface and on the plane z=0 at M=6.77 and angle of attack of 27 deg.

Conclusion

Numerical simulations of the aerodynamic coefficients of the hypersonic vehicles X–33 and X–34 as well as for spherically blunted cone have been performed with the aim of the validation of the author's computational code UST3D.

Experimental aerodynamic data used for the validation were obtained in wind tunnels of NASA Langley Research Center.

It is demonstrated that the numerical predictions obtained with the computational code UST3D are in acceptable agreement with the experimental data for approximate parameters of the geometry of the hypersonic vehicles and in good agreement with precise data for blunted cone.

The work was performed within the framework of the Program of basic research of the Russian academy of sciences and under support of the RFBR grant # 016 – 01 – 00379.

References

[1] Ewans M V and Harlow F H 1957 The particle-in-cell method for hydrodinamic calculations *Los Alamos Scientific Lab. Rept.* N LA–2139 Los Alamos
[2] Belotserkovsky O M and Davydov Yu M 1982 Method of the Large Particles for Gas Dynamics (Moscow: Nauka) p 391(in Russian)
[3] Zheleznyakova A L and Surzhikov S T 2013 Application of the Method of Splitting by Physical Processes for the Computation of a Hypersonic Flow over an Aircraft Model of Complex Configuration *High Temperature* **51** (6) pp 816–829
[4] Zheleznyakova A L and Surzhikov S T 2013 On the Way to Creation of Vortual Model of Hypersonic Vehicle (Moscow: IPMech RAS Press) p 159 (in Russian)
[5] Zheleznyakova A L and Surzhikov S T 2014 Calculation of a Hypersonic Flow over Bodies of Complex Configuration on Unstructured Tetrahedral Meshes Using the AUSM Scheme *High Temperature* **52** (2) pp 274–284
[6] Brauckmann G J 1999 X–34 Vehicle Aerodynamic Characteristics JSR **36** (2) pp 229–239
[7] Murphy K J, Nowak R J, Thompson R A, Hollis B R and Prabhu R 2001 X–33 Hypersonic Aerodynamic Characteristics *JSR* **38** (5) pp 670–698
[8] Neal L Jr 1963 Aerodynamic Characteristics at a Mach Number of 6.77 of a 9^0 cone configuration , with and without spherical afterbodies, at angles of attack up to 180^0 with various degrees of nose blunting. *NASA TN D-1606*
[9] Zheleznjakova A L 2016 A Unified Approach to Building Complex Virtual Surfaces and Computational Grids for the Comprehensive 3D Simulation of Aerospace Industry Products *Physical-Chemical Kinetics in Gas Dynamics* **17** (2) http://chemphys.edu.ru/issues/2016-17-2/articles/634/
[10] Zheleznyakova A L and Surzhikov S T 2013 Molecular dynamic-based unstructured grid generation method for aerodynamic application *Computer Physics Communication* **184** (12) pp 2711–2727
[11] Zheleznyakova A L and Surzhikov S T 2011 Triangular-Mesh Generation for Aerodynamic Problems by Molecular-Dynamics Simulation *Doclady Physics* **56** (7) pp 385–390
[12] Alter S J 1998 Grid Generation Techniques Utilizing the Volume Grid Manipulator *AIAA Paper 98–3012*

Approximate method for calculating convective heat flux on the surface of bodies of simple geometric shapes

V V Kuzenov [1,2] and S V Ryzhkov[2]

[1] N.L. Dukhov All-Russian Research Institute of Automatics (State Corporation «ROSATOM» company), Suschevskaya Street 22, Moscow 127055, Russia
[2] Bauman Moscow State Technical University, 2 Baumanskaya Street 5, 1, Moscow 105005, Russia

vik.kuzenov@gmail.com, svryzhkov@bmstu.ru

Abstract. The paper formulated engineering and physical mathematical model for aerothermodynamics hypersonic flight vehicle (HFV) in laminar and turbulent boundary layers (model designed for an approximate estimate of the convective heat flow in the range of speeds M = 6-28, and height H = 20-80 km). 2D versions of calculations of convective heat flows for bodies of simple geometric forms (individual elements of the design HFV) are presented.

1. Introduction

Development of the new aircrafts requires fundamental, experimental, theoretical and computational research and analysis of the aerodynamic characteristics, mass heat processes and flow characteristics of air breathing engine – the hypersonic rocket engine [1-4]. Experimental researches are expensive, in many cases modeling of physical and chemistry processes in the ground conditions accompanied with flying of hypersonic aircrafts in Earth's atmosphere is fundamentally impossible. In this case it is expedient to determine aerodynamics characteristics using theoretical and computational methods.

During HRE development lots of problems occur. Two of them we will distinguish:
calculation of viscous tangential tensions and friction coefficient C_f on the surface of the streamlined body; calculation of convective heat flux $q_{w,\ell}$ to the surface of the streamlined body.

It should be considered that the changes in the chemical composition of gas occur while aircraft moving in Earth's atmosphere with hypersonic velocities in shock and boundary (dynamic and heat) layers formed around HFV.

One of the ways to simplify the system of boundary layer equations, i.e. evaluation of heat flows and friction coefficient C_f, is the transition from the differential equations for a single particle calculation in a gas environment to satisfaction these equations in average by thickness of boundary layer. For heat flow assessment it is supposed that boundary layer can be determined using local similarity, it means that for boundary layer of the complex form surface creates an analogy with less complex form body, for example plates or cones (so called effective length method).

The purpose of this paper is to formulate of simplified method of convective heat flows for simple spatial form bodies. Wherein we should admit that heat flow computation is followed by external non viscous flow around surface of the streamlined body computation.

2. The main aspects of approximate computation method of convective heat flux approaching HFV surface

Consider the ways to define convective heat flow $q_{w,\ell}$ and local friction coefficient C_f (total friction coefficient can be found with integration of local coefficient through the whole body surface) at each point on the surface of an axisymmetric body (figure 1) according to effective length method proposed in Refs. 4 and 5. The effective length method's application area is high Reynolds number flow (Re > $10^4 - 10^5$). In that case the model of thin border layer is acceptable. Also it should be mentioned that the given method can be applied in flew areas with small pressure gradient along the line of the flow and without any gap zones.

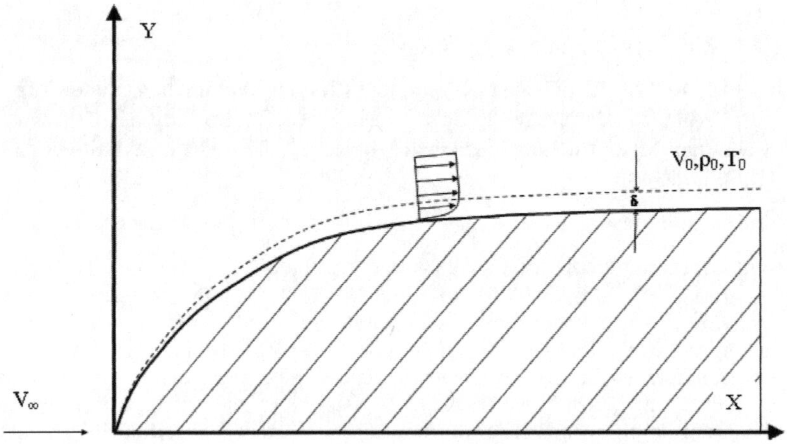

Figure 1. Scheme of the boundary layer near the surface of the axially symmetric body.

If the flow is laminar, the heat transfer around the critical point can be determined by the formula [5,6]:

$$q_{w,\ell} = 1,93 \times 10^{-4} V_\infty^{1,08} \left(H_0 - H_w\right)\sqrt{\frac{\rho_\infty}{R}}.$$

For the turbulent regime, we use formula:

$$q_{w,\ell} = 4,69 \times 10^{-4} V_\infty^{1,25} \left(H_0 - H_w\right)\left(1 + \frac{T_w}{T_0}\right)^{-\frac{2}{3}} \frac{\rho_\infty^{0,8}}{R^{0,2}}.$$

As a result the heat irradiation coefficient in critical point is inversely proportional to square root $\sqrt{R}$ (proportional to $R^{0,2}$ in case of turbulent flow) from the dulling radius R. Therefore, at high flight speeds and high braking temperatures with decreasing radius of blunting R the critical point values sharply increase convective and radiative fluxes.

Let's consider the application of effective length method in a point with Cartesian coordinate x^* for body rotation case with radius $R(x^*)$ (figure 2). l^* is the arc's length that forms from the beginning till the pending point. We consider that the heat border layer with thickness δ_T is formed in the point (x^*, R). The effective length x_{eff} is called a length of a flat plate that covered with the same border layer as on the length l^* of the considered body (figure 2) with the same options during the external flow.

For the axisymmetric case, the effective length x_{eff} will be the length of a cylinder with radius *R*.

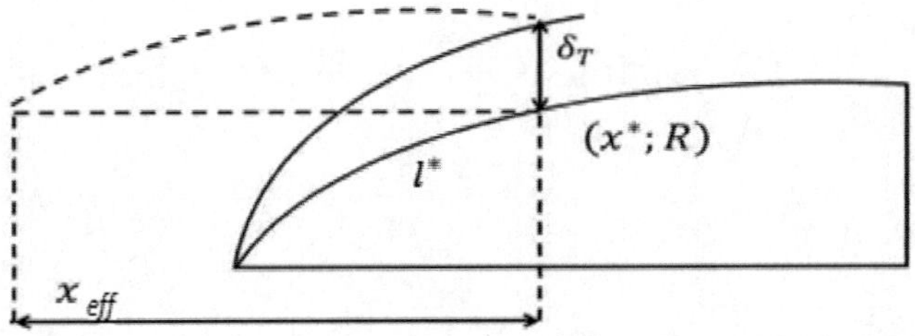

Figure 2. Scheme for the definition of the boundary layer characteristics by local similarity method.

For computing the heat and mass transfer of the classic bodies (plate, cylinder, cone, sphere and etc.) with gas flow which main parameters are variable with its length we use effective length method *l**.

According to this method (for laminar flow) effective length x_{eff} is determined by the ratio [7,8]:

$$x_{eff} = \frac{\int_0^x \left[R^2 K^2 K_1^2 \mu_0 \rho_0 U_0 C_{p,0}^2 (T_e - T_w)^2 \Pr_0^{-\frac{4}{3}} \right] dx}{\left[R^2 K^2 K_1^2 \mu_0 \rho_0 U_0 C_{p,0}^2 (T_e - T_w)^2 \Pr_0^{-\frac{4}{3}} \right]}.$$

The integral values are variable, they vary from the beginning of border layer formation (critical point) till the pending section with Cartesian coordinate *x*. *R*(*x*) is rotation radius of the axisymmetric body, $\mu_0, \rho_0, U_0, \mathrm{M}_0, \Pr_0$ are viscosity, density, velocity, Mach and Prandtl local numbers taken from the external border of the border layer (figure 1) in this section with Cartesian coordinate *x*; T_w, T_e are the streamlined body surface temperature and insulated wall temperature.

$$T_r = T_0\left(1 + r\frac{\gamma - 1}{2} M_0^2\right),\ r = \sqrt{Pr_0}\ ,$$

where r is recovery temperature coefficient (it shows what part of the kinetic energy of external flow spent on increasing the gas enthalpy near the streamlined body surface).

For accelerated and slightly slower flows coefficient K_1 introduces in effective length x_{eff}. It considers velocity gradient influence. This parameter is closed to a unit and depends on velocity gradient parameter and temperature factor [5- 8]:

$$K_1 = \left[1 + 0{,}16\left(1 + \frac{T_w}{T_0^*}\right)\left(\frac{2m}{m+1}\right)^{1/3}\right]^{1/2}.$$

Coefficient *K* is the compressibility factor:

$$K = \left(\frac{\rho_0 \mu_0}{\rho_w \mu_w} \right)^{1/3}.$$

The value of the dimensionless velocity gradient can be found from the formula:

$$m = \frac{x}{V_0} \frac{\partial V_0}{\partial x}.$$

The temperature T^* is defined as follows:

$$T_0^* = T_\infty + 0,5 \cdot (T_w - T_\infty) + 0,22 \cdot (T_r - T_\infty),\ T_r = T_0 + r \frac{V_0^2}{2 (C_p)_{cp}^*},\ r = \sqrt{Pr_0},$$

where $(C_p)_{cp}^*$ is the average heat capacity in the temperature range $[T, T^*]$.

The value of the Stanton number can be found using the ratios:

$$St_{x,eff} = 0,332 (m+1)^{1/2} \mathrm{Re}_{x,eff}^{1/2} \mathrm{Pr}_0^{-2/3} K \cdot K_1.$$

In this ratio the Reynolds number is defined with the formula:

$$\mathrm{Re}_{x,eff} = \frac{\rho_w V_0 x_{eff}}{\mu_w},$$

where ρ_w, μ_w are taken with temperature T_w of the streamlined body surface.

The convective heat flux $q_{w,\ell}$ under laminar flow can be found using ratio:

$$q_{w,\ell} = (C_p)_{cp}^* \rho_0 V_0 (T_e - T_w) St_{x,eff}.$$

In the presence of turbulent flow near the streamlined body surface we can use the following formula for calculation of effective length x_{eff} [5- 8]:

$$x_{eff} = \frac{\int_0^x \left[\left[R^{5/4} \left(1 + r \frac{\gamma - 1}{2} M_0^2 \right)^{0,1375} \rho_0 U_0 \mathrm{Pr}_0^{-0,7125} \mu_0^{0,25} C_{p,0}^{1,25} \right] (T_w/T_e)^{0,5} (T_e - T_w)^{1,25} \right] dx}{\left[\left[R^{5/4} \left(1 + r \frac{\gamma - 1}{2} M_0^2 \right)^{0,1375} \rho_0 U_0 \mathrm{Pr}_0^{-0,7125} \mu_0^{0,25} C_{p,0}^{1,25} \right] (T_w/T_e)^{0,5} (T_e - T_w)^{1,25} \right]},$$

where $r = \sqrt[3]{\mathrm{Pr}_0}$ is the coefficient of the temperature recovery, and heat transfer by convection can be determined by the formula:

$$q_{w,\ell} = \left(C_p\right)^*_{cp} \rho_0 V_0 \left(T_e - T_w\right) St_{x,eff}, \ \mathrm{Re}_{x,eff} = \frac{\rho_w V_0 x_{eff}}{\mu_w}.$$

The value of Stanton number for the turbulent regime can be found using the following ratio:

$$St_{x,eff} = 0,0296 \mathrm{Re}_{x,eff}^{-0,2} \mathrm{Pr}_0^{-0,57} \left(T_w / T_e\right)^{0,4} \left(1 + r\frac{\gamma - 1}{2} M_0^2\right)^{0,11}.$$

Using assumptions from work [8, 9], Stanton number $St_{x,эф}$ for transition area can be found using formula:

$$St_{x,eff.n} = St_{x,eff.turb} - \left(St_{x,eff.turb} - St_{x,eff.lam}\right) \exp\left[\frac{\mathrm{Re}_{x,eff} - \mathrm{Re}_{x,нп}}{\mathrm{Re}_{x,нп}}\right],$$

$St_{x,eff.turb}$, $St_{x,eff.lam}$ are determined using the formulas below; the Reynolds number $\mathrm{Re}_{x,нп}$ is corresponding to the transition from laminar to turbulent flow. For flow conditions $\mathrm{Re}_{x,eff} = 10^4 \div 10^6$ is agrees with experimental data [9].

If the gas consists of atoms then specific heat capacity (on mass unit) is (in that case there are only three degrees of freedom for translation motion):

$$C_{\mathrm{v},j} = \frac{3}{2} k, \ [\mathrm{J\ K^{-1}}]; \ C_{\mathrm{v},j} N_A, \ [\mathrm{J\ mol^{-1}\ K^{-1}}],$$

where $N_A = 6{,}022\ 141\ 10^{23}\ \mathrm{mol}^{-1}$ is the Avogadro number.

If the gas consists of diatomic molecules (N_2, O_2, H_2) then the molecule will have three translational, two rotation degrees of freedom ($i_2 = 2$), and one vibrational degree of freedom ($i_3 = 1$). The vibrational motion of a polyatomic molecule is much more complicated than the vibrational motion of a diatomic molecule: the number of vibrational degrees of freedom is $3n - 6$ for nonlinear polyatomic molecules and $3n - 5$ for linear polyatomic molecules. If the gas temperature is much higher than the characteristic temperature then $C_{\mathrm{вр},j} \to i_2 \frac{k}{2}$, [J K^{-1}], and $C_{\mathrm{кл},j} \to i_3 k$, [J K^{-1}]. The $C_{\mathrm{p},j}$ value can be found using ratio: $C_{\mathrm{p},j} = \gamma C_{\mathrm{v},j}$.

Thermodynamics properties approximate more accurate by polynomial [10]:

$$\Phi_j = \varphi_{1,j} + \varphi_{2,j} \ln x + \varphi_{3,j} x^{-2} + \varphi_{4,j} x^{-1} + \varphi_{5,j} x + \varphi_{6,j} x^2 + \varphi_{7,j} x^3,$$

$$\left(\frac{\mathrm{d}\Phi}{\mathrm{d}x}\right)_j = \left(\varphi_{2,j} - 2\varphi_{3,j} x^{-2} - \varphi_{4,j} x^{-1} + \varphi_{5,j} x + 2\varphi_{6,j} x^2 + 3\varphi_{7,j} x^3\right) \frac{1}{x},$$

$$\left(\frac{\mathrm{d}^2\Phi}{\mathrm{d}x^2}\right)_j = \left(-\varphi_{2,j} + 6\varphi_{3,j} x^{-2} + 2\varphi_{4,j} x^{-1} + 2\varphi_{6,j} x^2 + 6\varphi_{7,j} x^3\right) \frac{1}{x^2},$$

$$h_j = xT\left(\frac{\mathrm{d}\Phi}{\mathrm{d}x}\right)_j + \varphi_{8,j} \times 10^3 \text{, J/mol,}$$

$$C_{p,j} = 2x\left(\frac{\mathrm{d}\Phi}{\mathrm{d}x}\right)_j + x^2\left(\frac{\mathrm{d}^2\Phi}{\mathrm{d}x^2}\right)_j \text{, J/(mol·K),}$$

where $P_0 = 101325$ Pa, $x = T \times 10^{-4}$ K, Φ_j is the present Gibbs free energy, $C_p = \sum_i^{N_s} Y_i C_{p,i}$, N_s is the number of chemical components of the gas mixture, Y_i is the mass fraction of the i-th component in the mixture, $C_{p,i}$, h_i, ρ_i are the specific heat capacity at constant pressure, enthalpy and density of the i-th component. Approximation constants in the temperature range from 298 to 20000 K are calculated in [10]. Approximation constants over the range 298–20000 K are collected in [10, 12-16]

Notice that for temperature higher than ~ 10000 K in the shock layer there is more preferable using another formulation of the thermodynamic model, to separate translational, electronic, vibrational and rotational constituents of total and internal energy, in other words to apply the Born-Oppenheimer approximation.

3. Some results of analysis

A number of test (simulation) tasks are to be performed to justify quantitative analysis of engineering-physical-mathematical aerohydrodynamics model of HFV and supporting calculation solution. The following set of tasks are offered to be used: wedge linked to the plate; cone linked to the cylinder; cylinder linked to the plate; spherically blunt cone linked to the cylinder; supersonic flow of blunt axisymmetric bodies (in that case the calculation and experimental data of paper [11] has to be used).

The calculation results of the separate listed tasks (typical tasks of the mentioned above set): wedge linked to the plate; cone linked to the cylinder are provided below. The following incoming air flow parameters were used for verification:

Incoming flow pressure: $P = 2060$ Pa;
Incoming flow speed: $V = 1860$ m/s;
Incoming flow temperature: $T = 223$ K;
Incoming flow Much Number: $M = 6$;
Incoming flow gas mixture: Air;
Attitude from the Earth surface: $H = 25$ km.

Some calculation results of the flow of the wedge linked to the plate for provided initial data are illustrated in figures 3. The analytical estimation [4-7] of the effective length in the condition of flow streamlining the wedge under zero attack angle results in the expression $x_{eff} = z$, which corresponds to the graph shown at figure 3.

Similar outcome, received in the condition of air flow streamlining cone linked to the cylinder are shown at figures 4. That demonstrates the effective length of $x_{eff} \approx 4z/9$ and that result has very good correspondence with the theory [1-6].

IOP Conf. Series: Journal of Physics: Conf. Series **815** (2017) 012024 doi:10.1088/1742-6596/815/1/012024

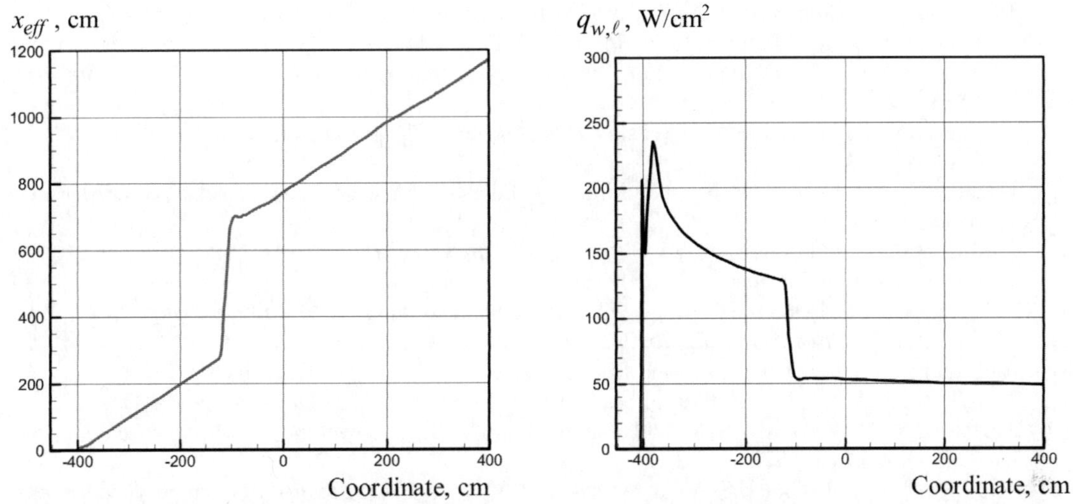

Figure 3. Effective length (*left*) and convective heat flow (*right*) dependence on the longitudinal coordinate (wedge coupled with the plate).

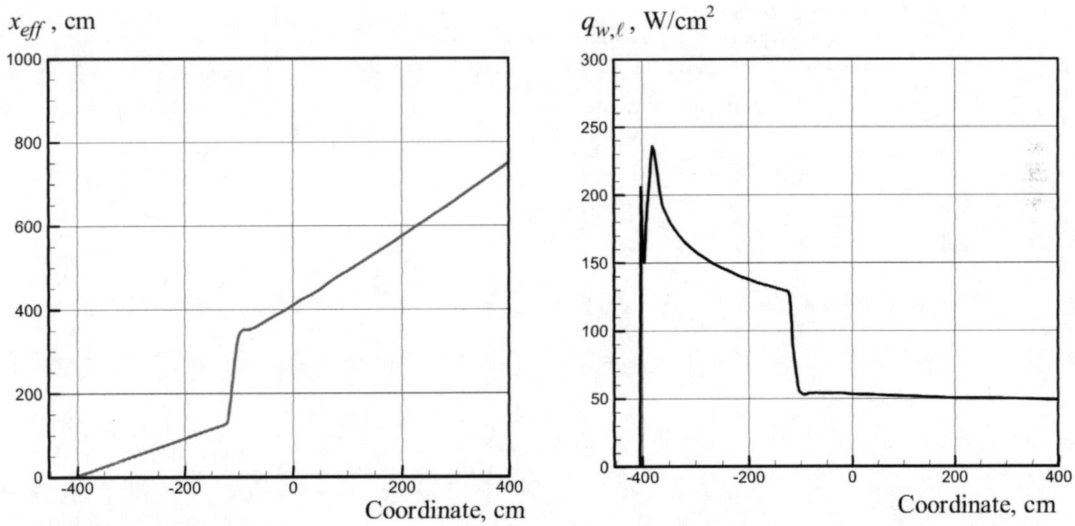

Figure 4. Effective length (*left*) and convective heat flow (*right*) dependence on the longitudinal coordinate (cone coupled with the cylinder).

4. Conclusion

The paper formulated engineering physical-mathematical model aerothermodynamics of hypersonic aircrafts in laminar and turbulent boundary layers (model designed for an approximate estimate of the convective heat flow in the range of speeds M = 6-28, and height H = 20-80 km). 2D model and calculation of the convective heat flow for simple geometric forms (individual elements of the design for hypersonic aerojet engine) are presented.

References

[1] Bocharov A N, Balakirev B A, Bityurin V A, Gryaznov V K, Golovin N N, Iosilevskiy I L, Evstigneev N M, Medin S A, Naumov N D, Petrovskiy V P, Ryabkov O I, Solomonov Yu S, Tatarinov, A.V, Teplyakov I O, Tikhonov A A and Fortov V E 2015 *J. Phys.: Conf. Ser.* **653** 012070
[2] Aksenov A A, Zhluktov S V, Savitskiy D V, Bartenev G Y and Pokhilko V I 2015 *J. Phys.: Conf. Ser.* **653** 012072
[3] Avduevsky V S, Marov M Ya, Noykina A I, Poleshaev V I and Zavelevich F S 1970 *Radio Science* **5** 369
[4] Avduevsky V S and Poleshaev V I 1991 *Advances in Space Research* **11** 365
[5] Znamenskii V V and Lunev V V 1981 *Fluid Dynamics* **16** 34
[6] Zemlyanskii B A, Lesin A B, Lunev V V and Shmanenkova G A 1982 *Fluid Dynamics* **17** 764
[7] Vulis L A 1972 *Combustion, Explosion, and Shock Waves* **5** 101
[8] Voloshchenko O V, Ostras' V N and Éismont V A 1983 *Fluid Dynamics* **18** 53
[9] Kobylyanskii A I, Veits I V, Gurvich L V and Suslov A A 1974 *High Temperature* **12** 1023
[10] Kuzenov V V, Ryzhkov S V and Shumaev V V 2015 *Problems of Atomic Science and Technology* **1** (95) 97
[11] Kuzenov V V, Ryzhkov S V and Shumaev V V 2015 *Problems of Atomic Science and Technology* **4** (98) 53
[12] Zheleznyakova A L and Surzhikov S T 2013 *High Temperature* **51** 816
[13] Zheleznyakova A L and Surzhikov S T 2014 *High Temperature* **52** 271
[14] Kotov M A, Kryukov I A, Ruleva L B, Solodovnikov S I and Surzhikov S T 2015 *33rd AIAA Applied Aerodynamics Conference*
[15] Dikalyuk A S and Surzhikov S T 2014 *High Temperature* **52** 35
[16] Shang J S, Andrienko D A, Huang P G and Surzhikov S T 2014 *Journal of Computational Physics* **266** 1

Experimental and numerical study of supersonic flow over two blunted wedges

M A Kotov, L B Ruleva, S I Solodovnikov and S T Surzhikov

A. Ishlinsky Institute for Problems in Mechanics of the Russian academy of sciences, Vernadsky prospekt 101(1), Moscow, 119526, Russia

E-mail: ma_kotov@mail.ru, ruleva@ipmnet.ru

Abstract. Shock-wave experiments with aerodynamic primitives in the form of a half wedge/double wedge with blunt edges are performed. Experimental studies were conducted using models, which are located at the exit of the supersonic nozzle of the hypersonic aerodynamic shock tube (HAST). Schlieren fragments of shock-wave formation over wedges were experimentally obtained. Calculations of the shock wave interactions over models carried out with the use of the authors' computer codes. The predicted results and their comparison with experiment showed a satisfactory agreement.

1. Introduction

Shock-wave interactions are characteristic feature of high-speed flows near the various elements of constructions of high-speed vehicles. The variety of arising shock-wave configurations makes the task of experimental and theoretical study very complex. However, the practical importance of the problem of the understanding of the mechanisms of shock-wave interaction between themselves and with surfaces of the vehicles puts this problem in a number of high priorities of experimental, theoretical and computational aerodynamics and physical mechanics.

In recent years experimental and computational studies [1 – 23] have allowed to classify the main problems, the study of which is of greatest interest and at the same time, the greatest difficulties. Among them:

- Calculation process of numerical evaluation of flow conditions in sections of high-speed ground test facilities and comparison with experimental data to understand the physics of hypersonic interactions in the facility sections [1 – 8] and hypervelocity flow interactions over sharp and blunt bodies with different incoming flow conditions [9 – 11];
- Ground test measurements for aerothermal and thermochemical effects on spherical capsule geometries with different gases for re-entry applications [12 – 15];
- Velocity/altitude simulation for scramjet engine testing at different flight conditions [4,16];
- The flow structure and heat exchange in the zone of interference between an inclined shock and the surface of a flat plate [17] and blunt bodies [18,19];
- Investigations of leading edge bluntness influence on flow in air-inlet model with rectangular cross-section [20,21];
- Transition between regular and Mach reflections of steady shock waves and its influence on flow characteristics [22,23].

In our previous works, the models of flat inlet sections were experimentally investigated with the use of the laboratorial hypersonic aerodynamic shock tube (HAST) [24-28]. Along with the experimental research some shock-wave configurations were calculated using authors' computational codes [25, 27, 29 – 31].

The present work continues the series of conjugated experimental and computational research. Technical capabilities of the HAST installation, as well as a short overview of previously obtained results are shown below.

2. Review of experimental and numerical results, previously obtained on the HAST

Detailed description of the HAST is presented in [24 – 28]. Scheme of the HAST function is shown in figure 1. Photo of the 15–meter facility with the receiver section is presented in figure 2.

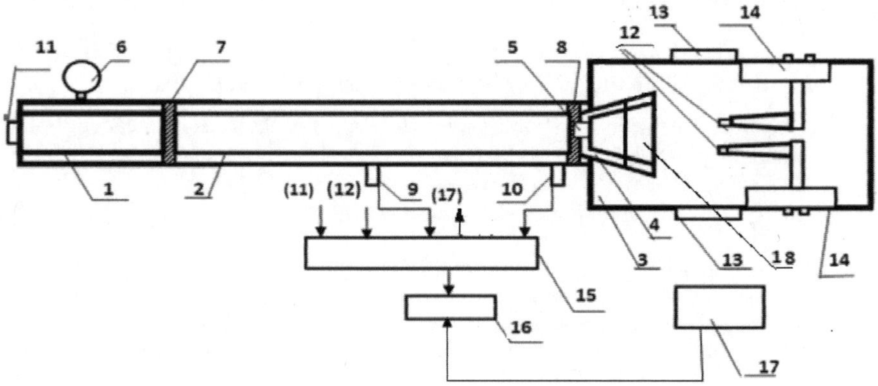

Figure 1. Scheme of the HAST facility. 1 – high-pressure section; 2 – low-pressure section; 3 – vacuum chamber; 4 – hypersonic nozzle; 5 – critical orifice; 6 – pressure gauge; 7 – high-speed valve device; 8 – diaphragm; 9, 10 – high-frequency dynamic pressure sensors; 11, 12, 14, 17 – sensors; 13 – observation windows.

Figure 2. Photo of the HAST.

When shock wave is initiated in a cylindrical channel it follows multiple shock reflections from the ends of the shock tube and portioned outflow through the crucial section of the supersonic nozzle onto the model at Mach numbers M = 6...7. The shock dynamic processes are registered and recorded by high-speed video camera and high-frequency dynamic pressure sensors.

Shock wave initiation parameters were as follows: pressure in the high pressure chamber was of 500 bar (air), pressure in the low pressure chamber was of 60 mbar; pressure in the vacuum chamber was of 10^{-3} mbar.

Figure 3 shows graphs of pressure registered by the sensors of pressure (the PCB sensors) at different distances from the exhaust nozzle. From the graph (figure 3) it follows that the initial shock velocity in shock tube is 757.7 m/s.

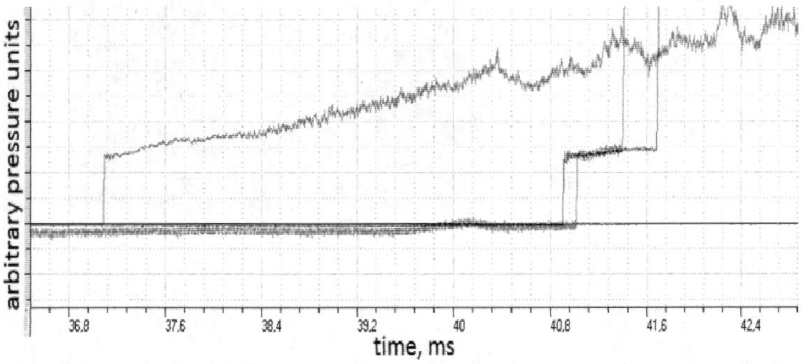

Figure 3. Pressure graphs: red line corresponds to the middle of the low-pressure section; green and pink correspond to distances of 200 mm and 100 mm before the nozzle entrance inside the low-pressure section.

With this experimental facility was obtained schlieren photography of M = 4.5...7 airflows over the model 10° half wedge [24], in channels with honeycombs [26], for spherical and conical models [27].

With the improvement of the equipment and experimental research methods [24, 25], several calculations of gas-dynamic characteristics have been produced using the developed computational codes [24 – 28]. Comparison of the experimental and computed results showed their satisfactory agreement [24 – 28]. Some examples of such a comparison are shown in figures 4, 5.

The upper parts of figure 4 and figure 5 correspond to the experimental data and the lower correspond to the calculations. Most of shock structures such as the incident waves, shear layers and expansion fans on the upper part of the figure coincide with the numerical simulation results.

Figure 4. The experimental and numerical images for the sharp wedge model with M = 7.0 [26].

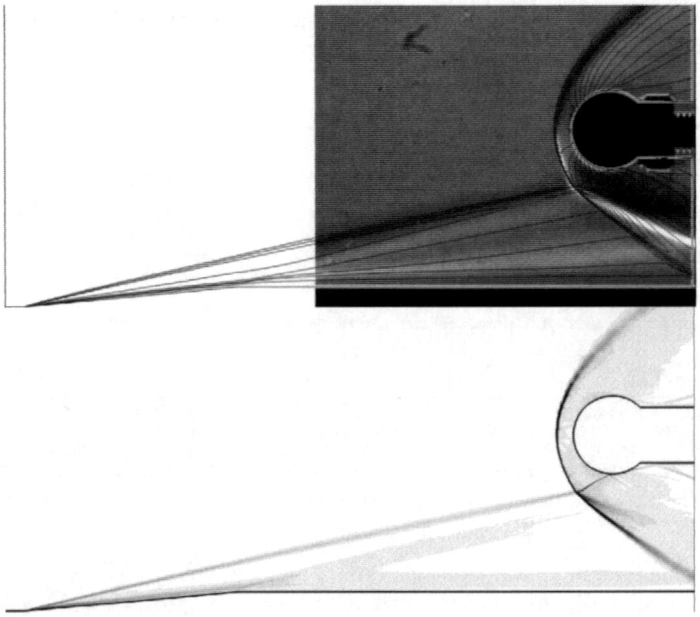

Figure 5. Comparison of computed density contours with the schlieren picture M=7.0 for the IV-th Edney configuration [27].

3. Experimental study of the double wedges configurations

The most detailed were investigated double wedges with sharp and blunt edges. For these two experimental configurations the CFD investigation was carried out with definition of gas-dynamic parameters which add information obtained in the experiments. At first we will analyze some experimental data and then the results of the experimental and corresponding CFD investigations will be considered. The double wedges were investigated for the three configurations of experimental models. These are described shortly below.

3.1. Double half wedge with blunt top and sharp bottom edge (configuration #1)

Experimental model is shown in figure 6. The distance between the front edges was changed. Initially, they were located in the same plane (see figure 7 from the left).

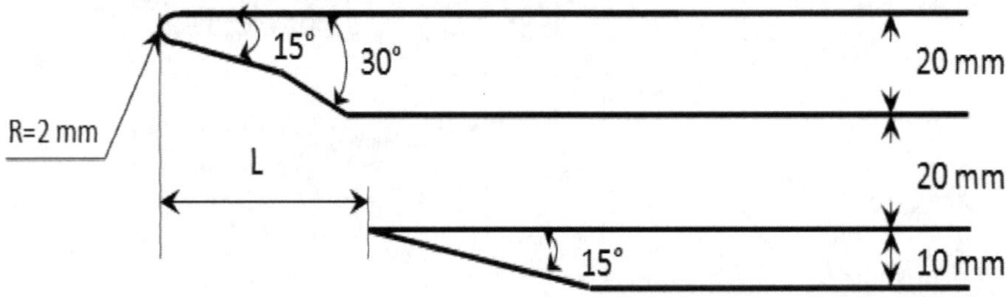

Figure 6. Configuration of the model #1: Blunt double wedge and sharp bottom wedge.

Then the bottom half wedge was moved to a distance at which the shock wave from the top wedge interacts with the front edge of the bottom model. In this case, the distance between edges is of 64 mm (figure 6, *L* parameter). Frames of high-speed video from these experiments are shown in figure 7. The goals of these experiments were obtaining the schlieren images of the shock wave configurations not only for validation the authors' computational codes, but also for receiving of the additional (indirect) information concerning parameters of input gas flow. It was established that these experiments corresponded to the following gas dynamic parameters on the test models (which were used in the calculations): $p_\infty = 9.977\times10^4$ erg/cm^3, $V_\infty = 1.402\times10^5$ cm/s, $T_\infty = 100$ K.

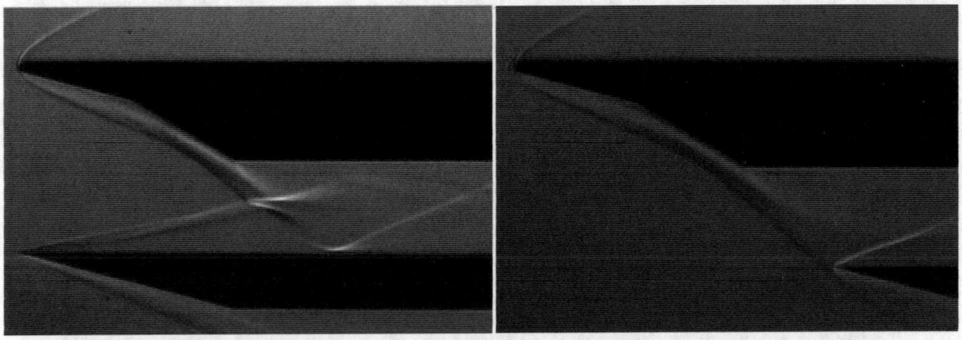

Figure 7. Schlieren image of M=7 air flow near the double wedge configuration #1.

3.2. Experiments with blunt bottom edge (configuration #2)

Experiments were carried out with models of blunt lower edge, as shown in figure 8.

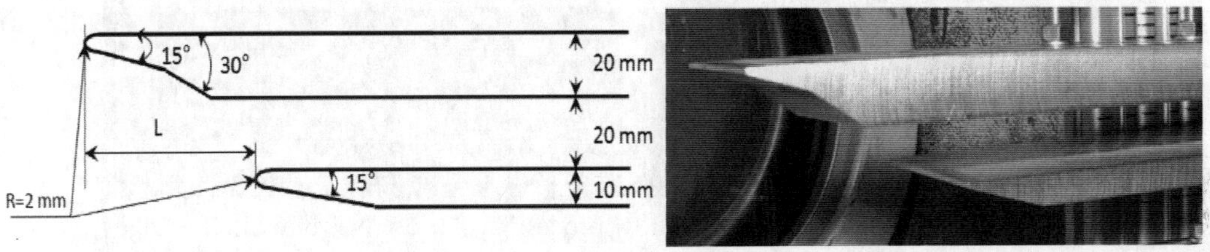

Figure 8. Configuration of the model #2: Blunt double wedge and blunt bottom wedge.

As expected, the concentration of reflected shock waves in the experiment was stronger with blunt bottom edge case. This may be due to the formation of high-enthalpy boundary layer and its effect on the flow behavior between models (figure 9).

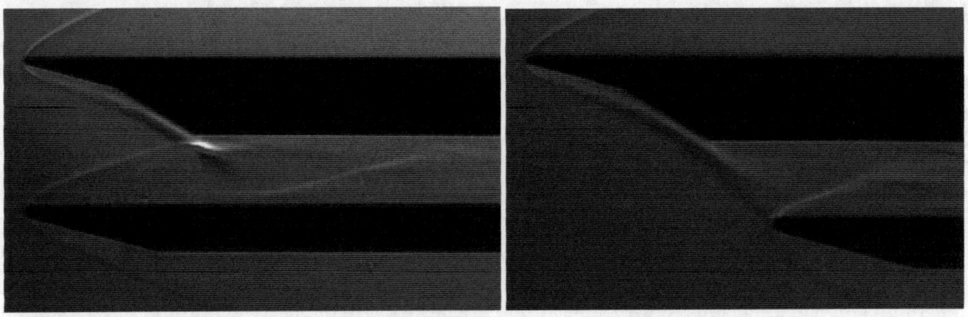

Figure 9. Schlieren image of M=7 air flow near the double wedge configuration with blunt edges #2 at different mutual location.

3.3. Experiments with two blunt wedges (configuration #3)

Two blunt half wedges with an angle 15° were tested at Mach number M=7. The radius of blunting a half wedge is of 2 mm, thickness of the upper half wedge is of 21 mm and lower one is of 11.5 mm, the distance between the models is of 20 mm. The shift of the lower half model relative to the upper half model was in range of 89 mm. It is just due to the mutual position of the models the detached shock from the upper blunted wedge reaches the front edge of the lower blunted wedge, what is shown in schlieren figure 10. Last frames in figure 10 illustrate the flow instability which is observed in experiments.

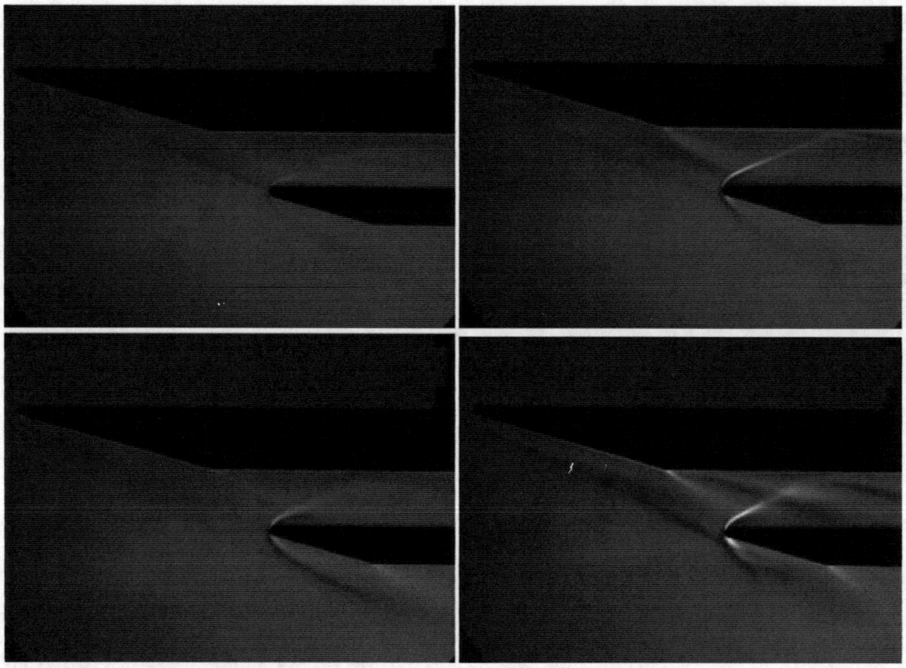

Figure 10. Shadow video sequence of airflow process evolution over blunt half wedges.

4. Numerical simulation results

4.1. Numerical investigation of the experimental configuration #1

A numerical simulation code NERAT-2D [30, 31] was used for this study. Used computational grid contains six blocks of the nonhomogeneous structured grids, which are shown in figure 11. Blocks #1 and #4 of the computational grid correspond to flowfield of blunted edges, block #3 describes compression flow, and block #5 corresponds to internal flow between two edges. Calculations were performed for the initial conditions presented in part 3.1. In these calculations instead of a bottom blunt wedge we used the blunt plate, because below the blunted nose of the bottom element of the model the flow field is not the point of our present investigation.

Distribution of gasdynamic functions for the case where the shock wave reflected from the upper edge of the model is incident on a blunt edge are shown in figures 12, 13.

Comparison of experimental and numerical data is shown in figure 13, where obtained numerical solution of shock-wave interaction for $L = 64$ mm combined with a schlieren photos for configuration of blunt double half wedge and blunt lower edge. Combined analysis of presented experimental and numerical data allows mark some significant peculiarities of the flow. Among those are:

- region of high pressure above blunt leading edge (see block 1 in figure 12);
- region of elevated pressure on bottom surface (see block 3 in figure 12);
- regions of elevated pressure on the bottom and top surfaces of the entrance region (see block 5 in figure 12);

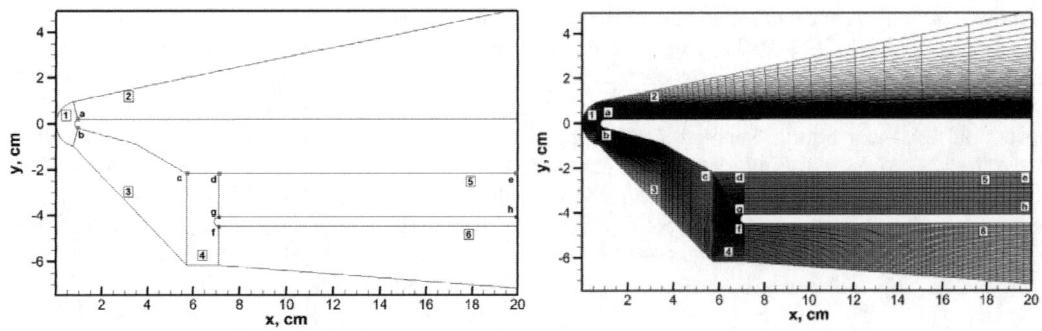

Figure 11. Six blocks of the grid for numerical simulation configuration #1.

- regions of braking of gas flow in all places along the surface with elevated pressure (see blocks 1, 3, 4, 5 in figure 13);
- in the point of interaction of shock wave with boundary layer (see block 5, top surface, figure 13) the separated flow is observed;
- waves rarefaction beyond the turn of the top surface are visible in experimental data (see figure 10);
- regions of elevated pressure are observed at intersections of shock waves and shock wave with boundary layer (figures 10,12).

Noticeable, that elevated pressure accompanied by the increasing of temperature.

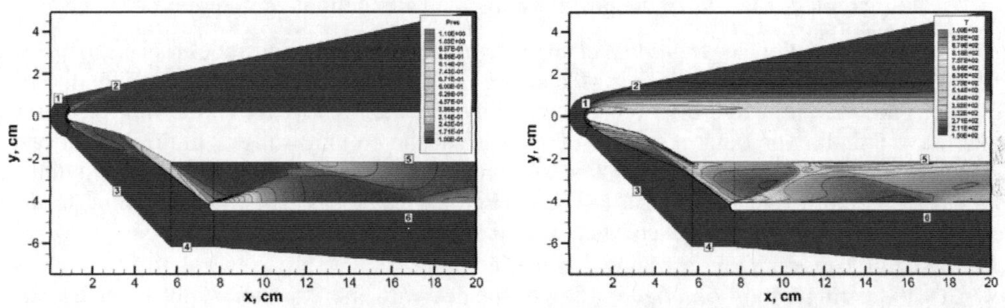

Figure 12. Pressure (left) and temperature for M=7, T=100 K.

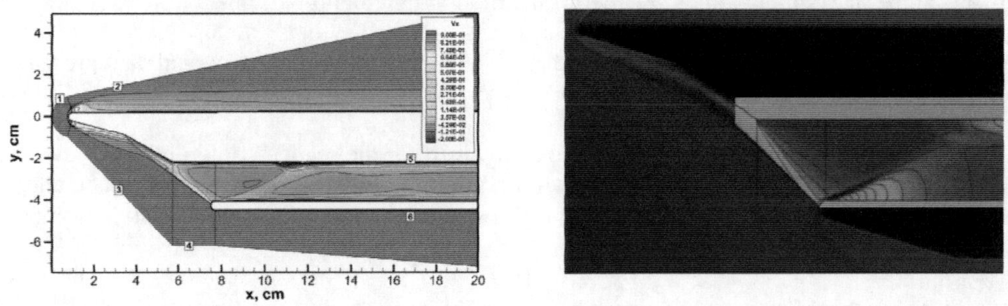

Figure 13. Longitudinal velocity and comparison of experimental and computational Schlieren images M=7; T=100 K.

So, these data demonstrate good agreement. Note that for precise control of parameters in reference points of given configuration, it is necessary to have experimental sensor readings.

4.2. Numerical investigation of the experimental configuration #3

As before, the NERAT-2D [30, 31] was used for these calculations. Used computational grid contains six blocks of the nonhomogeneous structured grids, which are shown in figure 14. Blocks #1 and #4 are introduced for numerical simulation of flowfield above blunt edges, block #3 describes compression flow near bottom surface of the upper wedge, and block #5 corresponds to internal flow between the two wedges. Calculations were performed for the initial conditions presented in part 3.1. One can see that the difference from the previous calculation case consists is in absent of turn of bottom surface for upper wedge. Four configurations of mutual location of upper and low wedges are investigated in the present numerical study, correspondingly: D_1=X_{R2}-X_{R1}=8.3 cm, D_2= 7.9 cm, D_3= 7.4 cm, D_4= 6.1 cm, where coordinates of X_{R1} and X_{R2} are shown in figure 14.

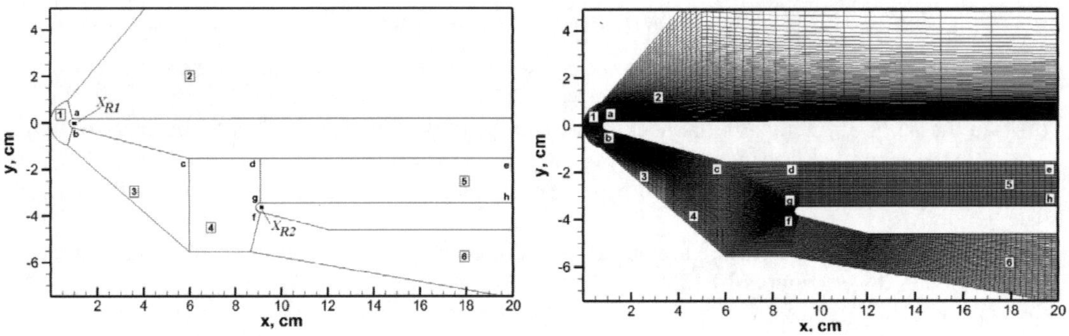

Figure 14. Six blocks of the grid for numerical simulation configuration #2.

The numerical simulation show gradual change of the location of falling shock wave relative to the nose of the bottom wedge at decreasing of distance *D*. Distribution of temperature for these four configurations are shown in figure 15 (correspondingly, figures 15,*a,b,c,d*). In the first case ($D=D_1$) the shock wave is falling on bottom surface of the nose of the bottom wedge. In the second case the shock wave is falling to the crucial point ($D=D_2$), while in the third case the shock wave is falling to the top part of the nose ($D=D_3$). In the last case (figure 15,*d*) the shock wave is falling to the top surface of the bottom wedge inside channel. Distributions of pressure along surfaces '*b-c-d-e*' and '*f–g*' for these four configurations are shown in figure 16.

Note that the configuration #2 (figure 15,b) coincides with the experimentally investigated case (see figure 10 (the upper left frame)).

So, the numerical simulation allow perform not only prediction of the possible shock wave configurations for different mutual locations of elements of tested models, but also to perform detailed comparison observed and numerical fields for the same initial conditions. Numerical data presented in figures 16 allow give prognostic decisions for further experimental study with the use of pressure transducers.

Finally we present numerical simulation results obtained for the second configuration of the wedge location (D_2 = 7.9 cm), but for another initial conditions, estimated by numerical calculations of flowfield in conical nozzle, namely p_∞ = 112.2 erg/cm^3, V_∞ = 1.242×10^5 cm/s, T_∞ = 50 K (figure 17).

5. Conclusion

Shock-wave experiments with wedge models with blunt edges have been performed. Experimental studies were conducted using models, which are located at the exit of the supersonic nozzle of the hypersonic aerodynamic shock tube (HAST).

Schlieren fragments of shock-wave interaction for different mutual location of two wedges were experimentally obtained at Mach number M=7.

Calculations of the shock wave interactions have been carried out with the use of the authors' computer codes. Numerical data and their comparison with experiment showed a satisfactory agreement.

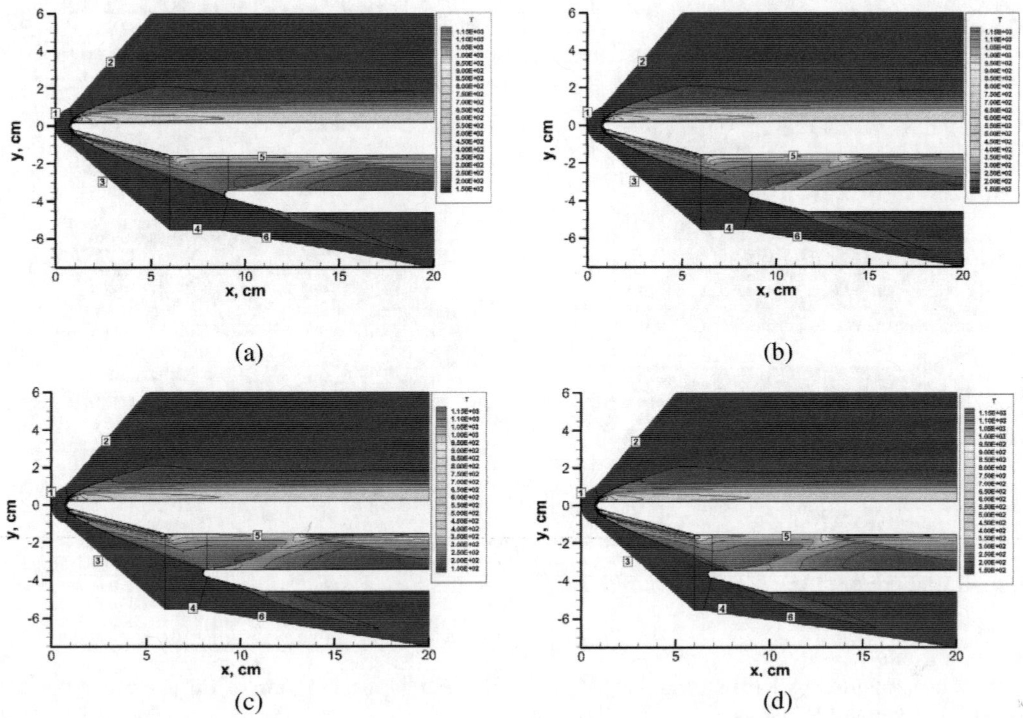

Figure 15. Temperature fields for four experimental configurations of mutual location of top and bottom elements of the model tested for M=7, *T*=100 K.

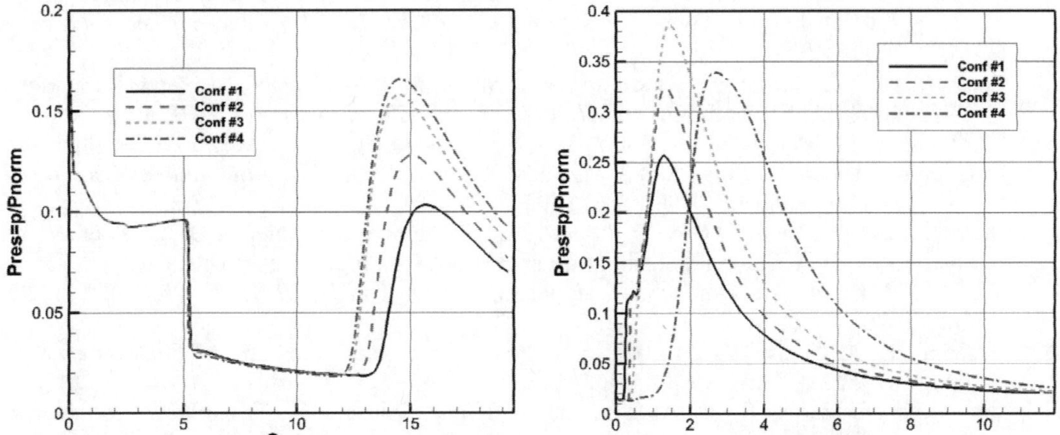

Figure 16. Distributions of pressure along top surface (points b-c-d-e from figure 14) and along bottom surface (points f-g-h) for four experimental configurations of mutual location of top and bottom elements of the model tested for M=7, T=100 K.

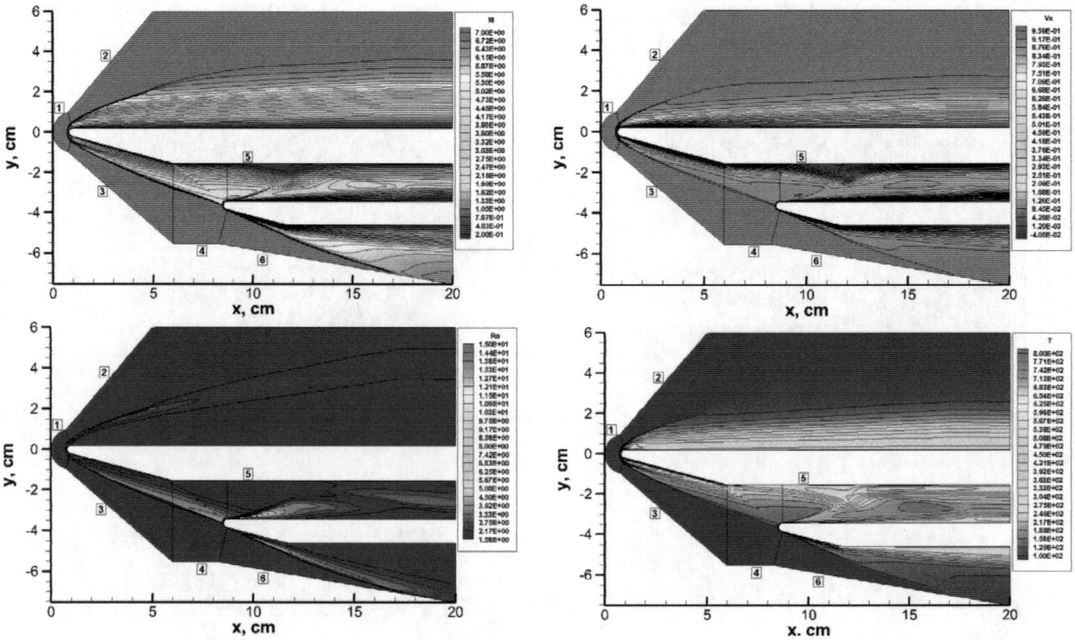

Figure 17. Mach number, longitudinal velocity ($V_x = u/V_\infty$), density (Ro=ρ/ρ_∞), and temperature (in K) M=7.5, p_∞=112.2 erg/cm^3, V_∞=1.242×10^5 cm/s, T=50 K.

Acknowledgments
This study was supported by RFBR grant No. 16-01-00379 and by the Program of Basic Researches of the Russian Academy of Sciences.

References

[1] MacLean M, Candler G and Holden M 2005 Numerical Evaluation of Flow Conditions in the LENS Reflected Shock-Tunnel Facilities *Proc. 43rd AIAA Aerospace Sciences Meeting* AIAA 2005–0903

[2] Hornung H G 2010 Ground Testing for Hypersonic flow, Capabilities and Limitations *CALIFORNIA INST OF TECH PASADENA GRADUATE AEROSPACE LABS*

[3] Cummings R M, McLaughlin T E and Barlow D N 2011 Supersonic Research Facilities and Opportunities at the United States Air Force Academy *Proc. 17th International Conference on Engineering Education* TU.A.SA 11.309

[4] Northam G B, Andrews E, Guy W, Pellett G L, Drummond P, Cutler A D and Rock K An Overview of Hypersonic Propulsion Research at NASA Langley Research Center

[5] Schneider S P 2008 Development of Hypersonic Quiet Tunnels *Journal of Spacecragt and Rockets* **45** 4

[6] Hadassah N 2010 Analysis and design of quiet hypersonic wind tunnels *PhD dissertation Rutgers The State University of New Jersey*

[7] Antón P S, Gritton E C, Mesic R, Steinberg P and Johnson D J 2004 Wind Tunnel and Propulsion Test Facilities: An Assessment of NASA's Capabilities to Serve National Needs. *RAND NATIONAL DEFENSE RESEARCH INST SANTA MONICA CA*

[8] MacLean M, Wadhams T, Holden M, Candler G and Nompelis I. 2009 Integration of CFD and Experiments in the CUBRC LENS Shock Tunnel Facilities to Understand the Physics of Hypersonic and Hypervelocity Flows *Proc. 4th Symposium on Integrating CFD and Experiments in Aerodynamics*, von Karman Institute, Belgium

[9] MacLean M, Wadhams T, Holden M and Hollis B 2005 Investigation of blunt bodies with CO2 test gas including catalytic effects *Proc. 38th AIAA Thermophysics Conference* AIAA 2005-4693

[10] MacLean M, Dufrene A, Wadhams T and Holden M 2010 Numerical and Experimental Characterization of High Enthalpy Flow in an Expansion Tunnel Facility *Proc. 48th AIAA Aerospace Sciences Meeting Including the New Horizons Forum and Aerospace Exposition* AIAA 2010-1562

[11] Hornung H G 2010 Selected Achievements and Discoveries Made in High-Enthalpy Flow Facilities *CALIFORNIA INST OF TECH PASADENA GRADUATE AEROSPACE LABS*

[12] MacLean M and Holden M 2006 Catalytic effects on heat transfer measurements for aerothermal studies with CO2 *Proc. 44th AIAA Aerospace Sciences Meeting and Exhibit* AIAA 2006-0182

[13] MacLean M and Holden M 2006 Numerical assessment of data in catalytic and transitional flows for Martian entry *Proc. 9th AIAA/ASME Joint Thermophysics and Heat Transfer Conference* AIAA 2006-2946

[14] MacLean M, Holden M, Wadhams T and Parker R 2007 A computational analysis of thermochemical studies in the lens facilities *Proc. 45th AIAA Aerospace Sciences Meeting and Exhibit* AIAA 2007-0121

[15] MacLean M, Mundy E, Wadhams T, Holden M and Parker R 2008 Analysis and Ground Test of Aerothermal Effects on Spherical Capsule Geometries *Proc. 38th Fluid Dynamics Conference and Exhibit* AIAA 2008-4273

[16] Holden M S 2005 Aerothermal and Propulsion Ground Testing That Can Be Conducted to Increase Chances for Successful Hypervelocity Flight Experiments *CALSPAN UB RESEARCH CENTER BUFFALO NY*

[17] Borovoi V Y, Egorov I V, Skuratov A S and Struminskaya I V 2005 Interaction between an inclined shock and boundary and high-entropy layers on a flat plate *Fluid Dynamics* **40** 6

[18] Edney B E 1968 Effects of shock impingement on the heat transfer around blunt bodies *AIAA Journal* **6** 1

[19] D'Ambrosio D 2003 Numerical prediction of laminar shock/shock interactions in hypersonic flow *Journal of spacecraft and rockets* **40** 2

[20] Borovoi V Y, Egorov I V, Noev A Y, Skuratov A S and Struminskaya I V 2011 Two-dimensional interaction between an incident shock and a turbulent boundary layer in the presence of an entropy layer *Fluid Dynamics* **46** 6

[21] Borovoy V Y, Mosharov V E, Radchenko V N, Skuratov A S and Struminskaya I V 2014 Leading edge bluntness effect on the flow in a model air-inlet *Fluid Dynamics* **49** 4

[22] Ivanov M S, Klemenkov G P, Kudryavtsev A N, Nikiforov S B, Pavlov A A, Fomin V M and Hornung H G 1997 Experimental and numerical study of the transition between regular and Mach reflections of shock waves in steady flows *Proc. 21st International Symposium on Shock Waves*

[23] Ivanov M S, Vandromme D, Fomin V M, Kudryavtsev A N, Hadjadj A and Khotyanovsky D V 2001 Transition between regular and Mach reflection of shock waves: new numerical and experimental results *Shock Waves* **11** 3

[24] Kotov M, Ruleva L, Solodovnikov S, Kryukov I and Surzhikov S 2014 Multiple Flow Regimes in a Single Hypersonic Shock Tube Experiment *Proc. 30th AIAA Aerodynamic Measurement Technology and Ground Testing Conference* AIAA 2014-2657

[25] Kuzenov V V, Kotov M A 2014 Analysis of gas-dynamic processes and development of model of flows in hypersonic shock tube *Herald of the Bauman Moscow State Technical University* **1** 94

[26] Kotov M A, Kryukov I A, Ruleva L B, Solodovnikov S I and Surzhikov S T 2015 Numerical and experimental investigation of the hypersonic flow structure in a complex flat duct *Herald of the Bauman Moscow State Technical University* **1** 100

[27] Kotov M, Kryukov I, Ruleva L, Solodovnikov S and Surzhikov S 2015 Supersonic Air Flows around Some Geometrical Primitives *Proc. 33rd AIAA Applied Aerodynamics Conference* AIAA 2015-3012

[28] Kotov M, Kryukov I, Ruleva L and Solodovnikov S 2017 The Investigation of Shock-Wave Interaction with Aerodynamic Models *Proc. 55th AIAA Aerospace Sciences Meeting* AIAA 2017-0262

[29] Seleznev R and Surzhikov S 2015 Quasi-One-Dimensional and Two-Dimensional Numerical Simulation of Scramjet Combustors Proc. 51st AIAA/SAE/ASEE Joint Propulsion Conference AIAA 2015-4166

[30] Surzhikov S T and Shang J 2012 Coupled radiation-gasdynamic model for stardust earth entry simulation *Journal of Spacecraft and Rockets* **49** 5

[31] Shang J S and Surzhikov S T 2012 Nonequilibrium radiative hypersonic flow simulation *Progress in Aerospace Sciences* **53**

IOP Conf. Series: Journal of Physics: Conf. Series **815** (2017) 012026 doi:10.1088/1742-6596/815/1/012026

Investigations of vibrational kinetics relaxation within air shock wave plasma

W Su[1], D Bruno[1] and Y Babou[2]

[1] Istituto di Nanotecnologia, Consiglio Nazionale delle Riecerche, 70126, Bari, Italy
[2] Departamento de Bioingeniería Ingeniería Aeroespacial, Universidad Carlos III de Madrid, 28911, Madrid, Spain

E-mail: ybabou@ing.uc3m.es

Abstract. A vibrationally detailed kinetic model is used to study the relaxation behind shock waves in air. The role of recently published data for the rate coefficients of the Zeldovich reactions of NO formation is studied in detail. Results allow to study the radiation emitted from the shock-heated gas. Comparison with some emission spectroscopy study performed in a shock tube facility shows only qualitative agreement with the model predictions but it allows to identify directions for model improvement.

1. Introduction

During the re-entry of a spacecraft into a planetary atmosphere, the relative stream velocity is so high that a bow shock is created in front of the vehicle. After the strong shock wave, in the relaxing flow, a massive amount of the free stream kinetic energy is converted through inelastic collisions, into internal energy of the surrounding gases, which results in intense convective and radiative heating over the spacecraft structure. Inter-particle collisions promote energy exchanges among the internal modes of atoms and molecules, and produce chemical reactions including dissociation and ionization. Since the time scale of energy exchange processes for internal levels and the chemical time scale are comparable with the fluid dynamics one, the flow behind the shock is in thermal and chemical nonequilibrium. Understanding such nonequilibrium processes in shock-heated flows is a key issue for the design of spacecrafts. Although progress has been achieved during the last decades, modeling of thermo-chemically nonequilibrium flows is still a challenging problem in both academia and industry.

The state of the gas, i.e. the concentration of the gas species and their internal energy (rotational, vibrational and electronic) level populations can be estimated by either multi-temperature (MT) models [1] or state-to-state (StS) approaches [2]. The MT models make the assumption that each energy mode is characterized by its own temperature, which is determined by solving an individual energy conservation equation. These models rely on the common assumption that the relaxation of the internal energy modes occurs through a series of Boltzmann distributions at their respective temperatures. However, in many flow conditions, the populations deviate from Boltzmann distributions [1]. Conversely, the StS approaches treat each molecule with different internal state as a separate pseudo-species, for which a continuity equation is solved. Without adopting any quasi-stationary distribution over the internal energy modes these approaches allow representing arbitrary nonequilibrium situations.

The StS approaches, however, require a large number of reaction rate constants corresponding to the different physico-chemical processes. In this study, the nonequilibrium relaxation in shock-heated air is simulated with a StS approach, since the vibrational state-specific rate coefficients from *ab initio* methods are becoming available to describe vibrational relaxation and reactive processes in N_2-O_2 mixtures. Particularly sensitive is the population of NO whose formation is governed by a strong coupling between chemical and vibrational kinetic and consequently tightly depending on the concentration of other molecular species.

Making use of the recent quasi-classical trajectory (QCT) results [3, 4, 5] for the Zeldovich exchange reaction of NO formation on theoretically calculated potential energy surfaces (PES), the thermal and chemical kinetics of NO is discussed in detail.

As an assessment of the modeling, the post shock radiative signature has been calculated under the adopted assumptions and critically compared to intensity spectra measured in shock tube experiments [6] in conditions representative of Earth re-entries ($u \sim$ 5-8 km/s).

2. Modelling

In this work, we consider an air gas mixture consisting of the five chemical species N_2, O_2, NO, N and O. For the shock conditions of interest, the ionization degree is expected to be small and the focus is on the coupling of translational, vibrational and chemical kinetics. Thereby, ionization is neglected and excitation of vibrational and electronic levels occurs mainly through heavy - heavy particle collisional impact.

A vibrational state-specific description is adopted for the diatomic molecules in their ground states. Vibrational levels are obtained from a RKR analysis on the low energy part of spectroscopically determined potential energy curves and the extrapolation to the dissociation limit [7]. This gives a total of 61 bound levels for N_2, 46 for O_2 and 48 for NO. The rotational mode is assumed in equilibrium at the gas temperature, T. In the following paragraphs, the physical model used for the simulation of the relaxation behind shock waves in air is presented.

2.1. Governing Equations

In the frame of reference moving with the shock wave, the flow is described by the stationary 1D Euler equations:

$$\begin{aligned} \partial_x(\rho u) &= 0, \\ \partial_x(\rho u y_k^v) &= 0, \qquad v = 0 \ldots v_{\max}^k,\ k \in \mathtt{H} \\ \partial_x(\rho u^2 + p) &= 0, \\ \partial_x\left[\rho u\left(h + \frac{u^2}{2}\right)\right] &= 0, \end{aligned} \tag{1}$$

with $\mathtt{H} = \{N_2, O_2, NO, N, O\}$. ρ, u, p and h are the density, velocity, total pressure and specific enthalpy of the gas mixture. y_k^v and $\dot{\omega}_k^v$ are the mass fraction and source term of the molecular species k with vibrational level v. $v_{\max}^k$ is the maximum vibrational quantum number for species k and $v_{\max}^k = 0$ for atoms. The species mass fractions and related chemical source terms are obtained by summing over all vibrational levels, i.e. $y_k = \sum_{v=0}^{v_{\max}} y_k^v$, $\dot{\omega}_k = \sum_{v=0}^{v_{\max}} \dot{\omega}_k^v$. The conservation equations are completed by closure relations obtained from ideal gas assumption and then solved by the ODEPACK library [8].

2.2. Vibration-Chemical Kinetic Mechanism and the Reaction Rates

The most important part of the modelling is the kinetic mechanism that strongly affects the thermal and chemical relaxation of the flow. The kinetic mechanism drives the evolution of the composition before equilibrium is achieved. Three different categories of physico-chemical processes are included in this work. By denoting AB and CD the diatomic molecules, and X

any of the 5 species, as well as v and w the specific vibrational levels, the processes include: 1) the vibrational-translational (VT) energy exchange that drives the vibrational level populations:

$$AB(v) + X \rightleftharpoons AB(w) + X, \tag{2}$$

2) the dissociation-recombination (DR) reactions, responsible for the production of atomic species:

$$AB(v) + X \rightleftharpoons A + B + X, \tag{3}$$

and 3) the two Zeldovich exchange (ZE) reactions [9] which are the essential mechanism for NO formation:

$$N_2(v) + O \rightleftharpoons NO(w) + N, \quad O_2(v) + N \rightleftharpoons NO(w) + O. \tag{4}$$

The generic chemical reaction i can be expressed in the general form:

$$\sum_{k=1}^{N_s} s_r^{k,i} X_k^v \rightleftharpoons \sum_{k=1}^{N_s} s_p^{k,i} X_k^w, \quad v \text{ } or \text{ } w = 0...v_{max}^k. \tag{5}$$

This allows to express the chemical source term $\dot{\omega}_{k,v}^i$ using the law of mass action:

$$\dot{\omega}_{k,v}^i = M_k \cdot \left(s_p^{k,i} - s_r^{k,i}\right) \cdot \left(k_f^i \cdot \prod_{k=1}^{N_s} [X_k^v]^{s_r^{k,i}} - k_b^i \cdot \prod_{k=1}^{N_s} [X_k^w]^{s_p^{k,i}}\right), \quad i = 1...N_r \tag{6}$$

where, $s_r^{k,i}$, resp. $s_p^{k,i}$ are the reactant, resp. product stoichiometric coefficients of species k in reaction i. k_f^i, resp. k_b^i are the forward, resp. backward reaction rate constants, which satisfy the detailed balance condition $k_b^i = k_f^i / k_{eq}^i$, k_{eq}^i being the equilibrium constant. M_k is the molar mass of species k. $[\cdot]$ denotes molar density. N_s is the total number of species in mixtures. N_r is the total number of chemical reactions.

Values for the state-specific rates are obtained, where available, from quasi-classical trajectory (QCT) calculations and the semi-analytical forced harmonic oscillator (FHO) model. For the remaining processes, where none of the above is available, the classical models of Landau-Teller (LT) relaxation and Park's two temperature model are used. Table 1 summarizes the state-specific reactions used in this work and the methods used to calculate the reaction rate constants.

2.2.1. State-specific rates from FHO and QCT calculations. In the present work, the state-specific rates for vibrational excitation and dissociation in diatom-diatom collisions are taken from the STELLAR Database [10]; it collects results from the work of Lino da Silva *et al.*, who applied the FHO model [11] to the production of a multi-quantum datasets of VT and DR rates [12] in a wide temperature range. The excitation and dissociation rates for $N_2 - N$ and $O_2 - O$ systems are from the work of Esposito et al., based on the QCT approach modelling the evolution of the vibrational and rotational states during a collision by means of classical mechanics [13, 14]. For the two Zeldovich exchange reactions, we have two alternative datasets using QCT calculations: one from Bose and Candler [15, 16] and one from the very recent work by Esposito [3, 4, 5]. The dissociation rates in $N_2 - O$ and $O_2 - N$ collisions are also from this recent work.

Table 1. Vibrational-chemical kinetic mechanism

Reaction	Model	Reaction	Model
$N_2(v) + N_2 \rightleftharpoons N_2(w) + N_2$	FHO	$N_2(v) + N \rightleftharpoons 3N$	QCT
$N_2(v) + O_2 \rightleftharpoons N_2(w) + O_2$	FHO	$N_2(v) + O \rightleftharpoons 2N + O$	QCT
$N_2(v) + N \rightleftharpoons N_2(w) + N$	QCT	$N_2(v) + NO \rightleftharpoons 2N + NO$	Park
$N_2(v) + X \rightleftharpoons N_2(w) + X$, $X \in \{NO, O\}$	LT	$O_2(v) + N_2 \rightleftharpoons 2O + N_2$	FHO
$O_2(v) + N_2 \rightleftharpoons O_2(w) + N_2$	FHO	$O_2(v) + O_2 \rightleftharpoons 2O + O_2$	FHO
$O_2(v) + O_2 \rightleftharpoons O_2(w) + O_2$	FHO	$O_2(v) + O \rightleftharpoons 3O$	QCT
$O_2(v) + O \rightleftharpoons O_2(w) + O$	QCT	$O_2(v) + N \rightleftharpoons 2O + N$	QCT
$O_2(v) + X \rightleftharpoons O_2(w) + X$, $X \in \{NO, N\}$	LT	$O_2(v) + NO \rightleftharpoons 2O + NO$	Park
$NO(v) + N_2 \rightleftharpoons NO(w) + N_2$	FHO	$NO(v) + N_2 \rightleftharpoons N + O + N_2$	FHO
$NO(v) + O_2 \rightleftharpoons NO(w) + O_2$	FHO	$NO(v) + O_2 \rightleftharpoons N + O + O_2$	FHO
$NO(v) + X \rightleftharpoons NO(w) + X$, $X \in \{NO, N, O\}$	LT	$NO(v) + X \rightleftharpoons N + O + X$, $X \in \{NO, N, O\}$	Park
$N_2(v) + N_2 \rightleftharpoons 2N + N_2$	FHO	$N_2(v) + O \rightleftharpoons NO(w) + N$	QCT
$N_2(v) + O_2 \rightleftharpoons 2N + O_2$	FHO	$O_2(v) + N \rightleftharpoons NO(w) + O$	QCT

2.2.2. State-specific rates from classical models. The classical models are used to define the state-specific reaction rates in a way to recover the macroscopic chemical and energy source terms. The LT model assumes vibrational energy relaxes as a consequence of VT energy transfer collisions $AB - X$ as:

$$\Omega^*_{\mathrm{VT}} = \rho_{AB} \frac{e^{AB}_{\mathrm{vib,eq}}(T) - e^{AB}_{\mathrm{vib}}}{\tau_{AB-X}}, \tag{7}$$

where, $e^{AB}_{\mathrm{vib,eq}}(T)$ is the specific vibrational energy in equilibrium at gas temperature T; τ_{AB-X} is the relaxation time, modelled from the Millikan-White's formula [17]. In a StS framework, this behaviour is recovered by choosing the state-specific rates for reactions $AB(v)+X \rightleftharpoons AB(w)+X$ as:

$$k_{\mathrm{f}}^{v\to w} = \frac{1}{N_{\mathrm{A}}[X]} \cdot \left(\frac{y^w_{AB}}{y_{AB}}\right)_{\mathrm{eq}} \cdot \frac{1}{\tau_{AB-X}}, \quad k_{\mathrm{b}}^{v\to w} = k_{\mathrm{f}}^{w\to v}, \tag{8}$$

where, N_{A} is the Avogadro's number; $\left(\frac{y^w_{AB}}{y_{AB}}\right)_{\mathrm{eq}}$ is the reduced equilibrium populations. Backward rates are calculated from the detailed balance condition, with $k_{\mathrm{eq}}^{v\to w} = \exp\left(-\frac{\epsilon^w_{AB} - \epsilon^v_{AB}}{k_{\mathrm{B}}T}\right)$.

Next, for reactions as eq. (5), according to Park's model [1], the chemical reaction rates are functions of different mode temperatures, depending on the type of reactions. The controlling temperatures T_{f}, T_{b} in forward and backward reactions is given by: $T_{\mathrm{f}} = \sqrt{T \cdot T_{\mathrm{vib}}}$, $T_{\mathrm{b}} = T$ for DR reactions, where T_{vib} is the energy-representative vibrational temperature for which a Boltzmann distribution would give the same vibrational energy as the non-equilibrium distribution. Then the macroscopic rate coefficients can be expressed by modified Arrhenius functions:

$$K_{\mathrm{f}}\left(T_{\mathrm{f}}\right)=\alpha_1\cdot T_{\mathrm{f}}^{\beta_1}\cdot\exp\left(-\frac{\gamma_1}{T_{\mathrm{f}}}\right),$$
$$K_{\mathrm{eq}}\left(T_{\mathrm{b}}\right)=\alpha_2\cdot T_{\mathrm{b}}^{\beta_2}\cdot\exp\left(-\frac{\gamma_2}{T_{\mathrm{b}}}\right),\tag{9}$$
$$K_{\mathrm{b}}\left(T_{\mathrm{b}}\right)=\frac{K_{\mathrm{f}}(T_{\mathrm{b}})}{K_{\mathrm{eq}}(T_{\mathrm{b}})},$$

where, the constants α, β, γ are adjusted to reproduce experimental determinations of K_{f} and K_{eq}. The total chemical and vibrational energy source terms are obtained by:

$$\dot{\omega}_k^*=M_k\cdot\left(s_{\mathrm{p}}^k-s_{\mathrm{r}}^k\right)\cdot\left(K_{\mathrm{f}}\cdot\prod_{k=1}^{N_{\mathrm{s}}}[X_k]^{s_{\mathrm{r}}^k}-K_{\mathrm{b}}\cdot\prod_{k=1}^{N_{\mathrm{s}}}[X_k]^{s_{\mathrm{p}}^k}\right),\tag{10}$$

$$\Omega_k^*=-M_kN_{\mathrm{A}}\left(\left\langle E_{\mathrm{f}}^k\right\rangle\cdot K_{\mathrm{f}}\cdot\prod_{k=1}^{N_{\mathrm{s}}}[X_k]^{s_{\mathrm{r}}^k}-\left\langle E_{\mathrm{f}}^k\right\rangle\cdot K_{\mathrm{b}}\cdot\prod_{k=1}^{N_{\mathrm{s}}}[X_k]^{s_{\mathrm{p}}^k}\right),\tag{11}$$

where, $\langle E_{\mathrm{f}}^k\rangle$, $\langle E_{\mathrm{b}}^k\rangle$ are the average vibrational energy lost/gained in reactions. In practice, it is usually set $\langle E_{\mathrm{f}}^k\rangle=\langle E_{\mathrm{b}}^k\rangle=e_{\mathrm{vib}}^k(T_{\mathrm{vib}})$ by assuming the reaction has a non-preferential character. In the current StS approach, the above behaviour is reproduced by choosing:

$$k_{\mathrm{f}}=K_{\mathrm{f}}\cdot\prod_{k=1}^{N_{\mathrm{s}}}\left(\frac{\exp\left(-\epsilon_k^w/k_{\mathrm{B}}T_{\mathrm{vib}}^k\right)}{\sum_w\exp\left(-\epsilon_k^w/k_{\mathrm{B}}T_{\mathrm{vib}}^k\right)}\right)^{s_{\mathrm{p}}^k},\tag{12}$$
$$k_{\mathrm{b}}=K_{\mathrm{b}}\cdot\prod_{k=1}^{N_{\mathrm{s}}}\left(\frac{\exp\left(-\epsilon_k^v/k_{\mathrm{B}}T_{\mathrm{vib}}^k\right)}{\sum_v\exp\left(-\epsilon_k^v/k_{\mathrm{B}}T_{\mathrm{vib}}^k\right)}\right)^{s_{\mathrm{r}}^k}.\tag{13}$$

Values for the macroscopic rate constants K_{f} and K_{b} are taken from Ref. [6].

2.3. *Post-shock radiative signature modelling*

In the following we present the approach adopted to rebuild the radiative signature measured in shock tube experiments (see [6, 18]). As in the experiments, the radiative signature of the post-shock plasma is considered in the range 200-400 nm and is dominated by transitions between bound energy levels of diatomic molecules. The spectral emission coefficient of the plasma is expressed as the sum of bound-bound transitions:

$$\eta_\sigma=\sum_{ul}n_u\frac{A_{ul}}{4\pi}hc\sigma_{ul}f(\sigma-\sigma_{ul})\tag{14}$$

where n_u is the population of the upper level, A_{ul} is the Einstein coefficient for spontaneous emission, σ_{ul} is the wavenumber of the transition, h is Planck constant, c is the light velocity and $f(\sigma-\sigma_{ul})$ is the line profile shape approximated by a Doppler profile to account for the thermal broadening (in our conditions the collisional broadening remains marginal).

The emission coefficient has been calculated accounting for the electronic systems of N_2, O_2 and NO molecules listed in Table 2. The spectroscopic data have been provided by the high resolution spectroscopic database HTGR considering an exhaustive set of vibrational bands, taken into account up to the last vibrational levels below the dissociation limit for each electronic state (for details see [19]). Also, although in practice the Second positive system of N_2^+ is a

Table 2. Diatomic molecule electronic systems considered in this study

Electronic system name	Upper state-lower state	Upper state energy (cm^{-1})	Calculated bands (0:v'_{max}, 0:v''_{max})
O_2 Schumann-Runge (SR)	$B^3\Sigma_u^- - X^3\Sigma_g^-$	49357.96	(0:19 - 0:21)
N_2 First Negative (FN)	$C^3\Pi_u - X^2\Sigma_g^+$	25566.077	(0:4 - 0:21)
NO γ	$A^2\Sigma^+ - X^2\Pi_r$	44198.94	(0:8 - 0:22)
NO β	$B^2\Sigma^+ - X^2\Pi_r$	45499.5	(0:37 - 0:22)
NO β'	$B'^2\Delta - X^2\Pi_r$	60019.0	(0:6 - 0:22)
NO δ	$C^2\Pi_r - X^2\Pi_r$	52377.5	(0:9 - 0:22)
NO γ'	$E^2\Sigma^+ - X^2\Pi_r$	60862.65	(0:4 - 0:22)

significant contributor above 300 nm, it is neglected in the present work which is focused on neutral molecules.

The population of molecular energy levels has been calculated in the frame of a two temperature model. For a well defined energy level of a diatomic molecule (n, v, J), characterized by the electronic state n and the vibrational, resp. rotational, quantum number v, resp. J, the population can be expressed as:

$$n_{nvJ} = N\frac{g_{nvJ}}{Q(T_{ve}, T_{rot})} \times exp\left(-\frac{E_{el}(n) + E_{vib}(n,v)}{k_B T_{ve}}\right) \times exp\left(-\frac{E_{rot}(n,J) + E_{inter}(n,v,J)}{k_B T_{rot}}\right), \quad (15)$$

where $Q(T_{ve}, T_{rot})$ is the two temperatures partition function [20], N is the total population of the considered molecule and g_{nvJ} is the degeneracy of the level (n, v, J); E_{el}, $E_{vib}(n, v)$ and $E_{rot}(n, J)$ are respectively the pure electronic, vibrational and rotational term and $E_{inter}(n, v, J)$ is the vibrational-rotational interaction term (in practice much smaller that the other terms). The total population density of the species, N and the temperatures T_{ve} and T_{rot} are output of the flow simulation discussed above.

The measurements dealt with in this study concern the intensity emitted by the flow in direction perpendicular to the its axis (for details on the data acquisition system and the methodology see [18]). Hence, the intensity collected by Optical Emission Spectroscopy corresponds to the local emission integrated along the flow diameter, where uniform properties are assumed in the transverse direction. Since the plasma medium is in practice optically thin, the transverse intensity $I_\sigma(x)$, at a position x behind the shock, is merely calculated as :

$$I_\sigma(x) = \eta_\sigma(x) \times D, \quad (16)$$

where the shock diameter D is assumed to be equal to the shock-tube test section diameter 50 mm and $\eta_\lambda(x)$ is the emission at the location x (along the flow axis).

Also, the acquisition time was set higher than the radiating flow lasting time, meaning that the collected radiation profile is not spatially resolved, but instead integrated along the flow axis. The intensity collected by the system is then calculated as:

$$I_\sigma^{coll} = \frac{1}{\Delta x}\int_{\Delta x} I_\sigma(x)dx \quad (17)$$

where $\Delta x = v_{shock} \times t_{rad}$ is the length of the plasma axial extent, and t_{rad} refers to the effective duration of the plasma radiation when passing through the measurement test section of 6 and 17 μs respectively for the 0.25 and 1 Torr cases. Finally, for comparisons with measured spectra, the so-called collected intensity is convoluted with the apparatus function approximated by a normalized triangle function of 2 nm full width half maximum.

3. Flow properties: results and discussion

We consider shock waves propagating at $u_1 = 5.0$ to 8.06 km/s in mixtures of N_2 and O_2 (80%-20% molar fraction composition) at $T_0 = 298$ K, $p_0 = 0.25$ or 1.0 Torr. The conditions just behind the shock are obtained assuming frozen chemistry and vibrational kinetics. The detailed State-to-State approach is used to solve for the evolution of the flow behind the shock waves. We employed two different models of state-specific rates for the Zeldovich exchange reactions, i.e. Esposito's and Bose's rates, which are labeled as 'Espo' and 'Bose' in the following figures respectively. Calculations show that the relaxation behaviors are similar in all cases. Therefore, in the following paragraphs, we firstly analyze the flow properties, vibration excitation, DR and exchange reactions, as well as non-Boltzmann vibrational distributions, and compare the Esposito's and Bose's rates based in the case with $u = 8.06$ km/s, $p_0 = 0.25$ Torr. Then we discuss the formation of NO as a function of the shock conditions.

3.1. Relaxation flow for $u_1 = 8.06$ km/s

Figure 1 shows the profiles of the gas temperature and representative vibrational temperatures of diatomic species (Figure 1(a)), and species mole fractions (Figure 1(b)) up to 1 m behind the shock. The solid lines are the results obtained by employing Esposito's ZE rates, and dashed lines are those by Bose's ZE rates. Differences are only noticeable for the vibrational temperature and abundance of NO. At the shock, the gas temperature increases rapidly and reaches about 31500 K. After the shock, the vibrational temperatures of N_2 and O_2 increase up to a maximum due to VT transitions. Then, once sufficient dissociation occurs, the loss of vibrational energy by dissociation takes over the energy gain due to excitation, the two temperatures fall and gradually equilibrate with the gas temperature. The two models for the ZE reactions result in large differences in the predicted evolution of the NO vibrational temperature.

The composition profiles show the dissociation of N_2 and O_2 and the formation of NO, N and O. At the early phase of the relaxation, the low vibrational temperatures of N_2 and O_2 prevent the dissociation and exchange processes to occur. Therefore, their mole fractions remain at the initial levels before 1×10^{-5} m. Then, due to VT excitation, the molecules gradually climb the vibrational ladder, which initializes the dissociation of upper levels. As the vibrational temperatures increase, the dissociation progressively goes deep into the lower energy levels. As a result, the molecules N_2 and O_2 are dramatically reduced, while the atoms N and O are abundantly produced. At about 1×10^{-1} m, the mole fraction of N_2 is reduced from 80% to 10%. Dissociation of O_2 is much faster so that already at 1×10^{-3} m, the mole fraction of O_2 falls below 0.1%. NO, instead, is produced in the Zeldovich exchange reactions. The onset of these processes requires enough atomic nitrogen and oxygen from dissociation of N_2 and O_2. Therefore, a long incubation distance is observed for the appearance of NO. Esposito's rates predict a relatively larger formation than Bose's rates. At about 2×10^{-5} m, Esposito's rates predict more than 0.1% mole fraction of NO, which reaches its maximum value of 0.65% at 8×10^{-5} m. Bose's rates obtained more than 0.1% at 4×10^{-5} m, and then get the maximum value of 0.24% at 1.7×10^{-4} m. After the peak, dissociation prevails over formation, and the mole faction of NO begins to curb. At 1.0×10^{-3} m, the mole fraction of NO falls back below 0.1%. Moreover, for the reaction $O_2(v) + N \rightleftharpoons NO(w) + O$, once NO is formed in sufficient amount, the reverse reaction takes over and leads to a production of O_2. This slightly increases the vibration temperature of O_2 and slows down the consumption of O_2 by dissociation, as observed between 3×10^{-4} m to 5×10^{-4} in Figure 1.

For the purpose of comparison, we also apply the MT method to simulate the relaxation flow; in this case, vibrational excitation is modeled by the Landau-Teller relation with Millikan-White's formula for the relaxation time, and the vibrational-chemical coupling is from Park's two-temperature model. The comparisons of resulting temperature and concentration profiles are illustrated in Figure 2. The MT method predicts different behaviors. All the vibrational

IOP Conf. Series: Journal of Physics: Conf. Series **815** (2017) 012026 doi:10.1088/1742-6596/815/1/012026

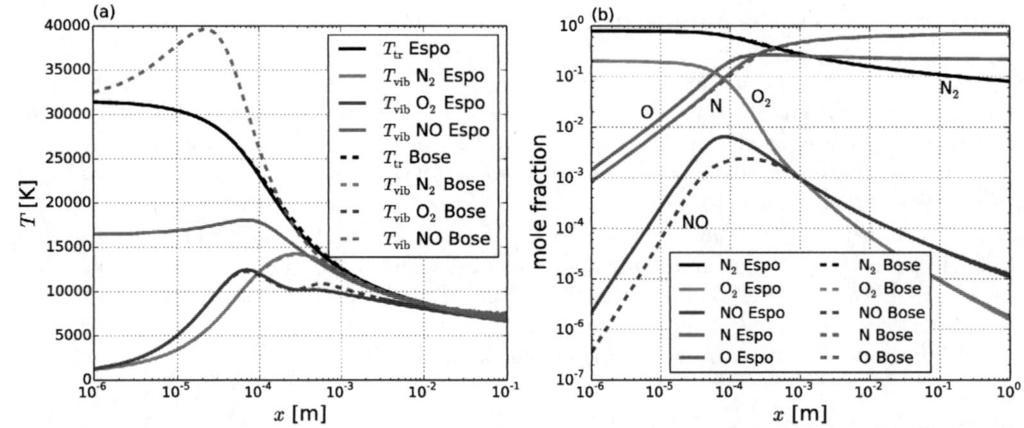

Figure 1. Comparison of Esposito's and Bose's rates: (a) Temperatures (b) Mole fractions ($u_1 = 8.06\ km/s$, $p_1 = 0.25\ Torr$).

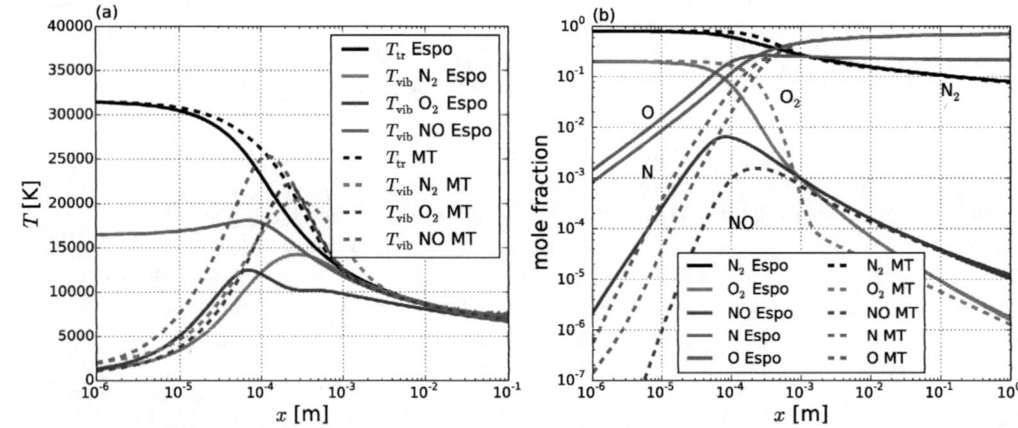

Figure 2. Comparison of StS and MT models: (a) Temperatures (b) Mole fractions ($u_1 = 8.06\ km/s$, $p_1 = 0.25\ torr$).

temperatures for diatomic molecules start from low temperatures. Then they increase up to the maximum values and undergo overshoot to the gas temperature during their decrease when dissociation is predominant. Dissociation of N_2 and O_2, as well as formation of NO are slower in the MT simulation, and longer incubation distances are observed. Finally, the MT model also predicts less amount of NO.

It is very interesting to plot the vibrational distribution functions (VDF) for diatomic species in the flow. Figure 3 shows the locations at which VDF for N_2 are plotted; $x_{0.79}$, $x_{0.7}$, $x_{0.4}$ and $x_{0.1}$ denote the locations at which the N_2 mole fraction is reduced to 79%, 70%, 40% and 10%, respectively. The calculated VDF together with the Boltzmann distribution functions at the equivalent vibrational temperatures are illustrated in Figure 4. At $x_{0.79}$, a location close to the shock, the VDF highly departs from the Boltzmann distribution with large overpopulation of the high-lying vibrational levels. Then, as excitation and dissociation proceeds, the VDF gradually tends towards a Boltzmann distribution at $x_{0.7}$. However, at $x_{0.4}$ when half of the N_2 dissociation has occurred, the VDF departs from the Boltzmann distribution again due to

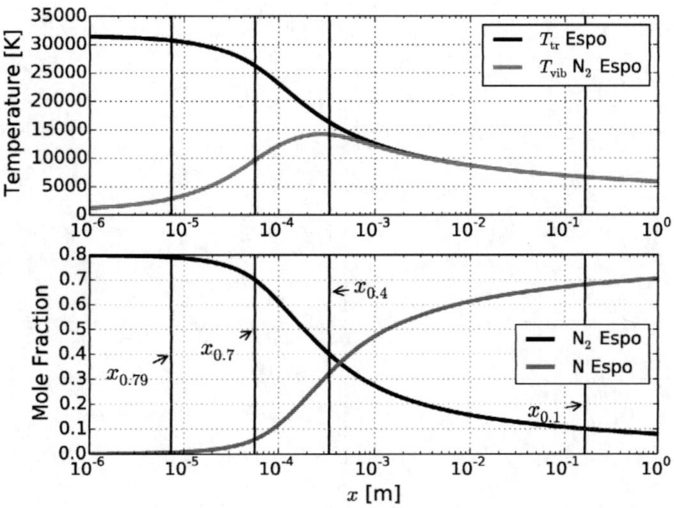

Figure 3. Locations at which the vibrational distribution functions for N_2 are plotted: $x_{0.79}$, $x_{0.7}$, $x_{0.4}$, $x_{0.1}$ is the position where the N_2 mole fraction is 79%, 70%, 40%, 10%.

the depletion of the high-lying levels caused by dissociation. Finally, when equilibrium has been reached at $x_{0.1}$, the VDF is close to the Boltzmann distribution.

Figures 5 and 6 show the locations and VDF for O_2; $x_{0.19}$, $x_{0.1}$, and $x_{0.001}$ are the locations at which the O_2 molar fraction is 19%, 10% and 1%. We also plot the VDF at $x_{\max O}$ where the O has reached its maximum. Similar behaviors are observed here as the ones of N_2. Interestingly, differences in the high-lying level populations predicted by Esposito's and Bose's rates are observed at $x_{\max O}$ and $x_{0.001}$. The Esposito's rates predict larger underpopulation of the high-lying vibrational levels. The discrepancy appears after the exchange reaction $O_2(v) + N \rightleftharpoons NO(w) + O$ starts proceeding predominatly in the backward direction.

Figure 7 and 8 show the locations and VDF for NO, and $x^F_{0.001}$ is where the NO mole fraction has reached 0.1% during formation; $x_{\max}$ is where the NO mole fraction has reached its maximum; $x^D_{0.001}$ is where the NO mole fraction has reduced to 0.1% during dissociation. It shows that both the Esposito's and Bose's results are close to the Boltzmann distribution functions at all three locations. However, VDF predicted by Esposito's rates shows underpopulation of the high-lying vibrational levels.

A detailed investigation shows that Bose's rates to form high-lying levels of NO are significantly larger than Esposito's. This explains the vibrational temperature overshoot predicted with the former model. On the other hand, Esposito's rates are larger for the formation of NO in intermediate-lying states. This leads to a larger production of NO. Note, however, that the original work [15, 16], only reported calculated rates for low and intermediate vibrational levels. Data for high-lying vibrational levels are obtained by extrapolation [7].

3.2. Formation of NO *as a function of shock speed*

We calculate shock waves propagating at velocities from 5.0 km/s to 8.06 km/s. All the upstream gas mixtures contain 80% N_2 and 20% O_2 in mole fraction at T_1 = 298 K. The upstream pressure and shock speed conditions considered are summarized in Table 3. Figure 9 illustrates the maximum NO mole fraction and the its position behind the shock waves. The position has been normalized by the upstream mean free path λ_1, which is estimated as 4.71×10^{-5} m and 1.88×10^{-4} m for 1 Torr and 0.25 Torr respectively. Esposito's rates produce the largest

IOP Conf. Series: Journal of Physics: Conf. Series **815** (2017) 012026 doi:10.1088/1742-6596/815/1/012026

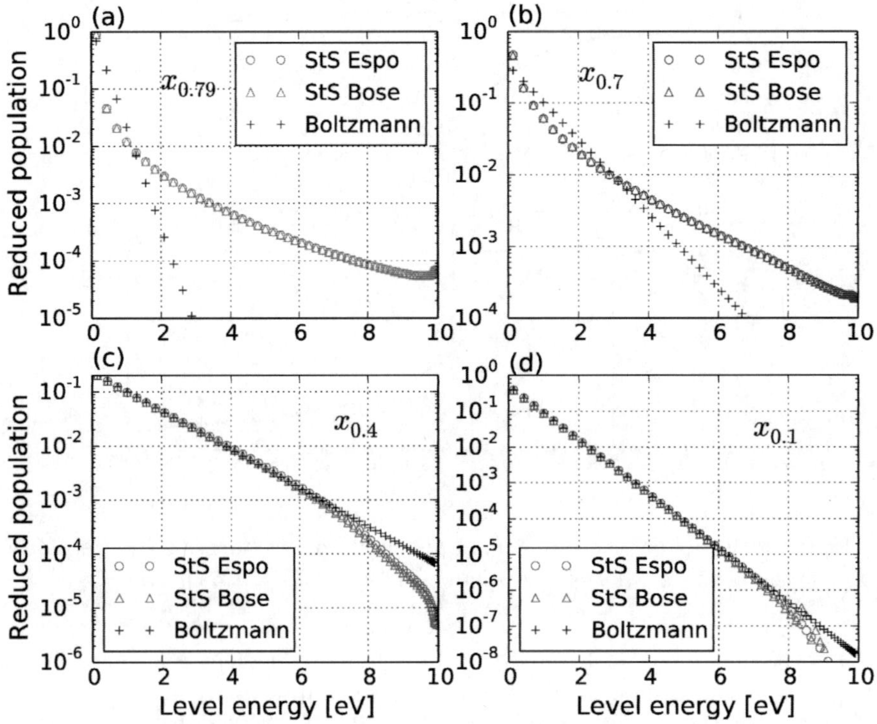

Figure 4. Vibrational distribution functions of N_2 and Boltzmann distribution functions at the equivalent T_{vib}: (a) at $x_{0.79}$; (b) at $x_{0.7}$; (c) at $x_{0.4}$; (d) $x_{0.1}$.

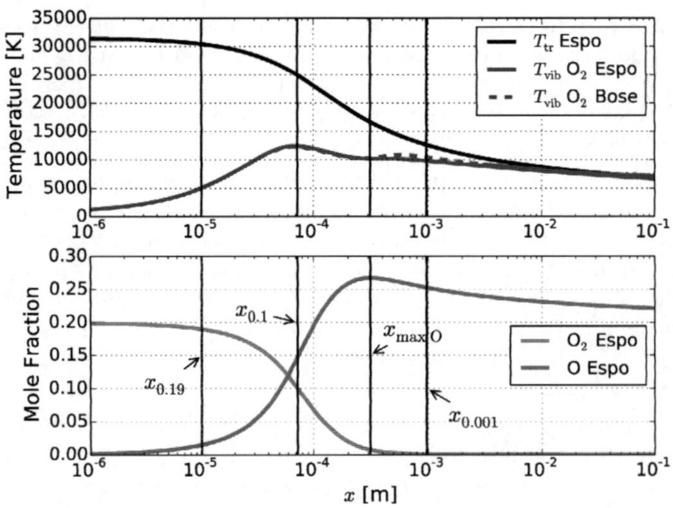

Figure 5. Locations at which the vibrational distribution functions for O_2 are plotted: $x_{0.19}$, $x_{0.1}$, $x_{\max\ \mathrm{O}}$, $x_{0.001}$ is the position where the O_2 mole fraction is 19%, 10%, maximum, 0.1%.

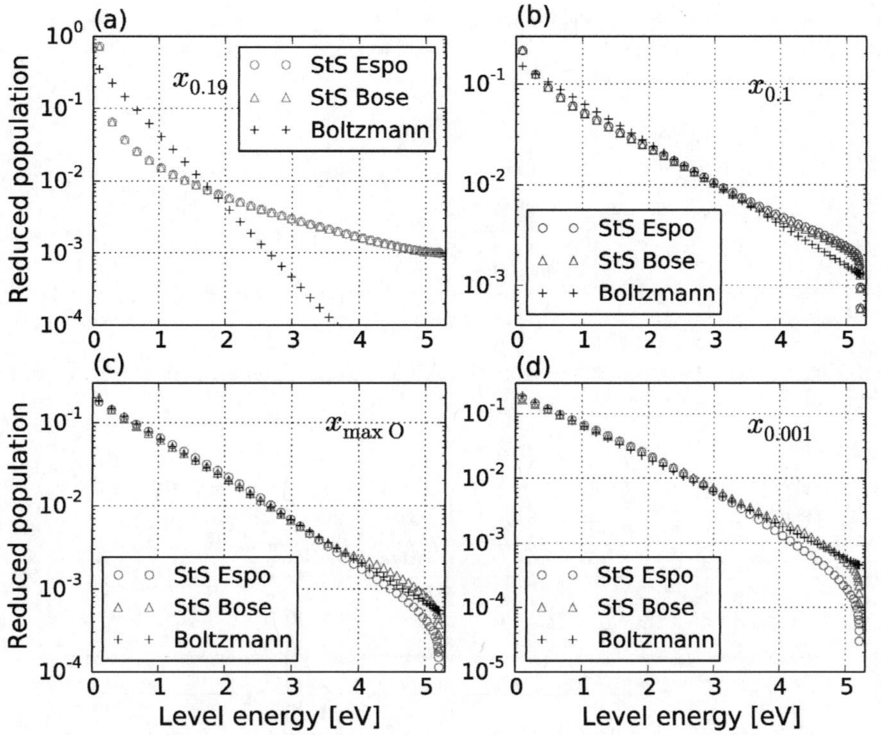

Figure 6. Vibrational distribution functions of O_2 and Boltzmann distribution functions at the equivalent T_{vib}: (a) at $x_{0.19}$; (b) at $x_{0.1}$; (c) at $x_{\max\ O}$; (d) $x_{0.001}$.

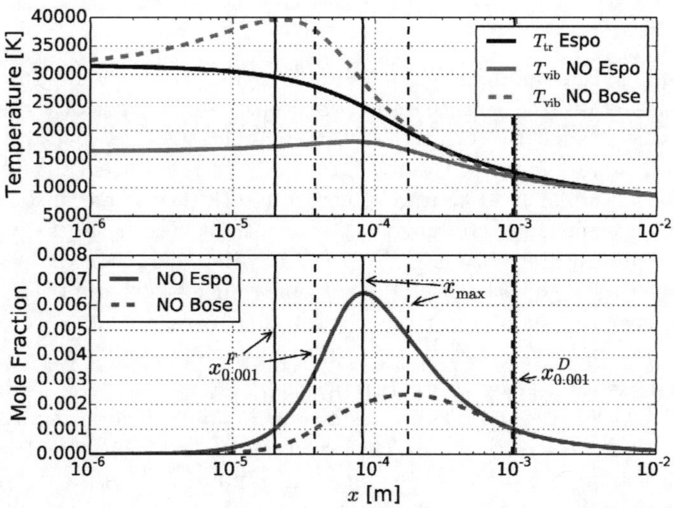

Figure 7. Locations at which the vibrational distribution functions for NO are plotted: $x^F_{0.001}$ is where the NO mole fraction is 0.1% during formation; $x_{\max}$ is where the NO mole fraction has its maximum; $x^D_{0.001}$ is where the NO mole fraction has reduced to 0.1% during dissociation.

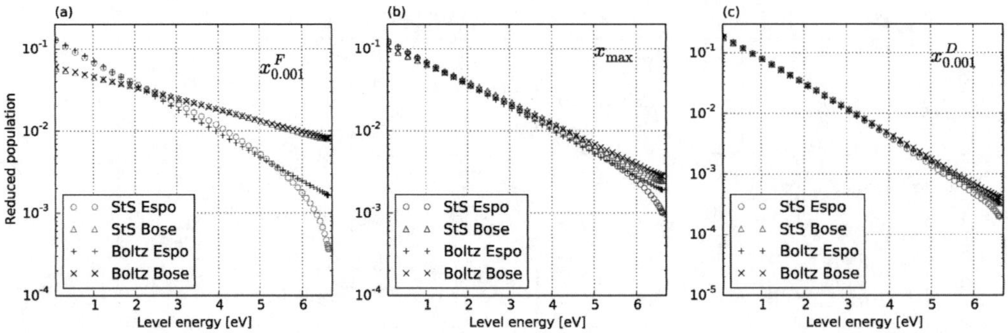

Figure 8. Vibrational distribution functions of NO and Boltzmann distribution functions at the equivalent T_{vib}: (a) at $x^F_{0.001}$; (b) at x_{max}; (c) at $x^D_{0.001}$.

Table 3. Upstream pressure and shock speed for the shock waves considered. Speed cases marked by * are considered for comparison with experimental data.

Case Index	u_1 km/s	p_1 Torr
1	5.00	1.00
2	5.38	1.00
3	5.56*	1.00
4	5.90	1.00
5	6.10*	1.00
6	6.41	1.00
7	6.49*	1.00
8	6.76*	1.00
9	6.94*	1.00
10	7.14*	0.25
11	8.06*	0.25

amounts of NO, while MT method has the smallest amounts. Besides, the predicted position of the maximum mole fraction by Esposito's rates is closest to the front of the shock wave, while the one by MT method is located farthest from the shock wave. Generally speaking, as the velocity of the shock wave increases, the formation of NO is reduced and the position of the maximum mole fraction shifts towards the shock front. Both effects can be explained by the strong temperature dependence of the dissociation reactions: faster shocks will result in larger temperatures, so the initial dissociation of N_2 and O_2 will occur faster, which in turn triggers NO formation closer to the shock front. At the same time, larger temperatures will promote NO dissociation, so that the maximum mole fraction decreases.

4. Assessment with shock tube experiments

The emission in the UV range, 200-400 nm, has been calculated at a dozen positions behind the shock using the three models investigated here referred to as MT, ES and BO. The resulting emission profiles for 5.56 and 6.94 km/s are plotted in Figure 10 with the detailed contributions of N_2, O_2 and NO radiating systems. At the lowest velocity, the differences between the O_2 and N_2 FN resulting profiles remain small. For all models the emission of O_2 SR dominates the peak emission, while the emission of N_2 FN remains weak. The cumulated emission shows that the peak emission contributes roughly to about half of the total emission. At higher velocity

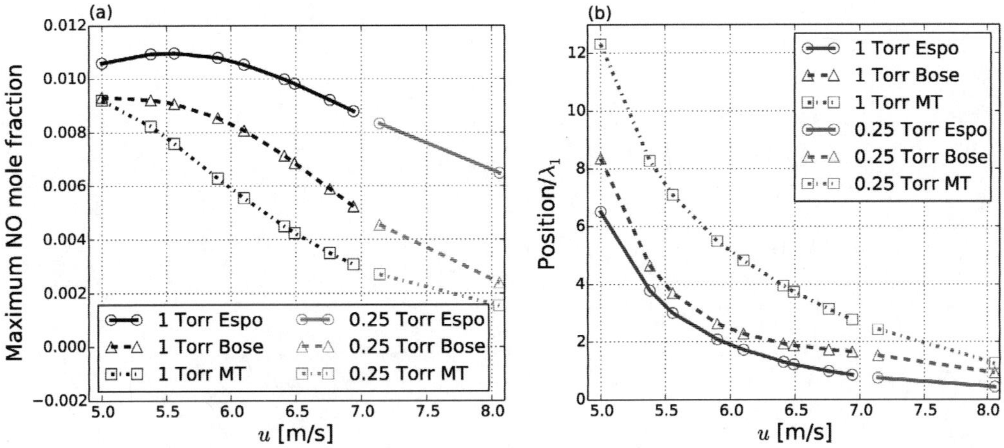

Figure 9. Formation of NO as a function of shock wave speed: (a) maximum mole fraction; (b) position of the maximum mole fraction (normalized by the upstream mean free path).

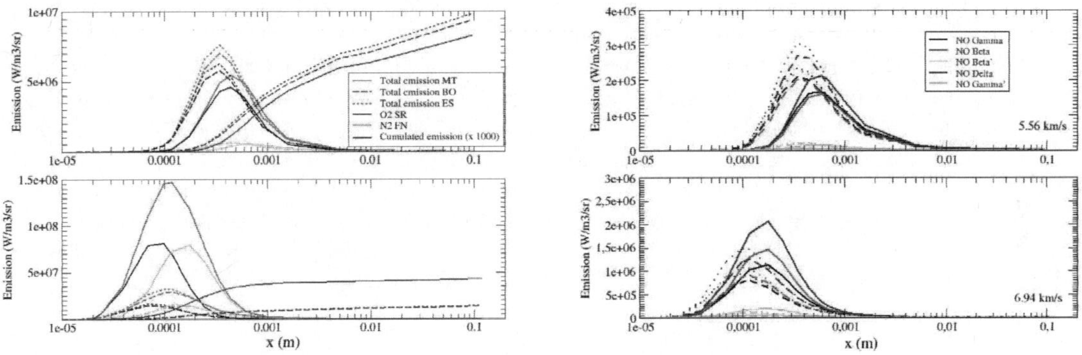

Figure 10. Detailed contributions of the electronic systems of Table 2 to emission profiles. Plain, dashed and dotted lines correspond to MT, ES and BO model results. (Left) Red, black, green and blue curves correspond respectively to total, O_2 SR, N_2 FN and cumulated emissions. (Right) Details of the NO systems.

the differences between MT and StS calculations become important, reflecting differences in the predicted flow properties and the emission intensity is dominated by the thermal equilibrium region. However, whatever the model, the relative contribution of N_2 FN has increased to the level of O_2 SR in the emission peak region. In all cases, the contribution of NO emission remains much smaller than the rest and has consequently been reported on a distinct plot for the sake of readability (see Figure 10, right plots), detailing the contribution of the NO systems listed in Table 2. With increasing velocity the contribution of NO radiation slightly increases, however more in the case of MT than in StS models.

Comparisons between rebuilt spectral intensities in the range 200-300 nm and measured spectra are gathered in Figure 11 for the 1 Torr case. Although the experimental and simulated

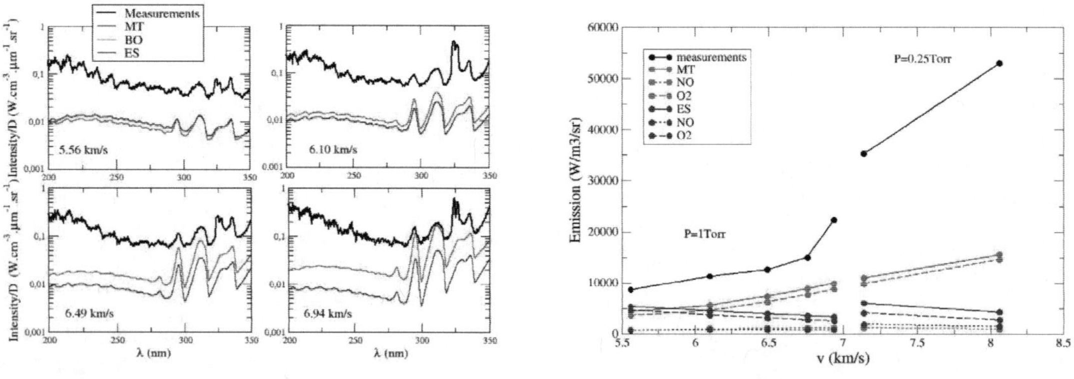

Figure 11. (Left) Measured and calculated spectra at $p_1 = 1\ Torr$. (Right) Total post-shock emission integrated in the range $200 - 300\ nm$ as a function of shock speed.

spectra show qualitative agreement, the absolute values differ of about one order of magnitude. Below 250 nm, the discrepancies between spectra increases with increasing velocity, whereas MT calculations tends toward a better agreement for the bands $\Delta v = 1$ of N_2 FN around 315 nm. For all the cases in Table 3, the total emission has been calculated in the range 200-300 nm and compared in Figure 11 to the experimental data to focus on the contributions to post shock emission due to NO systems and O_2 SR. As already appreciated from the spectral analysis in the 1 *torr* pressure case, all calculations under-estimate the measured total emission and the discrepancy increases with velocity. MT calculations seem to reproduce the main trend of the measurements, whereas StS calculations exhibit an opposite evolution. This is connected to the decay of the O_2 SR band which contributes up to 90% of the emission in the examined spectral range.

5. Conclusions

A detailed kinetic mechanism has been studied for describing the relaxation behind shock waves in air. The mechanism is vibrationally specific for the electronic ground state of the neutral molecules involved. In particular, two sets of reaction rate coefficients for the Zeldovich reactions of NO formation, obtained from accurate QCT calculations, have been critically analyzed and the results compared to a standard MT formulation. Results show that the NO formation kinetics is sensitive to the details of the adopted rate coefficients. Results from the calculations have been then used to estimate the radiation emitted assuming the electronic excited molecular states are populated according to a Boltzmann distribution at the N_2 vibrational temperature. This allows some comparison with recent optical emission spectroscopy measurements performed in a shock tube facility documenting the spectral intensity of the overall shock (without any spatial resolution). A poor agreement is found in all cases, and the contribution of O_2 SR in the range 200-300 nm was found to be dominant, calling for additional studies. However the lack of information on the spatial distribution of the intensity behind the shock makes it difficult to assess specific routes of improvement.

Also, indications on the merits of different kinetic models of NO formation are shown to be masked by the predominant contribution of O_2 emission in the same spectral range. Calculations with a 11 species air model, that take into account the ionization kinetics and the energy balance of the free electron component have also been carried out (not reported here), but results are seen not to differ appreciably. This is to be expected since, at these relatively low shock speeds, the free electron concentration is expected to be small and excitation processes are driven mainly by heavy particle collisions.

A better picture is expected from a detailed kinetic mechanism describing the population of the electronically excited states responsible for the emitted radiation. Future work will therefore be focused on the development of such a Collisional-Radiative model.

Acknowledgement

This work has been supported by the UC3M CONEX-Marie Curie fellowship and Italian project PON "Apulia Space" (03PE-00067-6).

References

[1] Park C 1990 *Nonequilibrium Hypersonic Aerothermodynamics* 1st ed (New York: Wiley-Interscience)
[2] Capitelli M, Armenise I, Bisceglie E, Bruno D, Celiberto R, Colonna G, D'Ammando G, De Pascale O, Esposito F, Gorse C, Laporta V and Laricchiuta A 2012 *Plasma Chemistry and Plasma Processing* **32** 427–450
[3] Akpinar S, Armenise I, Defazio P, Esposito F, Gamallo P, Petrongolo C and Sayos R 2012 *Chemical Physics* **398** 81–89
[4] Lagana A, Lombardi A, Pirani F, Gamallo P, Sayos R, Armenise I, Cacciatore M, Esposito F and Rutigliano M 2012 *The Open Plasma Physics Journal* **7** 48–59
[5] Capitelli M, Celiberto R, Colonna G, Esposito F, Gorse C, Hassouni K, Laricchiuta A and Longo S 2016 *Reactivity and Relaxation of Vibrationally/Rotationally Excited Molecules with Open Shell Atoms* (New York, NY: Springer New York) pp 31–56
[6] Dikalyuk A, Kozlov P, Romanenko Y, Shatalov O and Surzhikov S 2013 *44th AIAA Thermophysics Conference* pp 1–27
[7] Lopez B 2014 *11th AIAA/ASME Joint Thermoophysics and Heat Transfer Conference* (Atlanta) pp 16–20
[8] Radhakrishnan K and Hindmarsh A 1993 *Description and Use of LSODE, the Livermore Solver for Ordinary Differential Equations* Tech. rep. Lawrence Livermore National Laboratory Livermore, CA, USA
[9] Zeldovich Y B, Sadovnikov P Y and Frank-Kamenetskii D A 1947 *Oxidation of Nitrogen in Combustion* Tech. rep. Academy of Sciences of USSR, Institute of CHemical Physics Moscow-Leningrad
[10] Lino da Silva M STELLAR Database, http://esther.ist.utl.pt/pages/stellar.html
[11] Adamovich I V, Macheret S O, Rich J W, Treanor C E and Fridman A A 1996 *Molecular Physics and Hypersonic Flows SE - 5* (*NATO ASI Series* vol 482) ed Capitelli M (Springer Netherlands) pp 85–104
[12] Lino da Silva M, Lopez B, Guerra V and Loureiro J 2012 *5th Int. Workshop on Radiation of High Temperature Gases in Atmospheric Entry, ESA SP-714* (Barcelona, Spain)
[13] Esposito F, Armenise I and Capitelli M 2006 *Chemical Physics* **331** 1–8
[14] Esposito F, Armenise I, Capitta G and Capitelli M 2008 *Chemical Physics* **351** 91–98
[15] Bose D and Candler G V 1996 *The Journal of Chemical Physics* **104** 2825
[16] Bose D and Candler G V 1997 *The Journal of Chemical Physics* **107** 6136–6145
[17] Millikan R C and White D R 1963 *The Journal of Chemical Physics* **39** 3209–3213
[18] Kozlov P V, Romanenko Y V and Shatalov O P 2011 *ESA Special Publication* vol 689
[19] Perrin M Y, Riviere P and Soufiani A 2008 *Radiation database for Earth and Mars entry* Tech. rep. Centre National de la Recherche Scientifique Chatenay-Malabry, France
[20] Babou Y, Rivière P, Perrin M Y and Soufiani A 2009 *International Journal of Thermophysics* **30** 416–438

Vibrational-Chemical Coupling in mixtures $CO_2/CO/O$ and $CO_2/CO/O_2/O/C$

A A Kosareva[1], **E A Nagnibeda** [1]

[1] Saint-Petersburg State University, 7/9 Universitetskaya nab., St. Petersburg, 199034 Russia

E-mail: kos-hellen@yandex.ru

Abstract. The paper deals with modeling of vibration-chemical coupling in mixtures containing CO_2 molecules. The temporal vibrational and chemical relaxation in the five-component and three-component space-homogeneous mixtures $CO_2/CO/O_2/C/O$ and $CO_2/CO/O$ is studied taking into account vibrational energy transitions, dissociation, recombination and exchange chemical reactions. The vibrational-chemical coupling in considered mixtures is studied on the basis of three-temperature, two-temperature and one-temperature vibrational CO_2 distributions. Governing equations for macroscopic mixture parameters in the three-temperature, two-temperature and one-temperature approximations are solved numerically for different initial conditions. Comparison of the results obtained in three approaches showed the influence of non-equilibrium vibrational distributions on chemical reaction rates and on the temporal variation of the gas temperature, mixture composition and vibrational temperatures in two considered mixtures. The role of exchange chemical reactions in relaxation processes is also discussed in the paper.

1. Introduction

Modeling of non-equilibrium processes in mixtures containing CO_2 molecules is important for prediction of gas parameters near bodies moving in the Mars or Venus atmosphere [1, 2], in laser physics [3, 4, 5, 6, 7], chemical technology problems and other applications. During several last decades different models for non-equilibrium vibration-dissociation kinetics in CO_2 flows were elaborated. First of these models were proposed introducing a single vibrational temperature for three modes of a CO_2 molecule vibrations [8, 3, 9], then more accurate two-temperature CO_2 vibrational distributions were derived using the kinetic theory methods [10] and applied for simulations of a flow in a shock layer near a spacecraft Mars Sample Return Orbiter (MSRO) entering the Mars atmosphere [11]. In the present time, advances in numerical modeling made it possible to study the detailed vibrational CO_2 kinetics in the frame of the most accurate state-to-state approach [12, 13] solving the equations for populations of vibrational levels of all three (symmetric, bending and asymmetric) CO_2 modes. However, a practical use of the state-to-state kinetic model for prediction of CO_2 flow parameters calls some difficulties. First, practical realization of this model for mixtures containing CO_2 molecules requires to solve a huge number of equations for populations of vibrational levels of all three CO_2 modes. Another problem is connected with a lack of data for state dependent rate coefficients for vibrational energy transitions within three CO_2 modes and between different modes. Therefore simplified CO_2 kinetic models based on multi-temperature vibrational distributions remain attractive for practical applications.

In the present paper we study the vibrational-chemical coupling in the space-homogeneous five-component mixture $CO_2/CO/O_2/O/C$ with the use of three models proposed in [11], [10]. Within the first model, non-equilibrium vibrational distributions of CO_2 molecules are expressed in terms of two vibrational temperatures for combined symmetric-bending and asymmetric modes. In the second case the total vibrational temperature different from the gas temperature is introduced for all three CO_2 modes whereas the third model corresponds to the thermal equilibrium vibrational CO_2 distributions depending only on the gas temperature. Temporal relaxation of the considered mixture for different initial conditions is studied taking into account vibrational energy transitions of CO_2 molecules as well as dissociation, recombination and exchange reactions and the influence of CO_2 vibrational distributions on reaction rate coefficients and variation of macroscopic mixture parameters with time is analyzed. Calculations have been done also for a simplified case of vibration-dissociation relaxation in the three-component mixture $CO_2/CO/O$. The comparison of the results obtained for two considered mixtures showed a role of a chemical kinetics scheme in relaxation processes.

2. Vibrational distributions and basic equations

We consider vibrational-chemical kinetics in the high-temperature five-component mixture $CO_2/CO/O_2/O/C$ taking into account dissociation of CO_2, O_2 and CO molecules, recombination and exchange reactions:

$$CO_2 + M \leftrightarrow CO + O + M, \quad (1)$$

$$O_2 + M \leftrightarrow O + O + M, \quad (2)$$

$$CO + M \leftrightarrow C + O + M, \quad (3)$$

$$M = CO_2, CO, O_2, O, C,$$

$$CO_2 + O \leftrightarrow CO + O_2, \quad (4)$$

$$CO + O \leftrightarrow C + O_2, \quad (5)$$

$$CO + CO \leftrightarrow CO_2 + C. \quad (6)$$

Internal energy relaxation includes different processes. It is well known that the translational and rotational energy relaxation proceeds much faster comparable to vibrational relaxation, therefore we consider equilibrium distributions over translational and rotational energy with the gas temperature. The vibrational kinetics in the considered mixture includes the following channels [11]: intra-mode VV_m vibrational energy exchanges within the each CO_2 mode ($m = 1,2,3$) and in O_2 and CO molecules (VV_{O_2}, VV_{CO}); VT_m, VT_{O_2}, VT_{CO} vibrational- translational energy transitions of different CO_2 modes and of O_2 and CO molecules; VV vibrational exchanges between molecules of various species VV_{3-CO}, VV_{1-2-CO}, VV_{1-2-O_2}, VV_{CO-O_2}. Existing experimental data on relaxation times of different processes in mixtures containing CO_2 molecules [14, 15] show that under wide conditions intra-mode VV_m transitions and VV_{1-2} exchanges between the bending and symmetric modes in CO_2 molecules occur much faster than all other energy VV exchanges between two or three modes, $VT_m (m = 1,2,3, CO, O_2)$ exchanges of translation and vibrational energies and chemical reactions. In this case the following relation between relaxation times of various processes can be written [11]:

$$\tau_{tr} < \tau_{rot} < \tau_{VV_m} \sim \tau_{VV_{1-2}} \ll \tau_{VT_2} \sim \tau_{VV_{2-3}} \sim \tau_{VV_{1-2-3}} \sim \tau_{VT_3} \sim \tau_{react} \sim \theta, \quad m = 1,2,3, \quad (7)$$

here θ is the mean time of variation of macroscopic mixture parameters.

Rapid energy exchanges between bending and symmetric modes and within the asymmetric mode makes it possible to introduce the vibrational temperature T_{12} for combine symmetric-bending mode and T_3 for the asymmetric mode of CO_2 molecules. In this case vibrational level populations of two-temperature vibrational distributions take the form (see [16, 17]):

$$n_{i_1 i_2 i_3}(T_{12}, T_3) = \frac{n_{CO_2}(i_2+1)}{Z_{vibr}^{CO_2}(T_{12}, T_3)} \exp\left(-\frac{i_1\varepsilon_{100} + i_2\varepsilon_{010}}{kT_{12}} - \frac{i_3\varepsilon_{001}}{kT_3}\right). \quad (8)$$

Here n_{CO_2} is the CO_2 number densities, k is the Boltzmann constant, $Z_{vibr}^{CO_2}(T, T_{12}, T_3)$ is the non-equilibrium CO_2 vibrational partition function.

Following [11] we also suppose that CO and O_2 molecules equilibrate faster than CO_2 and therefore vibrational level populations of these molecules may be described by the thermal equilibrium Boltzmann distributions:

$$n_i^c(T) = \frac{n_c}{Z_{vibr}^c(T)} \exp\left(-\frac{\varepsilon_i^c}{kT}\right),$$

where $Z_{vibr}^c(T)$ is the vibrational partition function of spicies c, $\varepsilon_i^c = ih\nu_c$ is the vibrational energy of diatomic molecule c in the case of harmonic oscillations, ν_c is the frequencies of vibrations molecules c, h is the Plank constant. All distributions hereafter are considered for the harmonic oscillator model.

Under the conditions (7) closed description of vibrational-chemical relaxation in the space-homogeneous mixture $CO_2/CO/O_2/O/C$ is given by the set of macroscopic parameters including number densities of chemical species $n_{CO_2}, n_{CO}, n_{O_2}, n_O, n_C$, gas temperature T and two vibrational temperatures T_{12}, T_3 of coupled symmetric-bending and asymmetric CO_2 modes.

The system of equations for these macroscopic parameters includes equations of non-equilibrium vibrational-chemical kinetics coupled to the conservation equation for the total mixture energy:

$$\frac{dn_c}{dt} = R_c^{react}, \quad (9)$$

$$c = CO_2, CO, O_2, O, C,$$

$$\rho_{CO_2}\frac{dE_{12}}{dt} = R_{12} - m_{CO_2}E_{12}R_{CO_2}^{react}, \quad (10)$$

$$\rho_{CO_2}\frac{dE_3}{dt} = R_3 - m_{CO_2}E_3R_{CO_2}^{react}, \quad (11)$$

$$\frac{3}{2}nkT + (n_{CO_2} + n_{CO} + n_{O_2})kT + \rho_{CO_2}E_{vibr}^{CO_2}(T_{12}, T_3) + \rho_{CO}E_{vibr}^{CO}(T) + \quad (12)$$

$$+ \rho_{O_2}E_{vibr}^{O_2}(T) + n_{CO_2}\varepsilon_{CO_2} + n_{CO}\varepsilon_{CO} + n_{O_2}\varepsilon_{O_2} + n_O\varepsilon_O + n_C\varepsilon_C = const.$$

Here $n = n_{CO_2} + n_{CO} + n_{O_2} + n_O + n_C$ is the total numeric density, ε_c is the formation energy, E_{12}, E_3 are the vibrational energies of the coupled and asymmetric CO_2 modes:

$$\rho_{CO_2}E_{12}(T_{12}) = \frac{n_{CO_2}}{Z_{12}(T_{12})}\sum_{i_1,i_2}(i_1\varepsilon_{100} + i_2\varepsilon_{010})(i_2+1)\exp\left(-\frac{i_1\varepsilon_{100} + i_2\varepsilon_{010}}{kT_{12}}\right), \quad (13)$$

$$\rho_{CO_2}E_3(T_3) = \frac{n_{CO_2}}{Z_3(T_3)}\sum_{i_3} i_3\varepsilon_{001}\exp\left(-\frac{i_3\varepsilon_{001}}{kT_3}\right), \quad (14)$$

where $Z_{12}(T_{12})$, $Z_3(T_3)$ are the vibrational partition function of the coupled and asymmetric CO_2 modes.

In (12) $E_{vibr}^{CO_2}(T_{12}, T_3)$, $E_{vibr}^{CO}(T)$, $E_{vibr}^{O_2}(T)$ are the vibrational energies of molecules CO_2, CO, O_2 per the unit mass:

$$\rho_{CO_2} E_{vibr}^{CO_2}(T_{12}, T_3) = \frac{n_{CO_2}}{Z_{vibr}^{CO_2}(T_{12}, T_3)} \sum_{i_1, i_2, i_3} (i_1 \varepsilon_{100} + i_2 \varepsilon_{010} + i_3 \varepsilon_{001})(i_2 + 1) \cdot \tag{15}$$

$$\cdot \exp\left(-\frac{i_1 \varepsilon_{100} + i_2 \varepsilon_{010}}{kT_{12}} - \frac{i_3 \varepsilon_{001}}{kT_3}\right),$$

$$\rho_c E_{vibr}^c(T) = n_c \frac{h\nu_c}{\exp\left(\frac{h\nu_c}{kT}\right) - 1}, \quad c = CO, O_2.$$

Production terms R_{react}^c, R_{12}, R_3 in equations (9) – (11) describe variation of mixture composition due to chemical reactions and changing of vibrational energies E_{12}, E_3.

If all VV vibrational energy exchanges within and between three CO_2 modes proceed faster than VT_m transitions and chemical reactions the total vibrational temperature $T_v = T_{12} = T_3$ for CO_2 molecules may be introduced and vibrational distributions (8) take the form:

$$n_{i_1 i_2 i_3}(T_v) = \frac{n_{CO_2}(i_2 + 1)}{Z_{vibr}^{CO_2}(T_v)} \exp\left(-\frac{i_1 \varepsilon_{100} + i_2 \varepsilon_{010} + i_3 \varepsilon_{001}}{kT_v}\right). \tag{16}$$

In this case the system of governing equations contains, instead of two equations (10), (11), only one equation which follows from (15) if $T_{12} = T_3 = T_v$:

$$\rho_{CO_2} \frac{dE_{CO_2}^{vibr}(T_v)}{dt} = R_{CO_2} - m_{CO_2} E_{CO_2}^{vibr}(T_v) R_{CO_2}^{react},$$

where

$$\rho_{CO_2} E_{CO_2}^{vibr}(T_v) = \frac{n_{CO_2}}{Z_{vibr}^{CO_2}(T_v)} \sum_{i_1, i_2, i_3} (i_1 \varepsilon_{100} + i_2 \varepsilon_{010} + i_3 \varepsilon_{001})(i_2 + 1) \exp\left(-\frac{i_1 \varepsilon_{100} + i_2 \varepsilon_{010} + i_3 \varepsilon_{001}}{kT_v}\right)$$

is the vibrational energy of CO_2 molecules.

If the relaxation time of all internal energy modes of mixture components occurs considerably shorter than the mean time of chemical reactions, vibrational level populations are described by the Boltzmann thermal equilibrium distributions (16) with $T_v = T$. In this case non-equilibrium chemical reactions are simulated on the basis of maintaining thermal equilibrium vibrational distributions. The set of equations for macroscopic parameters n_{CO_2}, n_{CO}, n_{O_2}, n_O, n_C, T contains equations of chemical kinetics (9) with reaction rate coefficients depending only on the gas temperature and the energy conservation equation.

3. Reaction rate coefficients

Let us consider right hand part of equation (9):

$$R_c^{react} = R_c^{diss} + R_c^{exch}, c = CO_2, CO, O_2, O, C.$$

Dissociation-recombination processes are described by the following expressions:

$$R_{CO_2}^{diss} = \sum_M n_M \left(k_{rec,CO_2}^M(T) n_{CO} n_O - k_{diss,CO_2}^M(T, T_{12}, T_3) n_{CO_2}\right),$$

$$R_{CO}^{diss} = \sum_M n_M \left(k_{rec,CO}^M(T) n_O n_C - k_{diss,CO}^M(T) n_{CO}\right),$$

$$R_{O_2}^{diss} = \sum_M n_M \left(k_{rec,O_2}^M(T) n_O^2 - k_{diss,O_2}^M(T) n_{O_2}\right),$$

here $k^M_{diss,c}$, $k^M_{rec,c}$ are the coefficients of dissociation and recombination ratea of spicies c, M is the collision partner.

The dissociation rate coefficient for carbon dioxide in the three-temperature approximation may be written in the form of the generalized Treanor-Marrone model for multiatomic molecules [10]:

$$k^M_{diss,CO_2}(T,T_{12},T_3) = k^{eq,M}_{diss,CO_2}(T)\frac{Z^{CO_2}_{vibr}(T)}{Z^{CO_2}_{vibr}(-U)Z^{CO_2}_{vibr}(T_{12},T_3)}\sum_{i_1,i_2,i_3}(i_2+1)\cdot \quad (17)$$
$$\cdot \exp\left(\left(\frac{i_1\varepsilon_{100}+i_2\varepsilon_{010}}{k}\right)\left(\frac{1}{T}-\frac{1}{T_{12}}\right)+\frac{i_3\varepsilon_{001}}{k}\left(\frac{1}{T}-\frac{1}{T_3}\right)+\frac{\varepsilon_{i_1i_2i_3}}{kU}\right),$$

here $\varepsilon_{i_1i_2i_3} = i_1\varepsilon_{100}+i_2\varepsilon_{010}+i_3\varepsilon_{001}$, $k^{eq,M}_{diss,CO_2}(T)$ is the equilibrium dissociation rate coefficient defined by the Arrenius law, U is a model parameter, in calculations we had $U = D/6k$. The dissociation of diatomic molecules is assumed equilibrium and dissociation rate coefficients are calculated according to the Arrenius law as well as CO_2 dissociation in the one-temperature approach. The dissociation rate coefficients in two-temperature approach take the form (17) under conditions $T_{12} = T_3 = T_v$ and the dissociation rate coefficients in one-temperature approach are equilibrium $k^{eq,M}_{diss,CO_2}(T)$. For calculation of rate coefficients in the paper used the Park model [1].

Recombination rate coefficients are determined on the basis of detailed balance principle [10]:

$$k^M_{rec,CO_2}(T) = k^{eq,M}_{diss,CO_2}(T)\cdot$$
$$\cdot\left(\frac{m_{CO_2}}{m_{CO}m_O}\right)^{3/2}(2\pi kT)^{-3/2}h^3\frac{Z^{CO_2}_{vibr}(T)Z^{CO_2}_{rot}(T)}{Z^{CO}_{vibr}(T)Z^{CO}_{rot}(T)}\exp\left(-\frac{\varepsilon_{CO_2}-\varepsilon_{CO}-\varepsilon_O}{kT}\right),$$

$$k^M_{rec,CO}(T) = k^{eq,M}_{diss,CO}(T)\cdot$$
$$\cdot\left(\frac{m_{CO}}{m_Cm_O}\right)^{3/2}(2\pi kT)^{-3/2}h^3 Z^{CO}_{vibr}(T)Z^{CO}_{rot}(T)\exp\left(-\frac{\varepsilon_{CO}-\varepsilon_O-\varepsilon_C}{kT}\right),$$

$$k^M_{rec,O_2}(T) = k^{eq,M}_{diss,O_2}(T)\left(\frac{2}{m_O}\right)^{3/2}(2\pi kT)^{-3/2}h^3 Z^{O_2}_{vibr}(T)Z^{O_2}_{rot}(T)\exp\left(\frac{2\varepsilon_O}{kT}\right),$$

here m_c is the mass of species c, $Z^{CO_2}_{rot}(T)$, $Z^{CO}_{rot}(T)$, $Z^{O_2}_{rot}(T)$ are the rotational partition functions of CO_2, CO and O_2 molecules.

For accurate determination of exchange reaction rate coefficients state-dependent coefficients for rates of these reactions should be averaged over non-equilibrium vibrational distributions. However at present time analytical expressions for rate coefficients depending on vibrational levels of CO_2 modes are not available in the literature. Therefore in the present paper we consider the classical expression for R^{exch}_c with one-temperature rate coefficients:

$$R^{exch}_c = \sum_{dc'd'}\left(k^{d'\to d}_{c'\to c}(T)n_{c'}n_{d'} - k^{d\to d'}_{c\to c'}(T)n_cn_d\right).$$

Rate coefficients for the backward reactions are found using the detailed balance principle:

$$k_{C\to CO}^{CO_2\to CO}(T) = k_{CO\to C}^{CO\to CO_2}(T)\left(\frac{m_{CO}^2}{m_{CO_2}m_c}\right)^{3/2}\frac{\left(Z_{vibr}^{CO}(T)Z_{rot}^{CO}(T)\right)^2}{Z_{vibr}^{CO_2}(T)Z_{rot}^{CO_2}(T)}\exp\left(\frac{-D_{CO_2}+2D_{CO}}{kT}\right),$$

$$k_{O_2\to O}^{CO\to CO_2}(T) = k_{O\to O_2}^{CO_2\to CO}(T)\left(\frac{m_{CO_2}m_O}{m_{CO}m_{O_2}}\right)^{3/2}\frac{Z_{vibr}^{CO_2}(T)Z_{rot}^{CO_2}(T)}{Z_{vibr}^{CO}(T)Z_{rot}^{CO}(T)Z_{vibr}^{O_2}(T)Z_{rot}^{O_2}(T)}\cdot$$
$$\cdot\exp\left(\frac{-D_{CO}-D_{O_2}+D_{CO_2}}{kT}\right),$$

$$k_{O_2\to O}^{C\to CO}(T) = k_{O\to O_2}^{CO\to C}(T)\left(\frac{m_{CO}m_O}{m_Cm_{O_2}}\right)^{3/2}\frac{Z_{vibr}^{CO}(T)Z_{rot}^{CO}(T)}{Z_{vibr}^{O_2}(T)Z_{rot}^{O_2}(T)}\exp\left(\frac{-D_{O_2}+D_{CO}}{kT}\right),$$

where D_{CO_2}, D_{CO}, D_{O_2} are the dissociation energies.

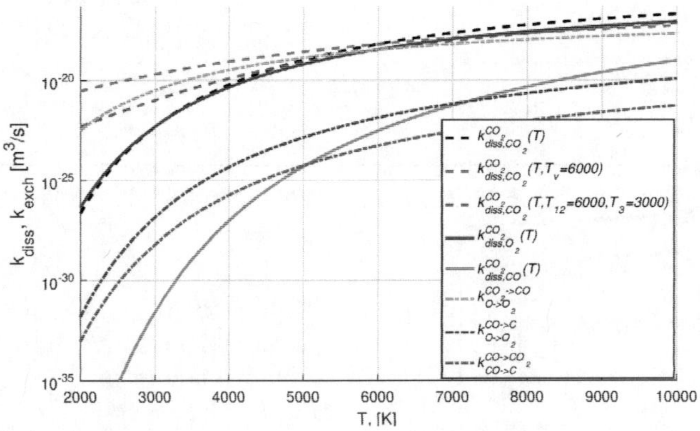

Figure 1. Reaction rate coefficients in dependence on the gas temperature T.

In figure 1 CO_2 dissociation rate coefficients calculated within three-temperature, two-temperature and one-temperature approaches are presented as well as one-temperature rate coefficients for dissociation of diatomic species (2), (3) and exchange reactions (4)-(6). Three-temperature and two-temperature coefficients are presented for fixed vibrational temperatures $T_{12} = 6000\ K$, $T_3 = 3000\ K$ and $T_v = 6000\ K$. It may be noticed that the one-temperature coefficients for CO_2 and O_2 dissociation (1), (2) as well as the forward exchange reaction (4) noticeably exceed coefficients for CO dissociation (3) and reactions (5), (6). In addition, comparing CO_2 dissociation rate coefficients calculated under considered conditions in three approaches one can conclude that one-temperature description provides underestimated values of coefficient $k_{diss,CO_2}(T)$ for the condition $T < T_{12} = 3000\ K$ and overestimated values of this coefficient for $T > T_3 = 6000\ K$ in comparison with the more accurate three-temperature coefficients $k_{diss,CO_2}(T, T_{12}, T_3)$. Similarly, we can notice that the one-temperature model leads to lower values of the coefficient $k_{diss,CO_2}(T)$ for $T < T_v = 6000\ K$ and higher its values for $T > T_v = 6000\ K$ in comparison with the two-temperature coefficient $k_{diss,CO_2}(T, T_v)$.

Now we consider the right hand parts of equations (10), (11) which describe changing of vibrational energies E_{12} and E_3 due to vibrational energy transitions. Following [18] we use approximate expressions of these parts via relaxation times:

$$R_{12} - m_{CO_2}R_{CO_2}^{react} = \rho_{CO_2}\sum_{\gamma}\frac{E_{12}^{eq}(T) - E_{12}(T_{12})}{\tau_\gamma}, \tag{18}$$

$$R_3 - m_{CO_2} R_{CO_2}^{react} = \rho_{CO_2} \sum_{\gamma} \frac{E_3^{eq}(T) - E_3(T_3)}{\tau_\gamma}, \tag{19}$$

where $\tau_\gamma, \gamma = VT_2, VV_{2-3}, VV_{1-2-3}$ were calculated on the basis of analytical approximations of experimental data [3, 14].

4. Results and discussion

In this section we present the results of numerical solutions of equations for number densities of species n_{CO_2}, n_{CO}, n_{O_2}, n_O, n_C, vibrational temperatures T_{12}, T_3, T_v and translation temperature T obtained for space-homogeneous five-component mixture in the three-temperature, two-temperature and one-temperature approaches. Calculations were done for the following initial conditions for temperatures and mixture composition: $T = 12000\ K$, $T_{12} = 6000\ K$, $T_3 = 3000\ K$, $T_v = 6000\ K$, $n_{CO_2} = 100\%$, $n_{CO} = n_{O_2} = n_O = n_C = 0$, $p = 1\ atm$. To solve stiff systems of ordinary differential equations we used the Gear method with a variable time step.

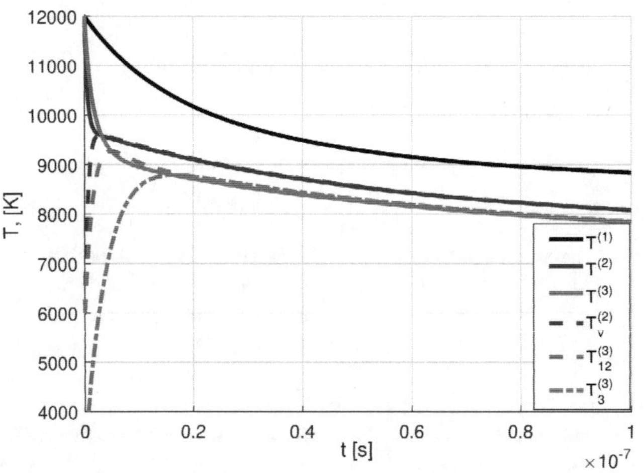

Figure 2. The gas temperature T and the vibrational temperature T_{12}, T_3, T_v in dependence on t; 1 – one-temperature, 2 – two-temperature, 3 – three-temperature approach.

In figure 2 temporal variation of the gas temperature T and vibrational temperatures T_{12}, T_3, T_v is shown whereas changing of the mixture composition in time for three approaches is demonstrated by figures 3 (a, b, c, d). It can be noticed that under considered conditions the use of the one-temperature approach provides noticeably overrated the gas temperature values than more accurate two-temperature and three-temperature kinetic models because in the one-temperature approach the mixture is considered in the thermal equilibrium and translation-vibrational energy transitions are not taken into account. The maximum difference between temperature values obtained in the three-temperature and one-temperature approaches reaches 12% whereas the two-temperature approach provides better consistency with T values obtained within the three-temperature model: the maximum discrepancy does not exceed 3%. In the considered cases of high initial values of the gas temperature vibrational relaxation proceeds much faster than chemical reactions which lead to chemical equilibrium establishing. The difference between parameters found in the frame of different kinetic schemes diminishes with time rising approaching to the equilibrium. Vibrational excitation of CO_2 molecules due to translational-vibrational energy transitions is seen in a short time interval (close to the initial moment of time) in the three-temperature and two-temperature approximations. Under considered conditions this process is found more fast within the two-temperature description, whereas in the frame of the three-temperature model increasing of the vibrational temperature of the combined symmetric-bending mode proceeds more rapid than increasing of the vibrational temperature of the asymmetric mode.

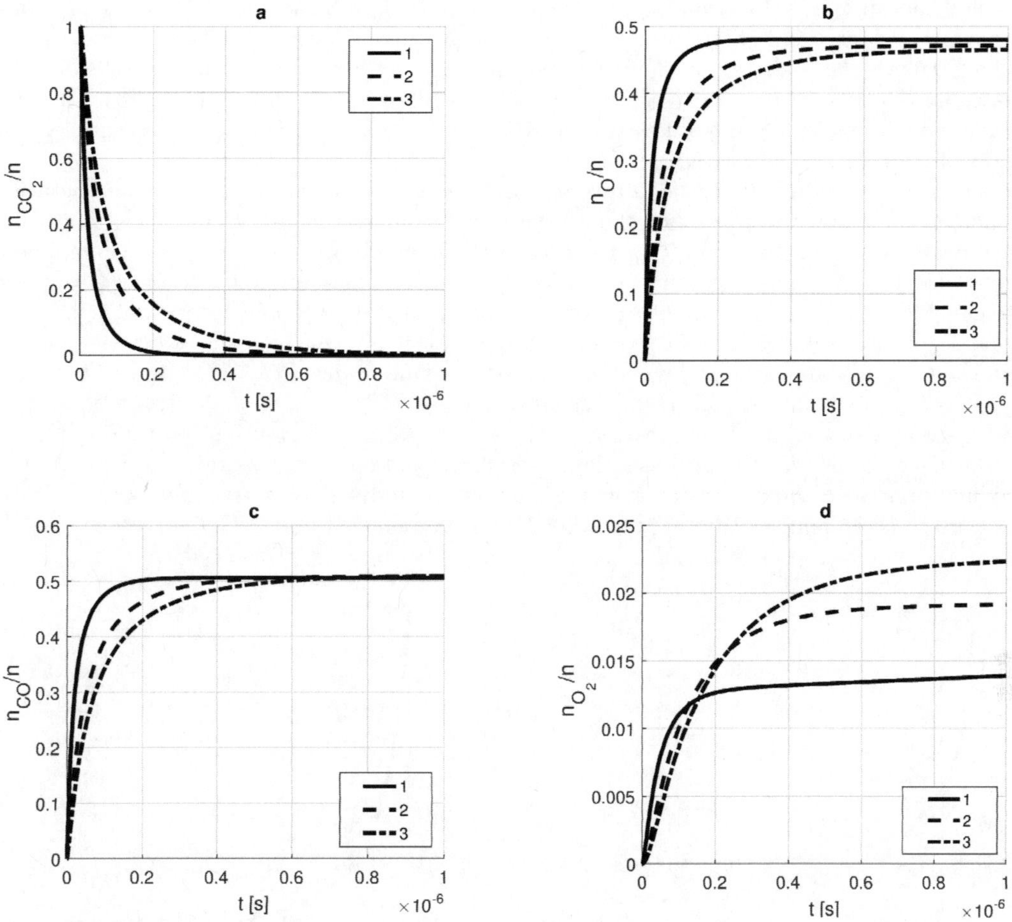

Figure 3. The relative number density of CO_2 (a), O (b), CO (c), O_2 (d) in dependence on t; 1 – one-temperature, 2 – two-temperature, 3 – three-temperature approach.

Figures 3 (a, b, c, d) show variation of number densities of mixture components CO_2 (a), O (b), CO (c), O_2 (d) with time. Again, the difference between parameters found within the non-equilibrium two-temperature and three-temperature CO_2 models occurs smaller than a discrepancy between the values obtained for thermal equilibrium and non-equilibrium CO_2 description. The values of number densities n_{CO_2}/n and n_{O_2}/n found in the three-temperature approach exceed equilibrium values more than in three times whereas the difference between the values found in the three-temperature and two-temperature approaches reaches 50%.

Rapid decreasing of CO_2 number densities due to dissociation reaction (1) (figure 3 a) resulting in increasing of number densities of O atoms (figure 3 b) and CO molecules (figure 3 c) is best seen in the one-temperature approach. Slower CO_2 dissociation within the two-temperature and three-temperature models than in the frame of one-temperature description is explained by a delay of dissociation for vibrational CO_2 excitation for small t values in a short time interval.

It is interesting that though diatomic species are considered in this paper in the thermal equilibrium, one can see a noticeable influence of CO_2 kinetic models on number densities of CO molecules and oxygen atoms. As it is seen in figures 3 (b, c) densities of these species are rather high because they form mainly as a result of CO_2 dissociation with high reaction rate coefficients as it is seen in figure 1. Oxygen molecules are formed in result of the active exchange reaction (4) with the high rate coefficient (see figure 1), however the density of O_2 molecules decreases fast due to active O_2 dissociation as it is also shown in figure 1. The impact of CO_2 kinetics on variation of O_2 molecules (figure 3 d) and C atoms is not so important because densities of these species are rather low.

Now let us discuss the influence of a mixture composition on variation of the gas temperature and number densities of species. For this purpose we compare the results presented above for the five-component mixture $CO_2/CO/O_2/O/C$ with ones obtained for the three-component mixture $CO_2/CO/O$ under the same initial conditions. The dissociation reaction (1) is taken into account with $M = CO_2, CO, O$, whereas dissociation of CO molecules is not included to the kinetic scheme. Vibration-dissociation relaxation of this mixture was studied numerically in [19] on the basis of the equations for number densities of molecules CO_2, CO, atoms O and the total energy conservation equation. In the present paper calculations of the gas temperature and three-component mixture composition were carried out for the same initial conditions and using the same models for CO_2 dissociation and vibrational energy transitions as were applied for the five-component mixture. Below we present the comparison of the results obtained in the most accurate three temperature approach for two mixtures.

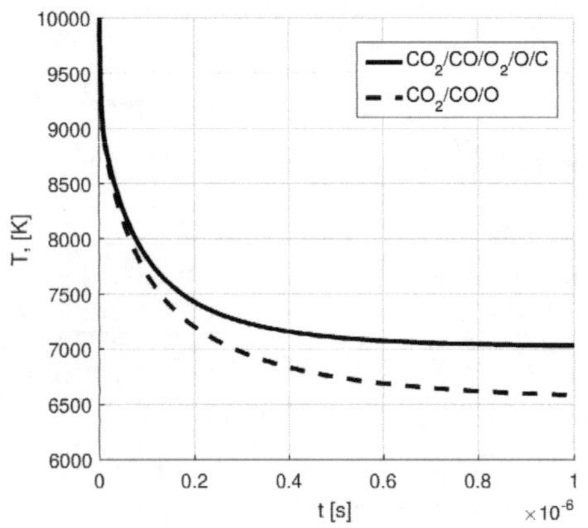

Figure 4. The gas temperature in dependence on t.

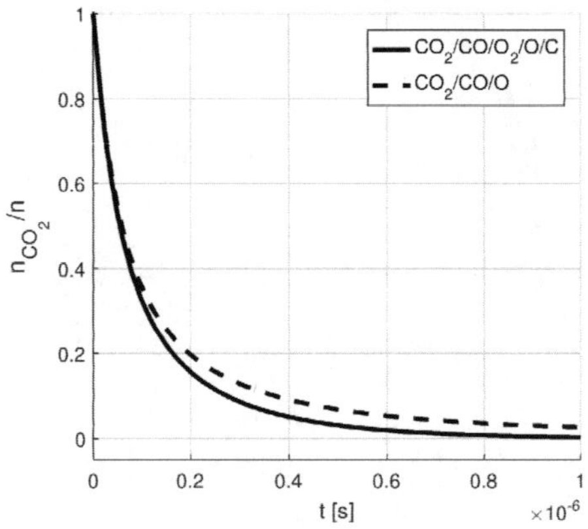

Figure 5. The relative number density n_{CO_2}/n, in dependence on t.

Figures 4, 5, 6 (a, b) plot the gas temperature T (figure 4) and number densities of species CO_2 (figure 5), CO (figure 6 a), O (figure 6 b) as functions of time for two considered mixtures. It may be noticed that in the beginning of the process, close to the initial moment of time, values of the gas temperature and number densities of species are found very close for both mixtures. It is explained by different characteristic times for vibrational and chemical relaxation. For small time values chemical reactions in both mixtures are frozen and the main process is the vibrational excitation of CO_2 molecules due to translational energy transitions to the vibrational energy. This process proceeds similarly in two mixtures and leads to close values of mixture parameters

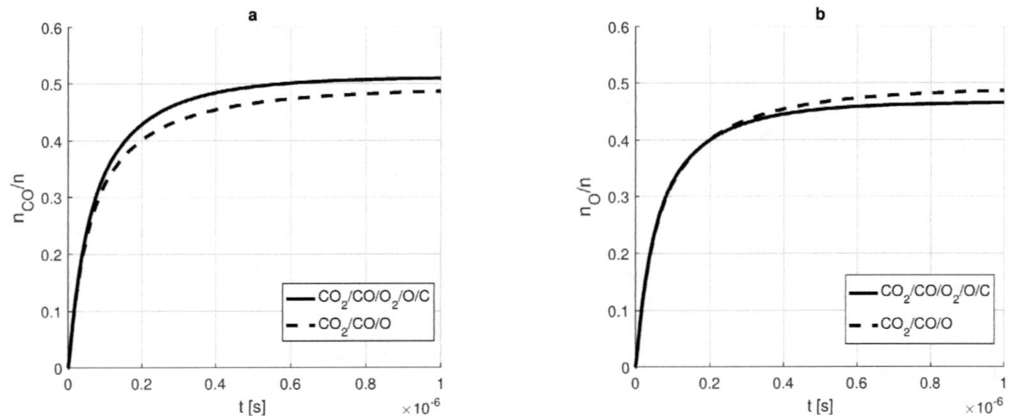

Figure 6. The relative number density n_{CO}/n (a), n_O/n (b) in dependence on t.

during a short time interval. Chemical relaxation and establishing of chemical equilibrium proceed differently in two mixtures. Calculations show higher temperature values in the five-component mixture (figure 4) and consequently more rapid dissociation and chemical reactions in this case than in the three-component mixture. Fast dissociation reaction (1) provides more intensive decrease of CO_2 number densities (figure 5) and increase of CO densities (figure 6 a) in the five-component mixture. Oxygen atoms form in result of the same dissociation reaction (1) in both mixtures. However values of n_O/n in the mixture $CO_2/CO/O_2/O/C$ (figure 6 b) are found lower than in the three-component mixture. It is explained by active O_2 dissociation reaction (2) in the five-component mixture.

5. Conclusion

In this paper vibrational-chemical kinetics in the high-temperature five-component mixture $CO_2/CO/O_2/O/C$ is numerically studied taking into account dissociation of molecular species, recombination, exchange reactions and vibrational energy transitions of CO_2 molecules. Calculations were carried out with the use of three-temperature, two-temperature and one-temperature description of CO_2 vibrational relaxation. Comparison of the results obtained in the frame of three CO_2 kinetic models showed the influence of CO_2 vibrational distributions on temporal variation of the gas temperature and mixture composition. The difference between mixture parameters obtained in the frame of the three-temperature and two-temperature models is found less than deviations from the data found using the thermal equilibrium one-temperature distributions. Calculations show that thermal equilibrium distributions establish under considered conditions much faster than the chemical equilibrium.

The results obtained for the mixture $CO_2/CO/O_2/O/C$ are compared with the data found for the three-component mixture $CO_2/CO/O$ taking into account vibration-dissociation CO_2 kinetics. The comparison showed the influence of a mixture composition on CO_2 vibration-dissociation relaxation, number densities of species CO_2, CO, O and the gas temperature.

Acknowledgments

This study is supported by the Russian Science Foundation (Project No. 15-19-30016).

References

[1] Park C, Howe J and Jaffe R 1994 *J. Thermophys. Heat Transfer* **8 (1)** 9-23

IOP Conf. Series: Journal of Physics: Conf. Series **815** (2017) 012027 doi:10.1088/1742-6596/815/1/012027

[2] Shevelev Yu D and Syzranova N G 2007 *Fiz.-Khim. Kinet. Gaz. Dinam.* **5** URL http://www.chemphys.edu.ru/Media/Files/2007-12-17-001.pdf [in Russian]
[3] Taylor R L and Bitterman S 1969 *Rev. Mod. Phys.* **41 (1)** 26-47
[4] Gordiets B F, Osipov A I and Shelepin L A 1980 *Nauka. Moscow* 512
[5] Likalter A A 1975 *Kvant. Elektron.* **5 (11)** 1307-1309 [in Russian]
[6] Kozak T and Bogaerts A 2015 *Plasma Sources Sci. Technol.* **24** 015024
[7] Aerts R, Martens T and Bogaerts A 2012 *J. Phys. Chem. C.* **116** 23257-23273
[8] Thomson R 1978 *J. Phys. D Appl. Phys.* **11** 2509-2516
[9] Brun R 1988 *AIAA paper* **88** 2655.
[10] Kustova E V and Nagnibeda E A 2006 *Chem. Phys.* **321** 293-310
[11] Kustova E, Nagnibeda E, Shevelev Yu and Syzranova N 2011 *Shock Waves* **21 (3)** 273-287
[12] Armenise I and Kustova E 2013 *Chem. Phys.* **415** 269-281
[13] Kustova E and Nagnibeda E 2001 State-to-state theory of vibrational kinetics and dissociation in three-atomic gases *22nd International Symposium on Rarefied Gas Dynamics, July 2001* AIP Conference Proceedings **585** 620-627
[14] Losev S A, Kozlov P V, Kuznezova L A, Makarov V N, Romanenko Yu V, Surzhikov S T and Zalogin G N 1988 Radiation of $CO_2 - N_2 - Ar$ mixture in a shock wave: experiment and modeling Proceeding of the Third European Symposium on Aerothermodynamics for Space Vehicles *ESTEC, Noordwiik, ESA* **426** 437-444
[15] Makarov V and Losev S 1997 *Khim. Phizika* **16 (5)** 29-43 [in Russian]
[16] Likalter A A 1976 *Zh. Prikl. Mekh. Tekh. Fiz* **4** 3-10 [in Russian]
[17] Kustova E and Nagnibeda E 1999 *Rarefied Gas Dynamics* **2** 289-296
[18] Kustova E and Nagnibeda E 2012 *Chem. Phys.* **398 (1)** 111-117
[19] Kosareva A A and Nagnibeda E A 2016 *Vestnik of Saint Petersburg University. Series 1. Mathematics. Mechanics. Astronomy* **3 (61)** 468-480 [in Russian]

Kinetic modelling of primary and secondary interstellar oxygen atom fluxes in the heliosphere

I I Balyukin[1,2], **V V Izmodenov**[1,2,3], **O A Katushkina**[1] **and D B Alexashov**[1,3]

[1]Space Research Institute of Russian Academy of Sciences, Profsoyuznaya Str. 84/32, Moscow, 117335, Russia
[2]Lomonosov Moscow State University, GSP-1, Leninskie Gory, Moscow, 119991, Russia
[3]A.Ishlinsky Institute for Problems in Mechanics, prosp. Vernadskogo 101, block 1, Moscow, 119526, Russia

E-mail: balyukin.ii@gmail.com

Abstract. The interstellar neutral (ISN) oxygen atoms penetrate into the heliosphere from the Local Interstellar Medium (LISM). Oxygen atoms that directly penetrate into the heliosphere are called primary, while secondary oxygen population is formed near the heliopause due to the process of charge exchange of interstellar oxygen ions with hydrogen atoms and its existence in the heliosphere was previously predicted theoretically (Izmodenov et al., 1997, 1999, 2004). In this work we study the distribution of both primary and secondary oxygen atoms in the heliosphere and its fluxes at 1 AU.

The analysis of the distribution of ISN O atoms in the heliosphere is only possible in the frame of a model that takes into account both filtration of the primary and origin of the secondary interstellar oxygen in the region of interaction of the solar wind (SW) with the LISM, as well as a detailed simulation of the motion of interstellar atoms inside the heliosphere. This simulation should take into account the temporal and heliolatitudinal dependences of ionization, the process of charge exchange with SW protons and the effect of the solar gravitational attraction. This paper presents the results of modelling ISN O atoms inside the heliosphere based on a new three-dimensional kinetic-MHD model of the SW interaction with the LISM (Izmodenov and Alexashov, 2015). Oxygen atoms are of a particular interest because they are strongly coupled with the protons through the charge exchange process and due to the fact that the fluxes of interstellar oxygen are directly measured by the Interstellar Boundary Explorer (IBEX). The first quantitative measurements of the interstellar heavy neutral atoms (O and Ne) obtained on the IBEX spacecraft were presented in (Park et al., ApJS, 2015). Qualitative analysis of these data shows that the secondary component of the interstellar oxygen atoms was also measured along with the primary interstellar atoms.

1. Introduction

Our Solar System is moving through the Local Interstellar Cloud (LIC) with the bulk velocity about $26\,km\,s^{-1}$ (see, e.g., McComas et al. [1]). The solar wind (SW) interacts with the ionized component of Local Interstellar Medium (LISM), which is surrounding our solar system, and the SW/LISM interaction region (heliospheric interface) is formed. It consists of two shocks, termination shock (TS) and bow shock (BS), and a tangential discontinuity – heliopause (HP). The region between the shocks is called a heliosheath.

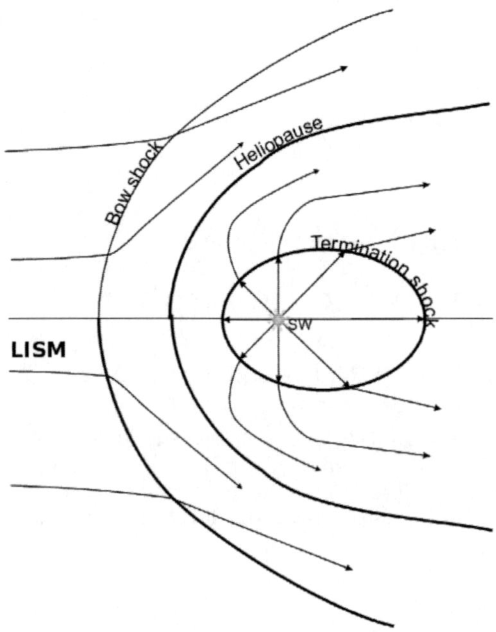

Figure 1. Qualitative picture of the SW/LISM interaction region: 2 shock waves (termination shock, bow shock) and tangential discontinuity (heliopause).

The LISM contains neutral elements such as H, He, O and Ne that have a large mean free path (by the process of charge exchange) that is comparable (for H and O) or larger (for He and Ne) than the character size of the heliosphere, so these atoms can penetrate into the heliosphere due to the relative motion of the Sun and LISM. In this work we study the distribution of the interstellar neutral (ISN) oxygen atoms in the heliosphere and particularly its fluxes at Earth's orbit.

Oxygen is the third element (after H and He) by its cosmic abundance in the LISM: $n_{O,LISM} = 0.78 \times 10^{-4} cm^{-3}$ (e.g., Izmodenov et al., [2]). Charge exchange cross section of interstellar oxygen atoms with protons is quite large (Izmodenov et al., [3]). Therefore, oxygen distribution inside the heliosphere is substantially modified by charge exchange in the heliospheric interface. For theoretical consideration it is convenient to divide interstellar oxygen atoms in two populations: primary and secondary. Atoms that directly penetrate into the heliosphere from the LISM we call primary. The secondary atoms are created via charge exchange between the oxygen ions and hydrogen atoms ($O^+ + H \rightarrow O + H^+$) in the region of disturbed interstellar plasma outside of the heliopause. Parameters of the secondary atoms depend on the local plasma properties in the region of their creation. Therefore, secondary oxygen atoms have smaller velocity and larger temperature compared to the primary ones because the interstellar plasma is decelerated and heated near the heliopause. Existence of the secondary oxygen population was theoretically predicted by Izmodenov et al. [3]. Filtration of interstellar oxygen at the heliospheric interface was also studied by Izmodenov et al. [4, 2].

Until recently, the interstellar neutral oxygen was measured only indirectly in the form of pick-up ions (Geiss et al., [5]) and anomalous cosmic rays (Cummings et al., [6]). First direct measurements of interstellar heavy neutral atoms (oxygen and neon) fluxes were performed by an Interstellar Boundary Explorer (IBEX) spacecraft. Qualitative analysis of this data performed by Park et al. [7] shows that the secondary component of the interstellar oxygen was detected for the first time. It is shown by Park et al. [7] that a sky map of interstellar heavy neutral atom fluxes contains a specific feature, which is called an extended "tail", associated with a contribution of secondary oxygen atoms.

In this work we perform a detailed numerical modelling of the ISN O fluxes at 1 AU. We consider both primary and secondary O atoms. Our model takes into account dynamics of atoms near the Sun under an influence of the solar gravitational force and local ionization processes as well as perturbations of the oxygen distribution at the heliospheric boundary.

The paper is organized as follows. In Section 2, the mathematical formulation of the problem and a detailed description of the model are provided. Section 3 describes the method of calculation of the model fluxes at 1 AU. Section 4 presents the results of the model numerical calculations. Finally, Section 5 is an overall summary of our work with conclusions.

2. Model of the ISN O distribution

In this section we introduce a brief description of the kinetic model of the ISN O distributions that we used in our calculations. Kinetic approach is only valid for description of the ISN atoms because of their large mean free path comparable to a character distance of the heliosphere. This model was proposed by Izmodenov et al. [8] and previously applied for modelling of He and H fluxes by Katushkina et al. [9] and Katushkina et al. [10], respectively. It is a 3D time-dependent version of the classical hot model (see, e.g., Izmodenov et al. [11]) with a specific boundary conditions at 90 AU based on the results of a global self-consistent model of the SW/LISM interaction. The numerical calculations were performed separately for the primary (O_{pr}) and secondary (O_{sec}) populations of the interstellar oxygen atoms.

2.1. Kinetic model

Kinetic approach is based on the conception of the velocity distribution function. Distribution of the ISN O atoms is described by a kinetic equation:

$$\frac{\partial f(\mathbf{r},\mathbf{v},t)}{\partial t} + \mathbf{v}\cdot\frac{\partial f(\mathbf{r},\mathbf{v},t)}{\partial \mathbf{r}} + \frac{\mathbf{F}}{m}\cdot\frac{\partial f(\mathbf{r},\mathbf{v},t)}{\partial \mathbf{v}} = -\beta(\mathbf{r},t)\cdot f(\mathbf{r},\mathbf{v},t), \quad (1)$$

where $f(\mathbf{r},\mathbf{v},t)$ is the velocity distribution function of O atom, $\mathbf{v}$ – individual velocity of an atom, t – time, m – mass of an atom, $\beta(\mathbf{r},t)$ – effective ionization rate. $\mathbf{F}$ is the force acting on each atom in the heliosphere. In general case, this force is the sum of a solar gravitational force $\mathbf{F}_g$ and a solar radiative repulsive force $\mathbf{F}_{rad}$, but we assume that $\mathbf{F}_{rad}$ is negligible ($\mu = |\mathbf{F}_{rad}|/|\mathbf{F}_g| \ll 1$) due to the O atom solidity. In view of this assumption resulting force $\mathbf{F}$ can be written in this form:

$$\mathbf{F} = \mathbf{F}_g = -m\frac{GM_\odot}{r^2}\frac{\mathbf{r}}{r},\ r = |\mathbf{r}|,$$

where G – gravitational constant, and $M_\odot$ – mass of the Sun.

Right side of the equation 1 represents the loss of atoms due to following processes: charge exchange on the SW protons ($O + H^+ \to O^+ + H$) and photoionization ($O + h\nu \to O^+ + e$). An electron impact ionization is not taken into account in our study because it is negligibly small as shown by Bzowski et al. [12]. Thus, the effective ionization rate can be written as: $\beta(\mathbf{r},t) = \beta_{ex}(\mathbf{r},t) + \beta_{ph}(\mathbf{r},t)$, where β_{ex} and β_{ph} are the rates of the charge exchange and photoionization, respectively.

Parameters β_{ex} and β_{ph} decrease with the distance r from the Sun $\propto 1/r^2$ in the region between the Sun and TS, since these values are proportional to the density of the SW protons and flux of the solar photons, respectively. Hereby, coefficient $\beta(\mathbf{r},t)$ can be rewritten as it follows:

$$\beta(\mathbf{r},t) = \beta^E(t,\lambda)\cdot\left(\frac{r_E}{r}\right)^2,\ \beta^E(t,\lambda) = \beta^E_{ex}(t,\lambda) + \beta^E_{ph}(t,\lambda),$$

where $r_E = 1$ AU, subscript "E" means that the value is taken at r_E. In general case, parameter $\beta^E(t,\lambda)$ depends on a time t and heliolatitude λ. We performed calculations with different ionization models ($\beta^E(t,\lambda)$; $\beta^E(t)$; $\beta^E = const$) and found that temporal and heliolatitudinal dependences of ionization rate has no qualitative effect on flux maps. Thus, in calculations we used a simplified model with constant ionization rate $\beta^E = 5.2 \times 10^{-7} s^{-1}$ (this value is obtained by averaging of the total in-ecliptic ionization rates from 2009 to 2011 presented by Bzowski et al. [12]).

2.2. Method of characteristics

The kinetic equation (1) is a linear partial differential equation that can be solved by a method of characteristics. A characteristic is a curve in the phase space $\mathbb{R}^6$ of coordinates and velocities and it is defined by the following equations:

$$\begin{cases} d\mathbf{r}/dt = \mathbf{v}, \\ d\mathbf{v}/dt = \mathbf{F}/m. \end{cases} \tag{2}$$

Along this characteristic the velocity distribution function $f(\mathbf{r}, \mathbf{v}, t)$ must satisfy the following equation:

$$\frac{df(\mathbf{r}, \mathbf{v}, t)}{dt} = -\beta(\mathbf{r}, t) \cdot f(\mathbf{r}, \mathbf{v}, t),$$

and the solution of that equation can be written in the following form:

$$f(\mathbf{r}, \mathbf{v}, t) = f_s(\mathbf{r}_s, \mathbf{v}_s) \cdot e^{-I}, \; I = \int_{t_s}^{t} \beta^E(t, \lambda) \cdot \left(\frac{r_E}{r}\right)^2 dt,$$

where $f_s(\mathbf{r}_s, \mathbf{v}_s)$ is the velocity distribution function of the O atoms at the outer boundary of the computational domain that is a Sun-centered sphere with radius $R_s = 90$ AU; $\mathbf{r}_s$, $\mathbf{v}_s$, t_s – position, velocity and moment of the time respectively when the atom crossed the outer boundary and enters the computational domain; I is a loss integral due to ionization processes. The integration is performed along the atom's trajectory that satisfies the system of equations (2) of the characteristic curve.

By solving the kinetic equation (1) with a specific boundary condition $f_s(\mathbf{r}_s, \mathbf{v}_s)$ we can find the value of the velocity distribution function everywhere in the computational domain. In the section below we describe a boundary conditions that are used in our modelling.

2.3. Boundary conditions

For the primary and secondary components of neutral oxygen atoms we used boundary condition on a sphere with radius $R_s = 90$ AU. To calculate the flow of oxygen atoms at this sphere we apply a new state-of-art kinetic-MHD model of the solar wind interaction with the LISM (hereinafter we call it a "global model"; this model is described in detail in the work of Izmodenov and Alexashov, [13]). In the present work parameters of the LISM were taken from the results of Witte et al. [14] that is based on Ulysses data: $V_{LISM} = 26.4\, km\, s^{-1}$; the LISM temperature $T_{LISM} = 6530 K$; direction of the interstellar wind in the ecliptic (J2000) coordinates – longitude $\lambda_{LISM} = 75.4°$ and latitude $\beta_{LISM} = -5.2°$; direction of the interstellar magnetic field in the ecliptic (J2000) coordinates – longitude 62.49° and latitude −20.79°. $\mathbf{B}_{LISM}\mathbf{V}_{LISM}$-plane coincides with the Hydrogen Deflection Plane (HDP) first proposed by Lallement et al. [15]. Hereinafter we also use a term "upwind" direction, which is defined by the opposite to the $\mathbf{V}_{LISM}$ vector direction, i.e., $(\lambda, \beta)_{upwind} = (255.4°, 5.2°)$.

At first, let us introduce a coordinate system, which we use to describe a boundary velocity distribution function: Z-axis pointing toward the "upwind" direction; XZ-plane (so-called BV-plane) is defined by two vectors – interstellar magnetic field vector $\mathbf{B}_{LISM}$ and the velocity vector of the interstellar wind $\mathbf{V}_{LISM}$; positive direction of X-axis have been chosen in the way that z-component of the $\mathbf{B}_{LISM}$ has a negative value, i.e., $(\mathbf{B}_{LISM})_x < 0$; Y-axis complements the left-handed triplet.

Then we calculated the following parameters of the O_{pr} and O_{sec} on the boundary sphere using the global model:

- zero velocity distribution function moment – number density n_O:

$$n_O(\mathbf{r}_s) = \int f(\mathbf{r}_s, \mathbf{v}) \cdot d\mathbf{v}; \tag{3}$$

- first velocity distribution function moments – components of bulk velocity (V_x, V_y, V_z):

$$V_i(\mathbf{r}_s) = \frac{1}{n_O(\mathbf{r}_s)} \cdot \int f(\mathbf{r}_s, \mathbf{v}) \cdot v_i \cdot d\mathbf{v}; \tag{4}$$

- second velocity distribution function moments – kinetic temperatures (T_x, T_y, T_z):

$$T_i(\mathbf{r}_s) = \frac{m_O}{n_O(\mathbf{r}_s) k_B} \cdot \int f(\mathbf{r}_s, \mathbf{v}) \cdot (V_i(\mathbf{r}_s) - v_i)^2 \cdot d\mathbf{v}. \tag{5}$$

In formulas (3, 4, 5) we use the following notations: $i \in \{x, y, z\}$, k_B – Boltzmann constant, $d\mathbf{v} = dv_x \cdot dv_y \cdot dv_z$ and integration performed all over the velocity phase space $\mathbf{v} = (v_x, v_y, v_z) \in \mathbb{R}^3$. Parameters ($n_O$, V_i, T_i) – velocity distribution function moments – depend on a position of the atom on a boundary sphere. Table 1 presents values of these parameters at the "nose" point $\mathbf{r}_N = (0, 0, 90\,AU)$. As we can see from this table that parameters V_x and V_y of primary oxygen are very small, which means that the direction of O_{pr} bulk velocity vector is almost identical to the V_{LISM} direction. The parameter V_y of the secondary oxygen is also very small, but the parameter V_x significantly differs from zero ($\approx -1.6\ km\ s^{-1}$). Thereby, the secondary oxygen bulk velocity vector is rotated in the BV-plane from the V_{LISM} vector to B_{LISM} by $\approx 4°$.

Table 1. Table of primary and secondary oxygen velocity distribution function moments at 90 AU in the "upwind" direction. These parameters were taken from the numerical calculations of the global model.

Component of ISN oxygen	n_O [$\times 10^{-4}\ cm^{-3}$]	V_x [$km\ s^{-1}$]	V_y [$km\ s^{-1}$]	V_z [$km\ s^{-1}$]	T_x [K]	T_y [K]	T_z [K]
Primary	0.272	-0.006	0.001	-26.877	6538	6522	6225
Secondary	0.373	-1.619	-0.005	-22.315	15541	6767	93982

To get a deeper understanding of properties of the oxygen flow at 90 AU let us introduce not only the moments of the calculated distribution functions but also its shape. Figure 2 shows $f_z(v_z) = f(V_x, V_y, v_z)$ functions of primary and secondary oxygen at $\mathbf{r}_N$ (or, in other words, the profiles in the "upwind" direction). The value of the $f_z(v_z)$ function is proportional to the amount of atoms that have z-component of velocity equals to v_z. Red curve of the figure 2 clearly shows that the primary oxygen can be fitted well with Maxwell – Boltzmann distribution function with appropriate parameters V_z and T_z. Green curve of the figure 2 presents the $f_z(v_z)$ profile of the secondary oxygen. Here we can see a strongly marked peak of atoms with the velocities $v_z \approx -26\ km\ s^{-1}$ and a wide range of slow atoms with approximately the same amount (hereinafter we will use a notion "plank" of slow atoms). This "plank" demonstrates that secondary oxygen component contains atoms that have a smaller velocities than the primary

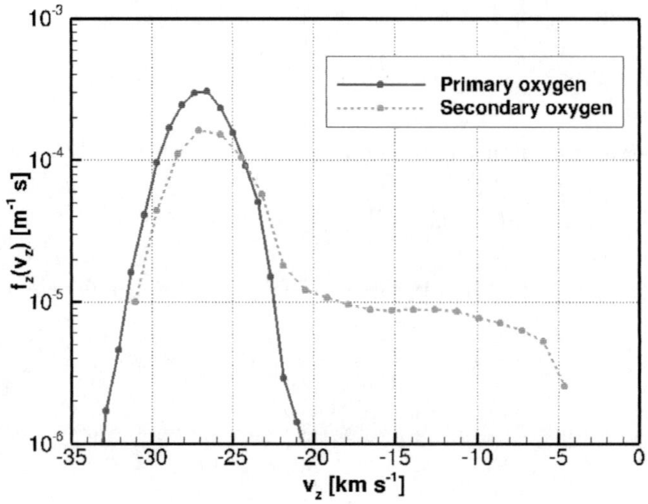

Figure 2. Profiles $f_z(v_z) = f(V_x,\ V_y,\ v_z)$ of the velocity distribution functions of the O_{pr} (red curve) and O_{sec} (green dashed curve) at $\mathbf{r}_N$ in the "upwind" direction. First profile is nearly a Maxwell - Boltzmann distribution function with parameters (V_z, T_z) for primary oxygen from table 1. Second profile has two main features: (a) a strongly marked peak of atoms with the velocities $v_z \approx -26\ kms^{-1}$; (b) a "plank" of slow atoms with velocities $-20 < v_z < -5\ km\ s^{-1}$.

ones. These slow atoms are created during charge exchange from O^+ ions decelerated near the heliopause. Oxygen atoms have small thermal velocities that is why all decelerated atoms with different velocities create such kind of "plank" on the total velocity distribution. This feature is absent for the secondary hydrogen atoms because they have much larger thermal velocities and the effect is suppressed.

Therefore, the velocity distribution function of secondary oxygen has a strong non-Maxwellian properties in the region near the TS. The charge exchange process in the heliosheath is responsible for these properties because it leads to disturbances of interstellar oxygen flow. A high value of T_z temperature (see table 1) also indicates that the $f_z(v_z)$ profile can not be properly fitted using a Maxwell – Boltzmann distribution function. Thereby, in the case of secondary oxygen it is necessary to use a specific form of velocity distribution function at 90 AU that takes into account the influence of the heliospheric interface.

As the next step we perform a fit of the velocity distribution functions of primary and secondary oxygen at 90 AU and present formulas of boundary distribution functions $f_s(\mathbf{v}_s)$ that we use in our numerical calculations. We start from the assumption that components (v_x, v_y, v_z) of vector $\mathbf{v}_s$ (atom velocity at 90 AU) are independent random variables, so the boundary velocity distribution function can be written in general form as:

$$f_s(\mathbf{v}_s) = n_O \cdot f_x(v_x) \cdot f_y(v_y) \cdot f_z(v_z), \quad \int_{-\infty}^{+\infty} f_i(v_i) \cdot dv_i = 1, \tag{6}$$

where $i \in \{x, y, z\}$, n_O is the number density of oxygen (primary or secondary). We assume $f_x(v_x)$ and $f_y(v_y)$ to be Maxwell – Boltzmann distributed with parameters (V_x, T_x) and (V_y, T_y), respectively:

$$f_i(v_i) = \frac{1}{\sqrt{\pi} c_i} \cdot exp\left(-\frac{(V_i - v_i)^2}{c_i^2}\right), \ c_i = \sqrt{\frac{2k_B T_i}{m}},$$

where $i \in \{x, y\}$. Parameters (V_x, T_x) and (V_y, T_y) and number density n_O of O_{pr} and O_{sec} were derived from the calculations of the global model and presented in table 1.

For primary oxygen we also assume $f_z(v_z)$ to be Maxwell – Boltzmann distributed with parameters (V_z, T_z), which were taken from table 1 as well. For secondary component of neutral oxygen atoms we use the velocity distribution function obtained numerically in the frame of the global model. The $f_z(v_z)$ factor in equation 6 mainly describes a non-Maxwellian properties of the secondary oxygen flow at 90 AU, so we use calculated values of the distribution function and a linear interpolation between them (green curve of the figure 2).

3. Calculation of the model fluxes at 1 AU

In this section we briefly describe the method of calculation of the model fluxes. Using the velocity distribution function we can calculate its integral characteristic, namely, an atom flux at 1 AU in a specified viewing direction. Every position of observer at 1 AU in the ecliptic plane can be defined by its ecliptic (J2000) longitude λ_{obs}. Plane of observations is perpendicular to a Sun – observer vector and can be consequently defined by its direction. This geometry of observations is similar to the IBEX observations. Line of sight (LOS), which lies in the observational plane, can be defined by only one angle and, for instance, it can be a α_{NEP}-angle that counted from the north ecliptic pole (NEP). Thereby, the flux at 1 AU in specified line of sight $\mathbf{S}_{LOS}$ can be calculated by the following formula:

$$Flux = \int_0^{+\infty} f(\mathbf{r}, \mathbf{v}, t) \cdot v^3 \cdot dv, \ \mathbf{v} = -v \cdot \mathbf{S}_{LOS},$$

where $[Flux] = s^{-1}cm^{-2}sr^{-1}$ and $\mathbf{v}$ is the absolute atom's velocity, which direction is determined by the LOS.

To calculate the model fluxes we use the following algorithm. A full sky map in ecliptic (J2000) coordinates $\{(\lambda, \beta) \in [0°, 360°] \times [-90°, 90°]\}$ was divided into $2° \times 2°$ bins in a form of 2D array. Thus, every bin is defined by the pair of indexes (i, j), where $i \in [0, 179]$ and $j \in [0, 89]$. By a specified longitude of observer λ_{obs} and a given value of α_{NEP} we can clearly define LOS and a bin of the sky map to which it belongs.

Total flux $F(i, j)$ in the (i, j)-bin was averaged over one year. As we have mentioned in the Section 2, the simplified ionization model with constant ionization rate was used in the present work, therefore the distribution function $f(\mathbf{r}, \mathbf{v}, t)$ and the value of the flux do not depend on time t. Thus, it does not matter what year is chosen for averaging of the flux $F(i, j)$. Simulations were performed for the lines of sight characterized by $\lambda_{obs} \in [0°, 360°]$ and $\alpha_{NEP} \in [0°, 360°]$ with steps of 1° (further steps reduction does not substantially change the value of the flux).

4. Results of the model numerical calculations

The results of the numerical calculations are presented in figure 3 in a form of the flux maps at 1 AU. The numerical calculations were performed for each sort of oxygen atoms separately: plot 3(A) presents fluxes of the primary oxygen and plot 3(B) presents fluxes of the secondary oxygen.

There are two peaks in the figure 3(A) that are obtained in the following directions of the sky map: (λ_1, β_1) $\approx$ (220°, 5°) and (λ_2, β_2) $\approx$ (290°, 5°). Due to the influence of the solar gravitational force the flow of oxygen atoms changes its direction, so these peaks in the flux map correspond to the opposite directions of the atoms flow at 1 AU. According to this figure, the primary oxygen flow at 1 AU is shifted by $\approx 35°$ from the "upwind" direction ($\lambda_{upwind} = 255.4°$). The symmetry of these two peaks is the result of the fact that $f_x(v_x)$ and $f_y(v_y)$ boundary functions of the primary oxygen at 90 AU are quite simillar (parameters (V_x, T_x) and (V_y, T_y) of these functions are about the same, see table 1).

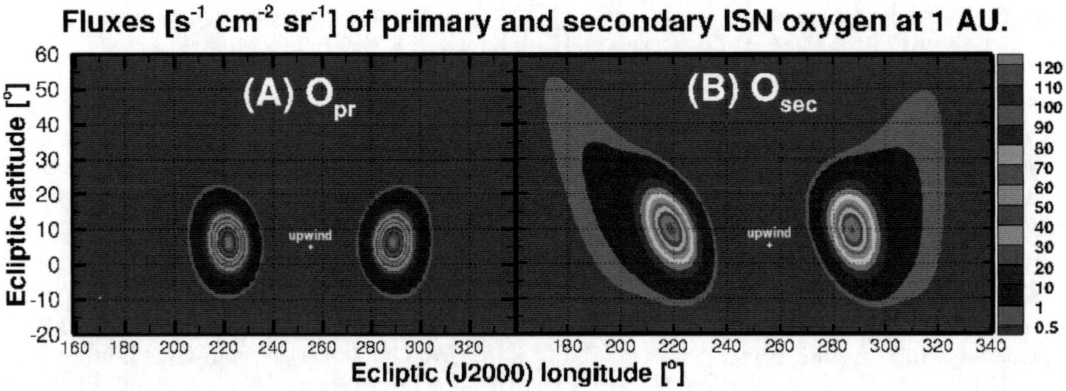

Figure 3. Numerical flux maps at 1 AU in the ecliptic (J2000) coordinates. Boundary conditions are based on the results of numerical calculations of the global model of SW/LISM interaction. Figure (A) presents fluxes of the primary oxygen, figure (B) presents fluxes of the secondary oxygen.

The O_{sec} flux map qualitatively differs from the O_{pr} flux map: on the one hand, there are also two visible peaks in figure 3(B), but on the other hand, these peaks have strongly pronounced "tails" and there is asymmetry between them. The left peak has a "tail" that lays in the region of the lower longitudes and higher latitudes from the main flow and the right "tail" lays in the region of the higher longitudes and higher latitudes. These "tails" are the product of the "plank" of slow atoms of the boundary function $f_z(V_z)$, which is discussed in Section 2.3 (see green curve of the figure 2). The asymmetry of the peaks in O_{sec} sky map of fluxes corresponds to the difference between the $f_x(v_x)$ and $f_y(v_y)$ boundary function of the secondary oxygen (see table 1). The structure of the "tails" strongly depends on the parameters of the boundary function $f_s(\mathbf{v}_s)$ and study of this dependence is beyond the scope of this paper.

The flux sky maps provided by IBEX contain only one visible "tail", which corresponds to the left one of the figure 3(B), due to the fact that these maps were obtained in the spacecraft reference frame. Nevertheless, according to the numerical modelling we can make a conclusion that the extended "tail" in IBEX data correspond to the decelerated and heated secondary oxygen population.

5. Summary and conclusions

In this work we have studied the distribution of the ISN oxygen atoms in the heliosphere and its fluxes at Earth's orbit. The main points of this work are:

(i) An advanced kinetic model that takes into account the secondary component of the interstellar neutral oxygen atoms was provided.

(ii) Numerical modelling of the ISN O (primary and secondary) fluxes at Earth's orbit was performed.

(iii) According to the results of the numerical calculations, we can make a conclusion that the extended "tail", which is observed in the IBEX data, is a product of the secondary component of ISN oxygen atoms, thereby the results of this work support the Park et al. [7] assumption that the "tail" is connected with the secondary oxygen population.

To perform a quantitative comparison between the IBEX data and numerical calculations we need to transform the model fluxes into the IBEX count rates. The method of transformation should make allowance for the motion and geometry of IBEX spacecraft observations, all technical features of the IBEX-Lo sensor and the results of its calibration. The matter of the future investigations is to consider all of these aspects and to perform a detail quantitative comparison of the results of numerical modelling with the IBEX data.

Acknowledgments

Work is supported by Program 7 of Presidium RAS (in part of studying of oxygen distribution in the heliosphere). Calculations of ISN oxygen distribution were performed by using the Supercomputing Center of Lomonosov Moscow State University.

References

[1] McComas D J, Bzowski M, Frisch P, Fuselier S A, Kubiak M A, Kucharek H, Leonard T, Möbius E, Schwadron N A, Sokół J M, Swaczyna P and Witte M 2015 *The Astroph. J.* **801** 28
[2] Izmodenov V V, Malama Y G, Gloeckler G and Geiss J 2004 *Astron. & Astroph.* **414** L29–L32
[3] Izmodenov V V, Malama Y G and Lallement R 1997 *Astron. & Astroph.* **317** 193–202
[4] Izmodenov V V, Lallement R and Geiss J 1999 *Astron. & Astroph.* **344** 317–321
[5] Geiss J, Gloeckler G, Mall U, von Steiger R, Galvin A B and Ogilvie K W 1994 *Astron. & Astroph.* **282** 924–933
[6] Cummings A C, Stone E C and Steenberg C D 2002 *The Astroph. J.* **578** 194–210
[7] Park J, Kucharek H, Möbius E, Galli A, Livadiotis G, Fusellier S A and Mccomas D J 2015 *The Astroph. J. Suppl. Ser.* **220** 34
[8] Izmodenov V V, Katushkina O A, Qumerais E and Bzowski M 2013 *Cross-Calibration of the Far UV Spectra of Sloar System Objects and the Heliosphere* ISSI Scientific Report Series vol 13 ed E Qumerais and M Snow et al (Bern:ISSI) p 7
[9] Katushkina O A, Izmodenov V V, Wood B E and McMullin D R 2014 *The Astroph. J.* **789** 80
[10] Katushkina O A, Izmodenov V V, Alexashov D B, Schwadron N A and Mccomas D J 2015 *The Astroph. J. Suppl. Ser.* **220** 33
[11] Izmodenov V V 2006 *The Physics of the Heliospheric Boundaries* ed. V V Izmodenov & R Kallenbach (ESA Publications Division, EXTEC) p 45
[12] Bzowski M, Sokół J M, Kubiak M A and Kucharek H 2013 *Astron. & Astroph.* **557** A50
[13] Izmodenov V V and Alexashov D B 2015 *The Astroph. J. Suppl. Ser.* **220** 32
[14] Witte M 2004 *Astron. & Astroph.* **426** 835–84
[15] Lallement R, Quemerais E, Bertaux J L, Ferron S, Koutroumpa D and Pellinen R 2005 *Sci.* **307** 1447–49

Shock-induced ejecta from a layer of spherical particles. Part I: SPH meso-scale simulation

M S Egorova[1], S A Dyachkov[1], A N Parshikov[1], V V Zhakhovsky[1], A A Serezhkin[1], I S Menshov[1,2], D B Rogozkin[1,3] and S E Kuratov[1]

[1]Dukhov Research Institute of Automatics (VNIIA), Sushchevskaya str. 22, Moscow, 127055, Russia

[2]Keldysh Institute of Applied Mathematics RAS, Miusskaya sq. 4, Moscow, 125047, Russia

[3]Moscow Engineering Physics Institute, National Research Nuclear University, Kashirskoe Shosse 31, Moscow, 115409, Russia

E-mail: egorova.maria.serg@gmail.com

Abstract. According to the data obtained by the Photonic Doppler Velocimetry (PDV) shock wave propagation through a layer of bulked metallic particles produces dust ejecta consisting of fragments with various velocities. While velocity distribution of dust particles can be measured by the PDV, the spatial mass distribution and size distribution of the fragments are hard to derive from the PDV experimental data.

To study the mechanism of ejecting and composition of ejected material, direct simulation of experimental conditions is performed with the massive-parallel SPH code. We find that the cumulative jets are produced by the collisions of neighbor spheres accelerated by a shock wave. Those of jets which are able to reach free boundary, leave the layer and decay with formation of many fragments.

The data we acquire via direct numerical simulation of the described process is used as initial data for further investigation of dust motion in the air with the finite differences method.

1. Introduction

When a layer of bulked metallic micron-size spherical particles is shock-loaded by pressures of from 60 to 100 GPa, a formation of the dust ejecta consisting of fragments with various velocities is observed [1]. Analysis of processes that lead to dust jetting and distribution of dust fragments by sizes and velocities is an unresolved problem up to now.

The spectrograms of particle velocities in the dust ejecta are one of the few sources of information available for analysis. The time evolution of PDV spectrograms enables us, based on the law of particle slowing-down in the air, to estimate the characteristic size of ejecting particles. There is an approach to solving the inverse problem of reconstruction of the ejected mass distribution with Doppler shift measurements [2, 3]. The inverse problem solution offers rough estimation of the size distribution of the fragments, but it does not demonstrate profound understanding of the physical processes that occur in a layer of spherical particles during shock wave propagation and release on the surface.

One can use, for example, optical registration of particles that detach from the layer when a shock wave hits its free surface. Numerical simulation either involves complex models of heterogeneous medium that operate averaged parameters of the layer (as it is shown in Part II), or completely corresponds to real pattern of the layer observed. The latter enables us to obtain sizes of the dust particles directly, without any averaging or simplifying, but appears to be time-consuming. Precise choice of an appropriate simulation method in combination with effective computational resource allocation can make a solution to be possible.

CSPH&VD^3, our massive-parallel software which implements SPH method with incorporated Riemann problem solution [4], allows us to simulate the experiment [1] with different packings of spheres.

Part I of the present work presents the simulation results which demonstrate the mechanism of dust ejecta in experimental conditions [1].

2. From experimental conditions to numerical simulation

Figure 1 demonstrates the scheme of experiment [1]. A 100 mm diameter high-explosive plane wave generator was used to generate a shock wave in a set of transmitters (consisting of a 1 mm thick plate of denal, which is a heavy tungsten alloy, and a 5 mm thick copper plate). The denal plate was machined with 8 notches. One of the notches was left empty for the transmitter free surface velocity measurement. The others were either filled with spheres with calibrated diameter (1 or 9 μm diameter golden spheres) or were containing a grooved plate.

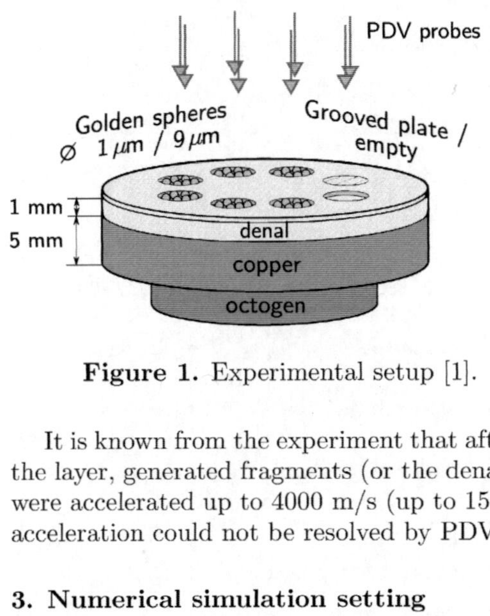

Figure 1. Experimental setup [1].

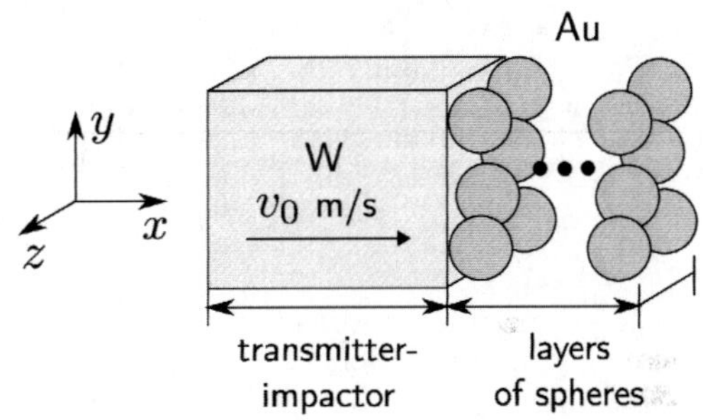

Figure 2. Simulation setup. Packings are illustrated further.

It is known from the experiment that after the shock wave broke through the free boundary of the layer, generated fragments (or the denal transmitter surface if considering the empty notch) were accelerated up to 4000 m/s (up to 1550 m/s for the empty notch) for less than 30 ns. The acceleration could not be resolved by PDV diagnostic used in the experiment.

3. Numerical simulation setting

Our simulation setting was slightly different from the experimental conditions. The scheme of the computational area is presented in figure 2.

In order to reproduce denal plate interaction with a bulked spheres layer, it was represented by a tungsten sample with periodical boundary conditions at y and z axes. We used pure tungsten instead of denal, because denal was a registered trademark for several tungsten heavy alloys and particular composition used in the experiment was left unknown. Thickness of the plate was 1.2 of height of the spheres layer and the length was estimated to be enough to avoid influence of the reflected rarefaction wave from the other side of the plate.

4. Numerical simulation technique

4.1. Modeling method

In order to solve the problem, the classic SPH method [5] was modified. The modification (CSPH) incorporates Riemann problem solution [6].

First of all, interparticle contact surface is introduced, which is similar to contact surface in a Riemann problem. Let us consider $ORST$ coordinate system, which is associated with two particles in contact, i and j. Let an OR axis point in $\vec{r}_j - \vec{r}_i$ direction, where $\vec{r}_i$ is a vector of a position of particle i. An interparticle contact surface is in the middle between particles i and j, normal to OR direction. We denote conditions of media to the left and to the right from the contact plane in direction of OR axis with $(.)^R$ superscript sign, so, for example, we have a Riemann problem for velocities $v_i^R = \vec{v}_i \cdot \vec{e}_{ji}$ and $v_j^R = \vec{v}_j \cdot \vec{e}_{ji}$. We denote the Riemann problem solution between particles i and j on OR axis with a $(.)^{*R}$ superscript. Then CSPH approximations of mass and impulse conservation laws take the following forms:

$$\frac{d\rho_i}{dt} = \frac{2}{\rho_i} \sum_j \frac{m_j}{\rho_j} \left(v_i^R - v_{ij}^{*R}\right) \frac{\partial W_{ij}}{\partial r_{ij}}, \tag{1}$$

$$\frac{d\vec{v}_i}{dt} = \frac{2}{\rho_i} \sum_j \frac{m_j}{\rho_j} \vec{\sigma}_{ij}^{*} \frac{\partial W_{ij}}{\partial r_{ij}}. \tag{2}$$

Here $\vec{\sigma}_{ij}^{*}$ is the Riemann problem solution for a tension vector on the contact plane after an inverse coordinates transform from $ORST$ to $Oxyz$.

Unlike [6], our code uses the conservative form of energy conservation law:

$$\frac{(e_i + \vec{v}_i^2/2)}{dt} = -\frac{1}{\rho_i} \sum_j \frac{m_j}{\rho_j} \vec{\sigma}_{ij}^{*} \cdot \vec{v}_{ij}^{*} \frac{\partial W_{ij}}{\partial r_{ij}}, \tag{3}$$

where $e_i + \vec{v}_i^2/2$ is a full energy in particle i. The Riemann problem solution for velocity $\vec{v}_{ij}^{*}$ and tension $\vec{\sigma}_{ij}^{*}$ are represented in the same coordinate system. To solve the Riemann problem the Dukowicz Riemann solver [7] is used in our code.

We consider a variable smoothing length, which is specified as $h_i = d_i = \sqrt[3]{m_i/\rho_i}$, assuming it to be equal to a length of cube side, where the cube is occupied by an SPH particle. Shape of an SPH particle is undetermined, so the assumption does not conflict with SPH theory. In order to assure conservation properties of particle interaction we need either to set $h_{ij} = (h_i + h_j)/2$ and then evaluate SPH sum assuming $W_{ij} = W_{ij}(|\vec{r}_{ij}|, h_{ij})$, or to set an SPH kernel to be equal to $W_{ij} = \big(W(|\vec{r}_{ij}|, h_i) + W(|\vec{r}_{ij}|, h_j)\big)/2$. We choose the first way without explanation. An interaction length depends on the chosen smoothing length, and in general $W(|\vec{r}_{ij}|, h_{ij}) > 0$ if $|\vec{r}_{ij}| < \kappa h_{ij}$. We use the Wendland C^2 kernel with $\kappa = 1/(2\sqrt{15}) \approx 1.94$ [8, 9].

4.2. Equation of state

Equation system (1)–(3) is enclosed by the equation of state (EoS) in Mie—Gruneisen form:

$$p - p_r = \Gamma_0 \rho (e - e_r). \tag{4}$$

Here e is the internal energy of medium, calculated from energy conservation law (3), Γ_0— Gruneisen parameter, $p_r(\rho)$ and $e_r(\rho)$ are the reference pressure and the internal energy curves, respectively.

The Shock Hugoniot curve $p_H(\rho)$ is used as reference pressure if density is higher than an initial density; elastic curves are used if density is lower than initial one:

$$p_r = \begin{cases} p_H, & \rho > \rho_0 \\ p_C, & \rho \le \rho_0 \end{cases}, \quad e_r = \begin{cases} e_H, & \rho > \rho_0 \\ e_C, & \rho \le \rho_0 \end{cases},$$

where $p_H = C_a^2(V_0 - V)/[V_0 - s_a(V_0 - V)]^2$, $e_H = p_H(V_0 - V)/2$, $p_C = K(V_0 - V)/V_0$, $e_C = p_C(V_0 - V)/2$, $V = 1/\rho$, K is an isothermal bulk modulus.

Elasto-plastic behavior of material follows Hooke's law (with constant shear modulus G) and von Mises yield criterion. Yield strength Y is considered constant.

Table 1 provides all the necessary parameters for our simulation.

Table 1. Material properties.

	ρ_0, kg/m^3	Γ_0	c_a, m/s	s_a	K, GPa	G, GPa	Y, MPa	T_m, K
Gold	$19.32 \cdot 10^3$	1.05	$3.07 \cdot 10^3$	1.540	166.4	28.5	40	1339
Tungsten	$19.6 \cdot 10^3$	1.17	$4.029 \cdot 10^3$	1.237	300.0	155.0	2400	3693

4.3. Solid—liquid phase transition

The preliminary modeling showed that experimental conditions [1] lead to pressures and temperatures that guarantee melting of materials. In order to conduct correct modeling of a phase transition phenomenon we added some code modifications. We used the following phase transition condition in form of Simon—Glatzel equation for gold [10]: if temperature exceeds

$$T > T_m \left(1 + \frac{P}{16.1}\right)^{0.57}, \tag{5}$$

then the metal is considered melt. Pressure in equation (5) is measured in GPa. For tungsten we have the following condition [11]:

$$T > T_m + m_c \left(1 - \frac{\rho_0}{\rho}\right), \tag{6}$$

where $m_c = 5000$ K is a Kennedy's constant.

If the conditions (5), (6) are not satisfied, respective materials are considered to be in solid state.

4.4. Initial packing of particles

Three-dimensional modeling of the shock wave propagation through a random metallic spheres pack requires development of a specific algorithm of packing of spheres in limited computational area with periodical boundary conditions (BC). Meanwhile the exact experimental golden spheres packing upon the denal transmitter is unknown (additional "shaking" may lead to packing density increase and may influence results of experiments, as will be shown further). Therefore two types of regular packing were used in our simulations: plain cubic and face-centered cubic (dense) without irregular vacancies. Considering the direction of shock wave propagation as Ox axis and applying periodic BCs by Oy and Oz, one can illustrate the packs by Ox axis in the following form (see figure 3).

It is important to identify SPH particles not only by a material they represent, but also by their initial belonging to one or another golden sphere (or the denal transmitter). It helps to apply correct BCs between SPH particles in the golden spheres in contact.

Number of layers can be calculated from the provided information about mass that was ejected from the notches per unit area (88 and 71 g/cm^3 for spheres of 9 μm and 1 μm respectively). We used 10 layers of 9 μm diameter spheres for cubic packing and 9 layers of 9 μm diameter spheres for dense packing; 70 layers of 1 μm diameter spheres for cubic packing and 60 layers of 1 μm diameter spheres for dense packing in our numerical experiments.

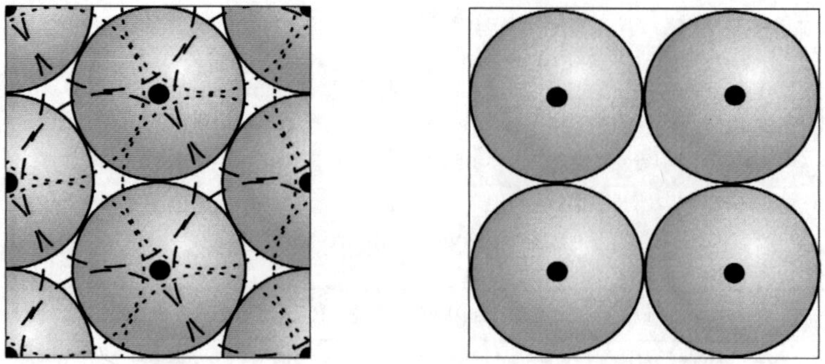

Figure 3. Dense and plain cubic packings of spheres.

4.5. Boundary conditions between spheres

The contact slip condition (non-viscous contact) and the no merging condition are applied between solid spheres. It means that between SPH particles that belong to different bodies there is no tangential stress. Contact pressure for such particles cannot become negative.

Absence of tangential stress in the SPH formulation we use means that for i and j particle

$$\vec{v}_{ij}^{*RST} = \left\{v_{ij}^{*R}, v_j^S, v_j^T\right\}^T, \tag{7}$$

$$\vec{\sigma}_{ij}^{*RST} = \left\{\sigma_{ij}^{*R}, 0, 0\right\}^T, \tag{8}$$

if calculating for particle i; non-negative contact pressure reads as

$$\sigma_{ij}^{*R} = \min(\sigma_{ij}^{*R}, 0) \tag{9}$$

Liquid phase is always non-viscous, so all the contacting pairs of liquid SPH particles employ conditions (7), (8). Condition (9) should not be applied to liquids (liquid materials can merge).

5. Simulation results

We choose the experiment in which the free surface was accelerated to 1500 m/s (as measured in the notch with no insertion) and golden spheres were of 9 μm diameter for discussion of physical phenomena which take place in layer of spheres. Our choice is driven by a low number of layers of spheres that are necessary for simulation in this case: 9–10 layers in contrast with 60–70 for 1 μm diameter spheres for a given mass per unit area (88 mg/cm^2). It is obvious that if we use the same number of SPH particles per a single golden sphere for the two cases, computational expenses grows greatly, but solving the harder problem will not help us to shed light on the ejecta phenomenon in the experiment under discussion. Considering the chosen computation area shape (in figure 3) it is convenient to show results for the 9 μm diameter case (spheres layer is not “stretched” in computational area).

5.1. Dense spheres packing

Here we consider a case of constructing initial computational area as shown on figure 4.

The layer of spheres consists of $2.488{\cdot}10^6$ SPH particles. A single sphere is comprised by $70{\cdot}10^3$ SPH particles. As it was already mentioned, we also considered the transmitter plate which acted like an impactor in our simulation so the whole number of particles reached $6.412 \cdot 10^6$. The number of SPH particles was estimated after they were placed at nodes of body centered cubic cells of size equal to the minimal computational area size divided by 70.

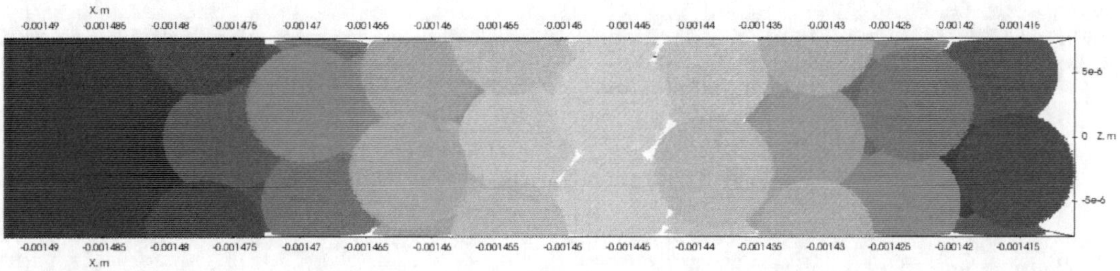

Figure 4. Initial pattern of spheres placement for dense packing (at $t = 0$). Here and below figures are provided in 3D perspective, and different spheres are shown in different colors.

At an initial time the transmitter plate is hit into layers of spheres with the experimental free surface velocity of 1500 m/s. Figure 5 shows flow evolution till and including ejecta formation. As a result of the impact of adjoining spheres cumulative jets are formed that hit shielding layers (figure 5, (a)). As shock wave propagates, the compaction of spheres and formation of typical Richtmayer—Meshkov instability structures are observed (figure 5, (b, c)). The eddy structures consist of SPH particles originating from different golden spheres. After shock wave arrival to the free surface of packing freely escaping jets are formed (figure 5, (d)). The jets are comprised by SPH particles, originating from upper layers of packing.

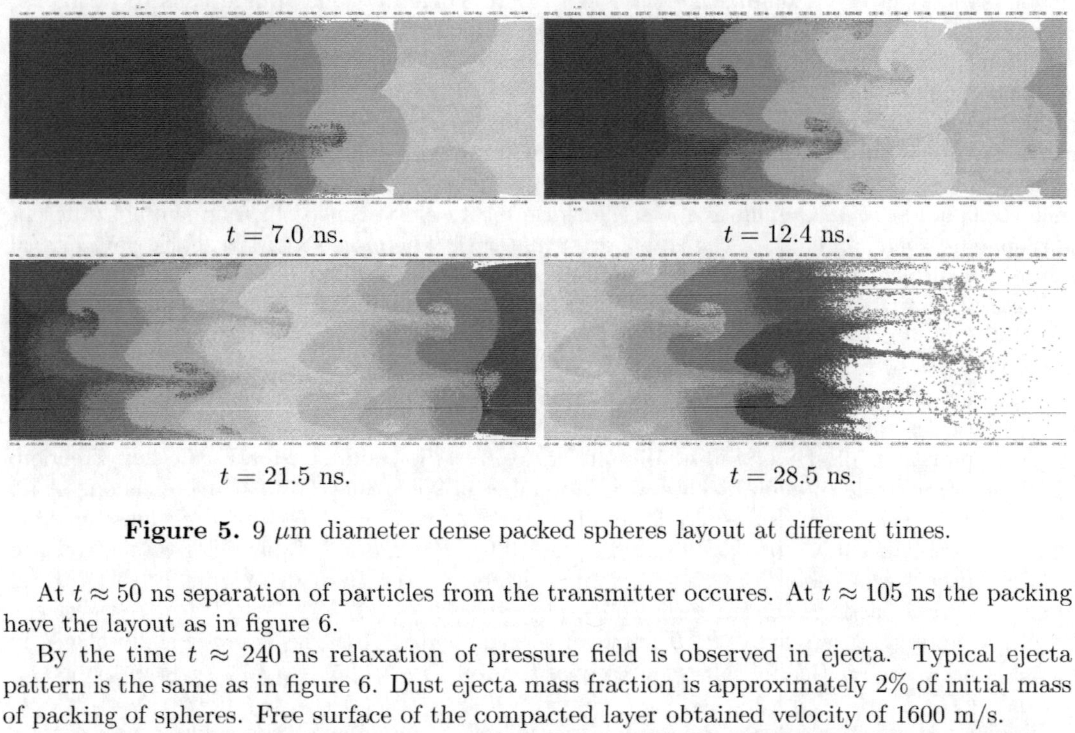

$t = 7.0$ ns. $t = 12.4$ ns.

$t = 21.5$ ns. $t = 28.5$ ns.

Figure 5. 9 μm diameter dense packed spheres layout at different times.

At $t \approx 50$ ns separation of particles from the transmitter occures. At $t \approx 105$ ns the packing have the layout as in figure 6.

By the time $t \approx 240$ ns relaxation of pressure field is observed in ejecta. Typical ejecta pattern is the same as in figure 6. Dust ejecta mass fraction is approximately 2% of initial mass of packing of spheres. Free surface of the compacted layer obtained velocity of 1600 m/s.

The results show that the flow pattern of the compacted spheres packing is determined by a cumulation phenomenon. The dust observed in experiments is in fact fragments of cumulative jets. Moreover, for the dense packing the fragments come from upper packing layers.

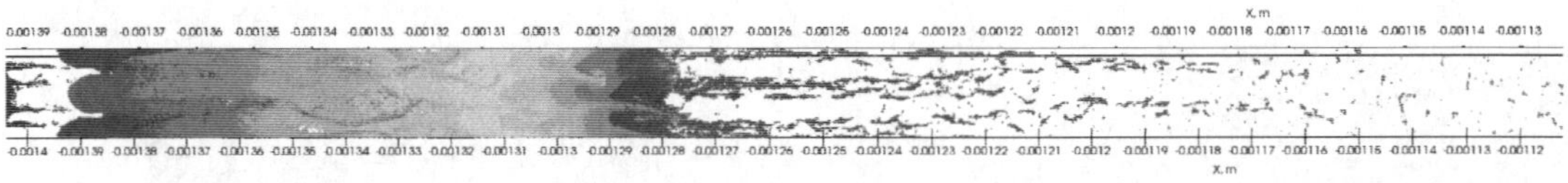

Figure 6. Layer layout at $t = 105.0$ ns.

5.2. Plain cubic spheres packing

Let us consider the case of initial computational area constructed as shown on figure 7.

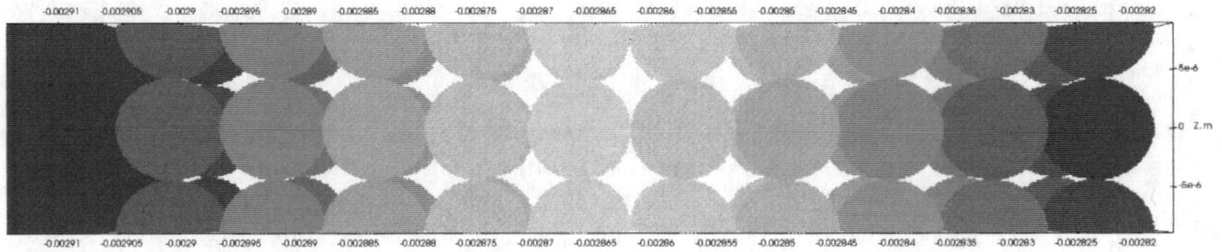

Figure 7. Initial pattern of the arrangement of spheres for plain cubic packing (at $t = 0$).

The spheres layer is comprised by $1.13{\cdot}10^6$ SPH particles. One sphere consists of $28{\cdot}10^3$ particles. Overall number of SPH particles is $4.77{\cdot}10^6$. The number was obtained after we placed them at nodes of body centered cubic cells (as in the previous case) of a size equal to minimal computational area size divided by 60.

Figure 8 shows evolution of packing while it undergoes shock wave loading by the transmitter plate. Figures show cumulative jets formation (figure 8, (a)). The jets have free space in upper layers and thus pass the free surface of the packing before the shock wave reaches upper layers of spheres (figure 8, (b)). Cumulative jets that form ejecta cloud consist of fragments of different (all) spheres layers, and also of the transmitter material. The moment of the shock wave arrival to the free surface of layers is shown in figure 8, (c).

Figure 9 shows jets that are observed in computational area some time after the shock wave breakage through the free surface. It is clear that the jet heads consist of the lower layers material and also from the transmitter plate material (tungsten in the simulation or denal, if interpreting the real experiments). Mass fraction of dust is 11% of overall ejecta mass. Tungsten mass fraction is $\approx$4% of the dust mass.

The simulations of shock loading of smaller spheres show similar small-scale flow trend to the described above. As mentioned above, a number of spheres of 1 μm diameter is one order greater than it is necessary to use for 9 μm diameter spheres. It took 60 layers for dense packing and 70—for plain cubic. It was required to use $30.5{\cdot}10^6$ and $26{\cdot}10^6$ particles respectively to obtain appropriate number of particles per single sphere ($44{\cdot}10^3$ and $30{\cdot}10^3$, respectively). Total number of SPH particles comprising golden spheres reached $14.9{\cdot}10^6$ and $7.9{\cdot}10^6$, respectively. As in previous cases, number of SPH particles was calculated after they were initially placed in nodes of body centered cubic cells of size equal to minimal computational area size divided by 60.

Similar to the larger spheres case, the formation of the cumulative jets is observed for plain cubic pack of spheres of 1 μm diameter each. Fragments of all layers of the pack and of the transmitter plate are found in the jets. Dense packing produces less jetting compared to plain

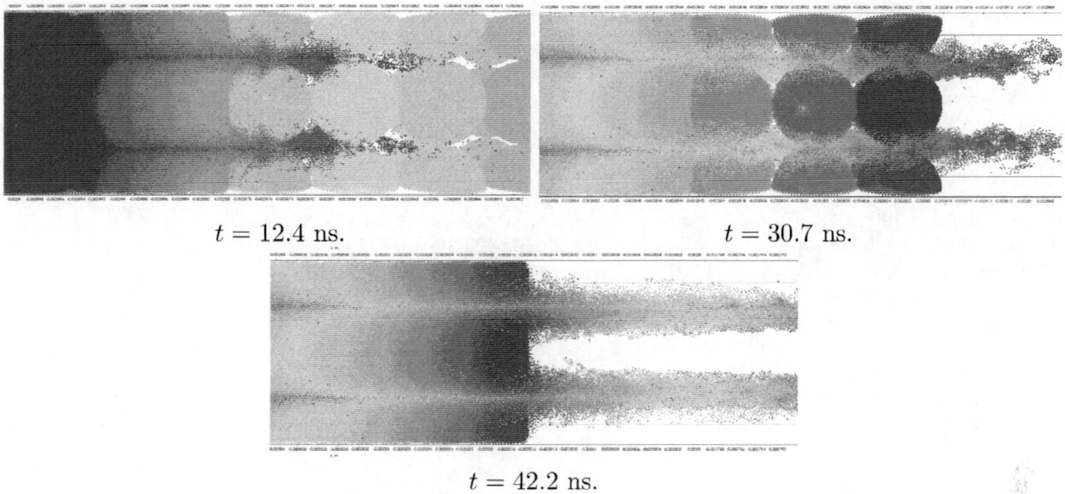

Figure 8. Spheres of a 9 μm diameter, plain cubic packing. Layout at different times.

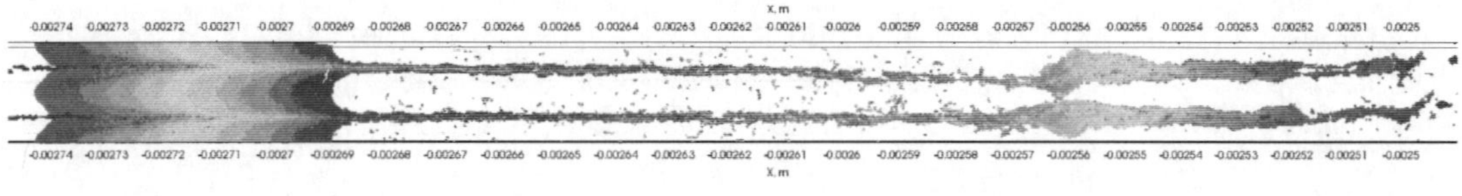

Figure 9. Layer layout at $t = 120.0$ ns.

cubic pack: the jets include only upper layers of the sphere pack. We do not provide full illustrations of the flow in this case due to the computational area length-to-width large ratio. The local flow has the same layout as shown above for the system of 9 μm diameter spheres.

6. Conclusion

The simulation results of shock-loaded regular layer of spherical particles have been presented. Two regular types of lattice have been discussed: plain cubic and faced-centered cubic one. Original results have demonstrated that the ejecta formed when a shock wave reaches the free surface of bulked spheres consist of the fragments of the cumulative jets. These fragments originate from different layers of spheres and also from the transmitter plate. We have found that ejecta structure is highly dependent on the initial packing of spheres. In the case of real random packing the ejecta appears to consist of the fragments both of the spheric particles and the transmitter plate.

CSPH&VD^3, the software we develop, enables us to estimate the distribution of fragments by their masses and velocities. The modeling results could serve as a reliable base for mesh methods and, as it is shown in Part II, is also suitable for further simulations of evolution of dust ejecta in the air.

References

[1] Prudhomme G, Mercier P and Berthe L 2014 *J. Phys. Conf. Series* **500** 142027

[2] Andriyash A, Astashkin M, Baranov V, Golubinskii A, Irinichev D, Kondratev A, Kuratov S, Mazanov V, Rogozkin D, Stepushkin S *et al.* 2016 *J. Exp. Theor. Phys.* **122** 970–983
[3] http://chemphys.edu.ru/issues/2015-16-4/articles/550/
[4] Dyachkov S, Egorova M, Mursov S, Parshikov A and Zhakhovsky V (submitted to Lobachevskii Journal of Mathematics)
[5] Monaghan J J 2005 *Rep. Prog. Phys.* **68** 1703
[6] Parshikov A N and Medin S A 2002 *J. Comp. Phys.* **180** 358–382
[7] Dukowicz J K 1985 *J. Comp. Phys.* **61** 119–137
[8] Wendland H 1995 *Adv. Comput. Math.* **4** 389–396
[9] Dehnen W and Aly H 2012 *Mon. Not. R. Astron. Soc.* **425** 1068–1082
[10] Errandonea D 2010 *J. Appl. Phys.* **108** 033517
[11] Kraut E A and Kennedy G C 1966 *Phys. Rev.* **151** 668

Shock-induced ejecta from a layer of spherical particles. Part II: modeling with the non-equilibrium two-phase model of a granular medium

A A Serezhkin[1], I S Menshov[1,2], M S Egorova[1], S A Dyachkov[1], A N Parshikov[1], V V Zhakhovsky[1], D B Rogozkin[1,3], S E Kuratov[1]

[1] Dukhov Research Institute of Automatics (FSUE VNIIA), Moscow, 127055, Russian Federation

[2] Keldysh Institute for Applied Mathematics RAS, Moscow 125581, Russian Federation

[3] Moscow Engineering Physics Institute, National Research Nuclear University, Kashirskoe Shosse 31, 115409 Moscow, Russian Federation

aaserezhkin@gmail.com

Abstract. In this work we consider the problem of high-velocity impact of a solid impactor on the layer of spherical solid particles. The impact initiates a wave-type process of compaction of particles in the layer, and the ejection of small solid particles from the compacted layer surface to the air. The first part of the paper addresses modelling the process of particle compaction and the initial stage of solid ejection with the SPH method. In the second part presented below, we consider the process of compaction and solid mass ejection using a new non-equilibrium two-phase model of the granular medium. This model makes it possible to describe mechanics of two-phase gas-solid medium in a wide range of the phase volume fraction, from regimes of isolated particle flows to regimes of dense packed particles.

Keywords: two-phase heterogeneous medium, numerical simulation, PDV.

1. Introduction

Process of high-velocity impact on a layer of spherical particles can be separated on two stages: the first stage is the compaction of particles and the second one is ejection of particles or fragments of particles from the free surface of the layer. Provided that we use continual approach, the choice of the physical model is of the first importance.

On the one hand, to simulate the compaction of particles, we need a model that should take into account mechanical interaction between the particles, plastic deformation, non-equilibrium in stress in the particles and in the gas filling the volume between the particles. The most appropriate model for this case is the 7-equation Baer-Nunziato model [1]. This model is based on the continuum theory of mixtures. Either the solid phase or the gas phase is considered as a continuum with proper physical and mechanical properties. This model works well for the regimes of flow when the volume fraction of the solid phase is greater than a critical value φ_S^* that corresponds to the volume fraction of the close-bed structure of the medium.

On the other hand, the simulation of the ejection stage is best fitted by the 5-equations model of a cloud of particles [2]. The solid volume fraction is assumed to be less then a critical value of φ_S^*. This model works well for the regimes of flow when solid particles are dispersive and do not interact with each other.

These two models form two classes of models of two-phase heterogeneous mixture mechanics. Models of one class differ by the closing relations and the model control parameters. If we want to simulate the both processes (compaction and ejection) we need to combine these two models in one numerical code, or develop a new model that could be capable to simulate the both regimes of flow in a unique way.

In the current work we use the second approach. We construct a new model for two-phase continuum dynamics that smoothly switches between the 5-equations model and the 7-equations model when the solid volume fraction crosses over the critical value φ_S^*.

One purpose of this work is interpretation of PDV measurements of disperse phase velocity in experiments [3], [4], [5]. The experimental data shows that the interaction of the compaction wave with the layer surface initiates ejection of a cloud of particles to the surrounding air. A simple analysis of velocity slowing-down given in [3] has shown that a typical diameter of particles in the cloud do not correspond to the diameter of particles in the layer.

In the first part of the present paper, the impact problem on a layer of particles is simulated with the SPH method. The results of SPH simulations show that the interaction of the compaction wave with the layer surface initiates ejection in the form of cumulative jets that are broken down in process. Jet fragments form a kind of fraction of solid particles that move with a velocity greater than the velocity of particles in the layer. Such a the velocity of the fragments is measured in experiments.

Simulating in framework of the continual model, one can not describe the formation of cumulative jets because this process occurs at the mesoscale level. Therefore, for simulation of the interaction of the compaction wave with the layer surface we use the results of SPH simulations as initial data.

Results of the SPH simulation show that the mass ejecting from the layer surface depends on the distribution of particles in the layer. Therefore, these results do not provide reliable data for the volume fraction in the jet domain because the initial arrangement of particles in the layer is chaotic. Nevertheless, the ejecting mass is small enough, so that we can use the drag coefficient for a single spherical particle for the simulation of slowing-down of fragments.

Along with simulating experiments with gold particles [3] we validate the proposed model in a similar problem of the impact with a smaller impact velocity on nickel particles [5]. Under these conditions, no cumulative jets appear, and therefore this experiment can be simulated by using the continuum model.

2. A wide-range model for two-phase gas-solid flows

The derivation of the system of governing equation is carried out with the continual approach proposed in [2]. We consider a small Euler volume that consists of two parts. The first part corresponds the solid phase (particles), the second one is the volume filled in gas. The ratio of the phase volume the whole one represents volume fractions (ϕ_S, ϕ_G). Here and after, indexes S and G indicate solid and gas phases respectively. Each phase is described by distributions of density (ρ_S, ρ_G), velocity (u_S, u_G), internal energy (e_S, e_G) and pressure (P_S, P_G). The evolution in time of conservative variables as mass, momentum and total energy ($E_S = 0.5u_S^2 + e_S$, $E_G = 0.5u_G^2 + e_G$) in the whole volume is defined by corresponding conservation laws written for corresponding sub-domains with taking into account change the volume fractions.

The system of governing equations that describes the evolution of conservative variables can be written in following form:

$$\begin{cases} \frac{\partial}{\partial t}\varphi_S\rho_S + \nabla(\varphi_S\rho_S u_S) = 0 \\ \frac{\partial}{\partial t}\varphi_S\rho_S u_S + \nabla(\varphi_S\rho_S u_S u_S) + \nabla(\varphi_S P_S) = P_I\nabla\varphi_S + K(u_S - u_G) \\ \frac{\partial}{\partial t}\varphi_S\rho_S E_S + \nabla(\varphi_S\rho_S E_S u_S) + \nabla(\varphi_S P_S u_S) = P_I u_S\nabla\varphi_S + A_{def} + K u_I(u_S - u_G) \\ \frac{\partial}{\partial t}\varphi_G\rho_G + \nabla(\varphi_G\rho_G u_G) = 0 \\ \frac{\partial}{\partial t}\varphi_G\rho_G u_G + \nabla(\varphi_G\rho_G u_G u_G) + \nabla(\varphi_G P_G) = -P_I\nabla\varphi_S - K(u_S - u_G) \\ \frac{\partial}{\partial t}\varphi_G\rho_G E_G + \nabla(\varphi_G\rho_G E_G u_G) + \nabla(\varphi_G P_G u_G) = -P_I u_S\nabla\varphi_S - A_{def} - K u_I(u_S - u_G) \end{cases} \quad (1)$$

where P_I and u_I are the average pressure and the average velocity on the interphase boundary, K is the drag coefficient [2], A_{def} is the work done for change the volume fraction of phases in the small Euler volume. The latter we determine in the following form:

$$A_{def} = -P_I\frac{d_S\varphi_S}{dt} + \lambda_S\nabla u_S + \lambda_G\nabla u_G$$

where λ_S and λ_G are parameters of model to be defined. The system of equations (1) is not closed. One need to define yet the following values: φ_S, φ_G, P_S, P_G, P_I, u_I, λ_S and λ_G.

First of all to do this, we use equations of state for each phase $e_S = e_S(\rho_S, P_S)$ and $e_G = e_G(\rho_G, P_G)$, and a condition $\varphi_S + \varphi_G = 1$. Then we employ the total entropy inequality [6]: $\varphi_S\rho_S\frac{d_S S_S}{dt} + \varphi_G\rho_G\frac{d_G S_G}{dt} \geq 0$ that must be valid for all regimes of flow. Using first the law of thermodynamics, equations of state, and the entropy definition we can the equation for the entropy evolution:

$$\begin{aligned} \varphi_S\rho_S\frac{d_S S_S}{dt} + \varphi_G\rho_G\frac{d_G S_G}{dt} = \varphi_S\varphi_G\frac{\left(\frac{d_G P_G}{dt} + (\rho_G C_G^2 + \lambda_G A)\nabla u_G\right)\left(\frac{P_S - P_I}{T_S} - \frac{P_G - P_I}{T_G}\right)}{Z} - \\ -\varphi_S\varphi_G\frac{\left(\frac{d_S P_S}{dt} + (\rho_S C_S^2 - \lambda_S A)\nabla u_S\right)\left(\frac{P_S - P_I}{T_S} - \frac{P_G - P_I}{T_G}\right)}{Z} - \\ -\left(\frac{\varphi_S\rho_G C_{GI}^2\left(\frac{P_S - P_I}{T_S}\right) + \varphi_G\rho_S C_{SI}^2\left(\frac{P_G - P_I}{T_G}\right)}{Z}\right)(u_G - u_S)\nabla\varphi_S + \\ + K(u_S - u_G)\left(\frac{\varphi_G\Gamma_S(u_I - u_S) + \varphi_S\Gamma_G(u_I - u_G)}{Z}\left(\frac{P_S - P_I}{T_S} - \frac{P_G - P_I}{T_G}\right) + \left(\frac{u_I - u_S}{T_S} - \frac{u_I - u_G}{T_G}\right)\right) \end{aligned} \quad (2)$$

where

$$A = \frac{(T_G - T_S)(\varphi_G \rho_S C_S^2 + \varphi_S \rho_G C_G^2) + (\varphi_G \Gamma_S T_S + \varphi_S \Gamma_G T_G)(P_S - P_G)}{\varphi_S \varphi_G ((P_S - P_I) T_G - (P_G - P_I) T_S)}$$

T_S and T_G are the temperatures of phases,

$$Z = \varphi_G \rho_S C_{SI}^2 + \varphi_S \rho_G C_{GI}^2, \qquad C_S^2 = \left(\frac{P_S}{\rho_S^2} - \frac{\partial e_S}{\partial \rho_S}\right)\left(\frac{\partial e_S}{\partial P_S}\right)^{-1}, \qquad C_{SI}^2 = \left(\frac{P_I}{\rho_S^2} - \frac{\partial e_S}{\partial \rho_S}\right)\left(\frac{\partial e_S}{\partial P_S}\right)^{-1},$$

$$C_G^2 = \left(\frac{P_G}{\rho_G^2} - \frac{\partial e_G}{\partial \rho_G}\right)\left(\frac{\partial e_G}{\partial P_G}\right)^{-1}, \quad C_{GI}^2 = \left(\frac{P_I}{\rho_G^2} - \frac{\partial e_G}{\partial \rho_G}\right)\left(\frac{\partial e_G}{\partial P_G}\right)^{-1}.$$

Note, that in above formulas C_S and C_G are the standard thermodynamic sounds velocities of the solid and gas phase. C_{SI} and C_{GI} are effective sound velocities at the interphase as proposed in [7]. In general C_{SI}^2 or C_{GI}^2 might be negative. In this case vacuum vapors arise nearby boundary. This situation is out of scope of the present study. Γ_S and Γ_G are Gruneisen coefficients of the phases.

To guarantee the entropy inequality to be valid, we suggest the following relations considered as additional closure equations:

$$\frac{d_G P_G}{dt} + (\rho_G C_G^2 + \lambda_G A)\nabla u_G = -\frac{P_G - P_I}{\tau T_G} \tag{3}$$

$$\frac{d_S P_S}{dt} + (\rho_S C_S^2 - \lambda_S A)\nabla u_S = -\frac{P_S - P_I}{\tau T_S} \tag{4}$$

$$u_I = \frac{\varphi_G \Gamma_S u_S + \varphi_S \Gamma_G u_G}{\varphi_G \Gamma_S + \varphi_S \Gamma_G} \tag{5}$$

where τ - is the pressure time relaxation in computations it could be takes equal the time of shock wave passing the distance of several particle size, $\tau = n d_p C_S^{-1}$. Providing (3)-(5), equation (2) can be rewritten as:

$$\varphi_S \rho_S \frac{d_S S_S}{dt} + \varphi_G \rho_G \frac{d_G S_G}{dt} = \varphi_S \varphi_G \frac{\left(\frac{P_S - P_I}{T_S} - \frac{P_G - P_I}{T_G}\right)^2}{\tau Z} + \frac{K(u_S - u_G)^2}{(\varphi_G \Gamma_S + \varphi_S \Gamma_G)}\left(\frac{\varphi_S \Gamma_G}{T_S} + \frac{\varphi_G \Gamma_S}{T_G}\right) - \left(\frac{\varphi_S \rho_G C_{GI}^2 \left(\frac{P_S - P_I}{T_S}\right) + \varphi_G \rho_S C_{SI}^2 \left(\frac{P_G - P_I}{T_G}\right)}{Z}\right)(u_G - u_S)\nabla \varphi_S \tag{6}$$

Due Gruneisen coefficient and temperature are positive, first two terms in the right-hand side of the equation (6) are positive. Therefore to ensure the entropy inequality for every thermodynamic process possible in the considered two-phase mixture, the last term must be vanish. This provides a quadratic equation for P_I:

$$P_I^2\left(\varphi_G T_S\Gamma_S+\varphi_S T_G\Gamma_G\right)-P_I\left(\left(P_S+\rho_G^2\frac{\partial e_G}{\partial\rho_G}\right)\varphi_S T_G\Gamma_G+\left(P_G+\rho_S^2\frac{\partial e_S}{\partial\rho_S}\right)\varphi_G T_S\Gamma_S\right)+$$
$$+\left(\varphi_S T_G\Gamma_G P_S\rho_G^2\frac{\partial e_G}{\partial\rho_G}+\varphi_G T_S\Gamma_S P_G\rho_S^2\frac{\partial e_S}{\partial\rho_S}\right)=0 \tag{7}$$

One can to prove that this equation always has one solution on an interval $[\min(P_S,P_G),\max(P_S,P_G)]$. If $P_I=P_G$, the left hand side of equation (7) equals to $\varphi_S\rho_G C_G^2\left(\frac{P_S-P_G}{T_S}\right)$, and has the same sign as $\left(P_S-P_G\right)$. If $P_I=P_S$, the left hand side of equation (7) equals to $\varphi_G\rho_S C_S^2\left(\frac{P_G-P_S}{T_G}\right)$, and has the same sign as $\left(P_G-P_S\right)$. This proves existing and uniqueness of the solution.

Using (3) - (5) and (7), we can get the equation for the volume fraction of the solid phase:

$$\frac{d_S\varphi_S}{dt}=-\frac{\varphi_S\varphi_G\left(P_G-P_I\right)}{\tau T_G\left(\varphi_G\rho_S C_{SI}^2+\varphi_S\rho_G C_{GI}^2\right)}+\frac{\varphi_S\varphi_G\left(P_S-P_I\right)}{\tau T_S\left(\varphi_G\rho_S C_{SI}^2+\varphi_S\rho_G C_{GI}^2\right)}+$$
$$+\frac{\left(\varphi_G\Gamma_S+\varphi_S\Gamma_G-\varphi_S\varphi_G A\right)\left(\lambda_G\nabla u_G+\lambda_S\nabla u_S\right)}{\left(\varphi_G\rho_S C_{SI}^2+\varphi_S\rho_G C_{GI}^2\right)}-\frac{\varphi_S\rho_G\left(u_G-u_S\right)C_{GI}^2}{\left(\varphi_G\rho_S C_{SI}^2+\varphi_S\rho_G C_{GI}^2\right)}\nabla\varphi_S \tag{8}$$

Equation (8) is not a closing relation but it allows us to write (1) as a closed system of governing equations:

$$\frac{\partial\mathbf{Q}}{\partial t}+\frac{\partial\mathbf{F}}{\partial x}+\overline{\mathbf{B}}\frac{\partial\mathbf{Q}}{\partial x}=\mathbf{H} \tag{9}$$

where $\mathbf{Q}$ is a vector of conservative variables $\varphi_S\rho_S$, $\varphi_S\rho_S u_S$, $\varphi_S\rho_S E_S$, ρ_S, $\varphi_G\rho_G$, $\varphi_G\rho_G u_G$, $\varphi_G\rho_G E_G$; $\mathbf{F}$ is the vector of conservative fluxes; the matrix $\overline{\mathbf{B}}$ is the matrix of the non-conservative part; $\mathbf{H}$ is non-differential right-hand side.

Closing relations (3) - (5) and (7) are not enough to close the system (1) because there are undefined parameters λ_S and λ_G. To estimate their values we study mathematical propertias of equation (9). To correctly describe evolutionary process, equation (9) must be hyperbolic. Let's recast equation (9) in a non-conservative form as follows:

$$\frac{\partial}{\partial t}\begin{pmatrix}\rho_S\\ u_S\\ P_S\\ \varphi_S\\ \rho_G\\ u_G\\ P_G\end{pmatrix}+\begin{pmatrix}u_S & \rho_S+\rho_S\varphi_G W\lambda_S & 0 & -\frac{\rho_S B_G\left(u_G-u_S\right)}{\varphi_S} & 0 & \rho_S\varphi_G W\lambda_G & 0\\ 0 & u_S & \frac{1}{\rho_S} & \frac{P_S-P_I}{\varphi_S\rho_S} & 0 & 0 & 0\\ 0 & \rho_S C_S^2-\lambda_S A & u_S & 0 & 0 & 0 & 0\\ 0 & -\lambda_S\varphi_S\varphi_G W & 0 & B_G\left(u_G-u_S\right) & 0 & -\lambda_G\varphi_S\varphi_G W & 0\\ 0 & -\rho_G\varphi_S W\lambda_S & 0 & -\frac{\rho_G B_S\left(u_G-u_S\right)}{\varphi_G} & u_G & \rho_G-\rho_G\varphi_S W\lambda_G & 0\\ 0 & 0 & 0 & \frac{-P_G+P_I}{\varphi_G\rho_G} & 0 & u_G & \frac{1}{\rho_G}\\ 0 & 0 & 0 & 0 & 0 & \rho_G C_G^2+\lambda_G A & u_G\end{pmatrix}\nabla\begin{pmatrix}\rho_S\\ u_S\\ P_S\\ \varphi_S\\ \rho_G\\ u_G\\ P_G\end{pmatrix}=\mathbf{h} \tag{10}$$

where **h** is the non-differential right-hand side:

$$\mathbf{h} = \begin{pmatrix} 0 \\ K(u_S - u_G)/(\varphi_S \rho_S) \\ -(P_S - P_I)/(\tau T_S) \\ \varphi_S \varphi_G \left((P_S - P_I)/(\tau T_S) - (P_G - P_I)/(\tau T_G)\right)/Z \\ 0 \\ -K(u_S - u_G)/(\varphi_G \rho_G) \\ -(P_G - P_I)/(\tau T_G) \end{pmatrix}$$

$$W = Z(\varphi_G \Gamma_S + \varphi_S \Gamma_G - \varphi_S \varphi_G A),\; B_S = Z\varphi_G \rho_S C_{SI}^2,\; B_G = Z\varphi_S \rho_G C_{GI}^2 .$$

The matrix of this system has 7 eigenvalues that depend on λ_S and λ_G. If $\lambda_S = \lambda_G = 0$, then the eigenvalues take the form:

$$\lambda_1 = u_S;\; \lambda_2 = B_S u_S + B_G u_G;\; \lambda_{3,4} = u_S \pm C_S;\; \lambda_5 = u_G;\; \lambda_{6,7} = u_G \pm C_G .$$

In this case all eigenvalues are real and the governing system is hyperbolic. Also values of $\lambda_1,...,\lambda_7$ show that small perturbations in the medium propagate with velocities of phases (or its linear combinations) and sound velocities. Therefore it can be assumed that this case corresponds to regimes of flow when solid directly interact each other. Particles forms the dense packed structure and the volume fraction of the solid phase is larger then the critical value, $\varphi_S \geq \varphi_S^*$.

If $\lambda_G = 0$, $\lambda_S A = \rho_S C_S^2$, then the eigenvalues are:

$$\lambda_{1,2} = u_S;\; \lambda_{3,4} = \frac{1}{2}(u_S + B_S u_S + B_G u_G) \pm \frac{1}{2}\left((B_S(u_G - u_S))^2 + 4\varphi_G C_S^2 (P_S - P_I)\frac{W}{A\rho_S}\right)^{1/2}$$

$$\lambda_5 = u_G;\; \lambda_{6,7} = u_G \pm C_G$$

In this case $\lambda_{3,4}$ might be complex, and hyperbolicity of the system violated. In order to overcome this difficulty we suggest to transfer to pressure equilibrium model by setting the pressure time relaxation parameter in (3) and (4), $\tau \to 0$. In this case $P_S - P_G \to 0$, and $\lim\limits_{P_S - P_G \to 0} (W/A) = Z\rho_S \varphi_G$, the system (10) is hyperbolical. Small perturbations propagate in this case with mass velocities of phases (or its linear combinations) and with sound velocity of the gas phase. Therefore we can assume that this case corresponds to a structure with separated particles which forms a cloud of particles in gas. We suggest apply this case when the solid volume fraction satisfies an inequality of $\varphi_S < \varphi_S^*$.

3. Numerical method

For numerical simulations we use in-house code named TIS. The basic idea of the numerical method used in this code is splitting the governing equations in accordance with main physical processes [8]. By using this approach, solving the problem can be reduced to more simple sub-problems without loss of fidelity. The dynamical process in the two-phase granular medium can be thought of as the combination of two sub-processes. One is hydrodynamic where the medium behaves like a two-phase fluid without the relaxation or drag process between the phases. Another sub-process is considered as Lagrangian one. It goes on the interface between phases and includes processes of drag and pressure relaxation. It means that calculation of one time step can be split in two stages. This splitting is mathematically described by 2 systems of equations: for the hydrodynamic stage

$$\frac{\partial \mathbf{Q}}{\partial t} + \frac{\partial \mathbf{F}}{\partial x} + \overline{\mathbf{B}} \frac{\partial \mathbf{Q}}{\partial x} = 0 \tag{11}$$

and for the Lagrangian stage:

$$\frac{\partial \mathbf{Q}}{\partial t} = \mathbf{H} \tag{12}$$

First, we solve non-conservative system of equations (11). For this we use a Godunov-type scheme with an approximate of Riemann solver HLLEM proposed in [9]. The method HLLEM is the extension of the HLL method [10] to non-conservative systems of the form (11). The solution is found on a moving eulerian grid. At the second stage system of ODEs (12) is solved with the Gear method for stiff systems.

4. Numerical results

4.1. Impact on a layer of nickel particles.

We consider a 0.35mm layer of nickel spherical particles with a diameter of 40 μm. The layer is posed on a 3 mm thickness copper plate. The process is initiated by an iron impactor of 10 mm thickness that runs over the copper plate with a velocity of 450m/s. The numerical domain is represented by 3 blocks that correspond to the impactor, the copper plate and the layer of particles. Each block is divided on 100 uniformly distributed numerical cells. Motion of the grid is defined by the velocity of block interfaces. Block boundaries correspond to positions of plate surfaces and boundary of the domain of solid particles. Maximal velocity of particles (km/s) it is shown in figure 1 vs time along with the experimental data [5].

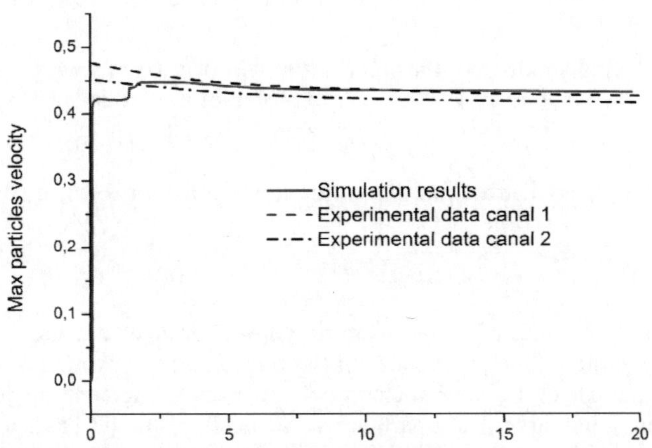

Figure 1. Comparison numerical and experimental data for the impact problem over the layer of nickel particles.

The agreement between numerical and experimental results confirms correctness of the numerical model in description process of compaction and ejection. Small difference in the results in an initial time period of 2 mks can be explained by mesoscale effects of the compaction wave at the particle layer surface.

IOP Conf. Series: Journal of Physics: Conf. Series **815** (2017) 012030 doi:10.1088/1742-6596/815/1/012030

4.2. Impact on a layer of gold particles

We simulate the experiment of ejecta from surface of spherical particles layer [3]. We consider a 58.3 μm layer of gold spherical particles with a diameter of 9 μm. The layer is posed on a denal plate. Impact is initiated by plane-wave generator. Part I of this work addresses the simulation of this problem by using SPH method. It is shown that the interaction of the compaction wave with the layer surface initiates ejection of the solid phase in the form of mesoscale cumulative jets that breaks down into separate fragments. The velocity of fragments is measured in experiments [3]. Simulation in the continual approach do not give adequate results because do not take into account mesoscale effects. However we can use results of SPH simulations as initial data (figure 2 and 3) and calculate slowing-down of expansion of cumulative jet fragments in air. Results of SPH simulation do not gives exact value of volume fraction of solid phase in area of cumulative jets, because in SPH simulations periodical position of particles in layer is proposed, and volume fraction depends on structure of particles position. Nevertheless we can use these results as a first approximation.

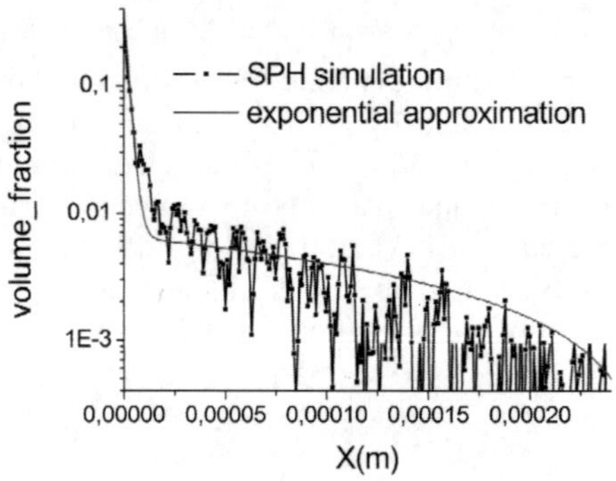

Figure 2. Results of SPH simulation: the solid phase distribution in the cumulative jets.

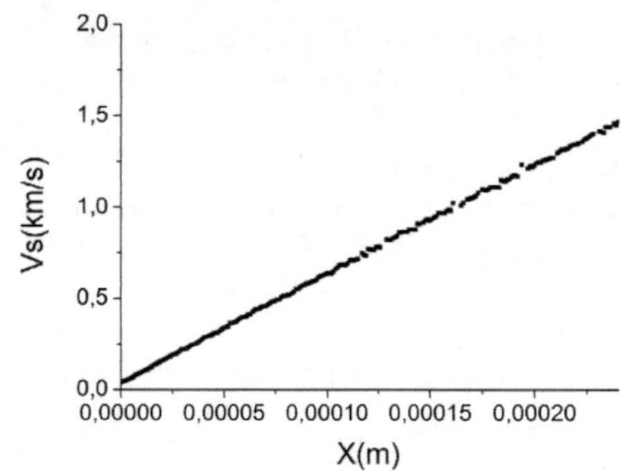

Figure 3. Results of SPH simulation: distribution of solid phase velocity in the cumulative jet.

We approximate data of the distribution of volume fraction from results of SPH simulation using following equation:

$$\varphi_S = 0.4e^{-5*10^{-5}x} - 25x + 0.0065 \tag{13}$$

The velocity distribution is nearly linear (figure 3). We assume that jet fragments spherically shaped. Moreover we need to define typical size of the fragments. We assume in the calculations two type of particles (fragments): of 1 μm diameter small particles - fragments of jets tips located at a distance of [50, 240] μm, and large 3 μm particles - fragments of the jet base, located at [0,50] μm. The calculation domain consists of three blocks: [0, 50] μm that corresponds position of the particles cloud from the jet base, [50, 240] μm that corresponds the cloud of particles from the jets tips, and [240, 1000] μm that corresponds air. Each block is divided into 100 uniform cells. The grid motion is defined by velocities of particles at blocks boundaries.

In figure 4 we present the comparison of experiment and numerical data for the maximal velocity of particles. Black line corresponds experiments, red line corresponds numerical results, and blue lines velocities at block interfaces. In the beginning small particles move faster than large particles. Due drag effects in air slowing-down of small particles is more than that of large particles, so that at a time moment of 2.7 mks the velocities became equal.

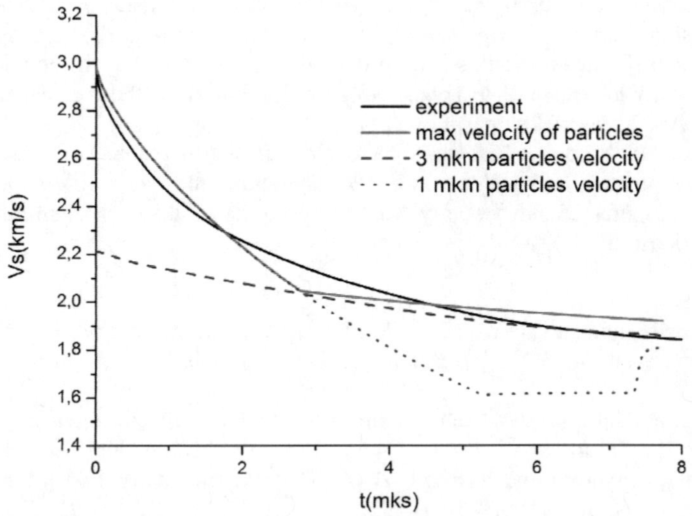

Figure 4. Maximal velocity of particles (km/s) vs time (mks). Comparison of experimental data and results of numerical simulations.

We can define the distribution of typical jet fragments more exactly by carrying out the following simulations. We assign any typical size of jet fragments and simulate expansion of only part of fragments cloud from the base of jets to any point in the middle of total cloud. In results we get curve of maximal velocity of particles. We are varying right boundary position of part of fragments cloud till tangent of the curves of maximal velocity of particles from experiments and from simulations. We can receive distribution of jets fragments typical size vs coordinate x or vs velocity of solid phase in jet. Results of particles size distributions definition are presented on figure 5 and 6.

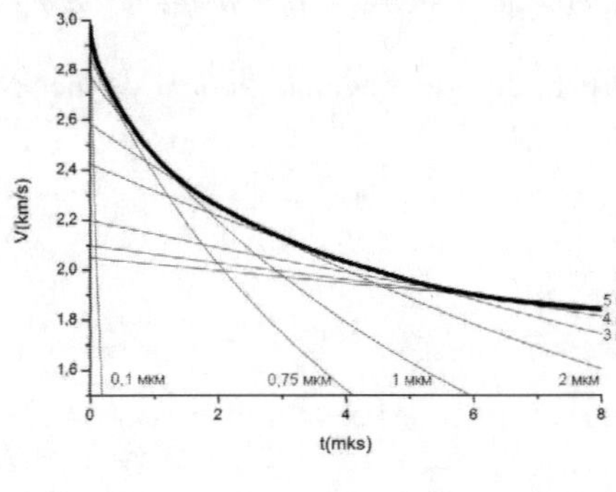

Figure 5. Estimation of the fragments size.

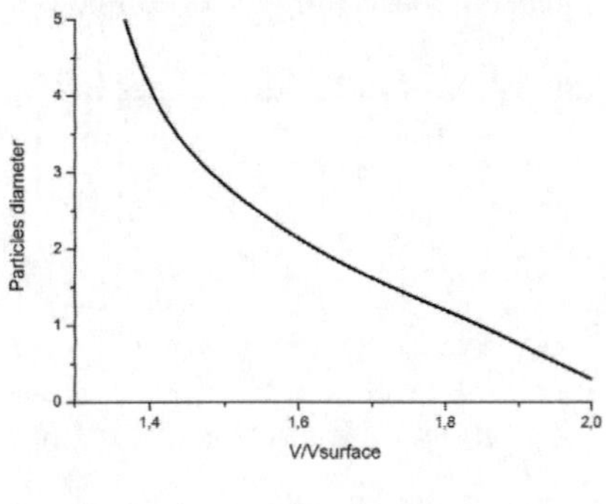

Figure 6. Fragments size (μm) vs the ratio of particles velocity to velocity of impactor surface.

5. Conclusion

In this work we have considered the problem of high-velocity impact of a solid impactor on the layer of spherical solid particles using the continual two-phase non-equilibrium approach. We have designed a new model that makes it possible to describe in a unique way mechanics of a two-phase gas-solid medium in a wide range of the phase volume fraction, from the regimes of isolated particle flows to the regimes of dense packed particles.

A continual approach model is not capable to simulate of mesoscale effects of cumulative jet ejection in the case of very high-velocity impact. Therefore, for the problem of interpretation of experimental PDV measurements of velocity, we have used the results of SPH simulation published in the part I of the work as initial data.

References

[1] Baer M R and Nunziato J W 1986 A two-phase mixture theory for the deflagration-to-detonation transition (DDT) in reactive granular materials *Int. J. Multiphase flow* vol 12 **6** pp 861-889

[2] Khomenko Y P, Ischenko A N and Kasimov V Z 1999 *Mathematical modelling of interior ballistic processes in barrel systems* (Novosibirsk: Publishing House of SB RAS) p 256

[3] Prudhomme G, Mercier P and Berthe L 2014 PDV experiments on shock-loaded particles *J. of Physics: Conf. Series.* vol 500 **14** 142027

[4] Bandurkin K V, Kamenev V G, et al 2015 Experimental (Laser-Heterodyne Method PDV) and Numerical Investigation of the Movement of the Dispersed Phase *Physical-Chemical Kinetics in Gas Dynamics* vol 16 **4**

[5] Fedorov A V, Mikhailov A L, et al 2014 Study of lead beahvior feachures at shock loading and further unloading. *Zababakhin Scientific Talks Int. Conf.*

[6] Nigmatulin R I 1991 *Dynamics of multiphase media* (Hemisphere Publ. Corp.) vol I

[7] Saurel R and Abgrall R 1999 A Multiphase Godunov Method for Compressible Multifluid and Multiphase Flows *J. of Computational Physics* **150** pp 425–467

[8] Menshov I and Serezhkin A 2013 Modelling non-equilibrium two-phase flow in elastic-plastic porous solids *11th World Congress on Computational Mechanics (WCCM XI) 5th European Conf. on Computational Mechanics (ECCM V) 6th European Conf. on Computational Fluid Dynamics (ECFD VI)*

[9] Dumbser M and Balsara D 2016 A new efficient formulation of the HLLEM Riemann solver for general conservative and non-conservative hyperbolic systems *J. of Computational Physics* vol 304 pp 275 - 319

[10] Harten A, Lax P D and van Leer B 1983 On upstream differencing and Godunov-type schemes for hyper-bolic conservation laws *SIAM Rev.* 25(1) pp 35–61

On using of gas detonation for spraying of biocompatible films onto the carbon nanocomposites

P A Tsygankov, A S Skriabin, E Yu Loktionov, V D Telekh and R I Chelmodeev

Department of Power Engineering, Bauman Moscow State Technical University, Baumanskaya 2-ya 5/1, Moscow, 105005, Russia

E-mail: terra107@yandex.ru

Abstract. The paper presents the experimental study of the using of gas detonation for spraying of hydroxyapapite (HOA) particles onto the carbon nanocomposites. The characteristic parameters of the heterophasic detonation flows were defined. The prepared films (with a thickness of 80…100 μm) were polycrystalline with Ca and P concentration ratio of ≈1.67 as for raw HOA. It was noted about the prospects of gas detonation for HOA spraying.

1. Introduction

The one of main problems of the modern engineering is designing of biocompatible materials for the bone substitutes [1]. Nowadays the films of hydroxyapatite (HOA) $Ca_{10}(PO_4)_6(OH)_2$ onto SiC and Ti substrates are under consideration [2]. Carbone nanoimplants (CNI) [3] are more perspective due to the structure features.

The commercial technologies [4] of HOA spraying are plasma and high velocity oxygen fuel spraying. The using of such methods is limited [2] because of poor adhesion, imperfection of structures etc. Therefore nowadays the application of gas detonation [5] for this aim is under studying [2]. The detonation provides pulsed action onto a substrate for minimization of abrasive wear and increasing of the adhesion.

2. Experimental set

Based on commercial CCDS 2000 equipment [5] the experimental set was designed (see figure 1). Gas detonation generated within a cylindrical barrel (1) with a diameter d_b of 1.6 cm in a stoichiometric acetylene-oxygen mixture by a spark (2). The mixture was supplied via controlled solenoid valves by a gas feed system (3). A portion of the mixture η within the barrel was varied from 0.6 to 0.95. Nitrogen flow was used for the blowdown among shots. The injection of fine ($d_p \leq 100$ μm) HOA particles was realized into the direction normal to the barrel axe by a pneumatic feeder (4). A volume flow rate of HOA was about of 0.3…1.0 mm^3 per a shot. Piezometric sensors were used for control of the pressure dynamics $p(t)$ and estimation of average velocity $\overline{D}$ at different distances from the spark. The one (5) was built in the barrel at the injection zone at the distance of a=56 cm from the spark and the other (6) was embedded into a substrate (7). A distance δ between the muzzle and the substrate was varied from 2.5 up to 23.0 cm. The length of the particle acceleration zone was of l=40 cm. The recording of data from the sensors was realized by a digital oscilloscope (8). A system of laser Schlieren photography was used for the visualization of the detonation flow (9). This system consisted from a Nd:YAG laser (10) and a high-speed camera (11). The red (λ=670 nm) and the green (λ=532 nm) lasers were used. The polarizing filter (12) regulated of an intensity of the laser radiation. The system consisting of an interference filter (13) with λ_{max}=532 nm and $\Delta\lambda$=11 nm for the green laser and with λ_{max}=670 nm and $\Delta\lambda$=11 nm for the red laser and a diaphragm (14) was used for cutoff of the flow (9) radiation and visualization of flow disturbances. The laser beam was collimated by lenses (15) and (16). Synchronizing unit BNC 575 (17) delayed a picture capturing by the camera relativity to the spark (2).

The study of the detonation dynamics was done both for the non-dusty flows and heterophasic flows with HOA particles with an average size of d_p=70 nm и 100 μm.

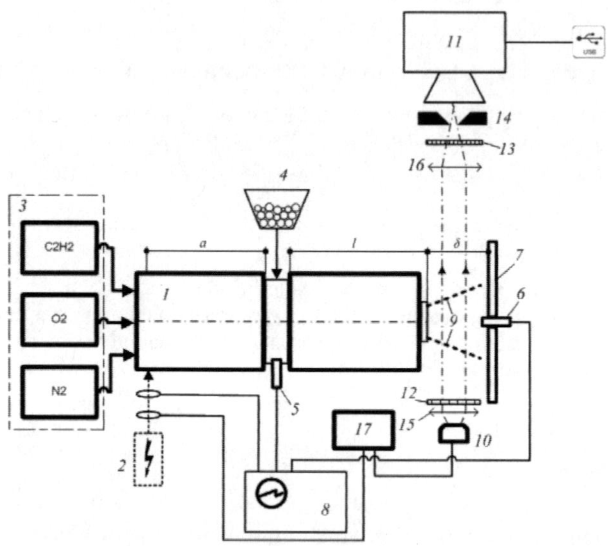

Figure 1. Experimental set: 1 – barrel; 2 – spark; 3 - gas feed system; 4 – pneumatic HOA feeder; 5 and 6 – piezometric sensors; 7 – substrate; 8 – oscilloscope; 9 – detonation flow; 10 – Nd:YAG laser; 11 – high-speed camera; 12 – polarizing filter; 13 – interference filter; 14 – diaphragm; 15 and 16 – lenses; 17 – BNC 575 synchronizing unit.

The prepared films and initial CNI were characterized by SEM (Zeiss Ultra plus 55), EDX (INCA Energy 350 XT), XRD (DRON 3M) and profilometry (FRT MicroSpy). The film adhesion was measured by ISO 13779–4 technique.

3. Results and discussion

The main features of the dynamics of non-dusty and heterophasic flows were defined. In figure 2 the dependences of pressure $\tilde{p}(\tilde{t})$ are presented. Here $\tilde{p} = p / p_0$ - dimensionless pressure and $\tilde{t} = t / t_0$ is dimensionless time, where $p_0 = 3.1$ MPa and $t_0 = 440$ μs.

The data evidenced about the presence of a fully developed detonation which was visualized as a strong shock wave (1) with amplitude of $\tilde{p}_{max} \approx 1...1.25$ and a further rarefaction (2). The main energy releasing occurred in a deflagration zone (3). The following pressure decreasing (4) was bound with the product cooling. The wide pressure peaks (5) were caused by a deflagration afterburning of the mixture. Further the rarefaction (6) caused by leaving of products from the barrel was recorded. Characteristic velocities $\overline{D}$ between the spark and the injection zone were varied from 0.7 up to 1.5 km/s. Such dispersion is explained by an irregularity of the valve operation, inhomogeneity of the mixture etc.

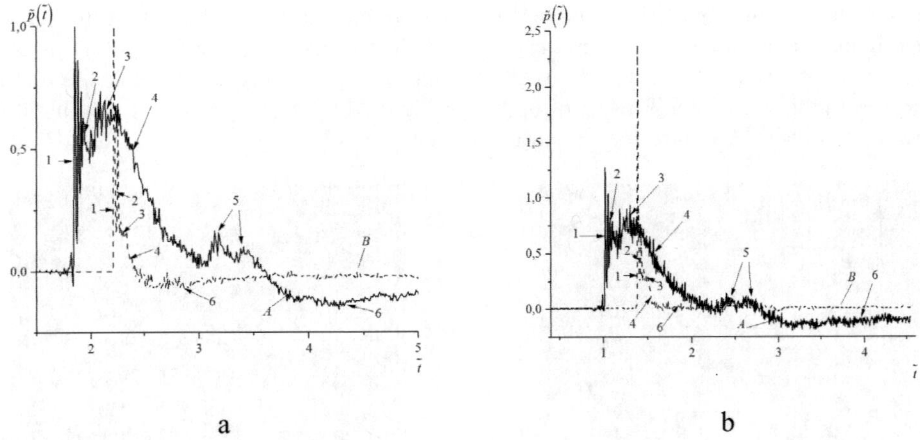

Figure 2. Typical $\tilde{p}(\tilde{t})$ data for non-dusty (a) and heterophasic flows (b) at the injection zone (A) and on the substrate (B) at η=0.95 and δ=2.5 cm

The impulse amplitude recorded by the sensor (6) and the flow structure were depended on δ (see figure 3). At δ≈2.5 cm the substrate was undergone by an action of strong shock wave (1) with amplitude of $\tilde{p}_{\max} \approx 1...3.2$. Maximal characteristic velocities (at η=0.95 and δ=2.5 cm) for the non-dusty and heterophasic flows outside the barrel were $\overline{D} \approx 2.30-2.76$ km/s. Further the shock amplitude was decreased due to δ increasing. Note that HOA injection with the relative weight content g of 1.25-2.50 % did not lead to a significant changing of a motion law for the flow. For this law it was established $x(t) \propto t^{0.15}$. On the other hand HOA injection caused a increasing of the transmitted on to the substrate impulse in ≈1.5…3.5 times than of non-dusty flows. This fact was linked with action of the dispersed phase.

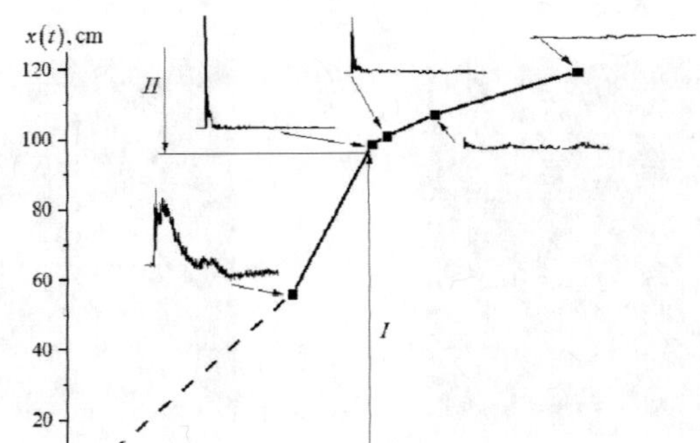

Figure 3. Typical *x-t* diagram of a detonation wave within (zone I) and outside (zone II) the barrel and characteristic $\tilde{p}(\tilde{t})$ patterns at η=0.95 and δ=2.5 cm

In figure 4 Schlieren photos of a motion of the detonation flow are presented at η=0.95 and varied delay time t_d. It was shown that the moving right to left the shock front was undergone diffraction at the muzzle (t_d≈650 μs). The shock (1) and combustion (3) fronts moved jointly at reaching of the substrate (2). At the periphery of the muzzle (4) the vortexes (5) and rarefaction waves (6) were registered. Further (t_d≈680 μs) the detonation deceleration occurred. The formation of high density shadowed region (7) and wave interference (8) between the substrate and the barrel were detected. At the larger times (t_d≈720 μs) the vortex growth was detected. This phenomenon resulted to the further «spreading» of the flow and stabilization of high pressure field over the substrate during of 50–70 μs. This fact was recorded by the piezometric sensor too (see figure 2).

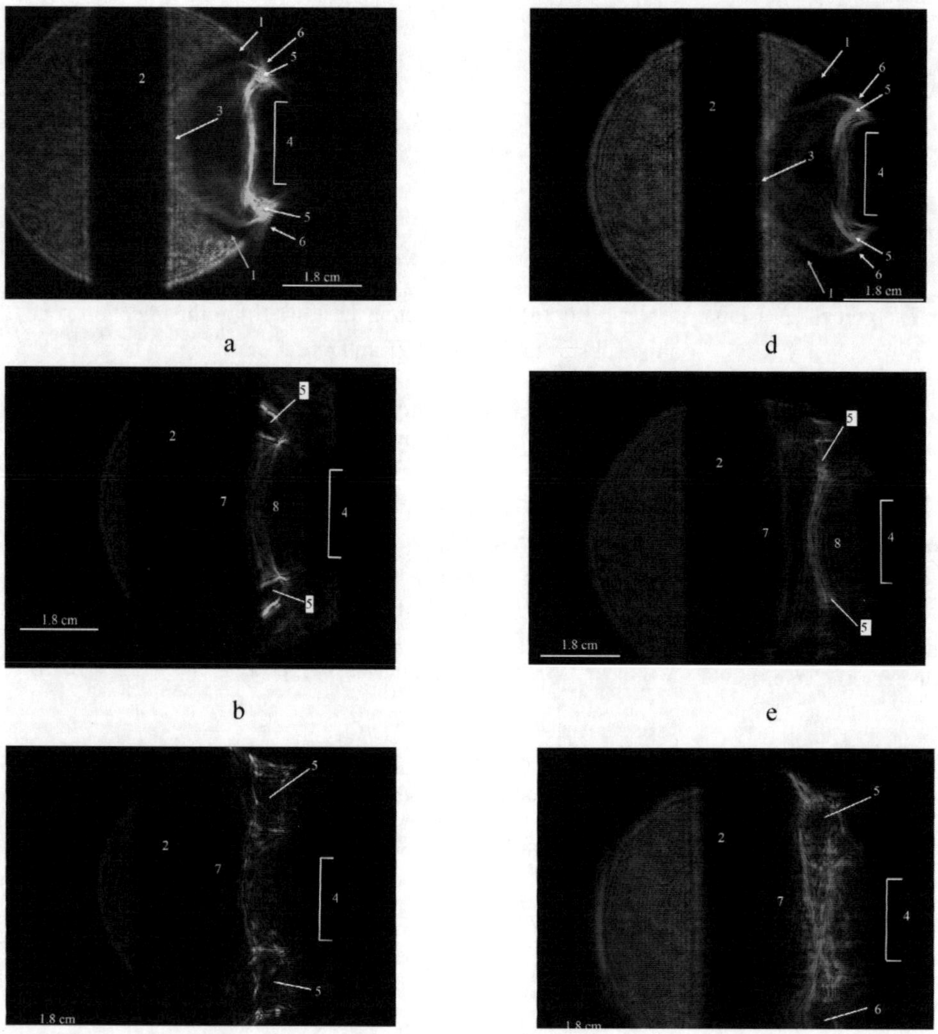

Figure 4. Schlieren photos of a motion of the detonation flows visualized by the green λ=532 nm (a–c) and the red λ=670 nm lasers (d–f)

Study of the HOA films on CNI allowed to define the main features. The films (see figure 5) had a non porous structure without visible defects. It was showed that the coatings are polycrystalline with Ca and P concentration ratio of ≈1.67 that corresponds to raw HOA.

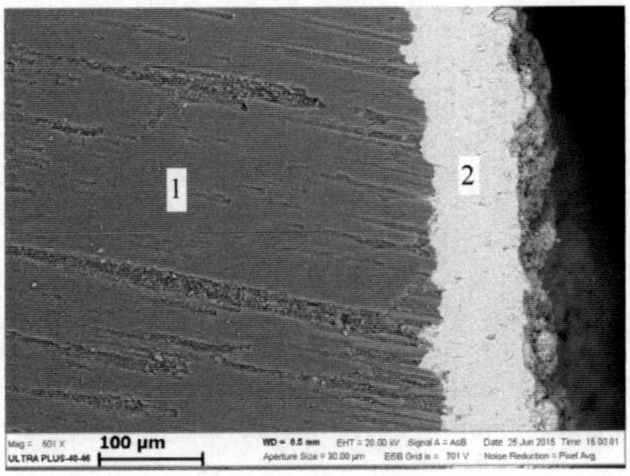

Figure 5. SEM image of the sample: 1 – carbon substrate; 2 – HOA film

The adhesion of the films depended non- monotonically on η and $\overset{\frown}{p}$ respectively (see figure 6). Here $\widetilde{A} = A / A_{max}$ is dimensionless adhesion and A_{max} is maximum measured adhesion at l≈5 cm and d_p≈70 nm. Initially the increasing of $\overset{\frown}{p}$ led to $\overset{\frown}{A}$ growth due to effective HOA acceleration but further increasing of $\overset{\frown}{p}$ (higher than of 0.85) caused the CNI micro crashing and $\overset{\frown}{A}$ decreasing.

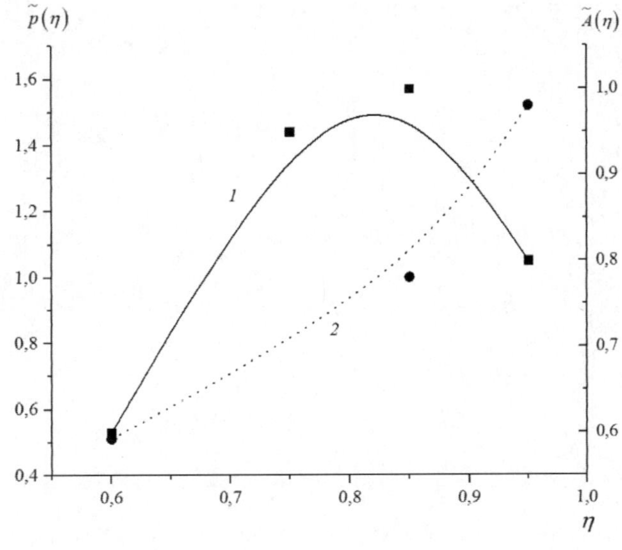

Figure 6. Dependences $\tilde{A}(\eta)$ (1) and $\tilde{p}(\eta)$ (2) at δ=5 cm and d_p≈70 nm

4. Conclusion

Gas detonation flows are suitable for spraying of HOA onto CNI. The prepared films (with a thickness of 80…100 µm) have the close packed structure as raw HOA. It was demonstrated that the flow dynamics has an effect on the adhesion. The optimal conditions of the film spraying were defined.

IOP Conf. Series: Journal of Physics: Conf. Series **815** (2017) 012031 doi:10.1088/1742-6596/815/1/012031

Acknowledgments
This work was performed at the facility "Beam-M" (RFMEFI59014X0001), supported by the State Task of the Russian Ministry of Education and Science. Authors thank Sc. Doctors V.Yu. Ulianitskiy, N.S. Gavryushenko and N.A. Medik for the discussion and «NTM+» Ltd for the equipment.

References

[1] Benson J 1971 *J. Biomed. Material Res.* **5** pp 41–48
[2] Nosenko V *et al* 2015 *Nanoscale Res. Lett.* **10** pp 464–469
[3] Gordeev S K *et al* 2002 Porouse composite for bone replacement and its preparation method *Russia patent* 2181600
[4] Gu Yu V *et al* 2003 *Biomaterials* **24** pp 1603–1614
[5] Nikolaev A Yu *et al* 2003 *Combust. Explo. Shock+.* **39**(4) pp 382–410

IOP Conf. Series: Journal of Physics: Conf. Series **815** (2017) 012032 doi:10.1088/1742-6596/815/1/012032

Plasma airflow jets diagnosis by means of time-resolved tomography

Y Babou[1], F Cannat[2], D Lequang[3] and Ch Rond[4]

[1] Departamento de Bioingeniería e Ingeniería Aeroespacial, Universidad Carlos III de Madrid, Leganés, Spain

[2] Mars Space Ltd, Basepoint Business Center, Southampton, United Kingdom.

[3] von Karman Institute for Fluid Dynamics, Rhode-St-Genèse, Belgium

[4] Laboratoire des Sciences des Procédés et des Matériaux, Université Paris XIII, France

E-mail: ybabou@ing.uc3m.es

Abstract. A simple optical arrangement combining a single high-speed CCD camera and a set of mirrors has been setup to record simultaneously light intensity of atmospheric air plasma jet issued from multiple directions. The adopted approach is presented and its potential benefit to examine plasma emission in 3D is outlined. Rebuild local emission distributions are briefly analyzed for two typical jet exhibiting different flow structures.

1. Introduction

Thermal plasma jets produced at atmospheric pressure are extensively used in a wide range of industrial applications necessitating high energy density medium or elevated thermal fluxes. Additionally to the inherent interactions between the hot jet submerged into quiescent surrounding gas [1],[2] (mixing, entrainment, transition), the technology adopted to warm up the working gaseous mixture can be a stringent source of undesirable plasma flow jet spatial and temporal disturbances potentially detrimental to thermal processing quality [3],[4]. The examination of the plasma jet dynamic can be easily performed by means of high speed imaging measurement techniques. Various processing can be applied to plasma intensity, recorded with a high speed CCD camera for example, to uncover underlying jet dynamic characteristics [5],[6],[7],[8]. Nevertheless, recorded intensity corresponds to the optical emission integrated along the optical path (in first approximation parallel to the camera viewing direction) therefore fine interpretation of plasma jet dynamics taking place locally is limited, and obviously local emission can no more be rebuilt with the classical Abel transform, applicable only to symmetric emission profile.

Tomography technique can provide a deeper understanding of the plasma jet temporal behavior, when analysis of line of sight measurements is no more sufficient, since tomography-based techniques are especially dedicated to rebuilt the local emission from intensity recorded at various directions. If tomography technique has already been used in the past to document thermal plasma jet local emission distribution [9], the present contribution addresses the development of tomography based technique using a single CCD camera. The new diagnostic method has been applied to the investigation of air plasma jets produced with a Microwave Plasma Torch (MPT). Principles of tomography and the numerical method adopted to rebuild local emission from experimental intensity profiles are described in Section 2. The experimental arrangement presented in Section 3 combines a single CCD camera with mirrors to produce simultaneous intensity records as if they were obtained with a setup of 5 CCD cameras. The rebuilt 3D emission distributions are shortly analyzed in Section 4 for two typical cases of plasma jets.

2. Local emission reconstruction method

Considering a two dimensional plasma medium of finite extent in the plane (x, y) and centered on the origin O as depicted in figure 1, the tomography technique consists in the reconstruction of the local emission distribution, referred as $J(x, y)$, starting from its orthogonal projection on plane perpendicular to direction passing by O and making an angle θ with x-axis. The line-of-sight intensity profile $I(x', \theta)$ corresponds to the plasma radiation integrated along optical path parallel to the viewing direction specified by θ. Knowing $J(x, y)$, projections at any angle θ are obtained by mean of Radon transform and reciprocally, the local emission distribution is reconstructed applying inverse Radon transform to side-on intensity profiles, known in practice for a limited number K of viewing directions.

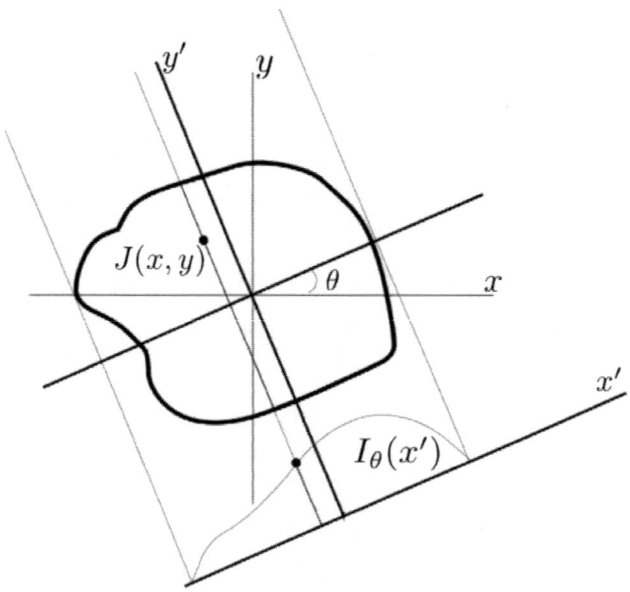

Figure 1. Tomography principle.

In the present study, the local emission distribution $J(x, y)$ is reconstructed adopting the approach described in [9], based on the iradon function under MATLAB environment to calculate the inverse Radon transform starting from K experimental intensity profiles recorded at equidistant angle. Assuming the plasma is optically thin, the intensity profiles recorded at two opposite directions are mutually flipped, reducing then the useful domain of angular scanning from $[0 - 2\pi[$ to $[0 - \pi[$. The approach consists in the generation of additional intensity profiles by means of trigonometric interpolation to extend the initial set of K measured profiles to enhance reconstruction procedure quality. Considering that profiles at any projection angle θ can be approximated by polynomials as $I(x, \theta) = \sum_n^N c_n(\theta)x^n$, polynomial coefficient are determined for $\theta_k = \frac{\pi(k-1)}{K}$, $k = 1, 2, ..., K$, by means of common least square fitting applied to recorded data. Also considering that $c_n(\theta)$ coefficients are smooth 2π-periodic functions, coefficients at intermediate angle $\theta_k + \frac{\theta_{k+1}-\theta_k}{2}$ can be expressed analytically for fast evaluation [9]. Such procedure leads to double the number of available intensity profiles, henceforth well determined at $\theta_k = \frac{\pi(k-1)}{2K}$, $k = 1, 2, ..., 2K$. Repeating the procedure P times with an initial set of K experimental profiles results in a set of intensity profiles determined for $\theta_k = \frac{\pi(k-1)}{2^P K}$, $k = 1, 2, ..., 2^P K$.

The effective enhancement when interpolated profiles are taken into account into the rebuilding procedure has been assessed in [9] through qualitative comparisons between a reference distribution J^{ref} and the distribution J^{sim} reconstructed with iradon applied to intensity profiles computed for $K = 4$ initial projection angles. The improvement with increasing numbers of input profiles was quantitatively estimated evaluating the Root Mean Square Error (RMSE) between a reference emission distribution J^{ref} and the distribution $J^{sim}(K, P)$ rebuilt using an initial set of K intensity profiles and doubling P times the initial set. Two synthetic emission distributions have been used to test the performance of the tomographic reconstruction. We have generated these planar distributions to be representative of the situations observed afterwards: a replete distribution and a hollow distribution that discloses intensity drop in the central region with irregular frontiers. The effect of the interpolated intensity profiles on the accuracy and quality of the rebuilt emission is clearly depicted in figure 2. From each reference distribution, $K = 5$ initial profiles have been obtained by means of the direct radon transform (from practical point of view, as if K detectors were disposed to record the intensity along $\{\theta_k\}$ directions, indicated with inserted white arrows). Then this initial set is processed with the inverse radon transform to rebuild the emission distribution. For both type of situation, the emission distribution rebuilt with $K = 5$ initial profiles ($P = 0$) exhibit a poor quality and are characterized by a significant numerical noise and artifacts. By extending the initial set of profiles, presently by doubling two times ($P = 2$) the quality is substantially enhanced the noise is significantly reduced at the expenses of sharp contours smoothing, accordingly to the trigonometric interpolation method.

Unlike classical tomography, the single CCD tomography technique presented in this paper involves a relationship between the spatial resolution and the number of intensity profiles taken into account. We evaluate the effect of the spatial resolution decay with increasing K, as constrained with arrangement combining a single linear detector array with $K - 1$ mirrors (analogous to the setup described in Section 3 for $K = 5$). With such arrangement, when K profiles are recorded with a detector of N_{pix} pixel array size, N_{pix}/K pixels are available to record a distinct intensity profile. The effect on the rebuilding procedure with increasing K is evaluated, for initial array size set to $N_{pix} = 800$ pixels, calculating the RMSE between a reference emission distribution J^{ref} and the distribution $J^{sim}(K)$ rebuilt with an initial set of K intensity profiles: $RMSE(K) = \frac{1}{N} \times \sqrt{\sum_{ij} (J^{ref} - J^{sim}(K))^2}$, with J^{ref} discretized according to the intensity profile array size reduction with increasing K as $N_{pix} = \frac{800}{K}$, i and j referring to pixel indexes. The effect of the spatial resolution diminution on the rebuilt data quality with increasing K can be assessed in figure 3 for the two reference shapes, where distributions rebuilt for K=3, 5 and 9 (and $P = 4$), are evinced with the reference distribution at finer spatial discretization. For $K = 3$, the insufficient number of sampling direction is responsible for shape symmetrization trends. For K=1 the rebuilt shape is perfectly symmetric. For $K = 9$, the shape is slightly altered, in comparison to the case $K = 5$, because of the too low spatial resolution. Especially, in the case of the hollow shape, the small jagged patterns at the center, still distinguishable for $K = 5$, are noticeably degraded and blurred for $K = 9$. RMSE values plotted in figure 4 show that increasing K enhances the rebuilding procedure up to optimal K (slightly varying with the reference distribution shape) after which rebuilt data quality is deteriorated following

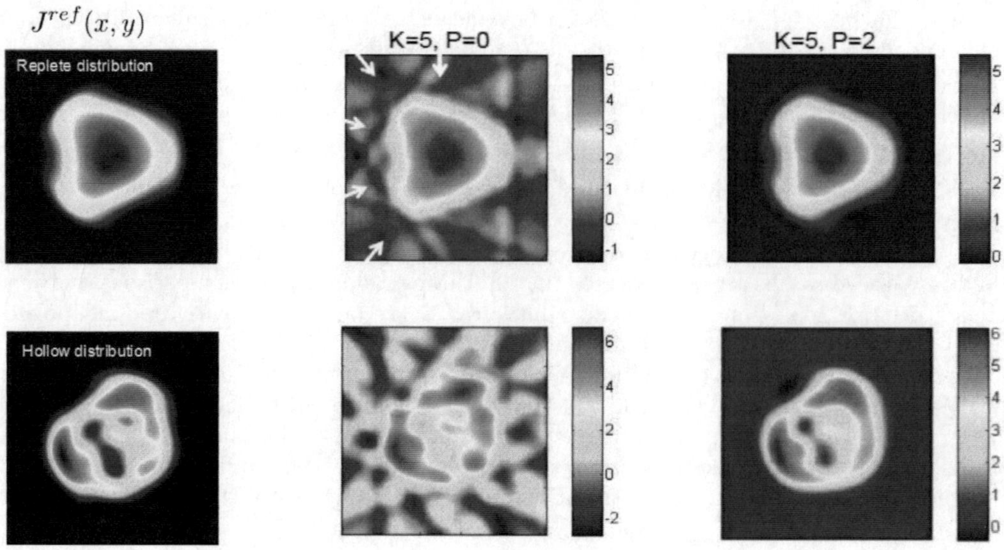

Figure 2. Effect of initial set doubling.

on insufficient spatial resolution. These considerations substantiate the number of projection angles considered in the implemented arrangement.

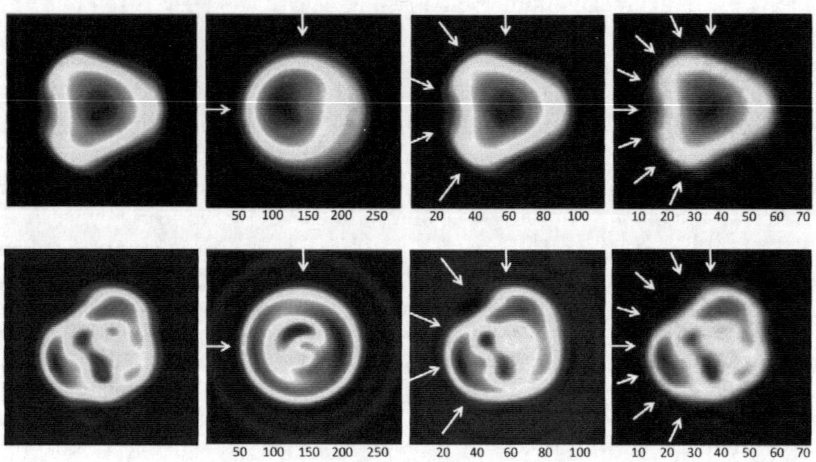

Figure 3. (From left to right) Reference distribution J^{ref}, and distributions $J^{sim}(K)$ rebuilt with increasing K, $K = 3$, $K = 5$ and $K = 9$, resulting in spatial resolution lowering. The upper, resp. lower, image series refer to the replete, resp. hollow, distribution case.

3. Equivalent N-camera arrangement

Various detection systems can be envisaged to sample simultaneously side-on intensity profiles emitted in multiple directions by a horizontal slice of plasma jet flowing along the direction z

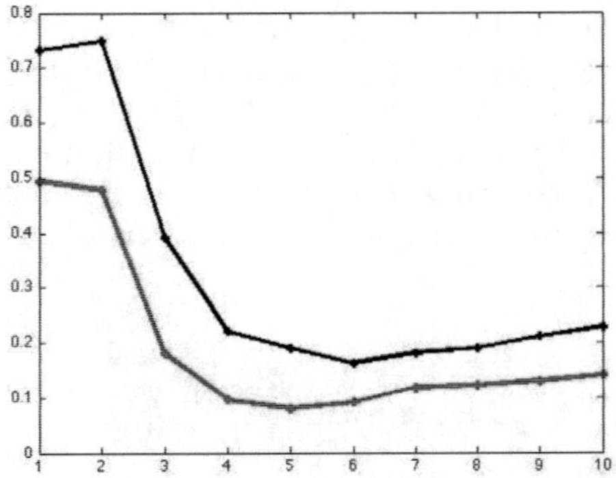

Figure 4. Evolution of the RMSE with increasing K. The black, resp. red, curve refers to the hollow, resp. replete, distribution case.

perpendicular to (x, y) plane. CCD high speed cameras are particularly well suited to provide, with simple arrangement, the required input data to rebuild, with the method exposed in Section 2, the temporal behavior of the local emission distribution $J(x, y, z)$ in the whole plasma jet volume. If K cameras are available, then the required side-on intensity spatial distribution $I(x', \theta_k, z)$, $k = 1, .., K$, can be straightforwardly acquired displaying the K cameras to record plasma intensity at K equidistant angles. Unfortunately, CCD high speed cameras remain still too expensive nowadays to easily access to more than one. To overcome the situation, an arrangement combing a single CCD camera with $K-1$ mirrors equivalent to K cameras setup is proposed. The arrangement is based on the simple idea that the direct front image of the plasma jet and additional images seen from the rear can be recorded simultaneously with a single CCD frame with the help of mirrors.

In the experiments presently reported, images are recorded with a Phantom v7.0 8 bit CCD camera, offering a large scale array of 800 $\times$ 600 pixels conductive to obtain five distinct images of the vertical jet with reasonable spatial resolution. The experimental arrangement has been designed to investigate the air plasma jet produced with a Microwave Plasma Torch consisting basically of a wave launcher operating commonly at 2.45 GHz coupled to cylindrical quartz tube of 30 mm inner diameter, the air flow is injected by means of a swirl injector from the bottom tube. In the present experiments the air plasma jet is produced in room ambient air at atmospheric pressure and the plasma temperature is of 5000 K. MPT properties have been detailed in previous contribution [8] and will be not repeated here. The CCD camera has been combined with four ordinary mirrors (20 cm length and 8 cm width) to record simultaneously side-on intensity spatial distributions $I(x', \theta, z)$ issued from five co-planar directions perpendicular to z-axis. The arrangement design is sketched in figure 5. The camera was positioned with the CCD chip vertical (parallel to z-axis), and perpendicular to and centered on the y-axis defined as the horizontal straight line crossing z-axis at the origin O located at 4.3 cm above the outlet. The mirrors have been displayed to collect the intensity issued from viewing direction of angles set to $\theta_k = \pi(k-1)/K + \frac{\pi}{10}$, $k = 1, 2, , K$, with K=5. With the adopted disposition, the CCD central image corresponds to the plasma jet direct front view at projection angle $\theta_8 = -\pi$ with $I(x', \theta_3) = I(-x', \theta_8)$. The four mirrors have been oriented to conserve, after reflection, the optical path parallel to y-axis. To do so, the angles between a given sampling direction θ_k and the normal to the mirror is set to $\frac{1}{2}(\frac{\pi}{2} - \theta_k)$. External mirrors (associated to sampling direction θ_1 and θ_5), resp. internal mirrors (associated to sampling direction θ_2 and θ_4), have been located at respective distance $r_{ext} = 14.7$ cm and $r_{int} = 11.9$ cm from z-axis to reduce width and depth covered by the arrangement in x and z directions (to minimize off-y-axis perspective distortions), letting however margin between adjacent images displayed on the frame. A common 35 mm lens NIKKOR-S was used to image the arrangement on the whole CCD frame, setting distance between camera lens and z-axis to 74 cm. The focus was adjusted using a simple ruler displayed vertically with printed numbers coinciding with z-axis. The focus was finely tuned *de visu* by balancing the blurring effect, albeit only slightly perceptible, to obtain for all viewing direction sharpest numbers. A typical raw CCD frame acquired for the reference case is shown in figure 6. The central image corresponds to the plasma direct view, the two images adjacent on the right and left to central image correspond to the rear plasma jet intensity reflected by internal mirrors and the two images on the edges of the frame correspond to rear intensity reflected by external mirrors.

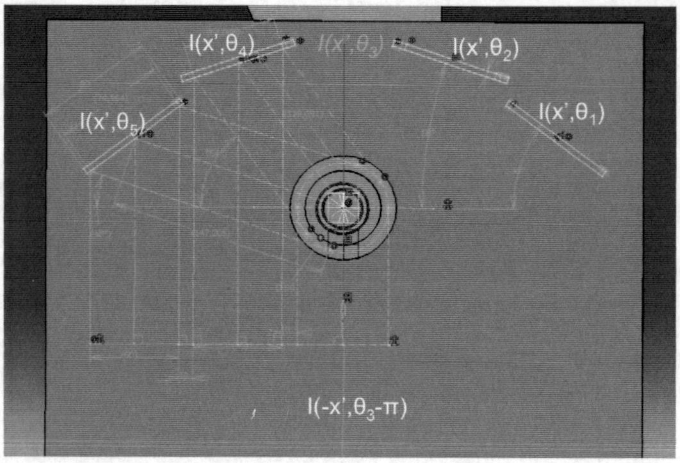

Figure 5. Top view drawing of the optical arrangement with four mirrors.

Since images are issued from apparent plasma jet at different location (especially at different depth), the intensity distributions imaged onto the CCD can be significantly distorted because of the actual perspective projections. Checkerboard images analysis with Camera Calibration Matlab tool box evinced that the distortion of images in the region of interest for the processing is negligible. The main effect of the perspective projection is the shortening of the images clearly visible in figure 6. Indeed, the images obtained for projection angles θ_k, with k=1, 2, 4 and 5, corresponding in fact to reflected images, are actually issued from intensity going along optical path lengthened of $d_{ext/int}$ compared to the direct image with $d_{ext/int} = r_{ext/int}(1+sin(\theta_{ext/int}))$. It comes out that the apparent relative depth between the plasma jet viewed directly and viewed through mirrors is of $d_{ext} = 19.2$ cm and $d_{int} = 21.5$ cm. Therefore the plasma jet images obtained with mirrors exhibit a foreshortening in comparison to the direct front image following on the apparent size reduction with increasing depth inherent to perspective projection. We checked with the checkerboard that the rear plasma images collected by the CCD chip result as a good approximation from intensity following optical path parallel to direction specified by θ_k before reflection and parallel to y-axis after reflection. Therefore, we reasonably assumed in the followings that distortions undergone by intensity distributions are weak enough to do not deteriorate reconstructed data reliability. However, such an approximation could be refined with a proper image geometrical calibration or using telecentric lens.

The raw frame records have been processed to elaborate five distinct temporal frame series corresponding to records as if they were obtained with a five cameras arrangement. Neglecting off-axis effects, the five plasma jet image series have been corrected as if they were obtained by mean of an equivalent five camera arrangement with cameras placed at different distances from z-axis (74 cm, 74 + d_{ext} cm and 74 + d_{int} cm). Hence, corrections have consisted in scaling intensity distributions obtained for different depths. The scaling procedure was performed using

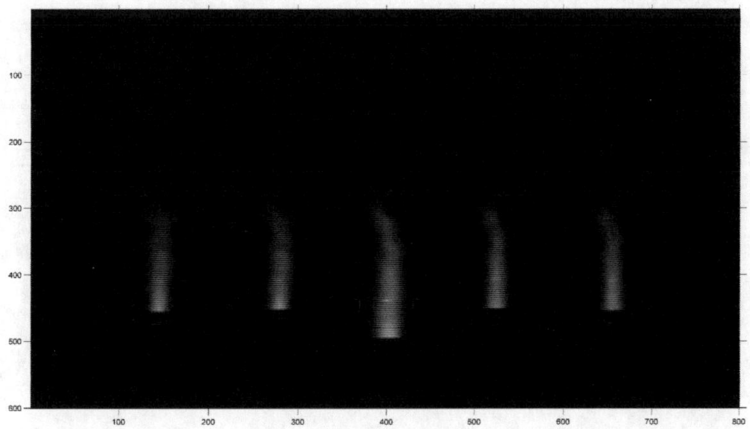

Figure 6. Single CCD frame recorded with the airflow plasma jet produced at 18 l/min.

the plasma jet itself as reference source, considering that for long acquisition time the plasma jet emission distribution is axisymmetric following on plasma jet oscillations statistical averaging. A plasma jet generated with the MPT at 18 l/min of air, to guaranty a fairly stable jet, has been used as reference. A single frame record is given in figure 6. Mean intensity distributions, obtained averaging 1000 raw frames recorded at 140 Hz exhibit a very good symmetry along the whole jet axial extent as seen in figure 7. For each projection angle, the spatial resolution is determined on the basis of the tube diameter to specify the array size in pixel of a window defining a region of 6 cm width and 8.2 cm height. Each window is centered on the jet axis, determined as the maximum intensity line (coinciding with the vertical z-axis).

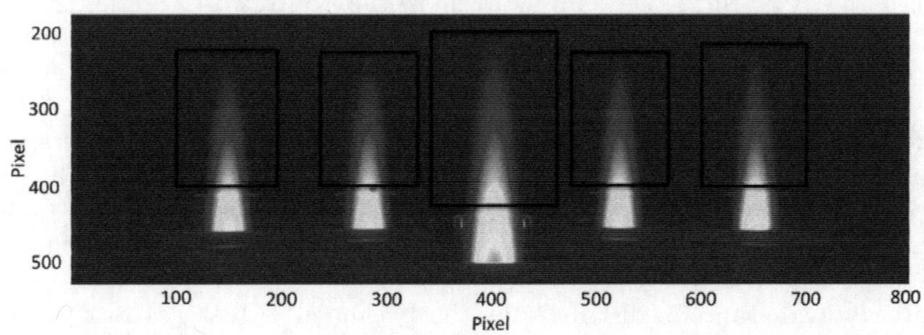

Figure 7. Sub-frames extraction.

Then, these sub-frames are extracted and discretized on a common grid data point size to have set of five images compliant in spatial resolution and data array size. The intensity profiles are scaled to a reference value to obtain images calibrated in relative intensity. Extracted and scaled instantaneous frames are displayed in figure 8 for two plasma jet cases discussed in Section 4. Such data process applied to time series of recorded images allow to elaborate five distinct movies equivalent to movies as if they were obtained with five distinct CCD cameras spread to record intensity distributions at viewing directions of angles $\theta_k = \frac{\pi(k-1)}{K}$, $k = 1, 2, ..., K$, with $K = 5$.

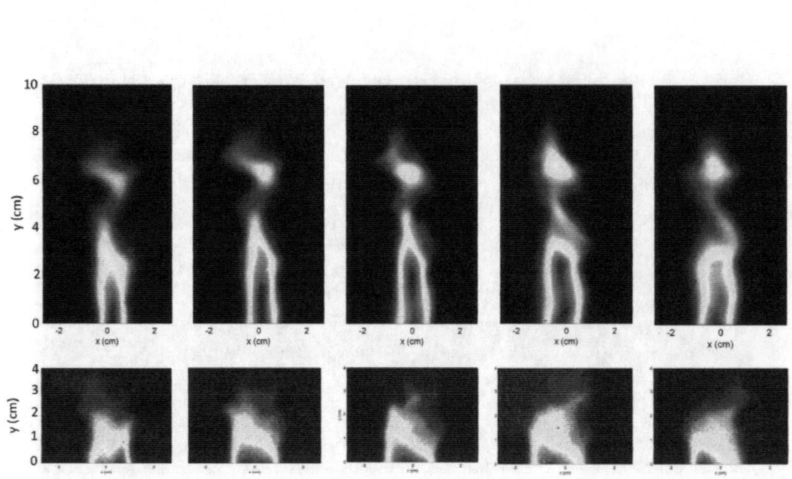

Figure 8. Five distinct frame series elaborated as input to tomography rebuilding procedure. Frames for $K = 1, ..., 5$ from right to left. First raw common plasma jet, second raw cyclonic plasma jet.

4. Results and discussion

Tomographic measurements have been performed with the MPT operating with air at two flowrates. In the following, a short analysis of the rebuilt emission distribution is proposed for two cases to outline the potential gain of the diagnostic. The first case (case:A) corresponds to a usual plasma jet case, it is produced at 46 l/min flowrate. The second case (case:B) corresponds to the plasma produced at 65 l/min exhibiting a cyclonic flow structure characterized by a pressure drop at jet center. For the two cases operating power of 2 kW was used and frames have been recorded with acquisition rate of 5 kHz with exposure time of 190 μs. The 3D distributions obtained using the approach described in Section 3 with the method outlined in Section 2, are examined hereafter for the so-called A and B cases. In the followings, resulting emission distributions are represented in two complementary manners to facilitate the appreciation of peculiarities taking place in the three space dimensions. Iso-emission envelops (*i.e.* $J(x, y, z) = constant$) are plotted for high, intermediate and low emission values, highlighting differences between hottest and coldest plasma regions. Horizontal cross sectional views at succeeding heights are shown together with iso-emission surface plots, to evince distribution characteristics in the plane (x, y). The intersections between the section plane and the envelops is represented by black lines displayed on the sectional views.

Looking at the instantaneous distributions the plasma flow features exhibits outstanding peculiarities. For case:A, evinced in figure 9, the situation is closely consistent with common cylindrical hot jet submerged into a cold quiescent gas [1],[2] characterized by mixing interactions established at the jet frontiers (turbulent region) disclosing a dominant potential core region (laminar region) very analogous to combustion and isothermal jet. However, the cross sectional view shows that the region that can be related to the core potential at $z = 0$ and 2 cm is closer to an elliptic distribution indicating a torsion-like along z-axis. Initial outer layer disturbances observed at the plasma jet base show that the cold air entrainment process has started, probably triggered earlier than with non swirling flows because the swirl enhances the entrainment of surrounding air. Moving away from the outlet, the initial disturbances seem to turn into large eddies stepping toward the jet centerline. The gradual increases of the cold air engulfment rate results in the strangulation-like phenomena at about $z = 4$ cm where eddies of entrained surrounding cold air have finally reached the centerline, resulting afterward in a detachment process of the jet termination. For case:B, shown in figure 10, the cyclonic

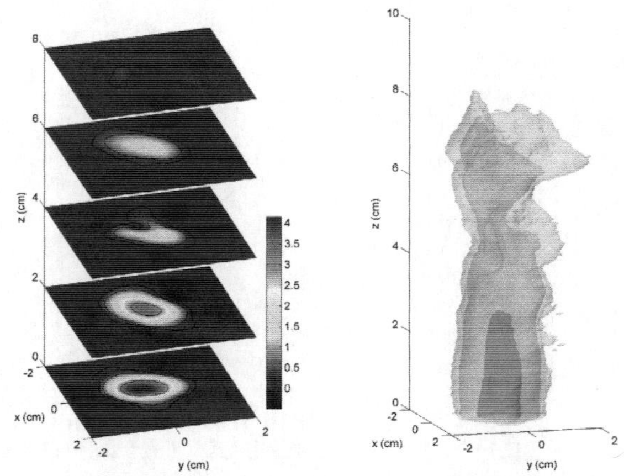

Figure 9. Instantaneous emission distribution for the common jet case (case:A). At left: cross sections of emission distribution in the plane (x, y). At right: iso-emission layers.

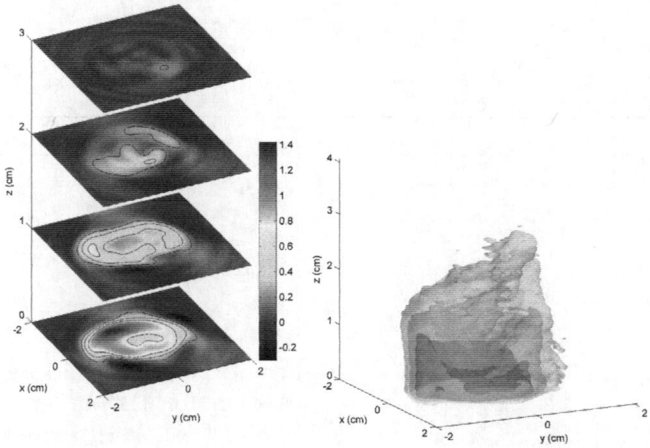

Figure 10. Instantaneous emission distribution for the cyclonic jet case (case:B). At left: cross sections of emission distribution in the plane (x, y). At right: iso-emission layers.

structure takes place following on the pressure reverse breakdown occurring as a rule for high swirl numbers. The plasma flow exhibits an outstanding structure recalling in a certain extent a hot tongue that sweeps on the tube inner wall. The mean emission, obtained averaging a series of 500 instantaneous emission values, is reported to highlight the differences between the plasma emission seen at low acquisition rate (as seen by a human eye or an ordinary camera) and the actual dynamic features accessible at 5000 KHz acquisition rate. The resulting mean emission distributions are displayed in figure 11 for case:A and in figure 12 for case B. For case A, the thermal jet emission appears to be structured in quasi-conical concentric layers uncovering a mean temperature decreasing monotonically away from the center. For case:B, the pressure collapse is well distinguishable with both representations of the mean emission. The emission distribution displays a symmetric annular shape in contrary to the instantaneous emission distribution, indicating that the dynamic of the hottest region is totally averaged for long recording time. For both cases, a fair symmetrical distribution was obtained with emission integrated over ~ 0.1 s.

Having access to the 3 D distribution of local emission, new features, that are inaccessible by means of image analysis, can be obtained. We consider presently the cyclonic jet (case:B), which

IOP Conf. Series: Journal of Physics: Conf. Series **815** (2017) 012032 doi:10.1088/1742-6596/815/1/012032

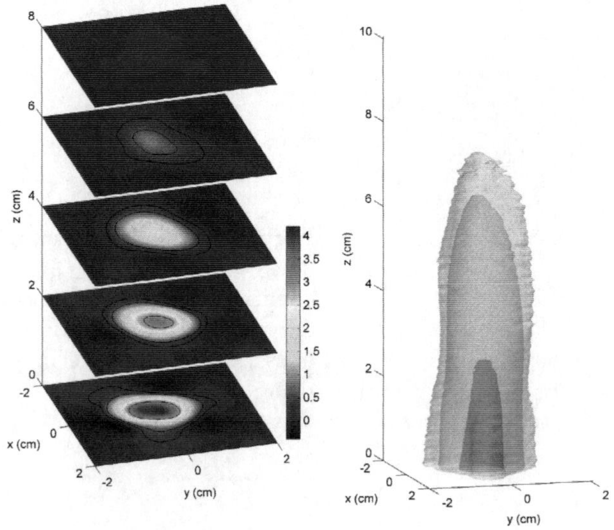

Figure 11. Mean emission distribution for the common jet case (case:A). At left: cross sections of emission distribution in the plane (x, y). At right: iso-emission layers.

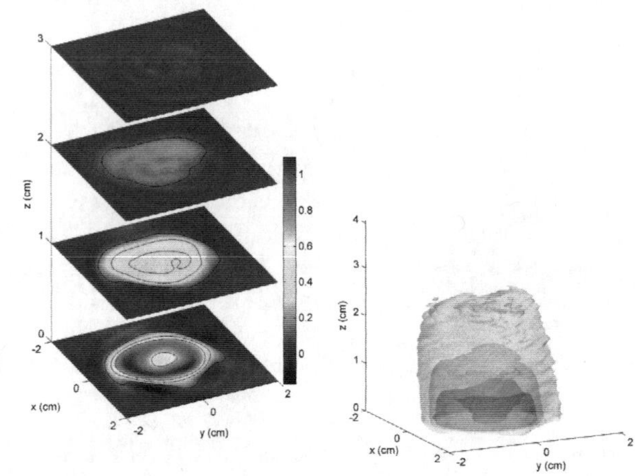

Figure 12. Mean emission distribution for the cyclonic jet case (case:B). At left: cross sections of emission distribution in the plane (x, y). At right: iso-emission layers.

is composed of a hot localized spot of plasma rotating in the plane transverse to the flow axis as depicted in the series of sectional views at the $z = 0$. This series represent well the rotation of the plasma spot shown here with time increment of $2.2{\cdot}10^{-3}$s. The angular position of this plasma spot is obtained as the angle θ defined by position of the maximum of the emission distribution with respect to x-axis. We consider first the temporal intensity collected through three distinct line-of-sight, namely following the direction parallel to y-axis and passing by $x = 0$ (the center), at $x = +1$ cm and $x = -1$ cm. The time evolution of the extracted intensities (as it have been collected by an array of three detectors) is reported in the upper plot of figure 13 together with the time evolution of the angular position. The associated FFT to each of these temporal information has been reported on the lower plot. The intensity acquired at $+/-1$ cm is characterized by a well distinct period of about ~150 Hz which is the period of the rotation of the plasma spot (period of θ). While the intensity recorded at $x = 0$ cm, is instead characterized by a frequency of twice the plasma rotation frequency (peak located at ~300 Hz). This feature is explained by the fact that during one period of rotation, the plasma spot is detected only once at $+/-1$ cm, while at $x = 0$ the plasma spot is seen two times per period : passing at $y \approx +1$ cm and at $y \approx -1$ cm (for each $x = 0$). In other words, considering only image analysis is insufficient to interpret the features displayed by the time intensity measurements since it can lead to misinterpret the frequency analysis. Without additional information (or a priori assumption), the two peaks at $\sim$150 and $\sim$300 Hz tell that the plasma dynamics is driven by two distinct periods. Only the local emission distribution analysis can reveal the underlying nature of the plasma dynamics.

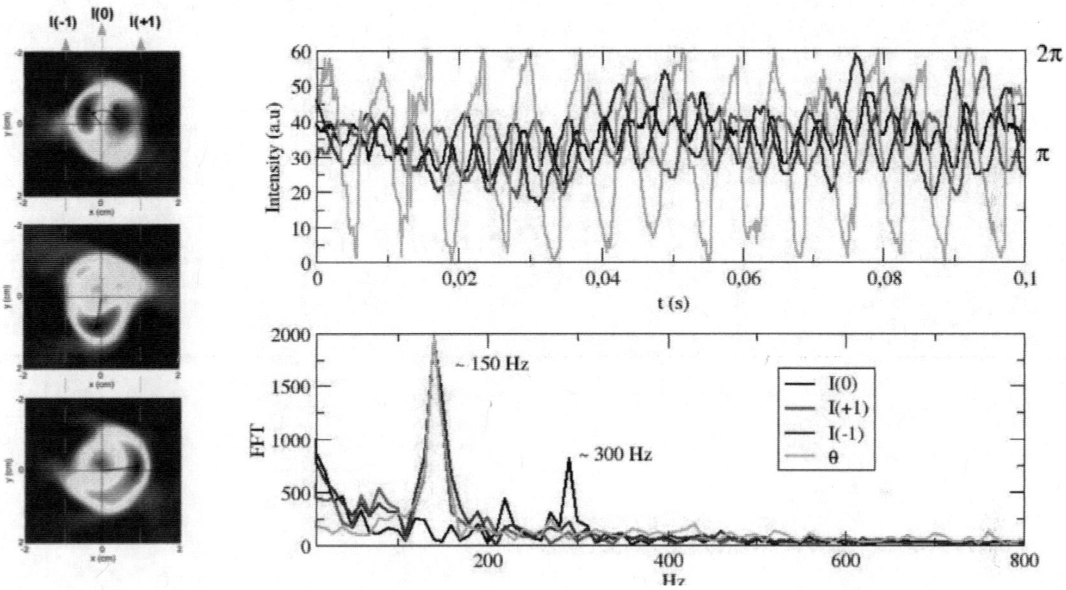

Figure 13. Left column: Sectional views of the plasma spot position at various instants. Right column: Time evolution of intensities and maximum emission angular position (upper plot). Associated FFT (lower plot))

5. Concluding remarks

A tomography approach based on a single CCD camera has been proposed to document the temporal behavior of the plasma jet local emission produced by a Microwave Plasma Torch. Assuming the plasma optical thin, the problem was abridged to the simultaneous sampling of radial intensity distributions issued from equidistant viewing directions in the range $[0-\pi[$. Allowing subsequently to propose a simple arrangement to record with the same frame the intensity issued from various direction with the help of ordinary mirrors. The technique was applied to the tomographic analysis of two plasma jet case exhibiting distinguishable structure to assess the potentiality of the adopted approaches. Starting from the 3 dimensions emission distribution, significant insights can be brought with numerical processing commonly used to examine 2D images (temporal and spatial analysis). Preliminary results showed that the technique is promising and it is worthwhile to evaluate its application to wide range of plasma jet facilities.

IOP Conf. Series: Journal of Physics: Conf. Series **815** (2017) 012032 doi:10.1088/1742-6596/815/1/012032

Acknowledgement
This work has been supported by the UC3M CONEX-Marie Curie fellowship and l'Institut Francais d'Espagne.

References

[1] Abramovich G N and Schindel L 1963 *General properties of turbulent jets* MIT press
[2] Spores R and Pfender E 1996 *Surface and Coatings Technology* **37**(3) 251–270
[3] Pfender E 1999 *Plasma Chem. Plasma P.* **19**(1), 1-31
[4] Fauchais P and Vardelle A 2000 *Plasma Physics and Controlled Fusion* **12** B, 365–383
[5] Nenicka V, Hlína J and Šonský, J 2005 *IEEE Transactions on Plasma Science,* **33**(2), 418-419
[6] Hlína J, Gruber J and Šonský, J 2006 *Measurement Science and Technology* **17** 918–922
[7] Benito I, Aleo M and Babou Y 2008 *VKI Plasmatron facility performances : I. Air plasma jet unsteadiness* Proceedings of the Third International Workshop on Radiation of High Temperature Gases in Atmospheric Entry Heraklion, Greece. European Space Agency, Noordwijk, The Netherlands
[8] Le Quang D, Babou Y and Andre P 2012 *Investigations of a microwave plasma source operating with air,* N_2*,* CO_2 *and argon gases* IOP Conference Series: Materials Science and Engineering, **29** 012009
[9] Hlína J, Chvála F, Šonský J and Gruber J 2008 *Measurement Science and Technology* **19** 015407

Experimental investigation of nitric oxide radiation in the shock-heated air

P V Kozlov

Institute of Mechanics Lomonosov Moscow State University, Moscow, Russia

E-mail: kalevala@mail.ru

Abstract. The experimental results of the absolute measurements of spectral radiation of air were obtained on the double diaphragm shock tube of Laboratory of Kinetic Process in Gases of Institute of Mechanics MSU. The radiative properties of nitric oxide in the spectral range 190-290 nm were investigated experimentally. The time dependence of the emission of molecular bands for NO shock wave velocities of 4-10 km / s were obtained. The article discusses the problems of absolute power measurement in the radiation behind shock wave front

1. Introduction

Estimation of effects of thermal and radiation fluxes on the aerospace vehicle surface in N_2–O_2 and CO_2–N_2 media during the aerocapture is currently of interest in connection with projects of mission to Earth, Mars and Venus. Up to now, some shock tube measurements of radiation intensities were carried out as a test data for validation the chemical kinetic models describing Earth, Mars and Venus atmosphere entry processes in [1]–[6].

Usually the optical designs of spectrographs are optimized to enable spatially resolved imaging over the length of the instrument's entrance slit. Spatial resolution is maintained at the exit plane, by this means providing dispersed spectra at multiple points simultaneously. The spectrally and spatially resolved image at the exit plane is recorded by an ICCD – at a short exposure time of 0,1–1 ms. Known as imaging spectroscopy, the technique has been implemented in the shock tube experiment in NASA Ames Research Center, University of Queensland and Chofu Aerospace Center [7]–[9]. This method has great potential in the study of radiative processes in shock waves. However, it features some difficulties, such as the deviation of intensity linearity from exposure time [7]. Also, if the length of the nonequilibrium radiation is less than the duration of the exposure of the ICCD, the investigation of nonequilibrium kinetics of radiation on the shock wave is virtually impossible. It is necessary to introduce a so-called non-equilibrium metric [10]. This is equivalent to the integration mode of the CCD measuring the shock wave radiation in the cross-section, as in our experiments.

In the present experiments, the duration of nonequilibrium radiation is in the range of hundreds of nanoseconds. Therefore, the use of the measuring channel, which registers the radiation of the shock wave through the PMT, provides additional information on the kinetic processes in the nonequilibrium zone. This type of data cannot be received for imaging spectroscopy in shock tubes experiments.

This paper represents description of a two-diaphragm shock tube and the results of measuring the absolute intensity of radiation of air heated in the shock waves with velocities V_{SW}=4–10 km/s.

2. Experimental Setup

Experimental investigations carried out in the double diaphragm shock tube (DDST) facility located at the Institute of Mechanics of Moscow State University. Schematic diagram of DDST given in figure 1. Inner diameter of all three seamless tube chambers is equal to 50 mm. Shock tube sections are provided with gateways for P1-P10 piezoelectric sensor mounting allowing to perform control of gas pressure and velocity of shock wave in each section. Transducer signal registered by 14–bit oscilloscope with 10MSa/s sample rate.

Burning mixture of $0.5(H_2+½O_2)+0.5He$ was used in the high-pressure chamber (HPC). The high-pressure chamber operated in the detonation-driven regime. The preparation of mixture produced in HPC. Hydrogen added to the preliminarily prepared mixture of helium and oxygen. HPC operational stability in the detonation regime substantially depends on duration of mixing of the burning gases. To ignite the mixture automobile spark plug used in the rear edge of the HPC. The discharge in the utilized spark plug appeared because of self-breakdown. The stored electrical energy on the high-voltage energy discharge capacitors did not exceed 100 mJ, the duration of discharge was approximately 300÷400 ns. Nanosecond discharge times are achieved by construction of the capacitors and spark gap forming a low-inductance circuit. To achieve rapid shaping of the plane front of combustion in the burning mixture the prechamber was arranged between the rear flange holding the spark plug and the section HPC. HPC filled by mixture $0.5(H_2+½O_2)+0.5He$ up to the total pressure of 400÷600 kPa.

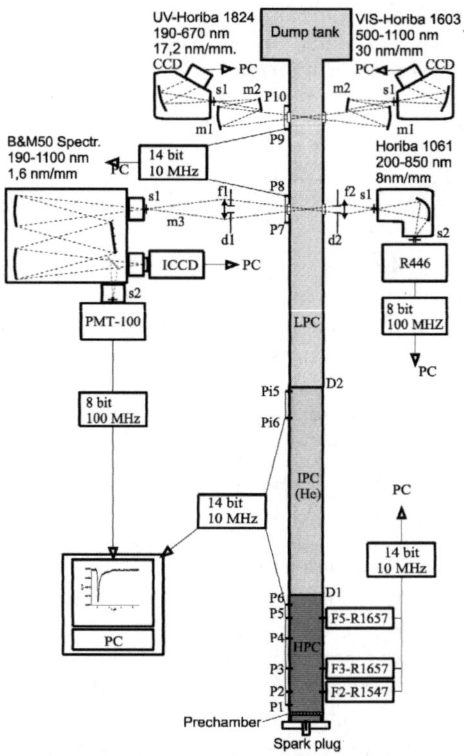

Figure 1. Experimental setup. Shock tube and system of registration.

Initial pressure of air P_1=25÷133 Pa in low pressure chamber (LPC). Changing the pressure and the gas composition (pure helium or a mixture of helium and neon) in the intermediary pressure chamber (IPC), it allows you to get a shock wave speed of 4÷10 km / s. Note that the use of the intermediate chamber with helium eliminates radiation from the contact area for all the shock wave velocity. Reproducibility of the operating parameters of the shock tube, from run to run are about 2% and basically determined by a quality of the mixture in HPC and a copper thickness of the diaphragm. Dump tank volume of 40 liters and LPC of shock tube evacuated up to 10^{-3} Pa by means of the dry pumping system with a leakage lower than 0.001 Pa/min (likely speed of degassing). To prepare the mixture of sample gas used pure gases produced by AGA Company. In the test experiments, the air gas fills up from the surrounding atmosphere in the low-pressure chamber. In this case, amount of water vapor and carbon dioxide was not controlled. Some works [11], [12] have shown a significant effect of driver gas on physicochemical processes in the driven shock-heated gas. This is because the diaphragm rupture occurs in finite time. The influence of driver gas diminished by means of intermediate section filled by helium. The pressure of helium in the buffer section was 10 kPa. Our work [13] describes in more details effects of driver gas on driven gas in a conventional shock tube operating in one and two diaphragm modes.

3. System Registration

Consider some parameters of optical registration system. Calibration of measured signal values to the absolute spectral radiance (W/cm^2/μm/sr) was executed with utilization of tungsten (SI–10) and deuterium (DDS–30) lamps calibrated by VNIIOFI. More detailed description of calibration procedure is presented in [4]. For better comparison of results obtained at different facilities, shock tube diameter should be considered, thus the experimental results are presented in the form of a spectral power density B_λ (W/cm^3/μm/sr). The spectral dispersion of the spectrograph with ICCD (ANDOR–520–18F–03) was 0.04 nm/pix. ICCD strobing was initiated by piezoelectric transducer P7. Minimal duration of strobe achieves up to 0.1 μs, and its location respective shock wave front can be determined with at least 0.5 μs accuracy. ICCD spectral sensitivity area is within 170÷850 nm. Measurements were carried out within 190÷850 nm. Spectrograph with ICCD allowed probing 25 nm of spectrum with high spectral resolution. Emission from the same direction as for the ICCD may be goes by mirror on PMT100 for investigation of temporal behavior of signal in spectral region 190÷850 nm. With splitter m3 the part of the radiation of the shock wave is directed to the spectrograph operating in the spectral range of 190-700 nm with PMT 100. Photomultiplier signal was registered by digital oscilloscope (8-bit, 100 MHz bandwidth). Temporal resolution of this scheme was 0.02 μs and was basically determined by the geometry of optical scheme. This PMT not calibrated in absolute units.

There is a radiation detection channel using photomultiplier R446 Hamamatsu. The photomultiplier is set at the output of the spectrograph Horiba–1061. The width of the recorded spectrum is 4 nm, and the temporary channel resolution of at least 0,04 μs. This PMT calibrated in absolute units in the spectral range 190÷800 nm. There are two channels that detect radiation in the spectral range of 190÷670 nm (UV) spectrograph Horiba CP140-1824 and range of 500÷1100 nm (VIS) spectrograph Horiba CP140-1603. The output spectrographs mounted gated CCD line Hamamatsu S11156. Both channels are calibrated in absolute intensity units. All channels (CCD and ICCD) that detect the spectral intensity distribution given the integral value during the emission across a shock wave passing by the measuring section. The integration time for the absolute values of the spectral distribution determined from the experimental temporary distribution of the radiation. Experimental procedure for measuring the intensity of the radiation emitted by the shock wave is given in [4].

4. Results of Experiment

Figure 2 shows the spectral distribution obtained with high spectral resolution on the ICCD camera in the experiment and the spectral distribution of $\gamma,\delta,\beta,\varepsilon$-bands nitric oxide calculated for equilibrium conditions at a temperature of 6000K. In the spectral region from 400 nm to 1100 nm are observed primarily atomic oxygen and nitrogen lines.

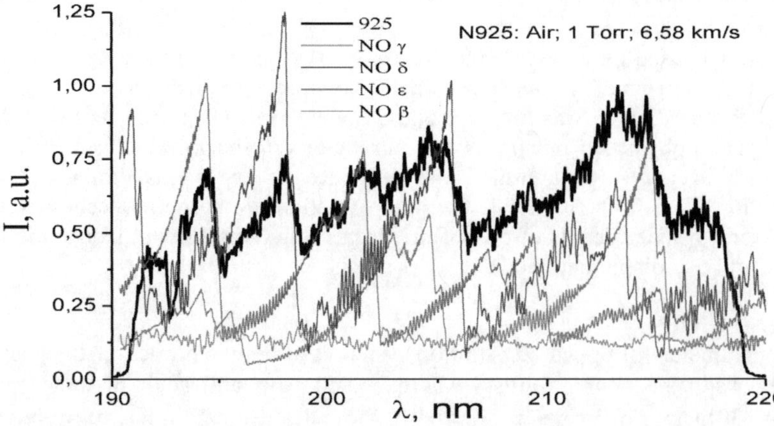

Figure 2. Spectral distribution of integral on time emissivity of air heated by shock wave (P_1=1 Torr, V_{SW}=6,58 km/s) and spectral distribution of γ,δ,β,ε-bands nitric oxide calculated for equilibrium conditions at a temperature of 6000K.

Temporal distributions of the intensities and spectral radiance for 10.4 km/s speed of the shock wave are presented in figure 3, 4. In the spectrum identified molecular nitrogen (the second positive system), nitrogen ion (the first negative system), nitric oxide (δ, γ , ε systems) and CN (violet band system) Observed carbon and hydrogen lines and CN violet bands in the spectra associated with the presence of impurities in the test air. As shown in [14] carbon impurity concentration of 1% in the test air contributes to the radiation from the violet CN bands compatible with of the radiating of $N_2^+(B)$ and $N_2(C)$ in the wavelength range from 290÷480 nm.

The integration of the experimental spectra recorded with a CCD camera was determined from the condition of the shock core length of 10 cm.

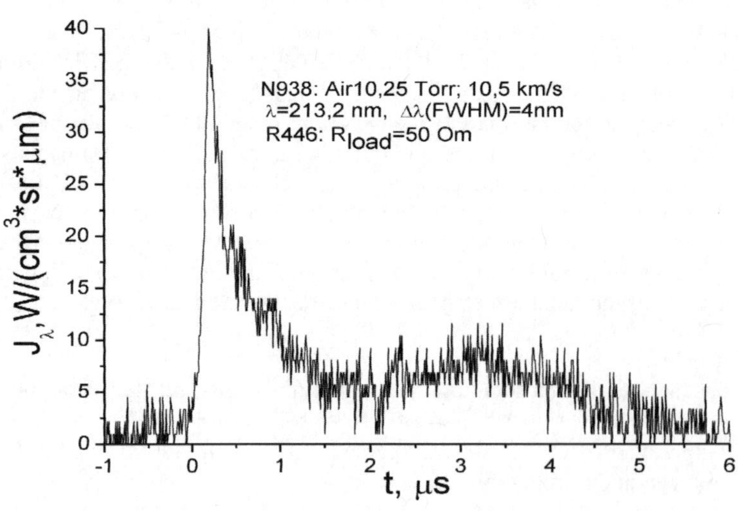

Figure 3. Experimental volumetric radiance profiles of air heated by shock wave P_1=0,25 Torr, V_{SW}=10,4 km/s.

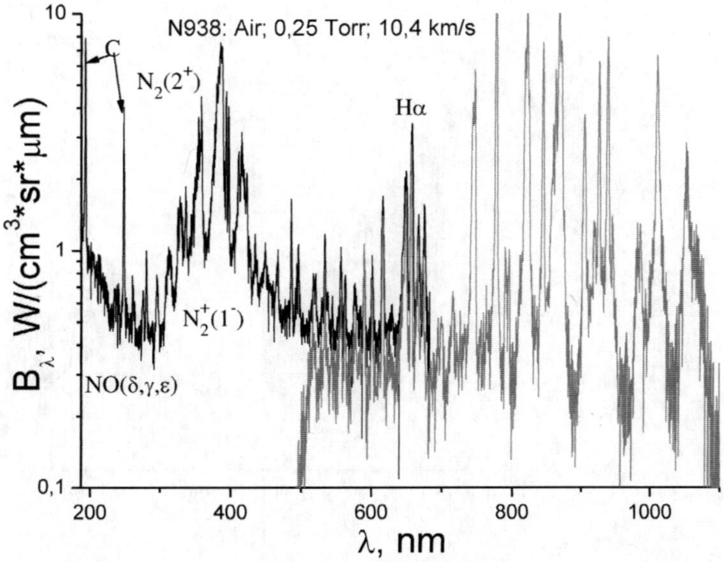

Figure 4. Spectral distribution of integral on time emissivity of air heated by shock wave P_1=0,25 Torr, V_{SW}=10,4 km/s.

Figures 5 ,7, 9, 11, 13 show examples of the volumetric radiance profiles of the nitric oxide emission obtained experimentally on PMT R446. The volumetric radiance profiles from PMT 100 in figure 13 obtained by conversion from the spectral distribution of emissivity with figure 14. The spectral distribution of integral on time emissivity in the wavelength range 190÷450 nm, recorded with a CCD camera shown in figures 6, 8, 10, 12, 14.

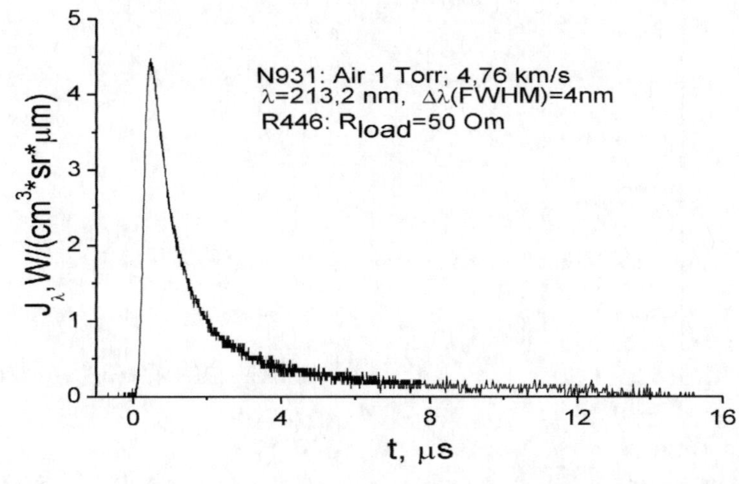

Figure 5. Experimental volumetric radiance profiles of air heated by shock wave P_1=1 Torr, V_{SW}=4,76 km/s.

IOP Conf. Series: Journal of Physics: Conf. Series **815** (2017) 012033 doi:10.1088/1742-6596/815/1/012033

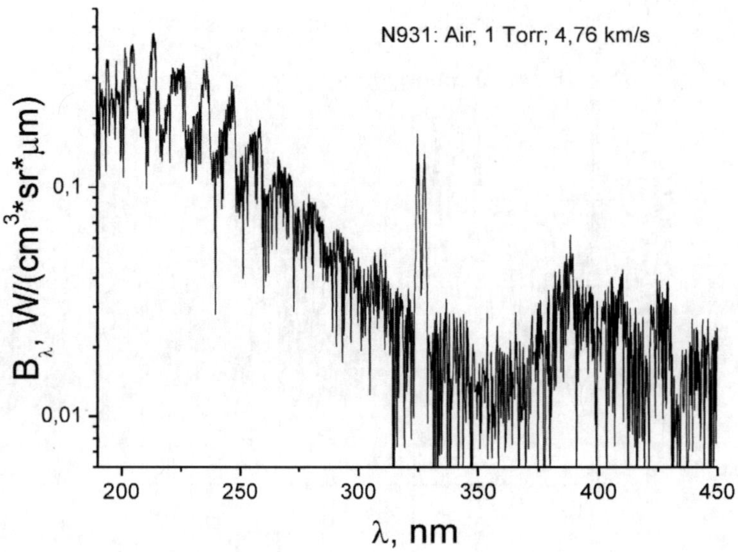

Figure 6. Spectral distribution of integral on time emissivity of air heated by shock wave P_1=1 Torr, V_{SW}=4.76 km/s.

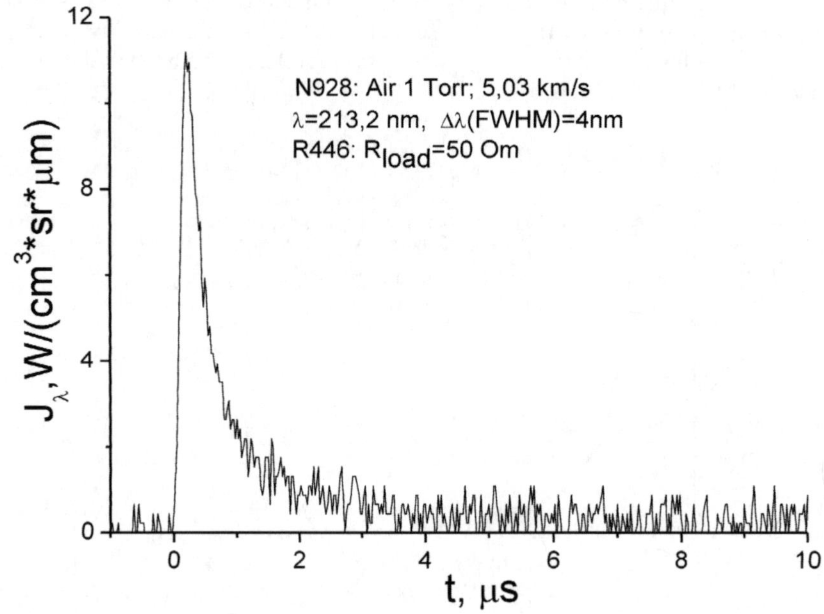

Figure 7. Experimental volumetric radiance profiles of air heated by shock wave P_1=1 Torr, V_{SW}=5.03 km/s.

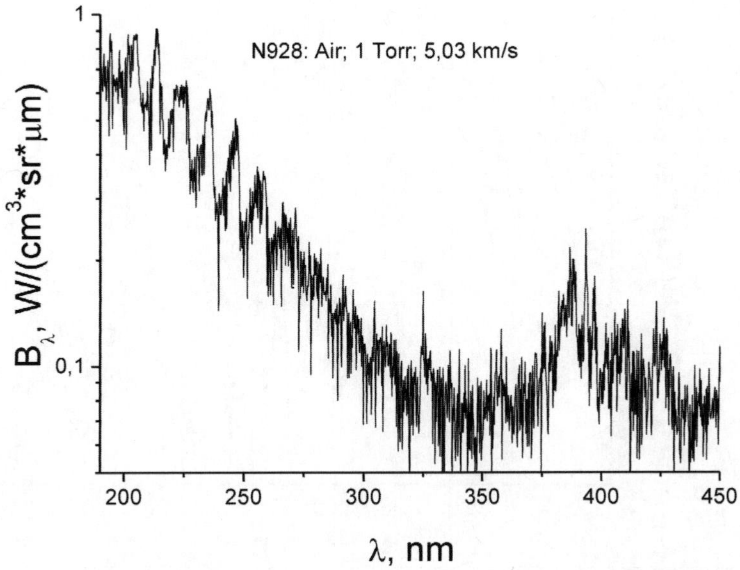

Figure 8. Spectral distribution of integral on time emissivity of air heated by shock wave P_1=1 Torr, V_{SW}=5.03 km/s.

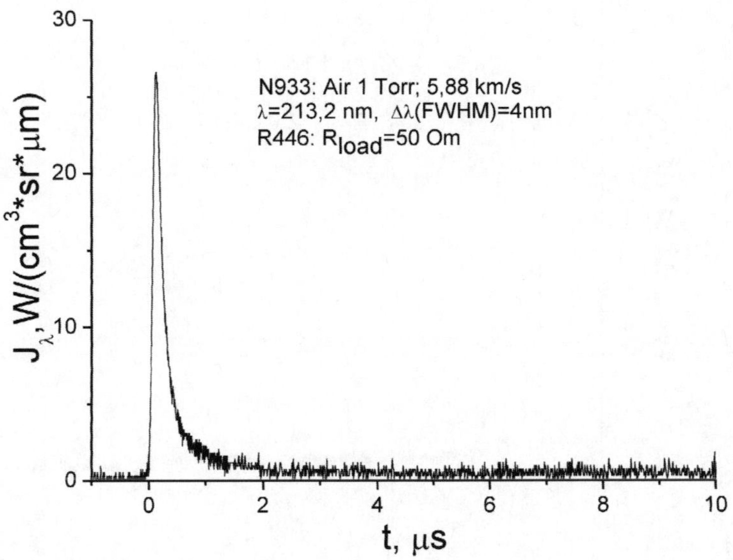

Figure 9. Experimental volumetric radiance profiles of air heated by shock wave P_1=1 Torr, V_{SW}=5.88 km/s.

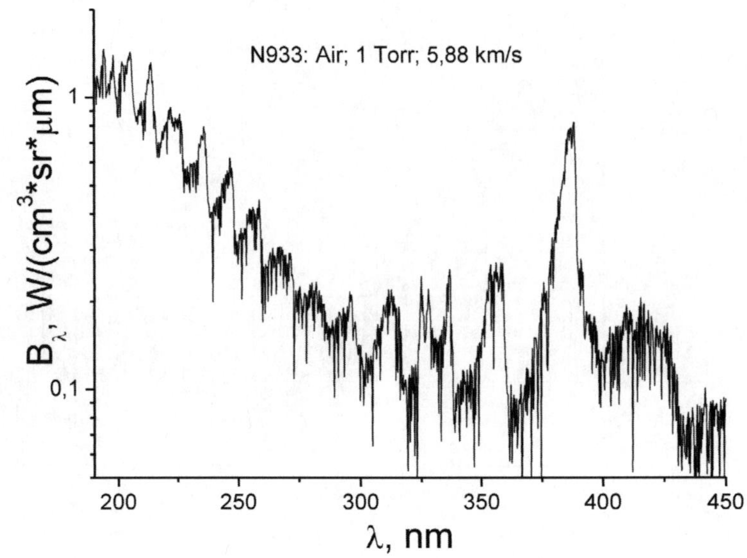

Figure 10. Spectral distribution of integral on time emissivity of air heated by shock wave P_1=1 Torr, V_{SW}=5.88 km/s.

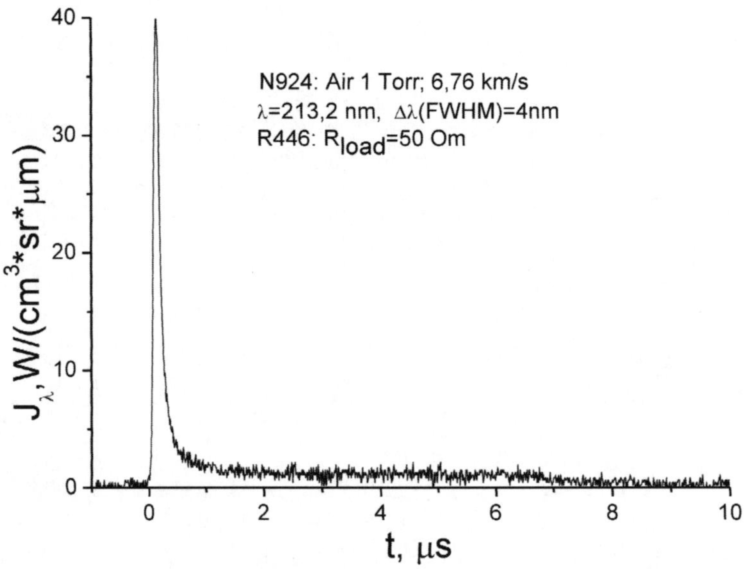

Figure 11. Experimental volumetric radiance profiles of air heated by shock wave P_1=1 Torr, V_{SW}=6.76 km/s.

IOP Conf. Series: Journal of Physics: Conf. Series **815** (2017) 012033 doi:10.1088/1742-6596/815/1/012033

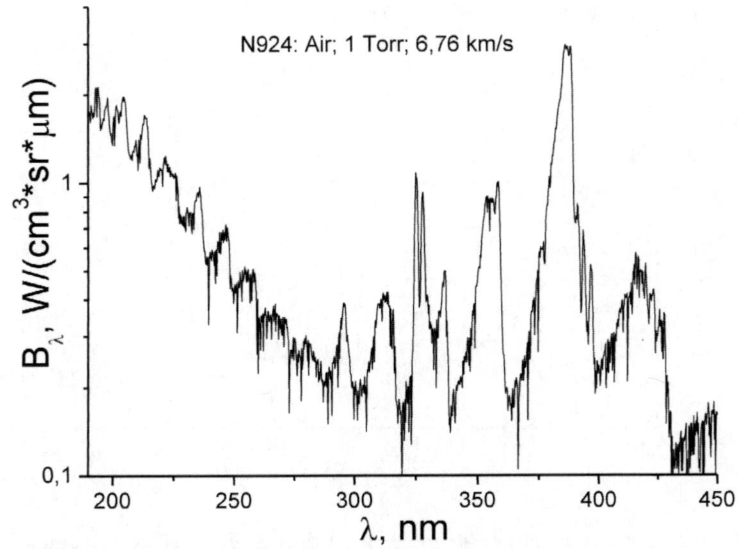

Figure 12. Spectral distribution of integral on time emissivity of air heated by shock wave P_1=1 Torr, V_{SW}=6.76 km/s.

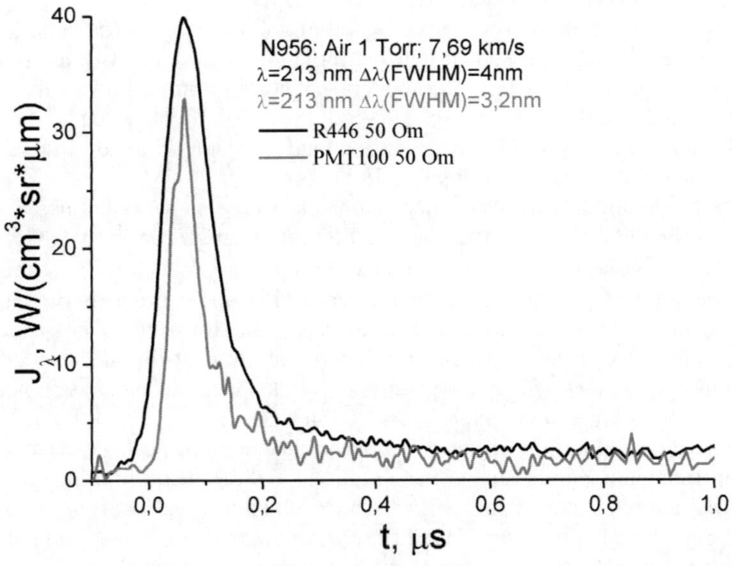

Figure 13. Experimental volumetric radiance profiles of air heated by shock wave P_1=1 Torr, V_{SW}=7.69 km/s.

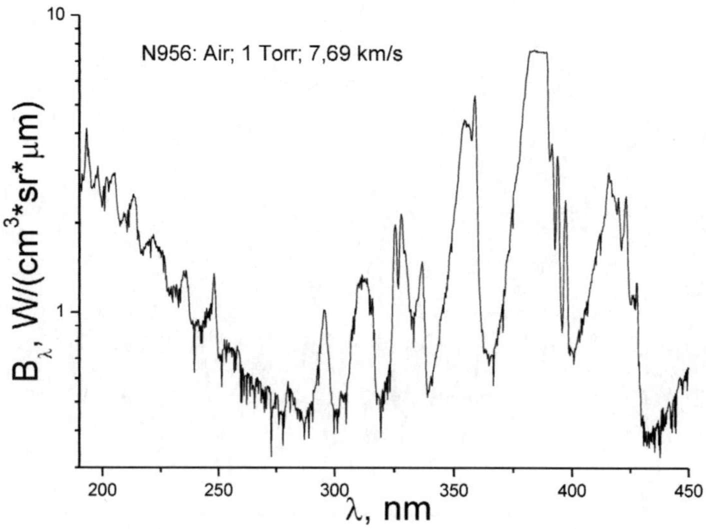

Figure 14. Spectral distribution of integral on time emissivity of air heated by shock wave P_1=1 Torr, V_{SW}=7.69 km/s.

Molecular bands (δ, γ, ε) of nitric oxide predominately emit in the UV spectrum with the shock wave velocity of 4-5.5 km/s. Nitrogen bands (the second positive system) and nitrogen ion (the first negative system) appear in the spectra with the increase of the shock wave speed. Also, atomic nitrogen and oxygen lines occur in the visible region of the spectrum.

The radiation peak depends on the speed and the initial pressure. Under the condition of constancy of the shock wave speed, it can be assumed that the radiation peak depends linearly on the pressure behind the wave (or the initial pressure). Duration of nonequilibrium radiation also depends on the speed and the pressure behind the shock wave. Typically, the peak duration leads to constant pressure behind the shock wave. In this study, the duration and the amplitude of the radiation peak is recalculated to the initial pressure of the test gas at P_1=1Torr.

The intensity of the NO bands increases approximately to the same extent as the increase in the shock wave speed, while the duration of the non-equilibrium radiation peak at a wavelength of 213 nm drops sharply. Figure 15 shows the experimental data of the radiation peak J_λ^{max} vs. the V_{SW}. The peak intensity (rhombic points) of RC-model calculated from [15] under our experiment conditions are shown for the comparison. Spectrally averaged radiations peak NO in the shock wave calculated in [16] are in good agreement with the experimental data of Gorelov [3] for shock speeds below 7km/s, but are lower than the experimental data in [3] at higher shock speeds. The radiation peak saturation is observed at speeds higher 7 km/s in our experiments, which is in agreement with [16].

Experimental data of the duration of the nonequilibrium radiation peak vs. speed is presented on the figure 16. The triangular points shows the duration of the nonequilibrium radiation peak at a wavelength of 236±7 nm recalculated for the pressure P_1=1 Torr, which was previously computed in [16] for the initial pressure P1 = 0.1 Torr. The difference between the calculation results of work [16] and our experiments is probably due to incorrect recalculation of peak for the pressure of 0.1 to 1 Torr. The duration of radiation peak of NO at a wavelength of 213 nm and 236 are practically identical. The time of population on FWHM of electronic states of NO molecule calculated within the RC-model [15] are shown in Figure 16 by rhombic points.

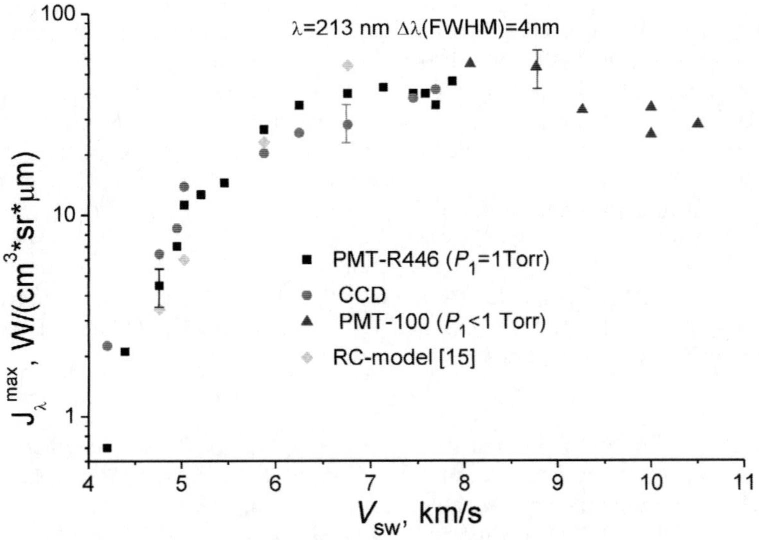

Figure 15. Radiance peak NO (213±2nm) of the shock-wave velocity at P_1=1 Torr.

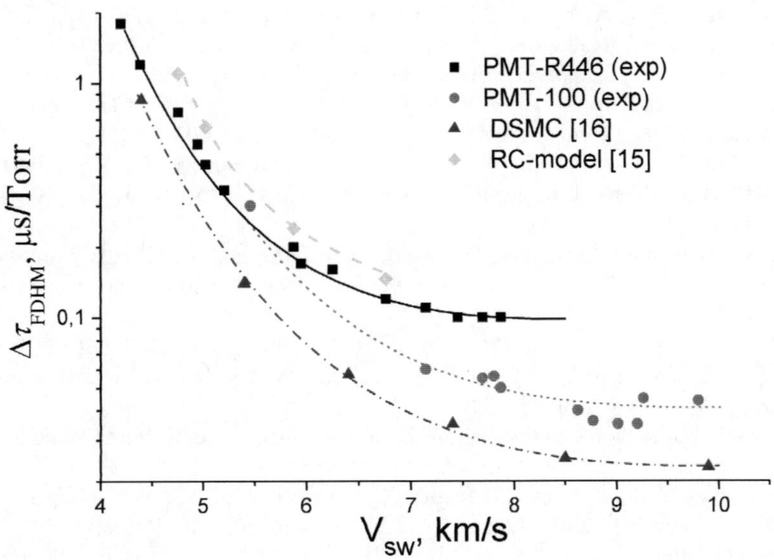

Figure 16 Duration radiance peak (FWHM) of the shock-wave velocity at P_1=1 Torr.

5. Conclusion

Nonequilibrium radiation is a major contributor in the radiation flux received by spacecraft at the entrance to Earth's atmosphere at speeds of 9-11 km / s. The duration of this radiation is very short. The paper shows the possibility of studying nonequilibrium kinetics in shock heated air with high spatial resolution. These experimental data were compared with the results of numerical calculation on the hybrid radiative-collisional model represented in Ref. [15]. Preliminary comparison of experiments described in the article with the results of DSMC simulation presented in Ref. [16] is performed, and proves them to be in a satisfactory agreement.

Acknowledgments

This study was supported by RFBR grant No. 16-01-00379 and by the Program of Basic Researches of Russian Academy of Sciences. The author would like to thank Drs. Shatalov OP and Surzhikov ST for their insight during many fruitful discussions.

References

[1] Dikalyuk A S, Kozlov P V, Romanenko Y V, Shatalov O P and Surzhikov S T 2013 Nonequilibrium spectral radiation behind the shock waves in Martian and Earth atmospheres *44th AIAA Thermophysics Conference AIAA 2013-2505* pp 1– 10

[2] Dikalyuk A S, Kozlov P V, Romanenko Y V, Shatalov and O P Surzhikov S T 2012 Nonequilibrium radiation behind the strong shock waves in Martian and Titan atmospheres: Numerical rebuilding of experimental data *50th AIAA Aerospace Sciences Meeting including the New Horizons Forum and Aerospace Exposition, AIAA* 2012–0795 pp 1–10

[3] Gorelov V A, Gladyshev M K, Kireev A Yu, Yegorov I B, Plastinin Y A and Karabadzhak G G 1998 Experimental and Numerical Study of Nonequilibrium Ultraviolet NO andN_2^+ Emission in Shock Layer *Journal of Thermophysics and Heat Transfer* Vol 12, No 2, pp 172–179

[4] Zalogin G N, Kozlov P V, Kuznetsova L A, Losev S A, Makarov V N, Romanenko Yu V and Surzhikov S T 2001 Radiation Excited by Shock Waves in a CO_2-N_2-Ar Mixture: Experiment and Theory*Technical Physics*, Vol 46, No 6, pp 654-661

[5] Surzhikov S T. 2012 Radiative-Collisional Models in Non-Equilibrium Aerothemodynamics of Entry Probes *Journal of Heat Transfer* **134** pp 031002-1 – 031002-12

[6] Brandis A M, Johnston C O, Cruden B A, Prabhu D K and Bose D 2012 Validation of High Speed Earth Atmospheric Entry Radiative Heating from 9 5 to 15 5 km/s *AIAA Paper 2012–2865* pp 1–23

[7] Cruden B A 2014 Absolute Radiation Measurements in Earth and Mars Entry Conditions *NATO TR RTO-EN-AVT-218* Von Karman Institute for Fluid Dynamics, Rhodes Saint Genese, Belgium pp 1–40

[8] Jacobs C M, McIntyre T J, Morgan R G, Brandis A M and Laux C O 2015 Radiative Heat Transfer Measurements in Low-Density Titan Atmospheres *Journal of Thermophysics and Heat Transfer* Vol 29 No 4 pp 835-844

[9] Takayanagi H and Fujita K 2012 Absolute Radiation Measurements Behind Strong Shock Wave In Carbon Dioxide Flow for Mars Aerocapture Missions *43rd AIAA Thermophysics Conference, Fluid Dynamics and Co-located Conferences AIAA 2012-2744* pp 1-12

[10] Brandis A M, Johnston C O and Cruden B A 2016 Non-equilibrium Radiation for Earth Entry *46th AIAA Thermophysics Conference, AIAA AVIATION Forum AIAA 2016-3690* pp1-19

[11] Kozlov P V, Makarov V N, Pavlov V A and Shatalov O P 2000 Experimental investigation of CO vibrational deactivation in a supersonic cooling gas flow *Shock Waves* Vol 10 pp 191–195

[12] Bogdanoff D W and Park C 2002 Radiative Interaction Between Driver and Driven Gases in an Arc-Driven Shock Tube *Shock Waves* Vol 12 pp 205–214

[13] Kozlov P V and Romanenko Yu V 2009 Experimental Investigation of Shock Heated Air

Radiation on the Double-Diaphragm Shock Tube *in Proceedings of the 3th All-Russian School-Workshop Aerophysics and Physical Mechanics* of *Classical and Quantum Systems* (in Russian) pp 21–29

[14] Johnston C 2008 A Comparison of EAST Shock-Tube Radiation Measurements with a New Air Radiation Model *46th AIAA Aerospace Sciences Meeting and Exhibit, Aerospace Sciences Meetings AIAA 2008-1245* pp 1–22

[15] Kozlov P V and Surzhikov S T Nonequilibrium Radiation NO in Shocked Air (2017) *55th AIAA Aerospace Sciences Meeting, AIAA SciTech Forum AIAA 2017-0157* pp 1–16

[16] Zhu T, Li Z and Levin D A 2014 Modeling of Unsteady Shock Tube Flows Using Direct Simulation Monte Carlo *Journal of Thermophysics and Heat Transfer* Vol 28 No 4, pp 623-634

IOP Conf. Series: Journal of Physics: Conf. Series **815** (2017) 012034 doi:10.1088/1742-6596/815/1/012034

Extinction and growth of cylindrical hotspots in AB model explosive: molecular dynamics studies

S A Murzov[2,1] **and V V Zhakhovsky**[2]

[1] Moscow Institute of Physics and Technology, Institutskiy Pereulok 9, Dolgoprudny, Moscow Region 141700, Russia

[2] Dukhov Research Institute of Automatics, Sushchevskaya 22, Moscow 127055, Russia

E-mail: murzovs@gmail.com

Abstract. Atomistic simulations of a sample with AB model explosive heated quickly up within a small central cylinder were performed with the goal of finding a critical radius of such hotspot. The heating time was chosen to be much shorter than an acoustic time required for rarefaction wave for arriving to the center of hotspot. An isochoric thermal decomposition of AB material evolves inside the hotspot before the arrival of rarefaction wave. We found that the critical radius of such hotspots is determined by the equality of the acoustic time and a characteristic time of isochoric thermal decomposition controlled by initial hotspot temperature. For supercritical hotspots the decomposition of AB material leads to deflagration. The earlier arrival of rarefaction wave extinguishes the chemical reactions in hotspots with the subcritical radius.

1. Introduction

The difference between scales of processes involved in simulation of hotspots (HS) in condensed phase explosives [1] makes a challenge to the modern computational facilities [2–4]. A characteristic time of reaction in PETN [5] or AB [6] energetic material are within picosecond timescale in molecular dynamics (MD) simulation. At the same time, experimental data provides several microsecond timescale for the detonation initiation and associated characteristic reaction times. The macroscopic kinetic models [7] are usually calibrated by experiments, then may be used in the hydrodynamic simulation. Run-distance to detonation [8] data is widely used in the calibration of such macroscopic kinetics. Another kind of models [9] are based on chemical principals and include information from the experiments, where the reaction time is about 1-100 seconds, for example, adiabatic explosion and one dimensional time to explosion.

The difference of "fast" and "slow" heating is in different characteristic times - the reaction time in AB material t_{react} (time to the acceleration of exothermal reactions), acoustic time of rarefaction immediately created HS t_a, duration of heating pulse t_T to create a local high-temperature domain (time of HS creation). "Fast" heating is characterized by almost instant thermal energy deposition into the media. After that, reactions may be occurred and dumping of pressure in the high temperature region due to thermal expansion. In this situation the parameter of pulse duration formally excluded. The simplified model of hotspot provides an understanding of critical size and presented in this paper. "Slow" heating is associated with a mechanical equilibrium of hotspot and surrounding media. In that case, hot gas is trapped within a sample and the absence of pressure gradient leads to the thermal diffusion mechanisms of hotspot growth. Also difference between chemical and heating time should be taken into account. There are two limit cases associated with the completed reactions in the end of heating or, if induction time is long enough, the reaction will start after the heating.

2. Timescale characterization of hotspot

An isochoric thermal decomposition time defines critical radius of the hotspots introduced in section 4. There is a possible analysis is presented based on characteristic times contributed to the different processes during extinction or growth of isolated HS.

Let's assume the heating duration is t_T. This time recognizes that hotspot created by a contact with warm piston or by a short laser pulse. Two other timescales are a characteristic time of exothermal chemical reactions t_{react} (time to acceleration) and an acoustic (hydrodynamic) time t_a. The acoustic time links the size of HS and velocity of perturbation propagation, which is equal to the sound in heated media for weak perturbations. Formal permutation of these time parameters provides $3! = 6$ relations:

1) a) $t_T < t_a < t_{react}$, b) $t_T < t_{react} < t_a$;
2) c) $t_a < t_{react} < t_T$, d) $t_{react} < t_a < t_T$;
3) e) $t_a < t_T < t_{react}$, f) $t_{react} < t_T < t_a$.

The variants in 1) associated with HS created by "fast" heating and described in the next chapter.

Another widely studied limit case is thermal hotspot. As it is mentioned in previous chapter, this HS is in mechanical equilibrium with surrounding cold media, therefore t_T is the biggest time. Diffusion processes, thermal conductivity and energy release of high-temperature media from the HS defines a characteristic time to explosion. It should be noted, the mechanical equilibrium blocks deflagration wave propagation (a pressure jump is needed), therefore an additional energy would be deposited to form the front. Variants of point 2) are associated to the thermal hotspots. Last cases in 3) are intermediate ones. If the least time is acoustic e), then a mechanical equilibrium will be observed after heating and hot unreacted material will be in low-density state, the reaction should start later. If the chemical time is shorter than both else as in f), hot products at high pressure will form a shock wave. This case is associated with hydrodynamic calculation without explicit reaction zone. Another specific case of HS is a dynamically formed during the collapse of pore or gas bubble [1,2].

3. Simulation techniques

Simulated samples are melted a priori at density 1.18 g/cm^3 and temperature T $\approx$ 77 K with the Langevin thermostat. After that a cylindrical hotspot is created within samples. Plane disk has a radius $\boldsymbol{L_r} = 120$ nm and width of 16 nm. Periodic boundary conditions are imposed in Z dimension. The creation of HS involves several steps:

(i) The Langevin thermostat equilibrate the sample at T = 77 K to melt initially and relax pressure to zero,

(ii) An additional Langevin thermostat in the central cylinder with radius of a specific HS,

(iii) Damping thermastat of waves coming to the free surface of the sample is realized by another Langevin thermostat with increasing rigidity as the free surface becomes closer and a target temperature $T = 77$ K. This thermostat acts only at ϕ and Z components of atoms velocity, which are perpendicular to the radial motion of the media (ϕ is a polar coordinates angle).

The time of the HS boundary arrival to the domain of damping thermostat defines an available molecular-dynamics simulation time.

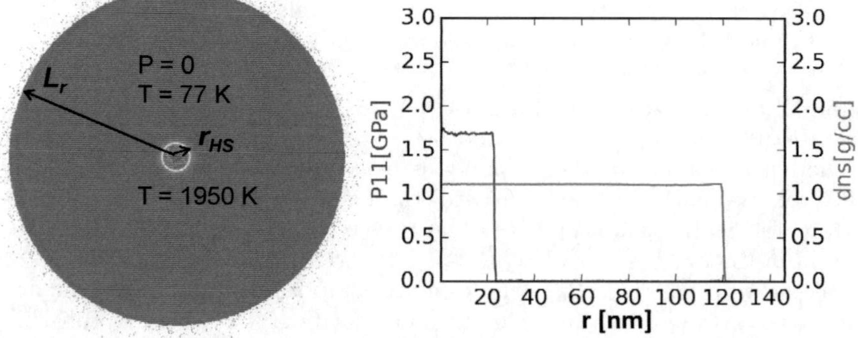

Figure 1. Radial profiles of pressure and density in the sample after heating. A steep ramp in a temperature profile without any products in heated domain after hotspot creation.

4. Simulation results and discussion

An initial perturbation of atoms velocity vector isotropically in a space with Maxwellian distribution equals to the target temperature of Langevin thermostat. The thermostat is turned on at the next integration step of MD simulation inside given cylinder of HS's radius. The time of heating is $t_T = 0.2$ ps and the target temperature is T = 2400 K. Characteristic time of thermostat is two times greater t_T. The temperature of HS is approximately 1950 K. Both temperature and pressure profiles have a "jump" on the boundary of heated and cold material (Figure 1) while density profile is plain.

The main requirement to preparation is absence of products molecules A_2, B_2 at the end of HS creating. Also an acoustic time t_a should be much larger then the time of heating t_T: $t_T << t_a$, where definitely t_a is a time of the first arrival of rarefaction wave into the center of HS. Target temperature of the thermostat is limited by fast dissociation of AB molecules. The last is motivated by the velocity perturbation independently to the whole molecules in the initial molecular crystal of AB. Therefore, a situation of relatively "too high" velocity of atoms inside one molecule would greatly overstep initial temperature and corresponded to the initial velocity permutation of atoms.

Different radii of HS provides different time of rarefaction wave arrival. In the center of sample, an isochoric thermal decomposition is observed before the arrival. We simulate initiation of reactions in case of HS with different radii $R_{HS} = 20, 23, 25$ nm, and on the figure 3, a products curve and density in the center of HS is shown, which is taken from fixed cylinder of 3 nm radius. It is observed, the HS with $R_{HS} = 20$ nm attenuates and the reactions of decomposition stop at times greater than $t \gtrsim 80$ ps after the start of simulation. In case of HS $R_{HS} = 23$ nm a slow production of A_2, B_2 molecules is observed before the rarefaction wave reflection from the center of cylinder and after begining of expansion in the center. This reflection induces additional increase of density and temperature in the center, therefore reactions accelerate and formation of deflagration front is observed. An additional maximum of density dependence in time is shown on the figure 3. In HS with $R_{HS} = 25$ nm, the reactions accelerate and form additional expansion of the gases outside heated hotspot with rise of the temperature and fall of the density in the center.

On the figure 2, an initial stage of products curves and associated temperature dependencies from figure 3 are shown. Dashed lines show temperature dependence and has a decrease after rarefaction wave arrival. However, before the arrival, this dependences of products mass fraction and temperature are similar to those in isochoric thermal decomposition simulation. The

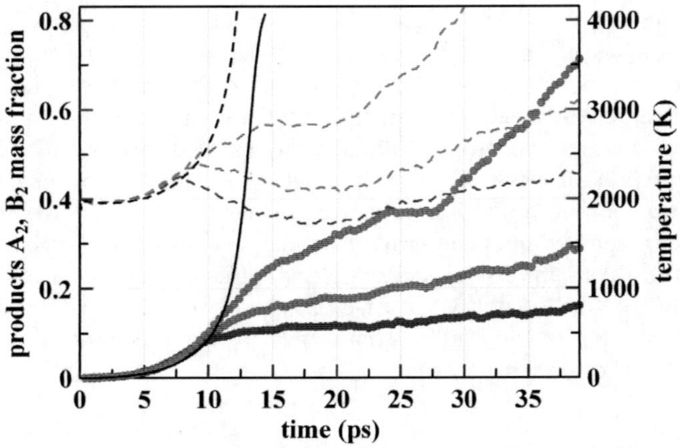

Figure 2. Temperature (dashed lines) and mass fraction of products (circles) in a small 3 nm central cylinder inside HS as functions of time are similar to ones (black lines) observed in the isochoric thermal decomposition, before the arrival of rarefaction wave. The next Figure 3 shows corresponding evolution of density at the center of HS.

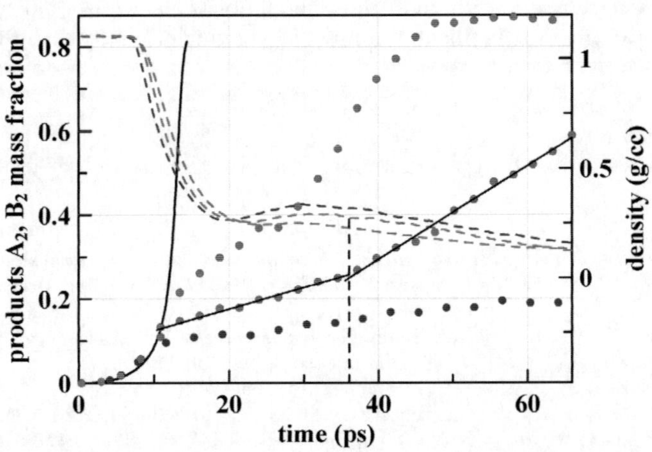

Figure 3. Density is shown by dashed lines and mass fraction of products by circles. Blue, red and green colored curves are associated to the HS of 20, 23, 25 nm, while black curves - to isochoric thermal decomposition. Black lines show a slope change when the reaction acceleration is observed in case of critical HS radius.

isochoric thermal decomposition simulation employs periodic boundary conditions in the box of $16 \times 16 \times 16$ nm^3. About 195 thousands of atoms in this box are simulated. It is shown, the fluctuation according to the induction time and local start of exothermal reactions is negligible. A distribution of the local reaction start may be assumed uniform relatively to the boundary of HS and cold media.

In that way, in the simulation of HS the sample width in Z direction should be equal to L_z = 16 nm. In case of system with $L_z = 8$ nm non-uniformity of reaction start inside cylindrical hotspot provides artificial uncertainty of critical hotspot.

Series of simulation with described "fast" heating in HS of cylindrical form provides condition for the critical radius.

The key characteristics of the "fast" HS are the size of HS and thermodynamic state of the explosive (temperature and density). A critical hotspot size is defined for given density and temperature. If radius of HS is less than the critical, exothermal reactions in the center are similar to the isochoric thermal decomposition, but the reactions decay and stop after rarefaction wave arrival. The rarefaction wave decreases temperature and density inside HS, slows energy release. After reflection of the rarefaction wave from the center, atoms move into the center again and additional spike of density and temperature should be observed. The condition of following HS's growth is transition to intensive energy release after the reflection of radial rarefaction wave. Otherwise, the HS should decay. In case of HS greater than critical, product molecules formation and expansion out of the center is observed, but the acceleration of reactions happens before the reflection of the rarefaction wave.

5. Conclusion

In this research we suggested a classification of hotspots according relations between the characteristic times of three main processes – chemical decomposition, material motion, and a time of hotspot formation. Molecular dynamic simulation of hotspots produced by the isochoric heating for a time much shorter than any processes involved in AB material evolution were performed with the aim to find a critical size of such hotspots. We determined the critical condition between extinction and growth of those hotspots as a simple equality of the rarefaction time and characteristic times of chemical decomposition, which is controlled by the initial radius, density and temperature of such hotspots.

Acknowledgements

This work is supported by the Russian Foundation for Basic Research grant 16-08-01181.

References

[1] Bowden F P and Yoffe A D 1948 *Third symposium on combustion, flame and explosion phenomena* 551–560
[2] Cherukara M J, Wood M A, Kober E M and Strachan A 2015 *Journal of physical chemistry C* **119**(38) 22008–22015
[3] Eason M R and Sewell D T 2015 *Journal dynamic behavior material* **1** 423–438
[4] Shan T R and Thompson A P 2014 *J.Phys.: Conference Series* **500** 172009
[5] Sergeev O V and Yanilkin A V 2014 *J.Phys.: Conference Series* **500** 172010
[6] Zhakhovsky V V, Budzevich M M, Landerville A C, Oleynik I I and White C T 2014 *Phys. Rev. E* **90** 033312
[7] Handley C A and James H R 2012 *AIP Conference proceedings SCCM - 2011* **1426** 519–524
[8] Lee E L and Tarver C M 1980 *Phys. Fluids* **23** 2362–2372
[9] Tarver C M, Tran T D and Whipple R E 2003 *Propellants, Explosives, Pyrotechnics* **28** **(4)** 189–193

Two-jet structure of the flow produced by magnetized hypersonic spherical source into the steady unmagnetized medium

E A Golikov[1,2], V V Izmodenov[1,2,3] and D B Alexashov[1,3]

[1] Space Research Institute, RAS Profsoyuznaya Street 84/32, Moscow, 117335, Russia
[2] Lomonosov Moscow State University, GSP-1, Leninskiye Gory, Moscow, 119991, Russia
[3] A. Ishlinsky Institute for Problems in Mechanics, RAS Vernadsky Street 101(1), Moscow, 119526, Russia

E-mail: golikov.e.a000@gmail.com

Abstract. In the present paper we consider the steady flow produced by the hypersonic magnetized spherical source into the steady unmagnetized medium. The magnetic field of the source is considered purely azimuthal. The basic dimensionless parameter of the problem is Alfvén number at the inflow boundary at the source equatorial plane, which we denote as ε. We first review the numerical results, that was already presented in our recent paper (Golikov et al, 2017). According to our numerical results there should be a tangential discontinuity dividing the flow and the ambient medium. In the present paper we analytically derive the asymptotic behaviour of the tangential discontinuity radius as epsilon approaches zero.

1. Introduction

An interest for the flow the present paper is devoted to, was inspired by the problem of stellar wind/interstellar medium interaction – an astrophysical problem, that rose up about half a century ago. We would like to briefly introduce the reader to the physical background of the problem first.

It was realized in the middle of the last century that the Sun constantly produces a flow of fully ionized hydrogen plasma (see historical review in Chapter 1 of [1]). First direct measurements showed that this flow is highly supersonic at the Earth orbit [2]. This flow, nowadays known as *the solar wind*, proceeds up to approximately 90 astronomical units, where it undergoes a shock transition, because of the ambient interstellar gas pressure. The region of the solar wind/interstellar medium interaction is commonly referred as *the heliospheric interface*. The structure of this region is strongly affected by a number of physical factors; for details, see, for example, reviews and papers [3], [4].

It should be noted now, that not only the Sun has *wind*: a number of images of the structures similar to the heliospheric interface in the vicinity of other stars where observed in recent years. These structures are called *astrospheres*. The shape of an astrosphere may greatly differ from that of the heliosphere.

In the simplest case a star can be considered as a steady hypersonic spherical source of the fully ionized hydrogen plasma – *the stellar wind*. Following [5], [6], we treat the flow hydrodynamically in our study. The flow is spherically symmetric and is governed by the system of Euler equations

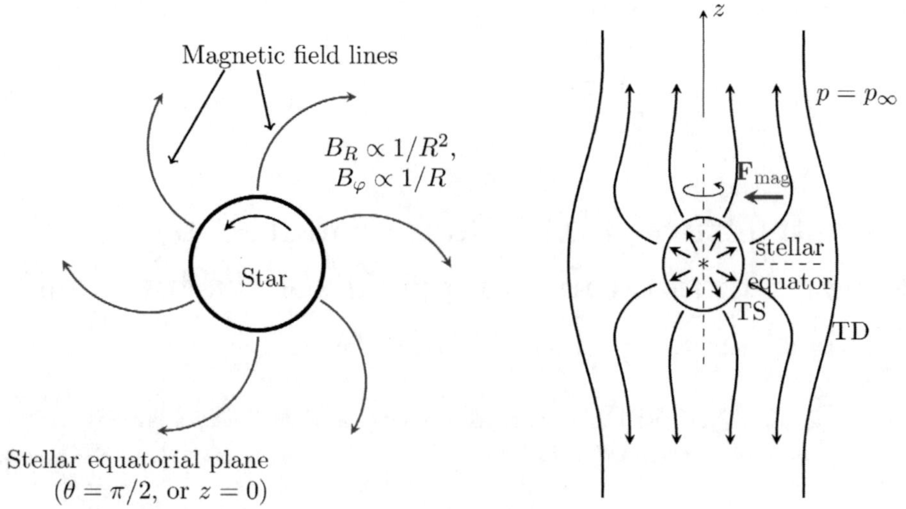

Figure 1. Left panel: schematic picture of the stellar magnetic field configuration in the equatorial plane $\theta = \pi/2$; the star is treated as a rotating "magnetic monopole". Right panel: schematic picture of the flow under consideration; TS is the termination shock, TD is the tangential discontinuity, dashed line is the stellar rotation axis.

together with the polytrope law, where we can omit the pressure gradient due to the hypersonic character of the flow. In this case, the solution is trivial:

$$\rho \propto R^{-2}, \quad V = \text{const}, \quad p \propto R^{-2\gamma},$$

where R stands for the distance from the origin, γ is the polytrope index (usually equals to 5/3, that corresponds to monoatomic gases).

However, the flow cannot proceed freely up to infinity due to existence of ambient interstellar gas. If we assume this gas to be at rest relatively to the star, then it causes the flow to overcome shock transition at certain distance R_{TS}. Index "TS" stands for *termination shock*; this term is traditionally used to distinguish this shock from the *bow shock*, that appears in ambient gas flow, if we assume it to be relatively moving with a supersonic velocity. The post-shock flow is subsonic, and if we treat it as incompressible, than the solution in the post-shock region is simple:

$$R > R_{TS}: \quad \rho = \text{const}, \quad V \propto R^{-2}, \quad p \sim p_\infty,$$

where p_∞ is an ambient gas pressure.

Let us now discuss the stellar magnetic field. The global magnetic field structure on the stellar surface corresponds to the magnetic field of a dipole. However, this magnetic field subsists not in vacuum, but in hydrogen plasma flow the star emits – the stellar wind. The stellar wind stretches the magnetic field lines, making their topology farther from the initial dipole field on the surface, but closer to the "split monopole" field (see Chapter 10 of [1]): such a field corresponds to the magnetic field of a (hypothetical) monopole, but with opposite polarities on opposite hemispheres. In our study the polarity of the field doesn't matter, since the magnetic field appears quadratically in equations of motion (see (7) and (8) below). Hence, we can assume the stellar magnetic field to be produced by a single "magnetic monopole".

There are no electric field in the stellar wind reference frame, due to its high electric conductivity. Since the magnetic dissipation is very low, the stellar magnetic field is completely "frozen" in the flow:

$$\nabla \times [\mathbf{V} \times \mathbf{B}] = 0 \tag{1}$$

It means that the magnetic flux through any fluid surface is constant. For purely radial field on the stellar surface this implies:

$$B_R \propto W_{\text{mag}}/R^2, \quad B_\theta = 0, \quad B_\varphi = 0,$$

where W_{mag} stands for "magnetic power" of the source.

Stars usually rotate, and this rotation affects their magnetic field. Let z-axis be the stellar rotation axis, and Ω be the stellar rotation frequency. Let us for now assume, that the magnetic field does not affect the flow (*kinematic approximation*). Applying the frozen-in condition (1) to the pre-shock flow, we get:

$$R < R_{TS}: \quad B_R \propto W_{\text{mag}}/R^2, \quad B_\varphi \propto W_{\text{mag}}(\Omega/R)\sin\theta, \quad B_\theta = 0 \tag{2}$$

Here θ is the angle counted from z-axis, φ is the azimuthal angle. This magnetic field configuration is nowadays referred as *Parker spiral*, after Parker (see [5]); see figure 1, left panel.

Kinematic approximation is not necessarily valid, and we generally have to add the magnetic force into the equation of gas motion:

$$\mathbf{F}_{\text{mag}} = \frac{1}{4\pi}\mathbf{B} \times (\nabla \times \mathbf{B}) \tag{3}$$

Magnitudes of magnetic force $\mathbf{F}_{\text{mag}}$ and of gas inertia $\rho(\mathbf{V}\cdot\nabla)\mathbf{V}$ are related to each other as squares of alfvén velocity and bulk velocity of the gas:

$$\frac{|\mathbf{F}_{\text{mag}}|}{|\rho(\mathbf{V}\cdot\nabla)\mathbf{V}|} \sim \frac{B^2/4\pi}{\rho V^2} = \frac{V_A^2}{V^2}$$

The square root of this quantity is usually referred as *Alfvén number*:

$$\text{Al} = \frac{V_A}{V}$$

So, one can say that the Alfvén number characterizes the dynamical impact of the frozen-in magnetic field on the flow. In the pre-shock region for kinemtically-treated magnetic field, we get:

$$\text{Al} = \sqrt{\frac{B^2/4\pi}{\rho V^2}} = \sqrt{\frac{B_\varphi^2/4\pi}{\rho V^2} + \frac{B_R^2/4\pi}{\rho V^2}} = \sqrt{\text{const} + O(R^{-2})}, \quad R < R_{TS}$$

Hence, as long as this quantity is small at the stellar surface, it remains so in the whole pre-shock region, and we can neglect the dynamical impact of the magnetic field there. However, it is not so in post-shock region; if we treat the magnetic field kinematically there, then:

$$R > R_{TS}: \quad B_R \propto W_{\text{mag}}/R^2, \quad B_\varphi \propto W_{\text{mag}}\Omega R\sin\theta, \quad B_\theta = 0 \tag{4}$$

From this follows, that $\text{Al} \sim R^3$ there, the magnetic force becomes very significant, and the kinematic approach is not valid in the post-shock region.

In the present paper we restrict ourselves to the case for which the "magnetic power" W_{mag} is vanishing, whereas the rotation frequency Ω is huge, so that the azimuthal magnetic field is finite:

$$W_{\text{mag}} \to 0, \quad \Omega \to \infty, \quad W_{\text{mag}}\Omega \to \text{const}$$

In this case the azimuthal component of the magnetic field is much larger, than the radial one (see 2), and the magnetic force (3) is directed towards the stellar rotation axis. This force collimates the flow in front of the axis, forcing it to form a *jet* (see schematic picture of the flow: figure 1, right panel). The radius of the jet appears to be finite, in other words, the tangential discontinuity (TD) dividing the stellar wind and the interstellar medium is formed.

Since the magnetic field is purely azimuthal in our consideration, it is convenient to denote the proportionality factor of the magnetic field B and $\sin\theta/R$ in the pre-shock region with a single symbol:

$$B = B_\varphi = \mathcal{F}_B \sin\theta/R, \quad R < R_{TS} \tag{5}$$

We have to note, that a stellar wind flow with magnetically-driven jets was proposed earlier by Drake (see [7]). However, the results presented in this paper, both analytical and numerical, are questionable. We would like to elaborate the main points of this paper in a tidier manner.

2. Mathematical formulation of the problem

As we have already noted in the Introduction, we shall omit the radial component of the magnetic field at the source surface. Then the magnetic field becomes purely azimuthal in the whole domain of the flow. From this follows, that there is no magnetic force acting in the azimuthal direction, and the flow becomes axisymmetric with no twist around the symmetry axis (z-axis).

The system of ideal MHD equations in cylindrical coordinates (z, r, φ) is then follows (see [8]):

$$\frac{\partial(r\rho V_z)}{\partial z} + \frac{\partial(r\rho V_r)}{\partial r} = 0, \tag{6}$$

$$V_z\frac{\partial V_z}{\partial z} + V_r\frac{\partial V_z}{\partial r} + \frac{1}{\rho}\frac{\partial}{\partial z}\left(p + \frac{B^2}{8\pi}\right) = 0, \tag{7}$$

$$V_z\frac{\partial V_r}{\partial z} + V_r\frac{\partial V_r}{\partial r} + \frac{1}{\rho}\frac{\partial}{\partial r}\left(p + \frac{B^2}{8\pi}\right) = -\frac{1}{4\pi r}\frac{B^2}{\rho}, \tag{8}$$

$$\left(V_z\frac{\partial}{\partial z} + V_r\frac{\partial}{\partial r}\right)\left(\frac{1}{\gamma - 1}\frac{p}{\rho}\right) = -\frac{p}{\rho}\left(\frac{\partial V_z}{\partial z} + \frac{\partial V_r}{\partial r} + \frac{V_r}{r}\right), \tag{9}$$

$$\frac{\partial(BV_z)}{\partial z} + \frac{\partial(BV_r)}{\partial r} = 0 \tag{10}$$

Here ρ, V_z V_r, p are the density, two components of velocity and pressure, respectively; B is the φ-component of the magnetic field (index φ is omitted since the two other components are assumed to be zero); γ is the ratio of specific heats.

We shall also use spherical coordinates (R, θ, φ) when it is convenient; spherical and cylindrical coordinates are connected as follows:

$$z = R\cos\theta, \quad r = R\sin\theta, \quad \varphi = \varphi$$

In order to set the supersonic inflow conditions at certain distance $R = R_{in}$ (index in stands for "inflow") we should specify the inflow velocity V_{in}, density ρ_{in}, pressure p_{in} and the azimuthal component of the magnetic field that depends on angle θ counted from the z-axis as $B = B_{in}\sin\theta$, where B_{in} is a constant.

However, instead of specifying density ρ_{in}, velocity V_{in}, pressure p_{in} and magnetic field $B_{in}\sin\theta$ at certain distance R_{in}, it is more convenient to specify the mass loss rate $\dot{M}$ =

$4\pi\rho_{in}V_{in}R_{in}^2$, the terminal velocity V_{term} (i.e. the velocity at infinite distance from the origin if we propose free supersonic unmagnetized spherical outflow into vacuum), the entropy $s = \log(p/\rho^\gamma)$ and the magnetic field multiplied by the distance from the origin at the equatorial plane $\mathcal{F}_B = B_{in}R_{in}$ (see (5) with $\theta = \pi/2$) there. It is so, because as long as we neglect the dynamical impact of the magnetic field on the flow, these quantities do not change with distance from the origin, hence, the flow does not depend on the inner boundary distance R_{in}. Whereas we do not neglect the magnetic field impact in the present consideration, the second set of quantities is still more convenient.

In the present paper we consider hypersonic outflow ($M \gg 1$) at the inflow boundary $R = R_{in}$, so the gas velocity equals to its terminal velocity $V_{in} = V_{term}$.

The outer boundary is a tangential discontinuity with an unknown shape and position. The total pressure $p + B^2/8\pi$ at this boundary equals to the ambient pressure p_∞, and the normal component of the velocity at the discontinuity is zero: $V_n = 0$. Notice that the purely azimuthal magnetic field is tangential to the outer boundary and the boundary condition $B_n = 0$ is satisfied automatically. (B_n is the normal component of magnetic field to the discontinuity.)

To finish the formulation of the problem we have to set the boundary conditions far from the source in the jets (at infinity, $z \to \pm\infty$). In the numerical model we assume the so-called soft boundary conditions there meaning that $\partial/\partial z = 0$ for all quantities.

2.1. Dimensionless formulation

In order to formulate the problem in dimensionless form, we choose the distance to the termination shock for the purely gas-dynamic case ($B = 0$) as a characteristic distance:

$$R_{TS}|_{B=0} = \left(\frac{\dot{M}V_{in}}{4\pi p_\infty}\frac{\gamma+3}{2(\gamma+1)}\right)^{1/2}$$

As the characteristic density and velocity we choose their values downstream from the termination shock for the purely gas-dynamic case:

$$\rho_{TS}|_{B=0} = \frac{2(\gamma+1)^2}{(\gamma-1)(\gamma+3)}\frac{p_\infty}{V_E^2}, \quad V_{TS}|_{B=0} = \frac{\gamma-1}{\gamma+1}V_E$$

We define the dimensionless flow parameters as $\hat{\rho} = \rho/(\rho_{TS}|_{B=0})$, $\hat{V} = V/(V_{TS}|_{B=0})$, $\hat{p} = p/\left(\rho_{TS}|_{B=0} \cdot V_{TS}^2|_{B=0}\right)$, $\hat{B} = B/\left(\rho_{TS}|_{B=0} \cdot V_{TS}^2|_{B=0}\right)^{1/2}$, so that the form of the governing equations system (6)-(10) remains unchanged after rescaling.

The pressure balance condition at tangential discontinuity then becomes:

$$\hat{p} + \frac{\hat{B}^2}{8\pi} = \hat{p}_\infty, \tag{11}$$

where

$$\hat{p}_\infty = \left(\left(\frac{\gamma-1}{2}\right)\left(\frac{\gamma-1}{2\gamma} + \frac{2}{\gamma-1}\right)^\gamma\right)^{\frac{1}{\gamma-1}}$$

is obtained from the solution of purely gas-dynamic problem (zero magnetic field).

As we have noticed previously, the inflow boundary conditions are formulated in terms of stellar mass-loss rate $\dot{M}$, stellar wind terminal velocity V_{term}, and the constant characterizing the azimuthal magnetic field $\mathcal{F}_B$. In the hypersonic outflow case we do not have to specify the entropy s, since the pressure at the inflow boundary do not affect the flow. In dimensionless form these quantities are given as follows:

$$\hat{\dot{M}} = 4\pi, \quad \hat{V}_{term} = \frac{\gamma+1}{\gamma-1}, \quad \hat{\mathcal{F}}_B = \sqrt{4\pi\varepsilon}\left(\frac{\gamma+1}{\gamma-1}\right)^{1/2},$$

where epsilon is a dimensionless parameter of the problem:

$$\varepsilon = \frac{\mathcal{F}_B}{\sqrt{\dot{M}V_{term}}} = \sqrt{\frac{B_{in}^2/4\pi}{\rho_{in}V_{in}^2}}$$

The second equality shows that epsilon is an Alfvén number at the inflow boundary $R = R_{in}$ at the equatorial plane $z = 0$.

From here and then we omit "hats" assuming all of the quantities to be dimensionless.

2.2. *First integrals*

First let us introduce a streamline function ψ, which is defined in axisymmetric case as follows:

$$\frac{\partial\psi}{\partial r} = -\rho V_z r, \quad \frac{\partial\psi}{\partial z} = \rho V_r r \tag{12}$$

At the inner boundary sphere ($R = R_{in}$) ψ can be expressed as following:

$$\psi|_{R=R_{in}} = \cos\theta$$

The problem formulated above has three first integrals along a streamline:

$$\frac{V^2}{2} + \frac{\gamma}{\gamma-1}\frac{p}{\rho} + \frac{1}{4\pi}\frac{B^2}{\rho} = \left(\frac{\gamma+1}{\gamma-1}\right)^2 \cdot \left(\frac{1}{2} + \varepsilon^2(1-\psi^2)\right), \tag{13}$$

$$\frac{B}{\rho r} = \sqrt{4\pi}\varepsilon\left(\frac{\gamma+1}{\gamma-1}\right)^{3/2}, \tag{14}$$

$$\frac{p}{\rho^\gamma} = S(\psi;\varepsilon) \tag{15}$$

The first integral (equation (13)) is the Bernoulli integral generalized for the case of MHD (see, e.g. [8]). Equation (14) is the integral that follows from eqs. (10) and (6). Equation (15) is the adiabatic condition.

The right-hand parts of these equations are constants along streamlines. The expressions for these constants have been easily deduced from the inner boundary conditions in their dimensionless form. Note, however, that the entropic integral (15) is not conserved at the TS. $S(\psi;\varepsilon)$ could be explicitly derived downstream the TS only if the pre-shock solution is known, and the exact shape of the TS is given *a priori*. The constants in the other two integrals (eqs. (13), (14)) do not depend on the TS, though.

Note that if the flow at the inflow boundary is *subalfvénic* (i.e. if $\varepsilon > 1$), than the termination shock is absent, and the flow differs qualitatively from the one for which it is superalfvénic. We consider only the superalfvénic case in the present paper: $\varepsilon < 1$. As for the case of the Sun, $\varepsilon \simeq 0.08$.

3. Results of numerical solution: parametric study

In this section we review results of numerical solution of the full two-dimensional problem (6)-(10) formulated in Section 2, that were presented previously in our paper [9]. In order to get the solution we have used the two-dimensional shock-fitting Godunov [10] scheme with Chakravarthy-Osher TVD limiter [11]. The numerical scheme is identical to that employed in [4].

We have used flow-adapting computational grid. The grid was adapted to the tube-like structure of the considered problem; it has (60×200) cells in the region between the termination

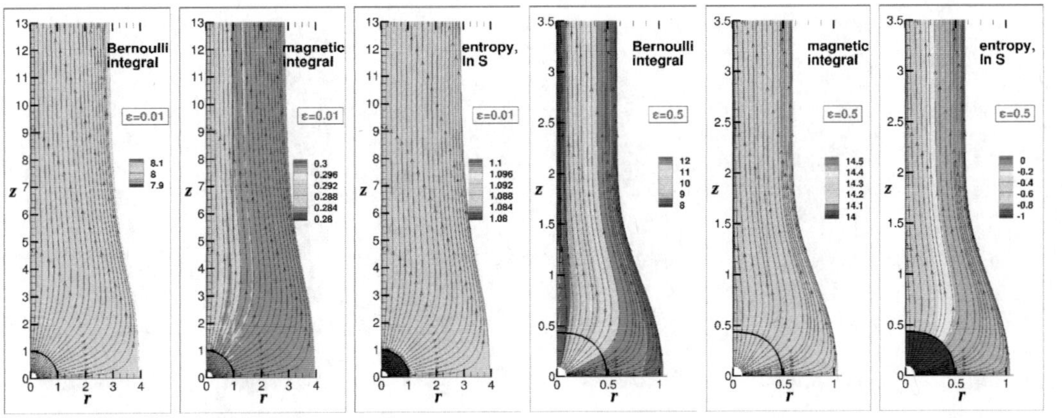

Figure 2. Three first integrals distribution for $\varepsilon = 0.01$ and for $\varepsilon = 0.5$. Each integral should remain constant along a streamline.

shock and the tangential discontinuity. Note that the grid has been constructed in such a way that cells are nearly squares everywhere, except the region far from the star, where the flow does not change much in the z-direction; hence there the cells are stretched along the z-axis.

The convergence of the result with respect to the spatial grid was verified via the cell bisection procedure. Also, in order to verify the quality of the numerical results, we have checked that the total mass flux through the outer boundary in the jet equals to the mass flux at the inner inflow boundary. We have also checked the conservation of the three first integrals (13), (14), (15) along the streamlines (see figure 2). The integrals are conserved in the numerical code within a few $(1-5)$ percent.

The results of the numerical calculations are shown in figures 3 and 4. Two-dimensional distributions of the flow parameters for three different values of ε: $\varepsilon = 0.01$, $\varepsilon = 0.25$ and $\varepsilon = 0.5$ are shown in figure 3, (top, middle and bottom rows respectively). Figure 4 shows one-dimensional distributions of the flow parameters in the pre-shock region (top row), and on the r- and z-axis of the post-shock region (middle and bottom rows respectively), for $\varepsilon = 0.05,\ 0.1,\ 0.25$, and 0.5. Note that on the latter figure the distance is normalized to the actual (i.e. ε-dependent) termination shock distance.

The distances to the TS and TD decrease with increasing parameter ε, meaning that the jet becomes more collimated. For $\varepsilon = 0.25$ the distance to the termination shock is 0.655 and the distance to the tangential discontinuity in the equatorial plane is 1.29. The jet radius is 0.851. For $\varepsilon = 0.5$ the distances are even smaller (bottom row of figure 3). It is interesting to note that for $\varepsilon = 0.25$ the termination shock is still very close to the spherically symmetric one (although, a slight deflection from a sphere could be seen in the figure). For $\varepsilon = 0.5$ the termination shock becomes distorted.

One can see from the plots of density that the flow between the TS and the TD is nearly incompressible in the case of small epsilon and the magnetic field grows linearly with the distance from the symmetry axis as it should follow from the integral (14) (see plots (a) and (d) on the top row of figure 3). At the same time for large epsilon the flow is sufficiently compressible (see plots (a) on middle and bottom rows of figure 3, and figure 4). Thus the behaviour of the magnetic field is not linear with respect to the distance from the symmetry axis (see plots (d) on middle and bottom rows of figure 3 and plots (b) of figures 4).

To understand the plasma and magnetic field distribution in figure 3 in detail it is very useful to plot (see plots (f) in figure 3) the r-projection of the sum of pressure and magnetic forces

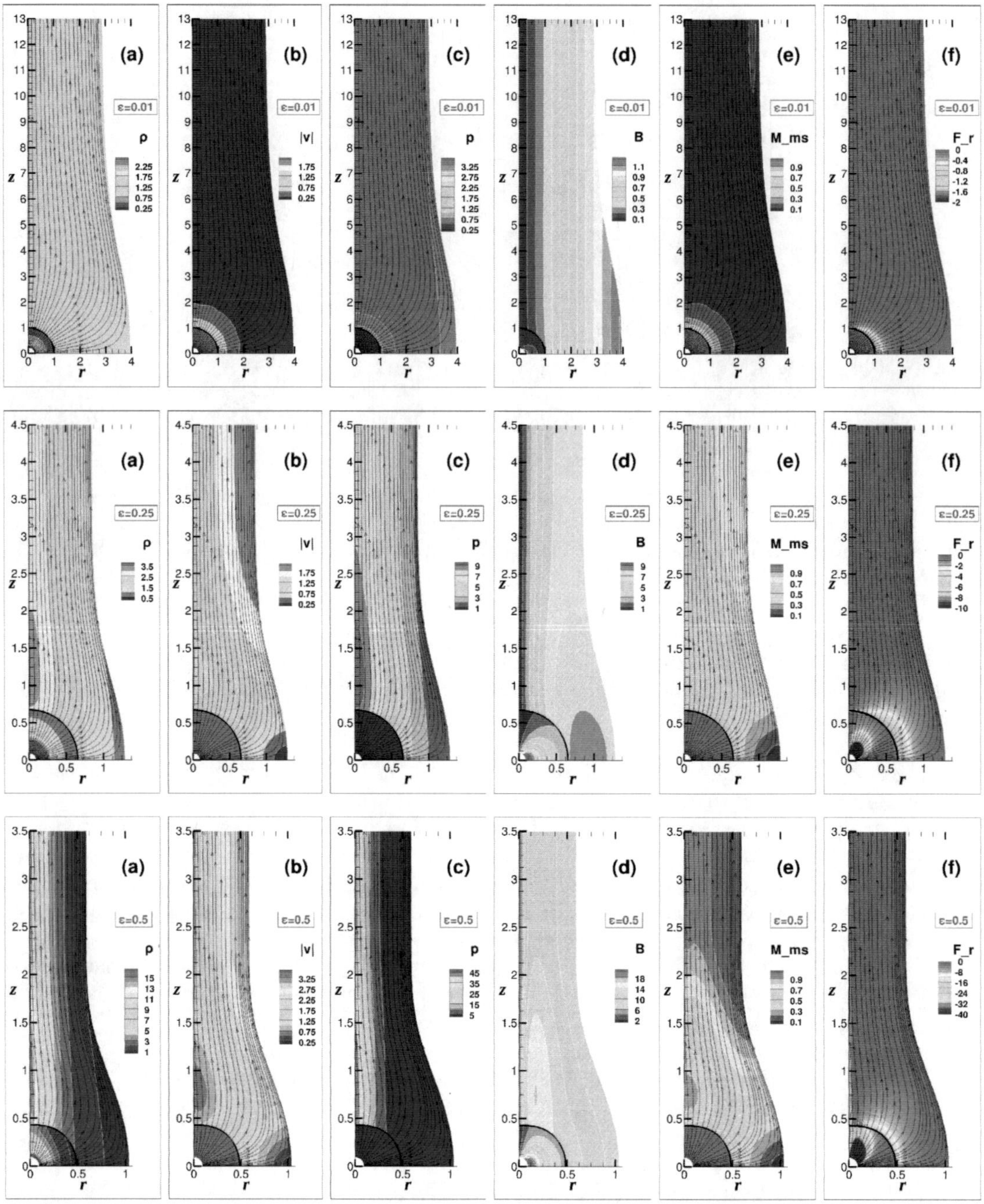

Figure 3. Two-dimensional distributions of the flow parameters. Top row: $\varepsilon = 0.01$, middle row: $\varepsilon = 0.25$, bottom row: $\varepsilon = 0.5$.

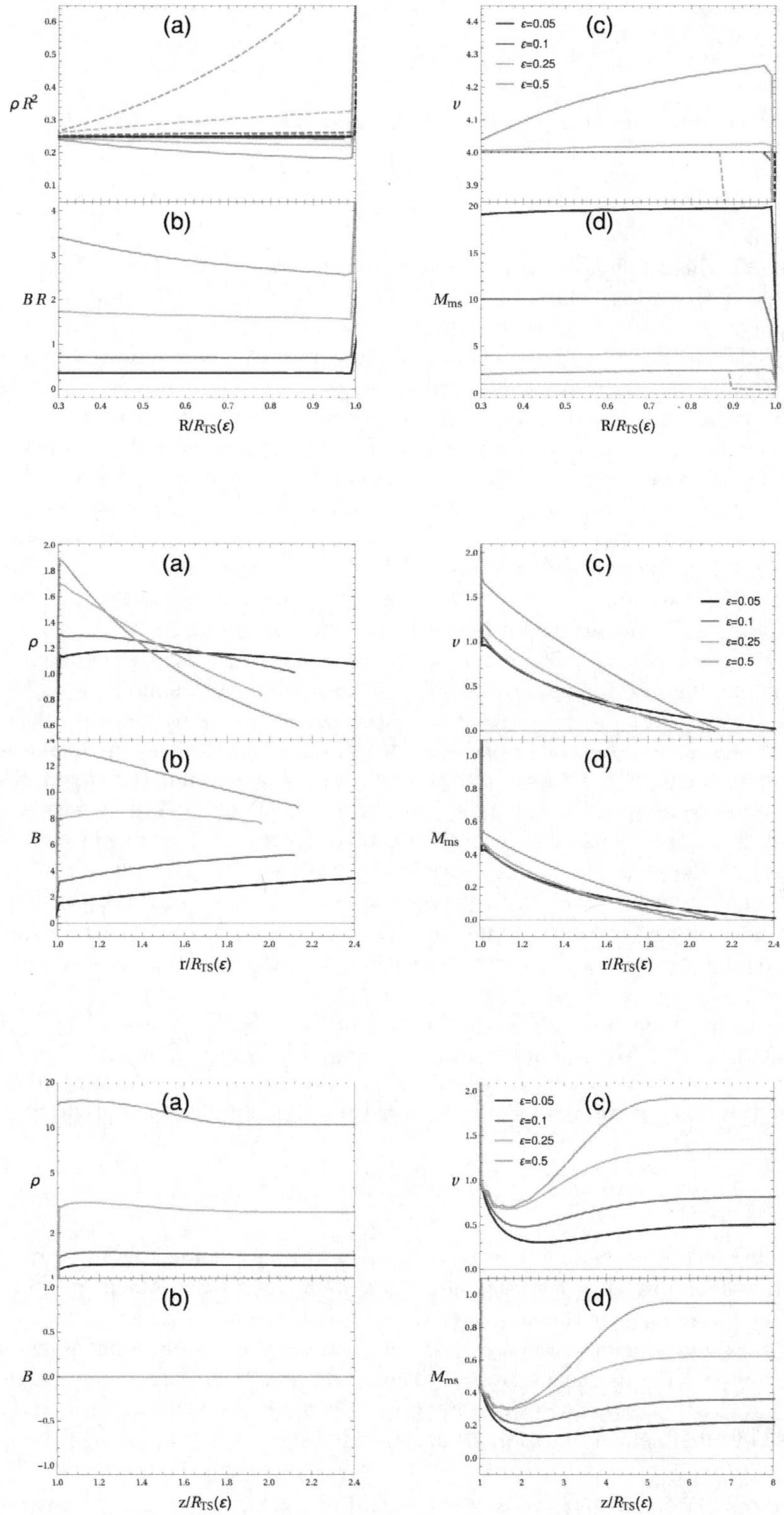

Figure 4. One-dimensional distributions of the flow parameters. Distance from the origin is normalized to the termination shock distance. Top row: pre-shock region $R < R_{TS}$. Solid lines correspond to r-axis distribution, dashed lines correspond to z-axis distribution respectively. Middle row: post-shock region, r-axis ($z = 0$). Bottom row: z-axis ($r = 0$).

acting on the flow in equation (8):

$$F_r = -\frac{\partial}{\partial r}\left(p + \frac{B^2}{8\pi}\right) - \frac{B^2}{4\pi r} \tag{16}$$

For all values of ε the maximum of the force (directed toward the axis of symmetry) in the subsonic region is in equatorial plane just after the termination shock because the magnetic field jumps significantly (by approximately the factor of four for $\gamma = 5/3$). Therefore the last term increases by the factor of 16. The magnetic force is partially compensated by the pressure gradient when the distance increases. F_r remains negative in the entire region. The flow is decelerated in the r-direction and the jet oriented toward the z-axis is formed.

While for $\varepsilon = 0.01$ the role of the magnetic force is restricted to the subsonic region, for $\varepsilon = 0.25$ and $\varepsilon = 0.5$ the magnetic force is also significant in the supersonic region. Similar to the subsonic region, the magnetic force acts toward the axis of symmetry (z-axis). The streamlines are slightly deflected towards the axis. This leads to a relative increase of the number density in the pole direction and a slight relative decrease of it in the equatorial direction as compared to the weak magnetic field case (e.g. $\varepsilon = 0.01$). The described effect is seen in the isolines of density in plots (a) on middle and bottom row of figure 3 and especially in plot (a) on top row of figure 4. Therefore for the cases of $\varepsilon = 0.25$ and $\varepsilon = 0.5$ the pre-shock density has the maximum at the pole ($r = 0$). The maximum remains beyond the shock for the nearly spherical (as for $\varepsilon = 0.25$) or ellipsoidal (as for $\varepsilon = 0.5$) shock. Further from the shock the density still slightly increases due to negative F_r force. Further small decrease of the density at the z-axis in the jet may be connected with the fact that the pressure gradient slightly overcomes the magnetic pressure. The slight pressure overcome may result in a small increase of the jet radius with z. Although the effect is small it can be recognized in bottom row of figure 3. It is also interesting to note that the pressure maximum at the z-axis (see plots (c) in bottom row of figure 3) corresponds to the minimum of the plasma velocity in this region, as it follows from the Bernoulli integral (see plots (b) in bottom row of figure 3 and the plot (c) in bottom row of figure 4). This local deceleration at the axis results in the larger velocities at the tangential discontinuity bounding the jet.

Since the velocity is larger at the TD, the Mach number is larger there. Plots (e) in figure 3 show the fast-magnetosonic Mach number $M_{\rm ms} = V/a_{\rm ms}$. It is also shown on one-dimensional plots (d) in figure 4. Note that fast magnetosonic wave speed is the only meaningful disturbance propagation speed in our case of axial symmetry and purely azimuthal magnetic field:

$$a_{\rm ms}^2 = \frac{\gamma p}{\rho} + \frac{B^2}{4\pi\rho}$$

For $\varepsilon = 0.5$ the flow in the jet becomes *transsonic* (plot (e) on bottom row of figure 3 and plot (d) on bottom row of figure 4) in a sense that the fast-magnetosonic Mach number in the jet becomes very close to unity near the z-axis and even greater near the tangential discontinuity.

For even larger values of ε (say, for $\varepsilon = 0.8$) the flow stops at some point on the symmetry axis downwind from the TS and turns around forming an eddy that causes sufficient numerical problems. The flow in the jet becomes completely supersonic. We do not show the results for this case because of possible computational uncertainties.

3.1. Radii of the tangential discontinuity at $z = 0$ and at $z = \infty$

Points on the figure 5 present the radii of the tangential discontinuity at $z = 0$ (crosses) and at $z = \infty$ (squares) with respect to parameter ε obtained from the numerical solution. We notice that in the large interval of the parameter ($\varepsilon \lesssim 0.1$) both dependencies are very well fitted with

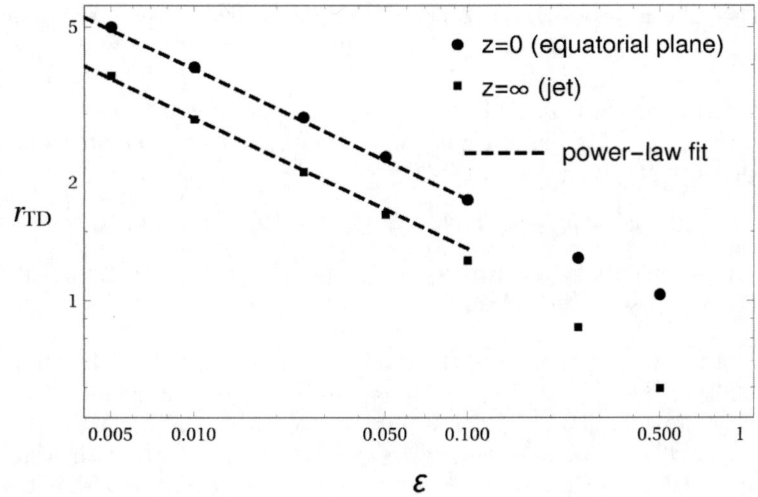

Figure 5. Radii of the tangential discontinuity at $z = 0$ (circles) and at $z = \infty$ (squares) with respect to parameter ε obtained from the numerical solution. Dashed lines correspond to the power-law fits for the numerical points (see (17)).

following power-laws:

$$\begin{aligned} r_{TD}(0)|_{fit}(\varepsilon) &= \frac{1}{3}4^{2/3}\varepsilon^{-1/3}, \\ r_{TD}(\infty)|_{fit}(\varepsilon) &= 4^{-1/3}\varepsilon^{-1/3} \end{aligned} \tag{17}$$

Note that exponents of both fits are the same and equal to $-1/3$. In the following section we explain it for the case of small epsilon without addressing our numerical results.

4. Analytical consideration

As we have already mentioned in Section 2.2, the entropic integral constant S is not conserved at the termination shock. In order to derive the expression for S downwind the TS we have to know the shape of the shock $R_{TS}(\psi)$ and the flow in the pre-shock region itself.

However, we are primarily interested in asymptotic behaviour of S as epsilon approaches zero. In the purely gas-dynamic case ($\varepsilon = 0$) the flow is known, and according to our problem rescaling (see Section 2.1) S in the post-shock region is expressed as follows:

$$S(\psi)|_{\varepsilon=0} = \frac{p_{TS}|_{\varepsilon=0}}{\rho_{TS}^{\gamma}|_{\varepsilon=0}} \equiv \frac{2}{\gamma - 1}$$

The expression for $S(\psi;\varepsilon)$ for a non-zero epsilon should approach this one as epsilon approaches zero, i.e:

$$S(\psi;\varepsilon) = \frac{2}{\gamma - 1} + o_{\varepsilon \to 0}(1)$$

At the moment we do not know what is the order of the first non-constant term in the series expansion. To denote it somehow, we shall introduce a new quantity – *kappa*:

$$S(\psi;\varepsilon) = \frac{2}{\gamma - 1} + O(\varepsilon^{2\varkappa}) \tag{18}$$

For now, kappa can be any positive number (if it is negative or zero, the O-term does not approach zero).

4.1. Radius of the tangential discontinuity at the equatorial plane (z = 0)
Let quantities with index "TD" and argument z denote the corresponding quantities at the tangential discontinuity at the height z, i.e.:

$$\rho_{TD}(z) = \rho(r_{TD}(z), z), \quad B_{TD}(z) = B(r_{TD}(z), z), \ldots$$

The radius of the tangential discontinuity at the plane $z = 0$ (i.e. at the equatorial plane) can be obtained if we consider the streamline $\psi = 0$ ($\theta = \pi/2$) that passes through the critical point at the TD.

At the critical point $V = 0$, therefore, the three first integrals (13), (14), (15) together with the boundary condition (11) provide four relations connecting five quantities. The five quantities are: 1) the dimensionless parameter ε, 2) the distance to the critical point $r_{TD}(0)$, and 3-5) the values of magnetic field, pressure and density $B_{TD}(0)$, $p_{TD}(0)$, $\rho_{TD}(0)$ at the critical point.

Using the four relations we can establish how $r_{TD}(0)$ depends on the parameter epsilon. The solution of the following system gives the dependency in the explicit form:

$$\begin{aligned} \left(\frac{\gamma}{\gamma-1} - 2\right) S|_{\psi=0}\rho_{TD}(0)^\gamma &= \left(\frac{\gamma+1}{\gamma-1}\right)^2 \cdot \left(\frac{1}{2} + \varepsilon^2\right) \rho_{TD}(0) - 2p_\infty, \\ \frac{1}{2}\varepsilon^2 \left(\frac{\gamma+1}{\gamma-1}\right)^3 r_{TD}(0)^2 &= \frac{p_\infty - S|_{\psi=0}\rho_{TD}(0)^\gamma}{\rho_{TD}(0)^2}, \end{aligned} \tag{19}$$

where p_∞ denotes the right-hand side of equation (11).

Using the expression (18) for entropy in the post-shock region, we find from the first equation of system (19), that:

$$\rho_{TD}(0;\varepsilon) = \rho_{TD}(0)|_{\varepsilon=0} + O(\varepsilon^{2\varkappa}) + O(\varepsilon^2), \tag{20}$$

where

$$\rho_{TD}(0)|_{\varepsilon=0} = \left(\frac{p_\infty}{S\Big|_{\substack{\psi=0\\ \varepsilon=0}}}\right)^{\frac{1}{\gamma}} = \left(\frac{(\gamma+1)^2}{4\gamma}\right)^{\frac{1}{\gamma-1}}$$

is obtained from the solution of purely gas-dynamic problem (zero magnetic field case).

We do not know which of the O-terms of expression (20) is dominant as long as we do not know whether kappa is below unity or not. For convenience we shall introduce another quantity – *bounded kappa* – as follows:

$$\bar{\varkappa} = \begin{cases} \varkappa, & 0 < \varkappa < 1, \\ 1, & \varkappa \geq 1 \end{cases}$$

Than, the expression (20) becomes:

$$\rho_{TD}(0;\varepsilon) = \left(\frac{p_\infty}{S\Big|_{\substack{\psi=0\\ \varepsilon=0}}}\right)^{\frac{1}{\gamma}} + O(\varepsilon^{2\bar{\varkappa}}) \tag{21}$$

Substituting (21) into right-hand part of the second equation of system (19) we get:

$$\frac{p_\infty - S_{\psi=0}\rho_{TD}(0)^\gamma}{\rho^2_{TD}(0)} = O(\varepsilon^{2\bar{\varkappa}}),$$

hence,

$$r_{TD}(0;\varepsilon) = O(\varepsilon^{\bar{\varkappa}-1}) \tag{22}$$

We immediately notice that if $\bar{\varkappa} = 1$ (or, equivalently, $\varkappa \geq 1$, see the definition of bounded kappa), then the radius of the TD at $z = 0$ does not approach infinity, which is impossible, since the solution with vanishing epsilon must approach a solution for zero magnetic field; this solution does not have a tangential discontinuity. Consequently,

$$\varkappa = \bar{\varkappa} < 1$$

4.2. Radius of the tangential discontinuity in the jet ($z \to \infty$)

Reasonable assumptions for the flow far from the source, i.e. for $z \to \infty$, are the following: 1) all parameters are independent of z coordinate, i.e. $\partial/\partial z = 0$, and 2) V_r component is negligible meaning that the inertial terms in equation (8) can be neglected. Therefore this equation can be written as

$$\frac{\mathrm{d}}{\mathrm{d}r}\left(p + \frac{B^2}{8\pi}\right) = -\frac{B^2}{4\pi r}, \tag{23}$$

and the streamline function satisfies the following ODE:

$$\frac{\mathrm{d}\psi}{\mathrm{d}r} = -\rho V_z r \tag{24}$$

Two differential equations (23) and (24) together with the first integrals (13), (14), (15) form closed system of equations for ρ, V_z, p, B, ψ as functions of r in the jet. This system can be reduced to the following system for ρ, V_z and r^2 as functions of ψ:

$$\begin{aligned}
\frac{1}{2}\frac{\mathrm{d}r^2}{\mathrm{d}\psi} &= -\rho^{-1}V_z^{-1}, \\
\frac{1}{2}\frac{\mathrm{d}V_z^2}{\mathrm{d}\psi} + \frac{1}{\gamma-1}\rho^{\gamma-1}\frac{\mathrm{d}S}{\mathrm{d}\psi} &= \left(\frac{\gamma+1}{\gamma-1}\right)^2(-2\psi)\varepsilon^2, \\
\frac{1}{2}V_z^2 + \frac{\gamma}{\gamma-1}S\rho^{\gamma-1} + \left(\frac{\gamma+1}{\gamma-1}\right)^3\varepsilon^2 r^2 \rho &= \left(\frac{\gamma+1}{\gamma-1}\right)^2\left(\frac{1}{2} + \varepsilon^2(1-\psi^2)\right)
\end{aligned} \tag{25}$$

System (25) is the second-order system of ordinary differential equations (ODE) for the functions $\rho(\psi)$, $V_z(\psi)$ and $r^2(\psi)$. The boundary conditions are posed as $r^2 = 0$ at z-axis ($\psi = 1$) and $p + B^2/8\pi = p_\infty$ at the tangential discontinuity ($\psi = 0$), where p and B are expressed in terms of r and ρ:

$$p(\psi) = S(\psi;\varepsilon)\rho(\psi)^\gamma,$$

$$B(\psi) = \sqrt{4\pi}\varepsilon\left(\frac{\gamma+1}{\gamma-1}\right)^{3/2}\rho(\psi)r(\psi)$$

Using the expression (18) for entropy in the post-shock region, we shall find from the second equation of system (25), that:

$$V_z^2(\psi;\varepsilon) = O(\varepsilon^{2\varkappa});$$

Then, it follows from the first equation of system (25), that:

$$r^2(\psi;\varepsilon) = O(\varepsilon^{-\varkappa})$$

The radius of the TD in the jet is defined as $r_{TD}(\infty) = r|_{\psi=0}$; hence,

$$r_{TD}(\infty;\varepsilon) = O(\varepsilon^{-\varkappa/2}) \tag{26}$$

IOP Conf. Series: Journal of Physics: Conf. Series **815** (2017) 012035 doi:10.1088/1742-6596/815/1/012035

5. Asymptotic behaviour of the tangential discontinuity at the equatorial plane and in the jet

Comparing expressions (22) and (26) we see that the asymptotic behaviour of the shape of the TD for the vanishing magnetic field of the source (i.e. for $\varepsilon \to 0$) is (generally) *different* at the equatorial plane and in the jet. For convenience, let us rewrite these expressions:

$$\begin{aligned} r_{TD}(0;\varepsilon) &= O(\varepsilon^{\varkappa-1}), \\ r_{TD}(\infty;\varepsilon) &= O(\varepsilon^{-\varkappa/2}), \end{aligned} \tag{27}$$

where we have already taken into account, that $\bar{\varkappa} = \varkappa$

We already know, that $0 < \varkappa < 1$. We also notice that in the case of $\varkappa = 2/3$ the asymptotics of $r_{TD}(0;\varepsilon)$ and $r_{TD}(\infty;\varepsilon)$ coincide. $\varkappa = 2/3$ is also consistent with power-law fits presented in Section 3.1 (see (17)). Below we introduce the proofs, that kappa could neither be greater, nor smaller than 2/3.

5.1. *Kappa cannot be greater than 2/3*

If we assume $\varkappa > 2/3$, then the radius of the jet approaches infinity *faster*, than the radius of the ecliptic, as epsilon approaches zero. From this follows, that for epsilon small enough, the TD is *wider* in the jet, than in the equator. However, below we introduce the proof that the radius of the tangential discontinuity should be maximal at the equatorial plane for any epsilon, hence, the assumption of $\varkappa > 2/3$ is contradictory.

Consider the zero streamline, $\psi = 0$, i.e. the streamline that passes through the ecliptic plane, and then, after reaching the stagnation point, throughout the TD. There are four relations for gas-dynamical quantities on the TD: namely, the three first integrals and the pressure equality condition. The entropic integral and the magnetic one allow to represent the pressure and the magnetic field in terms of density and the axial distance r, hence:

$$\begin{aligned} \frac{V^2}{2} + \frac{\gamma}{\gamma-1} S\rho^{\gamma-1} + \left(\frac{\gamma+1}{\gamma-1}\right)^3 \varepsilon^2 r^2 \rho &= \left(\frac{\gamma+1}{\gamma-1}\right)^2 \left(\frac{1}{2} + \varepsilon^2\right), \\ S\rho^{\gamma} + \frac{1}{2}\left(\frac{\gamma+1}{\gamma-1}\right)^3 \varepsilon^2 r^2 \rho^2 &= p_\infty \end{aligned}$$

The solution of the above system gives us V^2 and ρ as functions of r^2 at the TD.

Our goal is to prove that $r^2_{TD}(z)$ is maximal at the stagnation point ($z = 0$). In the following we first show that $d(V^2)/d(r^2) < 0$ at the TD. Then, as we show next, it implies $r^2_{TD}(z) \leq r^2_{TD}(0)$ for any z.

First we express both V^2 and r^2 in terms of ρ:

$$\begin{aligned} \frac{V^2}{2} &= \left(\frac{\gamma+1}{\gamma-1}\right)^2 \left(\frac{1}{2} + \varepsilon^2\right) - 2p_\infty \rho^{-1} - \left(\frac{\gamma}{\gamma-1} - 2\right) S\rho^{\gamma-1}, \\ r^2 &= 2\varepsilon^{-2} \left(\frac{\gamma-1}{\gamma+1}\right)^3 (p_\infty \rho^{-2} - S\rho^{\gamma-2}) \end{aligned}$$

Than, its derivatives with respect to ρ are written as follows:

$$\begin{aligned} \frac{\mathrm{d}V^2/2}{\mathrm{d}\rho} &= 2p_\infty \rho^{-2} + (\gamma-2) S\rho^{\gamma-2}, \\ \frac{\mathrm{d}r^2}{\mathrm{d}\rho} &= -2\varepsilon^{-2} \left(\frac{\gamma-1}{\gamma+1}\right)^3 (2p_\infty \rho^{-3} + (\gamma-2) S\rho^{\gamma-3}) \end{aligned}$$

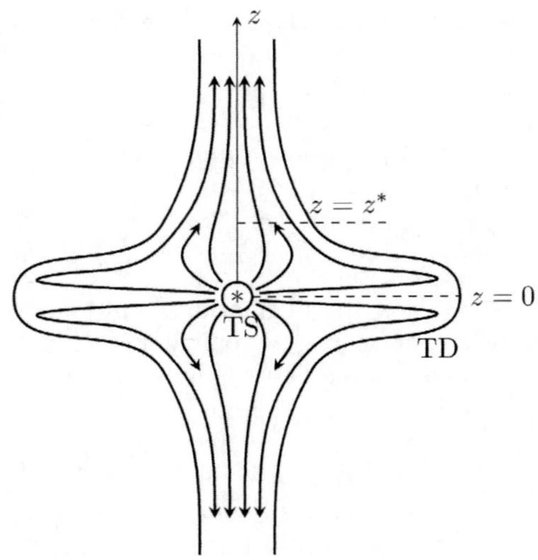

Figure 6. A sketch of the hypothetical flow for ε small enough if $\varkappa < 2/3$.

Consequently, the required derivative can be expressed as follows:

$$\frac{\mathrm{d}V^2}{\mathrm{d}r^2} = 2\frac{\mathrm{d}V^2/2}{\mathrm{d}\rho}\left(\frac{\mathrm{d}r^2}{\mathrm{d}\rho}\right)^{-1} = -\varepsilon^2\left(\frac{\gamma+1}{\gamma-1}\right)^3 \rho < 0$$

Next, let us integrate the last expression over r^2 from $r^2_{TD}(0)$ to $r^2_{TD}(z)$ for an arbitrary z:

$$\begin{aligned} V^2(z) - V^2(0) &= -\varepsilon^2\left(\frac{\gamma+1}{\gamma-1}\right)^3 \int_{r^2_{TD}(0)}^{r^2_{TD}(z)} \rho \, \mathrm{d}r^2 \\ &= -\varepsilon^2\left(\frac{\gamma+1}{\gamma-1}\right)^3 \rho^*(r^2_{TD}(z) - r^2_{TD}(0)), \end{aligned}$$

where we have used the mean value theorem for the integral. Since $V^2(0) = 0$ and $V^2(z) \geq 0$ for any z, this implies that:

$$r^2_{TD}(z) \leq r^2_{TD}(0) \quad \text{for any } z$$

This means, that assumption of $\varkappa > 2/3$ is incorrect, since as we noted in the beginning of current subsection, this implies that $r_{TD}(\infty) > r_{TD}(0)$ for epsilon small enough.

5.2. Kappa cannot be smaller than 2/3

For vanishing epsilon the tangential discontinuity radius can be expressed in the following way:

$$r_{TD}(z;\varepsilon) = O(\varepsilon^{-\alpha(z)}),$$

where for the exponent $\alpha(z)$ we already know that (see (27)):

$$\alpha(0) = 1 - \varkappa, \quad \alpha(\infty) = \frac{\varkappa}{2}$$

Let us assume that $\varkappa < 2/3$; hence, $\alpha(0) > \alpha(\infty)$. Note that $\alpha(z)$ is a function of real argument. If its limit at infinity is smaller, than its value at zero, then it follows from the very definition of limit, that there should be a *finite* value of argument $z = z^*$, such that $\alpha(0) > \alpha(z^*)$. More precisely,

$$\exists z^* \in (0, \infty): \; \alpha(0) - \alpha(z^*) > \frac{1}{2}(\alpha(0) - \alpha(\infty)) > 0$$

From this implies, that asymptotics of $r_{TD}(z^*; \varepsilon)$ differs from asymptotics of $r_{TD}(0; \varepsilon)$ as epsilon approaches zero, and:

$$\frac{r_{TD}(z^*; \varepsilon)}{r_{TD}(0; \varepsilon)} = O\left(\varepsilon^{\alpha(0) - \alpha(z^*)}\right) \xrightarrow[\varepsilon \to 0]{} 0$$

Hence, as epsilon approaches zero, the tangential discontinuity becomes infinitely narrower at some fixed finite "height" (z-coordinate), as compared with its equator; the geometry of the flow for epsilon small enough is sketched in figure 6. In such a flow there should be a huge vortex in the vicinity of the equatorial plane, since the flow has to nearly turn back after reaching the stagnation point. If we set epsilon to be even smaller, the hypothetical flow becomes even more exaggerated: the tangential discontinuity should nearly "lay" onto the stellar equator and the symmetry axis. This situation is mathematically unlikely and physically absurd, so, we conclude that the assumption of $\varkappa < 2/3$ is unsatisfactory.

6. Conclusion and discussion

In the present paper we have considered a flow produced by the hypersonic superalfvénic magnetized spherical source in the steady unmagnetized medium. An interest for such a flow was inspired by the problem of stellar wind/interstellar medium interaction. The flow is governed by the axisymmetric system of stationary ideal-MHD equations. The "magnetic power" of the source is assumed to be vanishing, whereas the rotation frequency is assumed to be huge, so that the azimuthal magnetic field produced by the source is finite and much larger than the radial one. The main results of the present paper can be summarized as following:

- The basic dimensionless parameter of the problem is ε (see the definition in Section 2.1), which describes the impact of the magnetic field on the flow. A zero magnetic field case corresponds to $\varepsilon = 0$.
- We have performed the numerical parametric study by varying the parameter ε in the range from 0.01 to 0.5; gamma was fixed at value of 5/3, that corresponds to monoatomic gases. The details of the numerical solution are shown on figures 3 and 4. The numerical results undoubtedly show the formation of tangential discontinuity (TD) dividing the flow and the ambient medium. These numerical results were presented earlier: see [9].
- According to our numerical results the radii of the tangential discontinuity at the equatorial plane ($z = 0$) and in the jet ($z \to \infty$) are both proportional to $\varepsilon^{-1/3}$ as long as epsilon is small enough: $\varepsilon \lesssim 0.1$, see figure 5 and fits (17).
- Using the three first integrals of the problem, (13)-(15), we compared asymptotic behaviour of the TD at $z = 0$ and for $z \to \infty$ as $\varepsilon \to 0$. We have found, see equation (27), that they are related to each other in a non-trivial way. We prove that the exponents of both asymptotics should be the same, or the flow is contradictory. We have found that there is a unique exponent for both asymptotics, that satisfies the relation between them. This exponent appears to be $-1/3$, which is fully consistent with our numerical results (17).

We have to note that besides we have explained the exponents of the tangential discontinuity asymptotic shape, we haven't explained the coefficients (see (17)). We are looking for the way to obtain them analytically, but at the moment these coefficients should be treated just as a fit.

Also, we do not present the results of our numerical calculations for $\varepsilon < 0.01$ and for $\varepsilon > 0.5$ because we are not completely sure that they are correct. Vortex flows beyond the termination shock are formed in the both cases. The extension of the presented parametric study will be elaborated in the future.

Additional important aspect of the considered problem is the stability of the considered flow and the tangential discontinuity. Theoretical considerations elaborated in the papers [12] and [13, 14] could be applied here.

Acknowledgments

This work has been supported by RSF grant No. 14-12-01096. Numerical calculations were performed using the Supercomputing Centre of Lomonosov Moscow State University (supercomputers Lomonosov and Chebyshev).

References

[1] Parker E N 1962 *Interplanetary Dynamical Processes* (New York: John Wiley & Sons)
[2] Gringauz K I, Kurt V G, Moroz V I and Shklovskii I S 1960 *Astron. Rep.* **37** 716
[3] Zank G P 2015 *Annu. Rev. Astron. Astrophys.* **53** 449
[4] Izmodenov V V and Alexashov D B 2015 *Astrophys. J. Sup.* **220** 32
[5] Parker E N 1958 *Astrophys. J.* **128** 664
[6] Parker E N 1960 *Astrophys. J.* **132** 851
[7] Drake J F, Swisdak M and Opher M 2015 *Astrophys. J.* **808** 44
[8] Kulikovskii A G and Lubimov G A 2005 *Magnitnaya Gidrodinamika* (Moscow: Moskow University Press)
[9] Golikov E A, Izmodenov V V, Alexashov D B and Belov N A 2017 *Mon. Not. R. Astron. Soc.* **464** 1065
[10] Godunov S K, Zabrodin A V, Ivanov M Y, Krayko A N and Prokopov G P 1976 *Chislennoye Resheniye Mnogomernyh Zadach Gazovoy Dinamiki* (Moscow: Nauka)
[11] Chakravarthy S K and Osher S 1985 american Institute of Aeronautics and Astronautics, Aerospace Science meeting
[12] Baranov V B, Ruderman M S and Fahr H J 1992 *Astron. Astrophys.* **261** 341
[13] Ruderman M S and Fahr H J 1993 *Astron. Astrophys.* **275** 635
[14] Ruderman M S and Fahr H J 1995 *Astron. Astrophys.* **299** 258